2807753161

AF443129

Springer Series in
MATERIALS SCIENCE   58

Springer
Berlin
Heidelberg
New York
Hong Kong
London
Milan
Paris
Tokyo

Physics and Astronomy   ONLINE LIBRARY

http://www.springer.de/phys/

Donglu Shi (Ed.)

Functional Thin Films and Functional Materials

New Concepts and Technologies

With 237 Figures and 48 Tables

  Springer

Professor Donglu Shi
University of Cincinnati
Department of Materials Science
and Engineering
Cincinnati, OH 45221-0012
USA
E-mail: dshi@uceng.uc.edu

Series Editors:

Professor R. M. Osgood, Jr.
Microelectronics Science Laboratory, Department of Electrical Engineering
Columbia University, Seeley W. Mudd Building, New York, NY 10027, USA

Professor Robert Hull
University of Virginia, Dept. of Materials Science and Engineering, Thornton Hall
Charlottesville, VA 22903-2442, USA

Professor Jürgen Parisi
Universität Oldenburg, Fachbereich Physik, Abt. Energie- und Halbleiterforschung
Carl-von-Ossietzky-Strasse 9-11, 26129 Oldenburg, Germany

ISSN 0933-033x

ISBN 3-540-00111-5 Springer-Verlag Berlin Heidelberg New York

ISBN 7-302-05903-9 Tsinghua University Press

Cataloging-in-Publication Data applied for
A catalog record for this book is available from the Library of Congress.
Bibliographic information published by Die Deutsche Bibliothek
Die Deutsche Bibliothek lists this publication in the Deutsche Nationalbibliografie;
detailed bibliographic data is available in the Internet at http://dnb.ddb.de
Library of Chinese Version Cataloging-in-Publication Data No.077184

Springer-Verlag Berlin Heidelberg New York
a member of BertelsmannSpringer Science+Business Media GmbH

http://www.springer.de

Cover concept: eStudio Calamar Steinen
Cover production: *design & production* GmbH, Heidelberg

Printed on acid-free paper   SPIN: 10838861   57/3141/di   5 4 3 2 1 0

Preface

In recent years, industrial applications have rapidly expanded the functional thin films into many new areas including membranes, optical devices, and microwave components in telecommunications. These applications typically employ materials with unique electronic, magnetic, acoustic, and thermal properties. Research on these emerging materials requires an extensive background in physical principles and characterization methods. The challenge for materials researchers is therefore to develop new technologies that can synthesize novel structures and characterize their corresponding unique properties. To achieve such a goal, research must be integrated with current trends in technological development and industrial needs, and must stimulate the interest of interdisciplinary groups.

We intend this book to provide an up-to-date introduction to the field of functional thin films with newly developed technologies and fundamental new concepts. The focus of the book is on the critical areas of novel thin films such as physical properties, structure, synthesis, phase equilibria, defect characterization, and novel applications. An important aspect of the book lies in its wide coverage of practical applications. Not only are the cutting-edge technologies in modern industry introduced, but also unique applications in many rapidly advancing fields. This book is written for a wide readership including university students and researchers from diversified backgrounds such as physics, materials science, engineering, and chemistry. Both undergraduate and graduate students will find it a valuable reference book on key topics related to solid state and materials science. It can also provide frontier researchers most up-to-date information on functional thin films, including novel syntheses methods, unique properties, and new applications.

The book devotes two chapters (Chaps. 1 and 2) to sol-gel science and technology. In these chapters, the authors address the most critical issues in membrane thin films from the perspectives of synthesis, structure, chemistry, and their unique applications. For instance, in the first chapter, the authors discuss in detail the advantages of polymer membranes in commercial applications. Compared with inorganic membranes, the polymer films can be fabricated in large quantities and with a desired geometry such as sheet, tube or hollow fiber. On the other hand, inorganic materials lack mechanical and fabricating flexibility, so that problems arise involving difficulty in housing, limited packing density relating to membrane area and higher cost. Many applications of organic membranes are introduced including microfiltration and ultrafiltration. The second chapter describes sol-gel synthesis and properties of several major ceramic adsorbents/catalysts or their support bodies for use in chemical reaction and adsorption processes. Nano-structured γ-alumina, zirconia and titania with uniform pore size distribution and an average pore diameter of about 3 nm are prepared by hydrolysis and condensation of corresponding metalorganic precursors. Experimental results on adsorption of ethylene and ethane and sulfation reaction on the sol-gel derived CuCl/alumina and CuO/alumina granules are reported to illustrate the advantages of the sol-gel derived adsorbents/catalysts as compared to those prepared by the conventional methods.

Chapters 3 and 4 focus on the fundamental concepts, novel synthesis methods, unique structures and properties of ferroelectric thin films. These chapters provide a brief summary of the current fundamental understanding of ferroelectricity and ferroelectrics and address critical material issues. One of the most actively pursued ferroelectric applications is highlighted: ferroelectric thin films for digital information storage as dynamic random access memories and non-volatile random access memories. The attention is mainly focused on three types of materials: barium strontium titanate, lead zirconate titanate, and strontium bismuth niobate tantalate, and on major thin film deposition processes including: sputtering, pulsed laser ablation deposition, chemical vapor deposition, liquid injection or mist, metal-organic decomposition, chemical solution deposition, and hybrid deposition processes. The challenges and current approaches are discussed.

Chapters 5 and 6 deal with nanostructured thin films and the mechanical behavior of fiber bridging in a ceramic matrix. They describe composite thin films with many unusual physical properties, which are not realizable with the pure single constituents. In recent years, ultrafine nanocrystalline materials have shown novel physical, chemical, magnetic, optical and electronic properties. Some interesting phenomena such as non-linear optical behavior and quantum confinement effects of carriers have also been reported in nanocomposite materials. Nanocrystalline and nanocomposite materials are believed to play an important role in explaining some fundamental physical problems, such as microstructural and property transitions between the molecular and bulk solid state and also may have technological applications in the future in structural engineering, optoelectronic devices, catalysts for chemical reactions, magnetic storage and optical coatings.

Chapters 7 and 8 are devoted to techniques developed recently in electron diffraction for the characterization of functional materials. They give details of sample preparation, diffraction principles, experimental procedures, and special methods for critical defect structure analysis. These chapters present many new characterization techniques using transmission electron microscopy. With these highly advanced techniques, defect structures including tweeds, twin boundaries, grain boundaries, stacking faults, and dislocations in a variety of materials are studied in detail.

Chapter 9 should be of most interest to researchers in computer simulation of novel structures and materials. This chapter gives detailed descriptions of novel methods in analyzing internal stresses in functional thin films. In this chapter, two main numerical simulation methods, namely, the finite element method (FEM) and the boundary element method (BEM) for macromechanical, and especially micromechanical analyses of various advanced materials, will be reviewed. Then the chapter will focus on the BEM approach, which offers many advantages over the FEM approach regarding the accuracy and efficiency for certain problems. The BEM formulations will be presented and some special issues associated with the BEM as applied to materials simulations will be addressed. Applications of the BEM to the micromechanical analysis of fiber-reinforced composites with the presence of the interphases and interfacial stress analysis of thin films and coatings will be presented.

We hope these chapters will provide timely and useful information for the progress of functional thin films and applications. We are grateful to all invited authors for their excellent contributions to this book.

Donglu Shi
Cincinnati
October 2001

Contents

List of Contributors

Donglu Shi
Dept.of Materials Science and Engineering, University of Cincinnati, OH
dshi@uceng.uc.edu

Naotsugu Itoh
Natl.Inst.of Adv. Ind. sci. & Tech., Tsukuba,Japan
itoh.n@aist.go.jp

Jerry Lin
Dept.of Chemical Eng.University of Cincinnati,OH
jlin@alpha.che.uc.edu

Guozhong Cao
Dept.of Materials Science and Engineering, University of Washington,WA
gzcao@u.washington.edu

Ren Xu
Dept.of Materials Science and Engineering, University of Utah,UT
renxu2020@yahoo.com

Feng Niu
Dept.of Materials Science and Engineering, Northwestern university,IL
fniu@amsuper.com

Yongjian Sun
Dept.of Materials Science and Engineering, Northwestern University,IL
ericsun@us.ibm.com

Xiaojin Wu
superconductivity Research Laboratoty
International superconductivity Technology Center Japan
wxjw@hotmail.com

Renhui Wang and Huamin Zou
Wuhan University,China
hmzou@whu.edu.cn

Yijun Liu
Dept. of Mechanical, Industrial and Nuclear Engineering, University of
Cincinnati,OH
yijun.liu@uc.edu

1 Membrane Thin Film Preparation and Applications

Naotsugu Itoh

1.1 Introduction

1.1.1 General Aspects

It is an undisputed fact that ever more organic polymer membranes are used in commercial applications. This is because these can be fabricated in large quantities as well as with a desired geometry (shape) like sheet, tube (1–10 mm in diameter) and hollow fiber (<1 mm), thereby having economical advantages. On the other hand, inorganic materials basically lack mechanical and fabricating flexibility compared to organic polymers, so that some problems such as difficulty in housing, limited packing density relating to membrane area and higher cost are encountered. However, inorganic membranes are used in particle and liquid filtration processes such as microfiltration (MF) and ultrafiltration (UF) since they are stable both chemically and thermally. In today's beer brewing industry, micro-organisms and yeast are removed by using porous ceramic membranes with submicrometer pore diameter to prevent beer spoil, while requiring no further pasteurization with heat treatment. This is one example of an application that became possible once inorganic membranes were successfully prepared.

As another aspect, growing interest has been focused on inorganic-based membranes for high-temperature gas separations and their applications to chemical reactions because of their capability of resisting at least the temperatures where softening begins. Among inorganic materials, including metals and metal oxides, i.e., ceramics, widely targeted compounds for separative and reactive membranes and the principal preparation methods are listed in Table 1.1, where whether porous or non-porous is the most important factor characterizing membranes in terms of structure.

1.1.2 Roles of Membrane

Membrane is a type of a barrier existing between two fluid streams, feed and permeate, but is required to separate a particular species from the others present

Table 1.1 Typical inorganic materials for membranes and the principal preparation methods

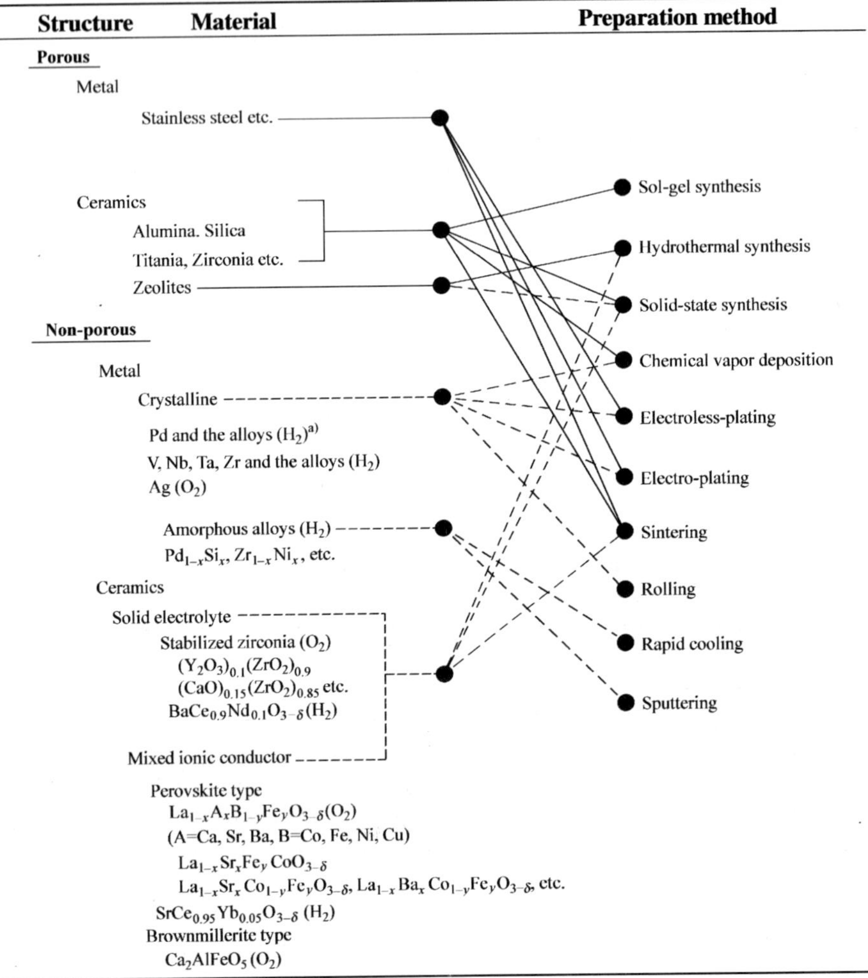

Metal
Crystalline
Pd and the alloys $(H_2)^{a)}$
V, Nb, Ta, Zr and the alloys (H_2)
$Ag\ (O_2)$
Amorphous alloys (H_2)
$Pd_{1-x}Si_x,\ Zr_{1-x}Ni_x$, etc.

Ceramics
Solid electrolyte
Stabilized zirconia (O_2)
$(Y_2O_3)_{0.1}(ZrO_2)_{0.9}$
$(CaO)_{0.15}(ZrO_2)_{0.85}$ etc.
$BaCe_{0.9}Nd_{0.1}O_{3-\delta}(H_2)$

Mixed ionic conductor
Perovskite type
$La_{1-x}A_xB_{1-y}Fe_yO_{3-\delta}(O_2)$
(A=Ca, Sr, Ba, B=Co, Fe, Ni, Cu)
$La_{1-x}Sr_xFe_yCoO_{3-\delta}$
$La_{1-x}Sr_xCo_{1-y}Fe_yO_{3-\delta},\ La_{1-x}Ba_xCo_{1-y}Fe_yO_{3-\delta}$, etc.
$SrCe_{0.95}Yb_{0.05}O_{3-\delta}\ (H_2)$
Brownmillerite type
$Ca_2AlFeO_5\ (O_2)$

a) indicates a gas species semipermeable

in feed, more selectively and more quickly. Let us look at the place where separation takes place. As shown by the schematic drawing in Fig.1.1, a thin layer directly related to separation, called the active layer governs the selectivity of C against the other species, A and B. The selectivity is strongly dependent on both membrane material and microscopic structure, as described later. In Fig.1.2, various separation methods are summarized, being closely related to the size of substance concerned. Microfiltration and ultrafiltration are used when suspended particles, micro cells, or large molecules in solutions are the objects of filtration. Reverse osmosis is very effective for desalination. Specific separation of hydrogen and oxygen is possible using palladium, ionic conductive membranes while usual gas separations by microporous membranes remain incomplete. Concerning the permeation flux, it can be considered that the flux is inversely proportional to the thickness of the active layer since the porous support usually has a very low resistance for permeation compared with the active layer. However, a decrease in the thickness leads to a decrease in the mechanical strength. In other words, to prepare a composite type of membrane as depicted in Fig.1.1, in which an active layer is formed on a porous support, is necerssary.

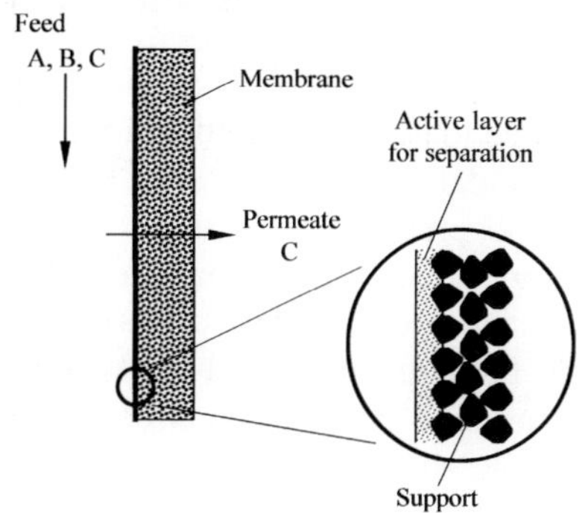

Fig.1.1   Separation by means of membrane.

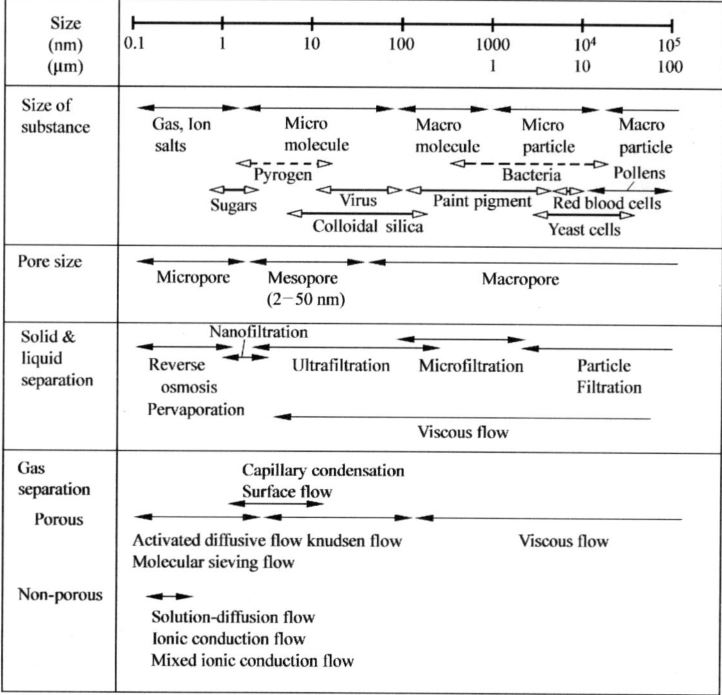

Fig.1.2   Membrane separations closely relating to the size of substance.

Therefore preparation methods for inorganic membranes including composite membranes will be described from a viewpoint of realizing a higher selectivity as well as a higher flux, which is indispensable to practical uses.

1.2 Preparation of Ceramic Membranes

1.2.1 Sol-Gel Process for Meso- and Micro-porous Membranes

Compared with several preparation methods of inorganic porous membranes, a sol-gel method is considered to be more flexible in terms of controlling pore size with various kinds of colloidal and polymer sols. Basically, the sol-gel process consists of four steps, that is, preparation of sols, coating, drying and firing steps. Each step can have a large affect on the characteristics of porous membranes obtained, such as pore size, pore size distribution, pore structure, defects, thickness, etc. Since the fundamental of sol-gel will be described in detail in Chap. 2, a concrete membrane preparation process by Asaeda et al. (1986, 1998), which is in principle similar to those attempted by other investigators (Yoldas et al., 1975; Uhlhorn et al., 1992; Brinker et al., 1993), will be presented below.

1) Preparation of sols

Figure 1.3 shows an example of preparing silica colloidal sols although there are many different routes with varying solvents, metal alkoxides, catalyst (acid or alkali), temperature, reaction time, pH, concentration and so on. In this case, the particle size of colloidal sols can be controlled mostly by the concentration and pH during the colloidal formation. The aging step, miscellaneous and dependent on know-how, is also important.

2) Coating, drying, firing

A procedure for dip-coating or slip-coating onto a porous support (substrate) is shown in Fig.1.4. The most important step in the sol-gel process is to prepare a coated layer without cracks. Drying is the step in which cracks most readily caused. Since with progress in drying of sols, a volumetric shrinkage of the coated layer follows, properly controlled drying and shrinking should offer a crack-free coated layer. The firing process to obtain a meso- or micro-porous layer firmly adhering on the support is also dependent on temperature, time and atmosphere.

Figure 1.5 shows an example of pore size distributions of a porous silica membrane determined by the dynamic humid air permeation method, which was recently proposed and based on the Kelvin capillary condensation theory (Tsuru et al., 1998). The well-known mercury porosimetry is also available for this purpose. Both methods can be readily applied to determine pore diameters larger than a few nanometers. However, as the radius of pores further decreases, the determination becomes difficult and ambiguous, in particular in the range below 1 nm.

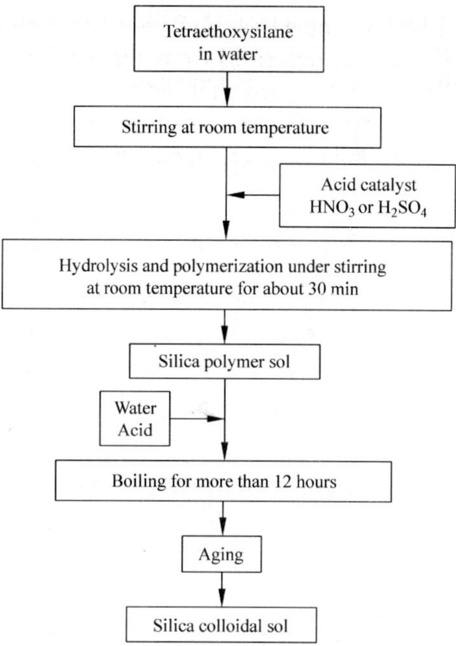

Fig.1.3 Preparation procedure of silica colloidal sol (Asaeda, 1998)

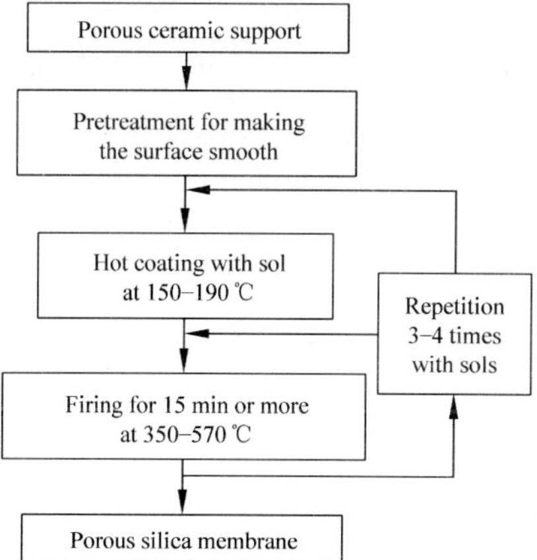

Fig.1.4 Coating procedure (Asaeda, 1998)

Porous layers prepared by means of a sol-gel technique, basically amorphous and mostly hydrophilic, are corroded by water. In addition, they dissolve easily in alkali solutions but are stable against acids. It is possible to change the amorphous material to a crystal line one with high-temperature treatment, whereby the most important problem is how to suppress the a

6

sintering process, which is accelerated with increasing temperature under coexistence of water vapor. Since such phenomena affect the performance of membrane negatively, careful control is required.

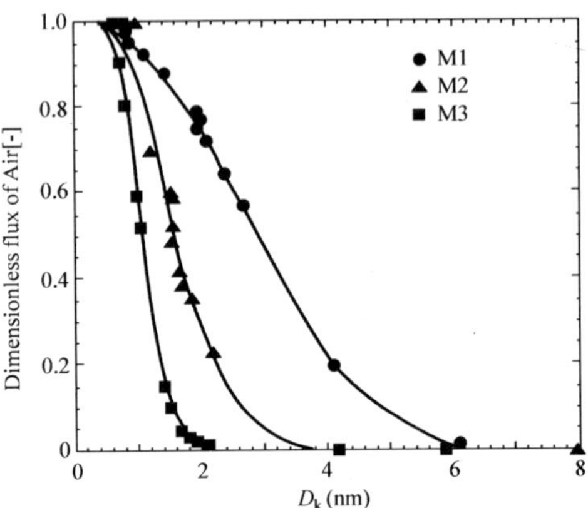

Fig.1.5 Relation between the dimensionless flux of air and the Kelvin diameter, D_k , for three types of silica-zirconia membranes (M1,M2,M3)(Tsuru et al., 1998).

1.2.2　Solid-State Synthesis for Mixed Conductor

It is well known that some perovskite-type oxides (ABO_3) containing transition metals at B-sites show high electronic and ionic conductivity at lower temperatures. The properties of these oxides can be easily modified by the substitution of A-site cations by other metals with lower valence. In a $La_{1-x}Sr_xCoO_3$ system, for instance, the substitution of Sr^{2+} for La^{3+} produces oxygen vacancies in the lattice to maintain the balance of electric neutrality, as schematically, drawn in Fig.1.6. Such vacancies produced in the lattice play an important role in diffusion of oxygen anions in the lattice matrix. Therefore, such compounds can allow oxygen to permeate without both the electrodes and the connecting lead, where the gradient of concentration of mobile ionic species becomes the driving force. Such a behavior is advantageous in fabricating a membrane separation unit because of its structural simplicity.

　　The disk samples of $La_{1-x}Sr_xCoO_3$ with varying Sr content, x, can be prepared as follows (Ohbayashi et al., 1974; Teraoka et al., 1988).

　　i) Starting materials.　The component metal acetates ($La(CH_3COO)_3 \cdot 1.5H_2O$, $Sr(CH_3COO)_2 \cdot 0.5H_2O$ and $Co(CH_3COO)_2 \cdot 4H_2O$) as starting materials are weighed exactly and then dissolved together in pure water to give a solution of the component metal acetates. This solution is a vaporized in rotating flask between 343 K and 363 K under a reduced pressure (200–300 mmHg[①]).

---

① 1 mmHg=133.322Pa

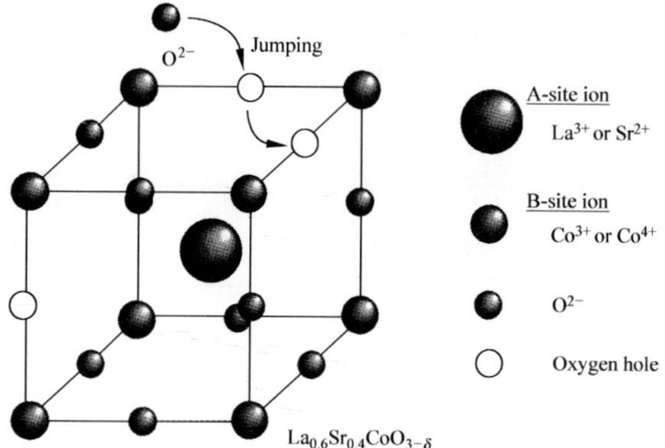

Fig.1.6 Oxygen anion diffusion in a perovskite type oxide (ABO_3)

ii) Preparation of precursor.   The residual powder is made to decompose at around 620 K in an electric furnace in an air flow. The product is then ground to obtain powder.

iii) Preparation of disk.   The obtained powder as the precursor of the perovskite-type $La_{1-x}Sr_xCoO_3$, is compressed into a disk with 13 mm in diameter and 1 mm thick under a hydrostatic pressure of 3500 kg/cm^{2}[1].

iv) Calcination.   The disk formed from the precursor powder is heated at a rate of 5 K/min in air and then kept at 1523 K for 5 h for sintering and changing to the perovskite-type oxide. Also, it is possible to prepare the disk from the powder having perovskite-type crystal structure, which had been obtained by calcining the precursor in air at 1123 K for 10 h. Thus a dense sintered disk can be obtained.

Generally, it is not easy to get a pore-free body from ceramic powder as starting material. Some ideas have been proposed to proceed the sintering process successfully: 1) to apply an extremely high temperature; 2) to control the distribution of particle diameter; 3) to use an additive such as polyvinylalcohol; 4) to utilize a chemical reaction; 5) to employ HIP (hot isotropic pressing), hot pressing, or molten particle deposition, etc. The sintering process of the precursor of $La_{1-x}Sr_xCoO_3$ can be observed by means of a high-temperature microscope, which made it possible to see changes in the shape of specimen when the temperature rose at a heating rate of 5 K/min (Itoh et al., 1994). The result is shown in Fig.1.7. It is obvious that a 2-mm-cube prepared by pressing the powder precursor transformed with increasing temperature, especially above 1473 K, and melted above 1793 K. Since it becomes evident that the specimen is significantly sintered in the range 1273–1530 K, the final sintering temperature is selected to be 1523 K in this case. Figure 1.8 shows an example of an X-ray diffraction pattern of the disk sample obtained together with that reproduced from the standard peaks in the ASTM card. Since both patterns are quite similar, the disk sample obtained is identified to have the

---

[1] 1 kg/cm^2=98066.5Pa

8

perovskite-type structure.

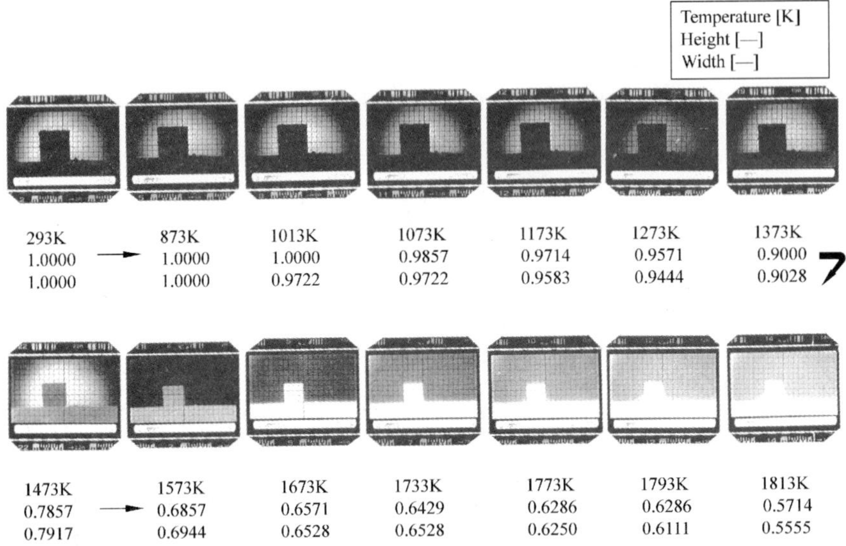

Fig.1.7 Shrinking behavior of the precursor of perovskite-type $La_{0.5}Sr_{0.5}CoO_3$ by means of high-temperature microscope (Itoh et al., 1994)

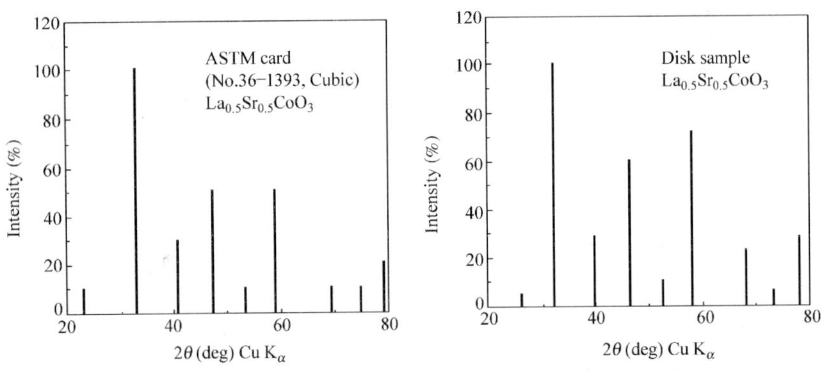

Fig.1.8 Comparison of the X-ray diffraction pattern of the sample with the ASTM standard peaks (Itoh et al., 1994)

Another route of solid-state synthesis without forming a solution can also be applied for $La_{0.2}Sr_{0.8}Fe_{0.2}Co_{0.8}O_{3-\delta}$ (Balachandran et al., 1995). $La(NO_3)_3$, $SrCO_3$ and $Co(NO_3)_2 \cdot 6H_2O$ and Fe_2O_3 are correctly weighed properly, mixed and milled in isopropanol for approximately 15 h. After drying, the mixtures are calcined in air at 850°C for 16 h. The powder is ground in an agate mortar, shaped in the desired form and heat-treated. During the heating, a solid-phase reaction as well as sintering progress, thereby producing a dense membrane without perforating pores. Such a synthetic process is also applied to the synthesis of solid oxide electrolytes, such as an yttria-stabilized zirconia (Y_2O_3-ZrO_2).

1.2.3 Hydrothermal Process for Zeolite Membrane

Zeolites, aluminosilicate hydrate represented by $M_{2/n}O \cdot Al_2O_3 \cdot xSiO_2 \cdot yH_2O$, are crystalline materials with rigid micropores of size 0.3–0.8 nm as illustrated in Fig.1.9. Since the pore diameter corresponds to the size of gas molecules, molecular sieving adsorptions and shape-selective reactions have been attempted. As an advanced application, fabricating zeolite in the form of membrane recently gained great interest and is a promising candidate for separation in a molecular-sieving manner. Although an in-situ synthesis of zeolite onto an asymmetric porous alumina is commonly employed, the whole procedure has been not yet established and there are many proposals for various types of zeolites on various supports (Ishikawa et al., 1989; Suzuki et al., 1990; Sano et

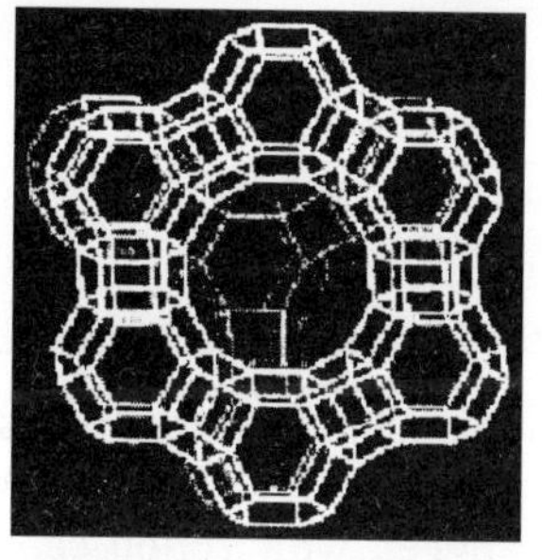

Faujasite
$(Na_2,Ca,Mg)_{29}[Al_{58}Si_{134}O_{384}] \cdot 240H_2O$
Channels: [111] 12ring 0.74nm

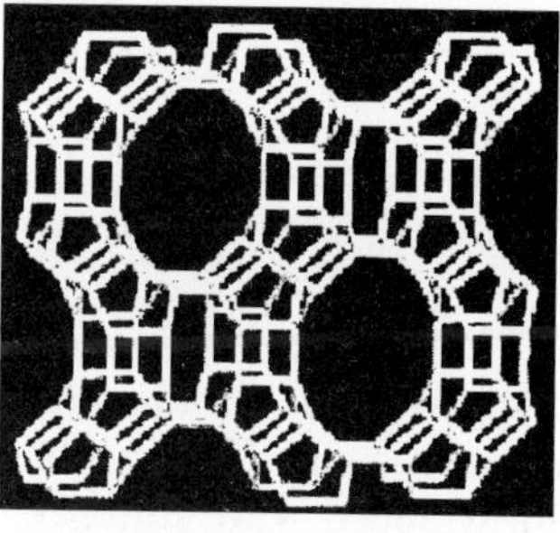

Mordenite
$Na_8[Al_8Si_{40}O_{96}] \cdot 24 H_2O$
Channels: [001] 12ring 0.65×0.7nm
[010] 8ring 0.26×5.7nm

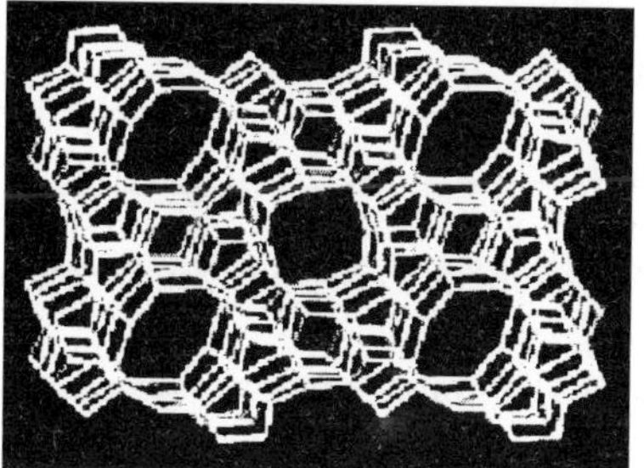

ZSM-5
$Na_n[Al_nSi_{96-n}O_{192}] - 16 H_2O$ ($n<27$)
Channels: [010] 10ring 0.53×0.56nm
[100] 10ring 0.51×5.5nm

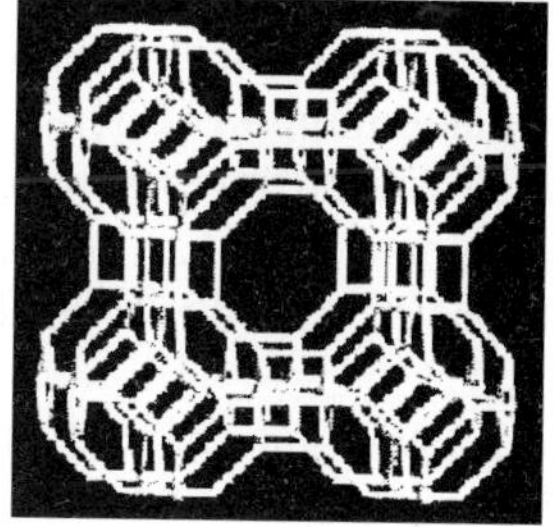

Linde Type A
$\{Na_{12}[Al_{12}Si_{12}O_{48}] \cdot 27 H_2O\}_8$
Channels: [100] 8ring 0.41nm

Fig.1.9 Molecular structure of zeolites (www-iza-sc.csb.yale.edu/IZA-SC/,The Structure Commission of the International Zeolite Association)

10

al., 1991; Dong et al., 1992; Jia et al., 1994; Matsukata et al., 1993; Yan et al., 1995; Kita et al., 1995; Kusakabe et al., 1996; Shimizu et al., 1996; Kanna et al., 1998). Here, a current typical way, a hydrothermal synthetic process for ZSM-5, is introduced in Fig.1.10 and explained below.

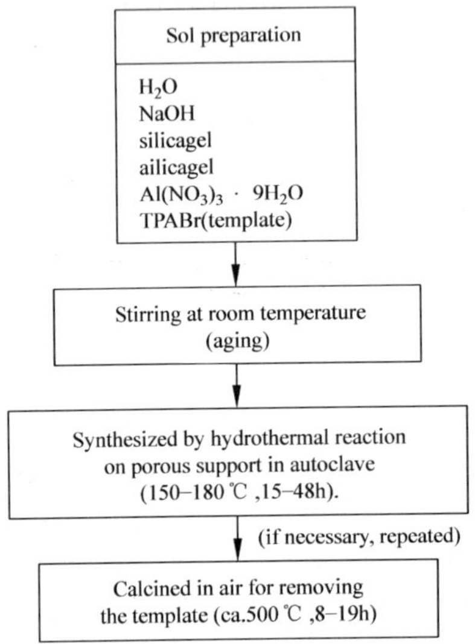

Fig.1.10 Synthesis procedure of ZSM-5 composite membrane

i) Support used. An alumina tube having submicrometer pores is usually employed as a support for zeolite synthesis. The support tube, both ends of which are wrapped with Teflon tape and plugged with Teflon caps, is placed vertically in a Teflon-lined autoclave.

ii) Composition. The raw material for zeolite synthesis is hydrogel or sol, and is basically composed of SiO_2, Al_2O_3, Na_2O and organic templating agent. For ZSM-5 synthesis, NaOH for the sodium source, colloidal silica or silica powder for the silica source, $Al(NO_3)_3$ for the alumina source, and tetrapropylammonium (TPA) for the templating agent, are commonly used. The ratios employed are over a wide range because ZSM-5 can be produced with various compositions, for instance, $Na_2O : SiO_2 : Al_2O_3 : TPA : H_2O = 0–4 : 0.5–10 : 0–0.06 : 1 : 40–1000$. Each is weighed according to the composition desired.

iii) Preparation of sol solution. The raw solution is prepared by successively adding NaOH, colloidal silica, $Al(NO_3)_3$, and TPA into the distilled water. Then the solution is stirred so as to mix well and age for several hours.

iv) Synthesis. Synthesis is first conducted in an autoclave at about 443 K for 15–24 h, and, if necessary, is repeated. Since the template fills in the zeolite pores during synthesis and thus blocks gas permeation, a defect free membrane should be impermeable. Normally, the membrane is impermeable after the second synthesis for γ-alumina supports, and after the second or third synthesis for α-alumina.

v) Calcination of template. After the synthesis is complete, the tube is carefully calcined in air to remove the template from the pores. The calcination is carried out in an electric furnace controlled to heat to around 770 K at approximately 0.01 K/s, holding there for 10 h, and cooling to room temperature at approximately 0.01 K/s. Such low heating and cooling rates are required not only to control the evolution rate of the template to be low enough not to break the zeolite layer, but also to avoid any thermal shock causing cracks.

1.2.4 Anodizing Process for Porous Membrane with Straight Pores

One of the difficulties in the practical use of ceramic membrane, especially at higher temperatures above 150°C, is how to house it in a membrane container, usually metal-made, where joining between ceramics and metal is not so easy because an organic adhesive agent and a polymer o-ring are no longer available. Instead of those, use of a high-temperature sealants like a graphite ferule (– 450°C) and an o-ring (–250°C) is attempted. A way of metallizing ceramics surface followed by welding is also proposed. Whatever the method, high-temperature sealing is very important step in the membrane separation at higher temperatures and is still under development. In this sense, a porous alumina membrane tube fromed by means of electrochemical anodic oxidation is thought to be very attractive because it can be obtained in a form adherent to metallic aluminum, which can be used as the sealing material as it is. Such a structure is also very useful as a support, on which a thin permselective layer formation by a sol-gel process, zeolite formation by hydrothermal synthesis (described above), palladium and its alloy deposition by means of electroless-or electro-plating (described below) and so on.

On the basis of the preparation method developed for a flat sheet of anodic membrane (Smith, 1973; Itaya et al., 1984), an anodic alumina membrane tube can be prepared by the following four steps (Itoh et al., 1998a). The last three steps are schematically shown in Fig.1.11.

i) Pretreatment of aluminum tube. The aluminum tube used was 45 mm long, 0.5 mm thick and 6 mm in outer diameter. First, the tube was ultrasonically cleaned in methanol for 20 min, and was annealed in air for 2 h at 300°C in an electric furnace. Chemical polishing was done with a solution of H_3PO_4 (85 wt%)-HNO_3 (61 wt%)-H_2O (8 : 1 : 1 by volume) for 3 minutes at 80–90°C. Subsequently, electrolytic polishing was done at a constant voltage of 20 V with a solution of H_3PO_4 (85 wt%)-H_2SO_4 (98 wt%)-H_2O (7 : 2 : 1 by volume), which contained 45g/l CrO_3 for 3 min at 80–90°C.

ii) Anodic oxidation. The whole inner side of the tube is oxidized on an anode in an aqueous solution of 4 wt% oxalic acid at 17°C using an experimental set-up, especially designed for anodizing specimens in the tubular form, as shown in Fig.1.12.

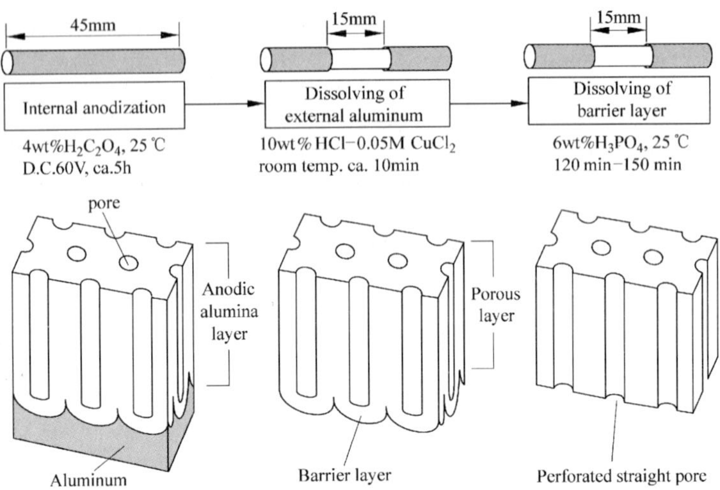

Fig.1.11 Preparation process of a porous alumina tube by the internal anodization technique (Itoh et al., 1998a)

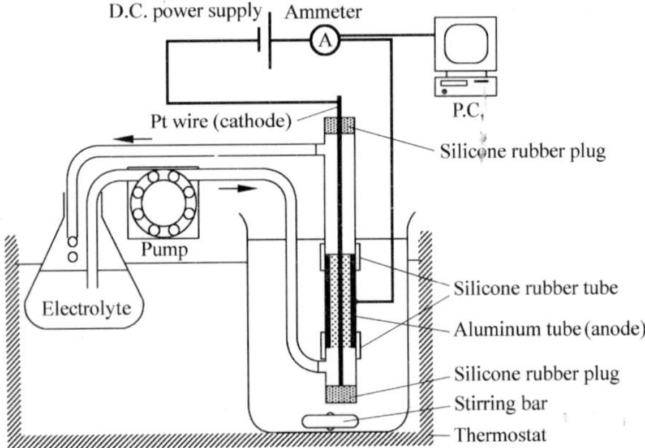

Fig.1.12 Schematic of the experimental set-up for internal anodization of aluminum tube (Itoh et al., 1998a)

iii) Dissolution of outer aluminum. The central part of the outer aluminum to be un-oxidized can be dissolved with an aqueous solution containing 20 wt% HCl and 0.1 mol/l $CuCl_2$ at room temperature for 10–20 min.

iv) Dissolution of outer alumina barrier. Subsequently, in the same manner as the above case, the thin barrier layer of aluminum oxide, formed at the bottom of pores, can be dissolved with an aqueous solution of H_3PO_4 for approximately. 90–120 min.

In Fig.1.13, a process to prepare the tubular membrane is shown in the sequence of 1) after polish, 2) after anodic oxidation and 3) after perforation.

The pore density of the alumina membrane prepared can be determined based on SEM observation. Figure 1.14 shows SEM images of the pores taken from the inner side when the anodizing voltage was changed. It is clearly seen that the pore density decreases with an increase in the voltage whereas the pore diameter increases. Using the number of pores photo counted in the SEM, pore density, N (number/m^2) is determined. The pore density is plotted against the voltage in Fig.1.15, where the result for a flat-type membrane estimated by means of small angle X-ray scattering (Itaya et al., 1984) is also presented for comparison.

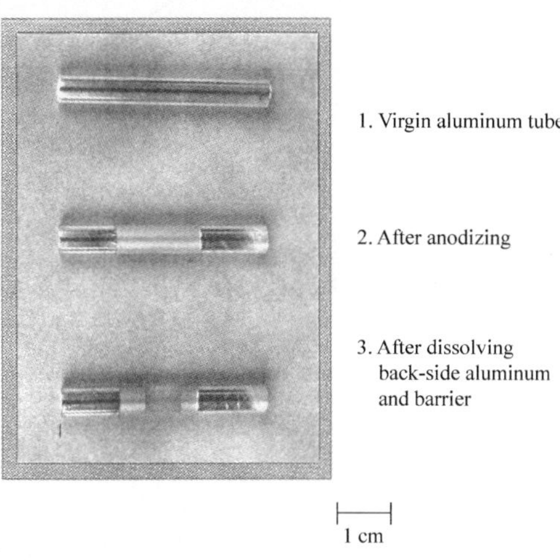

Fig.1.13 Outside appearance of the tube with progress of the preparation step: (1) virgin Al tube, (2) after anodizing, and (3) after dissolving Al and barrier layer (Itoh et al., 1996)

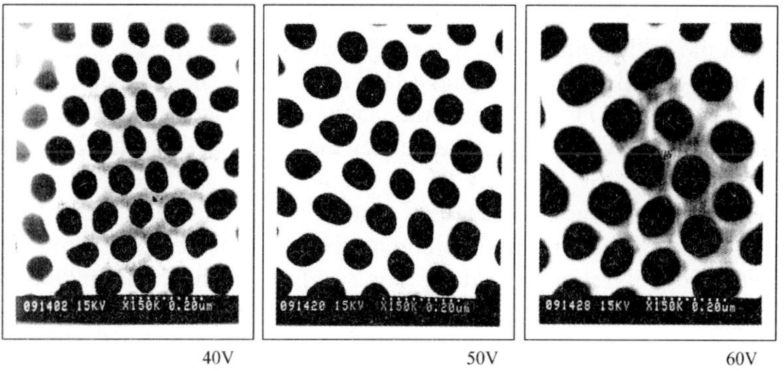

Fig.1.14 SEM images of the perforated pores with varying anodizing voltage (Itoh et al., 1996)

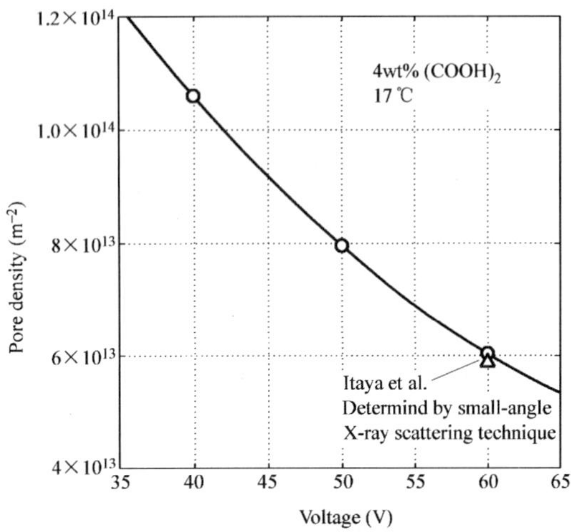

Fig.1.15   Relation between pore density and voltage (Itoh et al., 1996).

1.3   Preparation of Metal Membranes

The most important metal for hydrogen separation is palladium. As it has large hydrogen solubility and permeability, palladium foil has been used as a hydrogen purification membrane. In particular, since having a purity as high as 99.999%, the hydrogen produced by means of palladium membrane has been used in the semiconductor industry. However, there is a limitation in the use of pure palladium, that is, palladium membrane should be used above the critical temperature (approximately 310℃) in the pressure-composition (p-c) isotherms for the Pd-H system. This is because when Pd absorbs hydrogen at temperatures lower than the critical one, a phase transition from the hydrogen-lean α-phase to the hydrogen-rich β-phase occurs and leads Pd-membrane to distort and finally destroy (Lewis, 1967). To overcome such an inevitable proble, palladium membranes alloyed with other metallic elements such as Ag, Au, Ni, and so on, have been investigated. So far, a Pd-Ag membrane of all thos tested, having a permeability of hydrogen higher than that of pure palladium without the α-β phase transition (Knapton, 1962), is commercially available. In addition, the best composition of silver in the Pd-Ag binary alloy as the hydrogen-permeable membrane has been found to be within the range 23 at%–25 at%. Empirically, when the total quantity of the valence electrons of additive elements in any palladium alloy system is near 0.23–0.25, for instance, $Pd_{0.77}Ag(\text{I})_{0.23}$, $Pd_{0.92}Y(\text{III})_{0.08}$, $Pd_{0.92}Gd(\text{III})_{0.08}$, $Pd_{0.89}Gd(\text{III})_{0.06}Ag(\text{I})_{0.05}$ etc., optimum permeablities of hydrogen may be obtained (Sakamoto et al., 1990). Simultaneously with exploitation of such superior alloys, many efforts (introduced below) have been made to fabricate as thin a palladium film as possible for producing higher permeability.

1.3.1   Electroless Plating

Electroless plating is a type of thin film deposition technique utilizing chemical redox reactions, as represented by the following scheme:

$$2Pd^{2+} + N_2H_4 + 4OH^- = 2Pd^0 + N_2 + 4H_2O$$

As reducing agents, hydrazine (N_2H_4), sodium borohydride ($NaBH_4$), sodium phosphinate ($NaHPO_2$), etc. can be used. A typical procedure is illustrated in Fig.1.16 (Uemiya et al., 1988; Kikuchi, 1996).

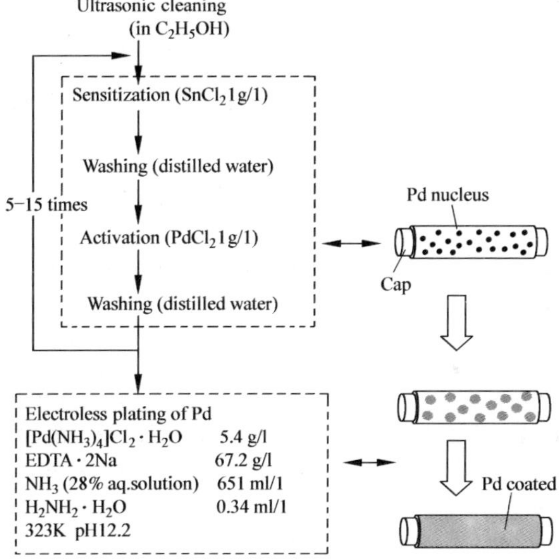

Fig.1.16   Electroless plating of palladium on porous support tube (Kikuchi, 1996)

   i) Sensitization. Prior to electroless plating, palladium nuclei should be deposited on the outer surface of the cylindrical alumina support to obtain dense film. This sensitization process is performed by treating with Sn(II) chloride solution followed by activation with Pd(II) chloride solution, and is repeated 2–10 times.

   ii) Electroless deposition. A solution consisting of [Pd(NH$_3$)$_4$] Cl$_2$, EDTA 2Na, ammonia, and hydrazine (pH 12.2, 323 K) is used for electroless plating of palladium. The Pd nuclei on the support catalyzes the reduction of Pd cations to metallic palladium, which will progresses over the surface to form Pd film. It takes about 50 min for an increment of 1 μm in thickness.

   In order to attain extremely high selectivity for H_2 permeation, palladium membrane should have no pinholes, resulting in a lower limitation of the

16

membrane thickness, which depends on the average pore size of the support used. Empirically, when a support with pore diameter of 200–300 nm is used, the thickness of Pd layer needed for pinhole-free membranes is said to be at least 5 and 15 μm, respectively. Generally, the rate of H_2 permeation through the membrane increases in inverse proportion to its thickness, as demonstrated in Fig.1.17, indicating that the rate-determining step in this range of thickness is the diffusion process of hydrogen in the palladium film.

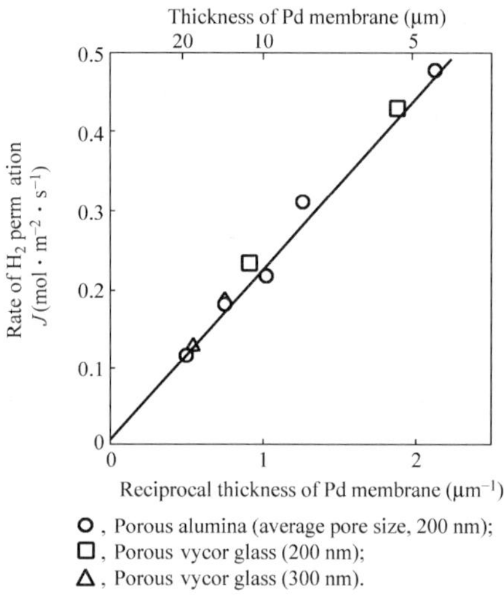

Fig.1.17 A linearity between the hydrogen permeation rate at 773 K and the reciprocal of thickness of palladium film deposited on porous supports (Kikuchi and Uemiya, 1991)

Further improvement of H_2 permeability is possible by alloying palladium with silver, where a successive deposition technique (Uemiya et al., 1991; Kikuchi and Uemiya, 1991) or   a simultaneous one (Shu et al., 1993) can be employed. In the former, after successive electroless plating of palladium and silver on the support, the metallic double layer is treated at high temperature, alloying the two metals by the mutual thermal diffusion, as seen in Fig.1.18. It becomes clear that above 1073 K the alloying can go to completion to obtain a uniform distribution profile of Ag and Pd. The permeability of H_2 through the obtained Pd-Ag alloy membrane, dependent on the content of Ag, attains a maximum at 23 wt% Ag in Pd. Such an increase in H_2 permeability is due to an increased solubility of H_2, while the H_2 diffusivity should decrease somewhat on addition of Ag. As another approach to improve the permselectivity and the adhesion to the support, a method of repairing the defects including pin holes left, that is, an electroless plating combined osmosis technique, has been proposed (Lin et al., 1999).

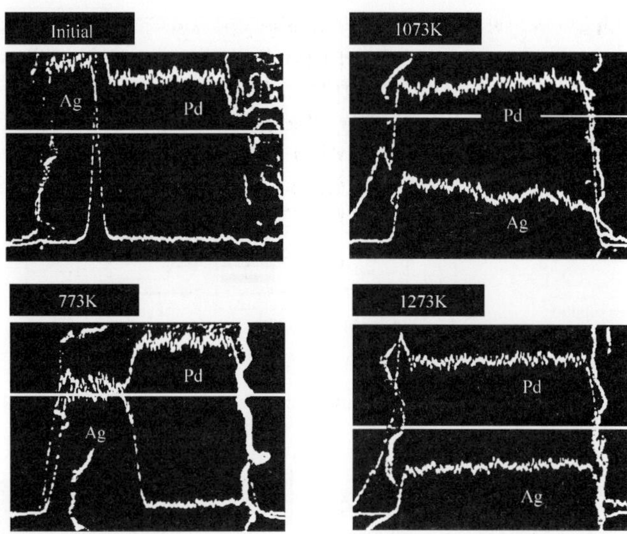

Fig.1.18 Progress of mutual diffusion between palladium and silver-deposited films according to high-temperature treatment, followed by means of electron probe microanalysis (Kikuchi et al., 1991)

1.3.2 Electroplating

The electrolytic process is controllable and fast compared to the electroless plating because the deposition rate of palladium is in the range of 0.2–0.4 μm/min while 0.01–0.04 μm/min is achieved in the electroless plating. A $Pd_{78}Ni_{22}$ alloy film of approximately. 1 μm can be deposited on a porous stainless steel disk (an average pore diameter of 0.5 μm), where the plating solution is forced to penetrate inside the pores by evacuating the opposite side of the disk as shown in Fig.1.19 (Nam et al., 1999). The hydrogen permeability (Pe) at 550℃ is found to be as high as 250 $cm^3/(cm^2 \cdot min \cdot atm)$ with a selectivity (S) of 4700 given by a ratio of H_2 to N_2 permeance (cf. S=1100, Pe=70 at 450℃, S=450 and Pe=30 at 350℃).

When ceramic materials are used as the support, direct electroplating is no longer possible as mentioned before. Prior to that, an electrode must be attached by any technique.   As one of the many methods conceivable for a such case, a process of sputtering followed by electroplating is shown in Fig.1.20(a) (Itoh et al., 2000). A 4 μm-thick palladium layer packed in the pores of anodic alumina tube (Fig.1.20(b)) can resist shape changes due to hydrogen absorption and desorption, and has S=1700 and Pe=25 at 350℃.

1.3.3 CVD Process

An increase in the resistance to hydrogen embrittlement may also be realized by

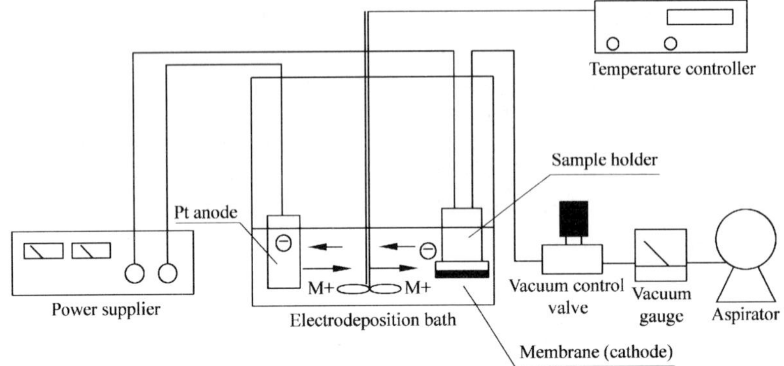

Fig.1.19 Shematic diagram of vacuum electrodeposotion setup (Nam et al., 1999).

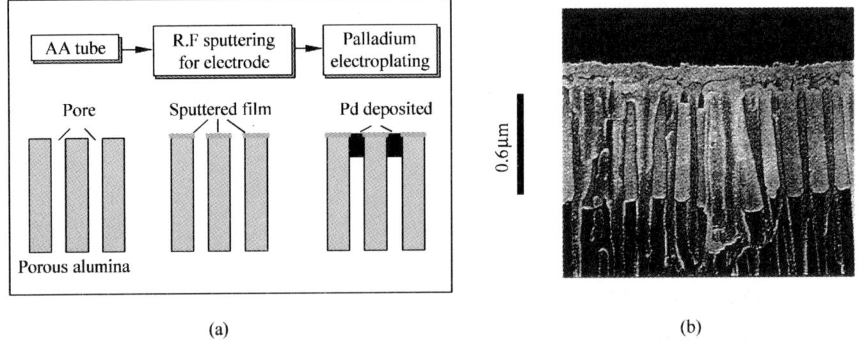

(a)   (b)

Fig.1.20 **(a)** Deposition process of palladium inside the straight pores of anodic alumina (AA) by electroplating method, and **(b)** SEM photo of the crosssection of palladium deposited inside the straight pores of AA layer (Tomura, 1999)

a low-temperature metal-organic chemical vapor deposition (MO-CVD) process, which possibly impregnates palladium in macropores of an α-alumina support tube. As the metal source palladium (II) acetate and the like can be used, and the deposition in the pores is carried out as follows (Yan et al., 1994).

i) Support and palladium salt. A porous α-alumina tube (2.6 mm O.D., 2.0 mm I.D., void fraction of approximately. 0.40, and average pore size of approximately 150 nm) can be used as the support. Fine particles of palladium (II) acetate are used as the palladium source, decomposing between 195℃ and 240℃ as follows.

$$Pd(CH_3COO)_2 = Pd + \text{volatile}$$

ii) Experimental set-up. Deposition of palladium is carried out in a glass reactor (Pyrex, 250 mm long and 32 mm O.D.) as shown in Fig.1.21. The porous

α-alumina tube is fixed inside the reactor with O-ring seals at the bottom outlet. Except for the central part of about 10 mm length, the support tube is coated with glass sealant paste, prepared by mixing a Na_2O-B_2O_3-SiO_2 powder (Nippon Electric Glass, GA-4) with α-terpineol and ethyl cellulose at a mass ratio of 0.7 : 0.27 : 0.03. After drying at 120℃ for 20 min, the support tube is treated at 750℃ for 1 h in air. The upper end of the support tube is also closed with the sealant. Both the shell and tube sides of the reactor are evacuated with rotary vacuum pumps and controlled to be, respectively, 100–200Pa and 6–9Pa by introducing argon in to the reactor.

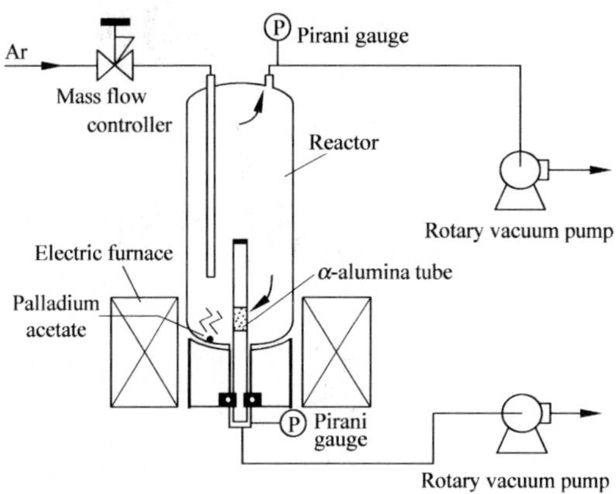

Fig.1.21 Schematic diagram of a CVD apparatus for palldium deposition onto a porous tube (Yan et al., 1994)

iii) Deposition of palladium. About 20 mg of $Pd(CH_3COO)_2$ placed near the bottom of the reactor is decomposed, where some of the palladium acetate sublimes before decomposition. This kind of sublimation is somewhat advantageous since there is a possibility that a part of the vapor is transported through the support tube, trapped in pores, and decomposed there.

From microscopic observation, a finely packed structure can be observed in the specimen prepared at 300℃. Figure 1.22 is the depth profile of the membrane obtained by the reaction at 300℃, showing that the palladium layer extends as far as 8 μm inside of the support. As a result of testing the resistance to hydrogen embrittlement, the hydrogen and nitrogen permeabilities are mostly unchanged after the sixth repetition of heating and cooling between 100℃ and 300℃, as shown in Fig.1.23, in which the separation factor is maintained to be higher than 1000 at 300℃.

1.3.4 Sputtering

Sputtering, categorized as a physical vapor deposition, PVD, is widely used for preparation of thin films. When argon is used as a discharge gas, a glow discharge can be generated at argon pressures of 10^{-2}–10^{-1} torr and current

20

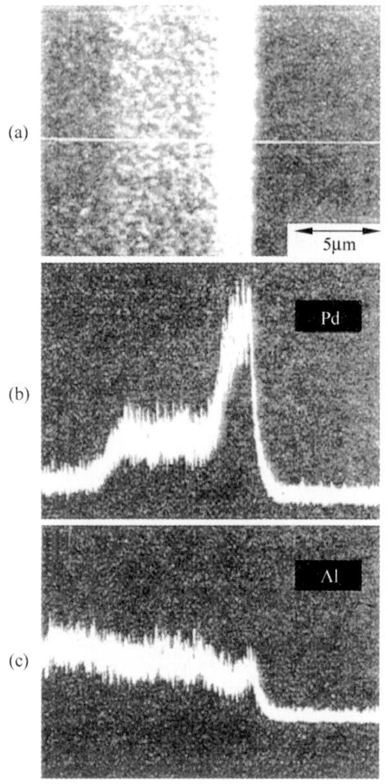

Fig.1.22　SEM image of palladium film deposited on a porous support (**a**) and EDX line profiles of palladium (**b**) and aluminum (**c**) (Yan et al., 1994)

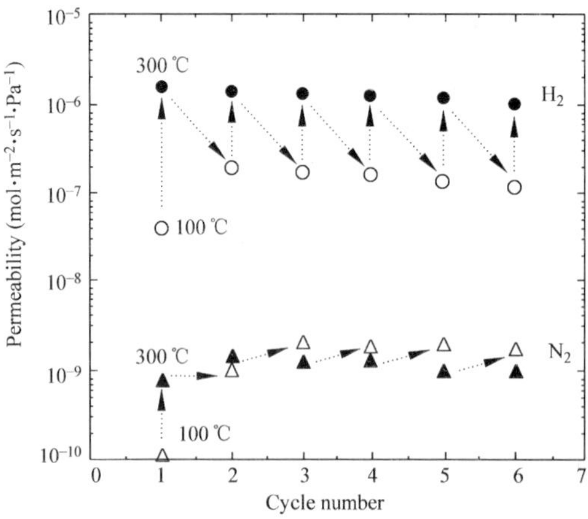

Fig.1.23　Temperature repetition test between 100℃ and 300℃ for checking the stability of palladium film (Yan et al., 1994)

density of 0.1–10 mA/cm^2. When argon ions produced in argon plasma hit the surface of target, atoms or molecules of the target surface evaporate and deposit on the support. Accordingly, even materials with a high melting point, such as metals, alloys or metal oxides can be used as targets. The growth rate of thin film can be controlled within approximately 0.005–5 μm/min.

Figure 1.24 shows a RF-diode glow discharge sputtering apparatus designed to apply for tubular supports, which are attached to a shaft of a motor and rotated at 20 rpm (Itoh et al., 2000). Table 1.2 compares the sputtering rates. It is clear that the growth rate of thin film depends on the kind of target and the temperature of the support.

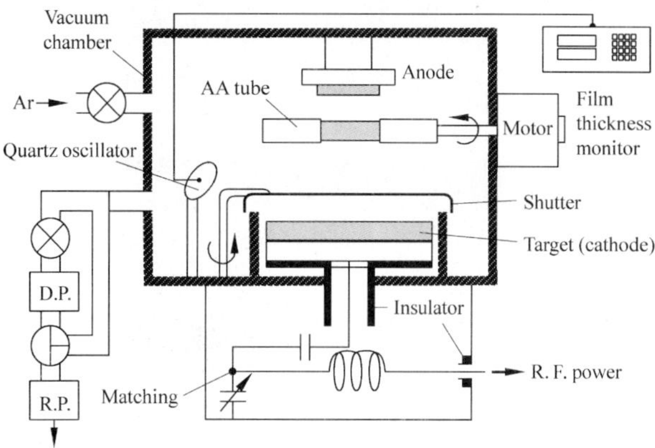

Fig.1.24 Schematic diagram of a R. F. sputtering apparatus for a tubular sample (Itoh et al., 2000)

Table 1.2 Deposition rates of Pd$_{85}$Ag$_{15}$ and Pt (Itoh et al., 2000)

	Target	Substrate temp. (℃)	Thickness(Å)	Deposition rate (Å /min)
1	Pd$_{85}$Ag$_{15}$	26	269	53.8
2	Pd$_{85}$Ag$_{15}$	150	310	62
3	Pd$_{85}$Ag$_{15}$	300	466	93.2
4	Pt	23	288	57.6

RF Plasma power   50W
Deposition time   5min
Deposition pressure 0.07Torr   (1torr=133.322 Pa)
Target-substrate distance 5cm

In the case of palladium film being exposed to a hydrogen atmosphere, the palladium lattice expands and then a number of cracks appear. Such a serious problem will, according to conditions, be solved by making a multi layered film. Figure 1.25 compares a palladium mono-layered film with a palladium/platinum double-layered film after exposure to a hydrogen atmosphere, where a glass plate is used as support. Platinum has a very low hydrogen solubility, so very

22

little shape change occurs in hydrogen. In the mono layer, many cracks can be seen while less cracks occur in the double layer. This is probably because the adhesive force between the palladium and platinum films becomes larger than that between palladium film and glass. The adhesion between the metals belonging to the platinum group is expected to be larger than the metal-ceramics.

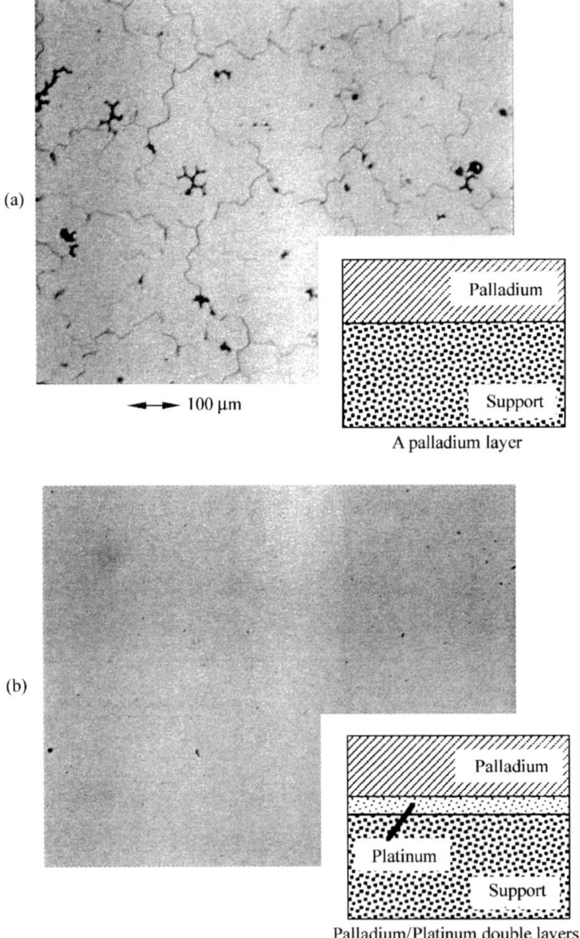

Fig.1.25 A multi-layer structuring, that can relieve an inner stress generated when a palladium layer is exposed in a hydrogen atmosphere, thereby depressing an outbreak of cracks

1.3.5 Rapid Quenching

Some amorphous alloys are known to show a high capacity of dissolving hydrogen, comparable to or exceeding crystalline metals. The high hydrogen solubility in amorphous alloys is considered to be closely related to their disordered atomic structure. The solubility of hydrogen in various amorphous alloys composed of Pd-Si or Pd-Si-M (M: other metallic elements) systems has been studied whereas most measurements were carried out at ambient temperatures (Stolz and Kirchheim, 1984, Szokefalvi-Nagy et al., 1987, Yoshinari and Kirchheim, 1991) The pressure-composition (p–c) isotherms in the amorphous $Pd_{1-x}Si_x$ alloys have been investigated intensively over a wide range of hydrogen pressures at room temperature, and have been elucidated by a theoretically derived equation (Szokefalvi-Nagy et al., 1987).

1) Preparation

The amorphous $Pd_{1-x}Si_x$ alloys with three compositions ($x = 0.15$, 0.175, 0.20) used are prepared by a rapid quenching method using a single-roller melt spinning apparatus as schematically drawn in Fig.1.26 (Itoh et al., 1997). The roller, made of copper (20 mm wide and 200 mm in diameter) is regulated to rotate at 1500 rpm. The alloy sample is put into a quartz tube, which has a thin slit (6 mm × 0.5 mm), heated to melt by an induction coil connected to a high-frequency generator, and then sprayed with 1.8-atm pressurized argon. The ribbon samples thus prepared have a mean thickness of about 40 μm and are approximately 5 mm in width. Of course, a wider ribbon would be obtained by making use of a wide-slit nozzle.

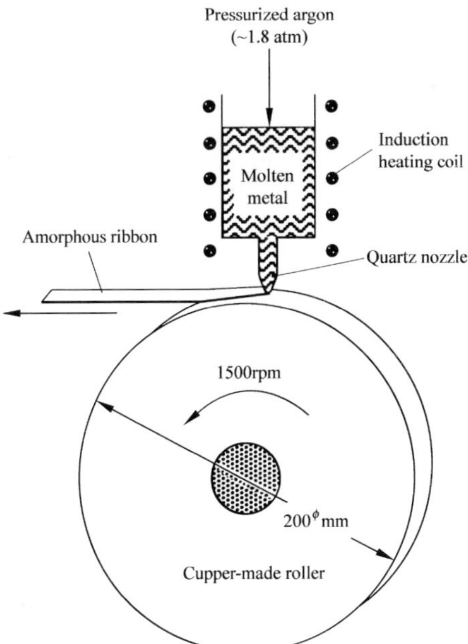

Fig.1.26 Single-roller melt-spinning technique for preparing amorphous ribbon (Itoh et al., 1997a)

24

2) Characterization

X-ray diffraction (XRD) patterns of the amorphous $Pd_{0.8}Si_{0.2}$ ribbon using Cu-Kα radiation are taken to see the crystallographic structure, Fig.1.27, in which that for the crystallized one and the ASTM reference peaks are also presented. A broad peak characterizing an amorphous structure can be seen. On the other hand, the crystallized one shows many sharp peaks reflecting a crystalline structure, the diffraction angles, 2θ, of which mostly correspond to the reference peaks from the ASTM card of the crystalline $Pd_{0.8}Si_{0.2}$.

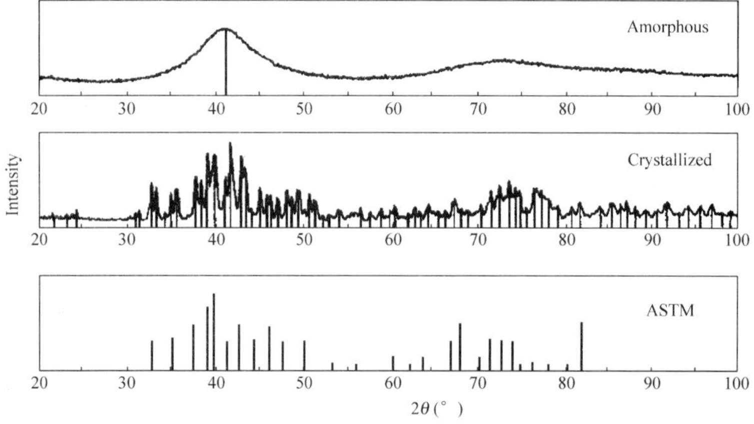

Fig.1.27   X-ray diffraction patterns of the $Pd_{0.8}Si_{0.2}$ alloy (Itoh et al., 1997a). The X-ray radiation used: Cu-Kα.

Differential scanning calorimetry (DSC) is used to determine the crystallization temperature of the amorphous alloy at a heating rate of 10°C/min in He or H_2 flow. DSC charts for the three amorphous are shown in Fig.1.28. The initial peak temperatures of crystallization in He flow for the three compositions are found to be 364.1°C (x=0.15), 380.1°C (x=0.175), 387.8°C (x=0.20), respectively. The crystallization temperatures become about 30°C lower in H_2 atmosphere.

Thermo-mechanical analysis (TMA) is carried out to observe thermal changes in dimensions of the ribbons at 5°C/min in N_2 or H_2 flow. Figure 1.29 shows a TMA chart for the amorphous $Pd_{0.85}Si_{0.15}$ (11.04 mm long and 1.75 mm wide). It can be seen that around 350°C a sudden contraction of 118 μm occurs. Since this temperature range corresponds to that of the crystallization process, as shown in Fig.1.28, it can be regarded that the total contraction rate of 1.07% is caused by the crystallization. Such a phenomenon would infer that the amorphous material should have a lower density than the crystallized. In fact, the densities measured by the Archimedes method of the three amorphous samples are 1.4%–2.4% lower than the corresponding crystallized one. The latter, in general, has a packed structure with the constituent atoms closer than the former.

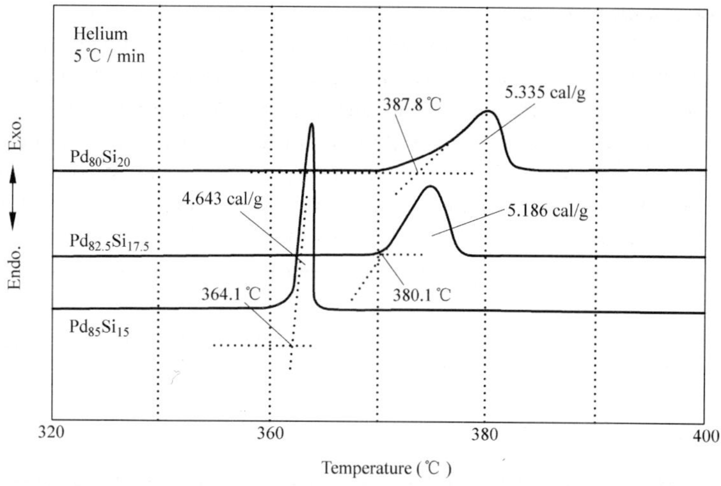

Fig.1.28   DSC charts of the Pd-Si amorphous alloys (Itoh et al., 1997a)

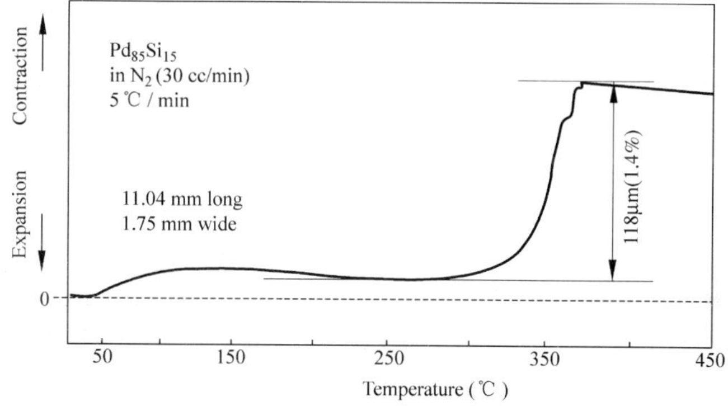

Fig.1.29   TMA chart of the amorphous $Pd_{0.85}Si_{0.15}$ Alloy (Itoh et al., 1997a)

1.4   Applications to Separations and Reactions

In the 1960s, a water desalination process by means of reverse osmosis was established. This was not turned into reality until an asymmetric cellulose membrane was developed by Loeb and Sourirajan (1962). The asymmetric structure, composed of a thin dense layer, effective for separation, and its thick and porous support layer, could drastically lower the resistance for water transport through the membrane and result in the high flux of water necessary for application to a commercial plant. Since then, such a structure has been in widespread use. Another progress in the 1960s was a development of producing asymmetric   membranes   from   synthetic   polymers   that   are   durable   in concentrated acids, alkalis as well as organic solvents. Then, it was pointed out that such solvent-resist membranes had the possibility of being employed in any system where a liquid-phase chemical reaction took place (Michaels, 1968). However, unfortunately there was no organic membrane possessing a long-term stability in the temperature range above $200\,^{\circ}\mathrm{C}$. Accordingly, a variety of separations by ceramic membranes, which can offer a high-temperature use, specifically selective separations and so on, will be increasingly adopted in the future chemical processes.

In this section, the principles of the flows of fluid through the various types of membranes introduced above, which are presented schematically in Fig.1.30, will be described. Furthermore, some applications of membranes to reactions will be explained concisely.

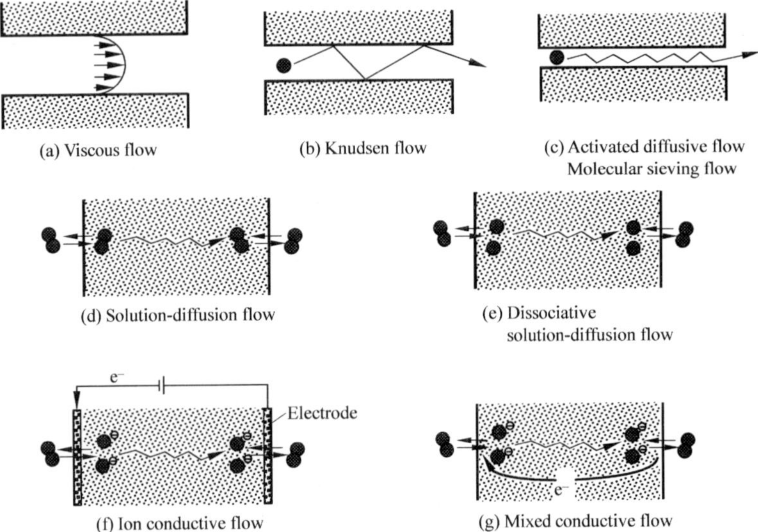

Fig.1.30　Mechanisms of gas flow through various materials

1.4.1　Microfiltrations and Ultrafiltrations

Principally, filtration is to separate liquid and suspended or colloidal solids from each other, where liquid mainly passes through membranes. Particles in the range　50 nm–2 μm of average diameter are processed with microfiltration. Paint pigment, bacteria, red blood cells, yeast cells, etc. are involved in this range. Ultrafiltration targets particles ranging from 2 nm to 100 nm such as viruses, proteins, colloidal particles etc..

In these filtrations, the flux of permeate, J_v, is given by the following equation,

$$J_v = \frac{\Delta P}{(R_B + R_G + R_M)} \tag{1.1}$$

where ΔP is the transmembrane pressure (TMP), and R_B, R_G and R_M are the resistances due to permeation through the boundary layer, the gel layer and the membrane, respectively. R_B and R_G, respectively, arise from the concentration polarization and the gel formation, so that those phenomena may be depressed by choosing the operatiry conditions and improving the flow pattern around membranes. R_M, relating to liquid flow through mesopores of membrane, can be written using the Darcy or the Hagen-Poiseuille equation.

$$R_M = \frac{\mu t_m}{k_D} \qquad \text{(Darcy equation)} \tag{1.2}$$

$$R_M = \frac{\mu t_m}{k_H r^4} \qquad \text{(Hagen-Poiseuille equation)} \tag{1.3}$$

where k_D and k_H are the constants, μ is the viscosity, t_m is the thickness, and r is the pore radius. As ΔP is increased, the flux proportionally increases at low ΔP, but gradually slows down and finally saturates at a certain ΔP (called the limited operation pressure). This is because increasing the flux simultaneously makes the gel layer grow thicker, which becomes a resistance for liquid permeation. Therefore, as a result, the flux neither increases nor decreases and remains a limited flux as shown in Fig.1.31 (Defrance and Jaffrin, 1999).

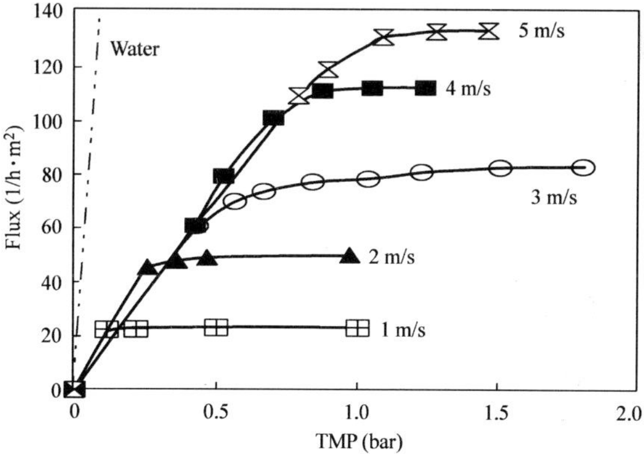

Fig.1.31 Behavior of microfiltration flux with varying transmembrane pressure (TMP) as well as linear velocity of feed (Defrance and Jaffrin, 1999)

$$1 \text{ bar} = 10^5 \text{Pa}$$

1.4.2   Gas Separations

1.4.2.1   Knudsen Regime

When a gas molecule (atom) passes through a straight pore of 2–100 nm to diameter, which is comparable or smaller than the corresponding mean free path, the flow of gas obeys the Knudsen law, and the permeance of the gas, Q_K (mol/s), is given by the following equation (Kennard, 1938).

$$Q_K = \frac{4}{3}\pi N r_p^3 \sqrt{\frac{2}{\pi MRT}}\frac{\Delta P}{t_m} \tag{1.4}$$

where N (m^{-2}) is the pore density, r_p (m) is the pore radius, M (kg $\cdot$ mol^{-1}) is the molecular weight, R(m$^3 \cdot$ Pa/(mol $\cdot$ K)) is the gas constant, T(K) is the absolute temperature, and ΔP(Pa) is the transmembrane pressure. The above equation can be rewritten using permeability, $\overline{P_K}$ (mol/(Pa $\cdot$ m $\cdot$ s)), porosity, $\varepsilon(-)$, and membrane area, A(m^2) as follows.

$$Q_K = \frac{4}{3}r_p \sqrt{\frac{2}{\pi MRT}}\left(\frac{N\pi r_p^2}{A}\right)A\frac{\Delta P}{t_m} = \overline{P_K}\varepsilon A\frac{\Delta P}{t_m} \tag{1.5}$$

Therefore, as temperature is fixed, the permeance ratio of gas i to gas j becomes,

$$\frac{Q_i}{Q_j} = \frac{\overline{P_i}}{\overline{P_j}} = \sqrt{\frac{M_j}{M_i}} \tag{1.6}$$

This is called an ideal separation factor, corresponding to the maximum separation performance attainable in this regime.

Using the tubular anodic alumina membrane (Itoh et al., 1998a), gas permeabilities of H_2, He, N_2, O_2, Ar and CO_2 at 40℃ are measured and plotted against the inverse of the square root of molecular weight in Fig.1.32, in which one can find a linearity through the origin. Also, in Fig.1.33, the temperature dependency of the permeabilities is shown. The permeabilities of all gases tested increase with an increase of $1/\sqrt{T}$. From such results, it is made clear that the gas permeation through the anodic alumina membrane obeys the Knudsen flow mechanism.

Based on this principle, ^{235}U (0.72 wt%), which is very valuable as a nuclear fuel, in ^{238}U (>99 wt%) is separated using a multi-stage membrane module because of a quite low separation factor between their gaseous forms, $^{235}UF_6$ and $^{238}UF_6$, 1.0043. In this case, more than 1000 stages are needed to enrich up to 3 wt% ^{235}U. Excluding such special purposes, gas separations by the Knudsen-flow-controlled porous membranes may not be feasible.

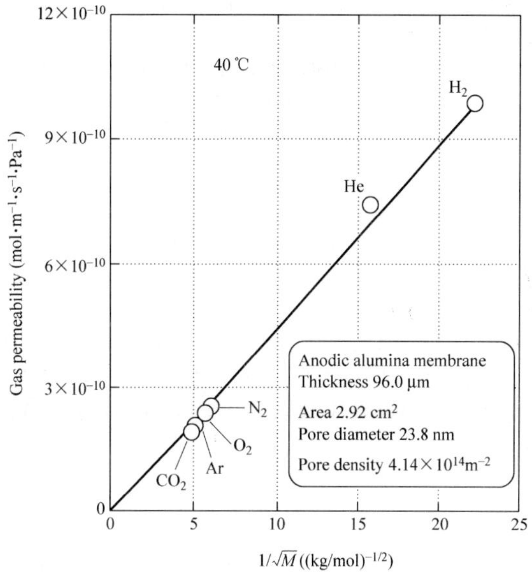

Fig.1.32 Plot of gas permeability vs. $1/\sqrt{M}$ (Itoh et al., 1996)

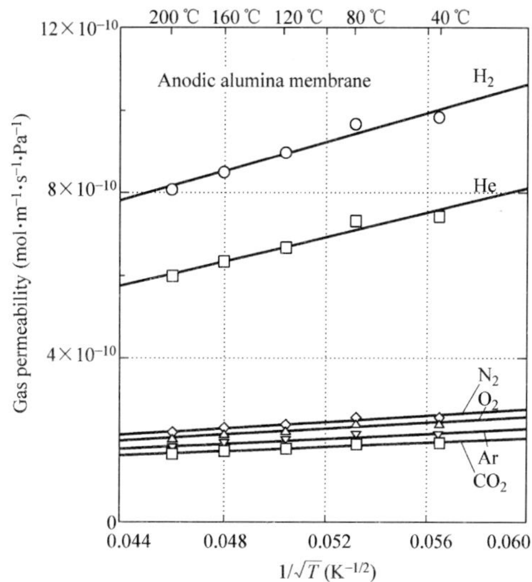

Fig.1.33 Plot of gas permeability vs. $1/\sqrt{T}$ (Itoh et al., 1996)

1.4.2.2 Surface Diffusion Regime

Gas molecules can more or less adsorb on the pore wall and migrate in the adsorbed layer. This flow, called surface diffusion, is observed particularly when gases having higher boiling points are used. The surface diffusion flux, Q_s(mol/s), is usually represented in the form of the Fick equation, that is, proportional to the concentration gradient of adsorbed layer as follows.

$$Q_s = -DSA\frac{dC}{dt_m} \tag{1.7}$$

Condensable gases like organic vapor, mainly pass through according to this mechanism. For instance, a fluorocarbon ($CClF_2CClF_2$, molecular weight 171 g/mol, b.p. 276.7 K) can permeate through porous vycor (silica) glass with 4 nm pores much more than expected from the Knudsen plot as shown in Fig.1.34a (Kato, 1994). However, as the temperature increases, the permeance approaches the value estimated from the Knudsen flow: this is explained by considering that both the Knudsen and the surface flows contribute to the total flux, Q_T, as follows.

$$Q_T = Q_K + Q_s \tag{1.8}$$

Namely, the difference obtained by subtracting the Knudsen permeance from the total permeance in Fig.1.34b corresponds to the permeance due to the surface flow. The permeability, $\overline{P_S}$, that is, Q_s will decrease with increasing temperature since the amount of adsorption decreases. It is also obvious that the surface flow is closely related to the boiling point of permeate.

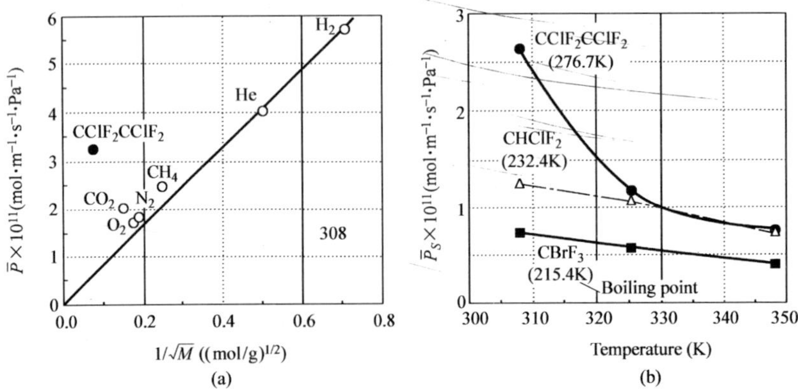

Fig.1.34 Permeability of fluorocarbons through vycor glass with 4nm-pore diameter: (a) permeability of ClF$_2$ ClF$_2$ largely deviated from the Knudsen plot;and (b) temperature dependency of the permeability attributable to the surface flow (Kato, 1994)

1.4.2.3 Capillary Condensation Regime

As the diameter of pore becomes small, condensation of gas inside the pore can occur more easily. This is quantitatively explained by use of the Kelvin equation as follows.

$$\ln\left(\frac{P_{\mathrm{c}}}{P_{\mathrm{sv}}}\right) = -\frac{2V_{\mathrm{m}}\gamma\cos\theta}{rRT} \tag{1.9}$$

where P_{c}(Pa) and P_{sv}(Pa) are the condensation and saturated vapor pressures, V_{m}(m^3/mol) is the molar volume of condensate, γ(N/m) is the surface tension of condensate, θ is the contact angle (approximated to be zero in usual calculations), r (m) is the pore radius, R(Pa $\cdot$ m^3/(K $\cdot$ mol)) is the gas constant, and T(K) is the temperature. Namely, vapor existing within pores with a small radius can condense more readily than that placed in a test tube. Condensed molecules due to the capillary condensation can prohibit the other molecules from permeating. Organic gases, inherently having high boiling temperatures, are more condensable, compared with inorganic gases.

1.4.2.4 Activated Diffusion Regime

It can be supposed that as the pore size approaches the molecular size, an energy barrier, over which a gas molecule must climb for permeation, becomes larger. In other words, an excess kinetic energy is necessary for a molecule to go through a narrow path. Such a behavior seems to be observed clearly when the pore diameter is smaller than approximately 2 nm. Since the kinetic energy is originated from molecular motion, increasing temperature should help this kind of diffusion process, which may be called an activated diffusion. The temperature dependencies of permeation rates of gases, i.e., the activation energy, differ from each other, depending roughly on the size of molecule. Figure 1.35 proves this, that is, H_2 has an activation energy larger than He in the case of a microporous membrane prepared by the sol-gel method (Kitao et al., 1990). On the other hand, N_2 is found to show an opposite dependency. This is inferred to be due to some pin holes left in the membrane.

1.4.2.5 Molecular Sieving Regime

Further, when the pore diameter becomes small and is comparable to molecular size, gas molecules with diameters larger than the pore can no longer enter into the pore. Sifting gases by their dimensions is one of our ultimate goals, although not realized so far, in membrane separation technology. Of all approaches made so far, fabricating zeolites in the form of membrane is considered to be promising. Zeolites, well known as adsorbents and catalysts, have a regulated pore structure, whose diameter is in the range 0.3–1 nm and controllable by changing the ratio of aluminum to silica in the starting raw materials for synthesis. The obtained film is polycrystalline and unfortunately involves many spaces between crystals, while an ideal film covered with a single crystal is

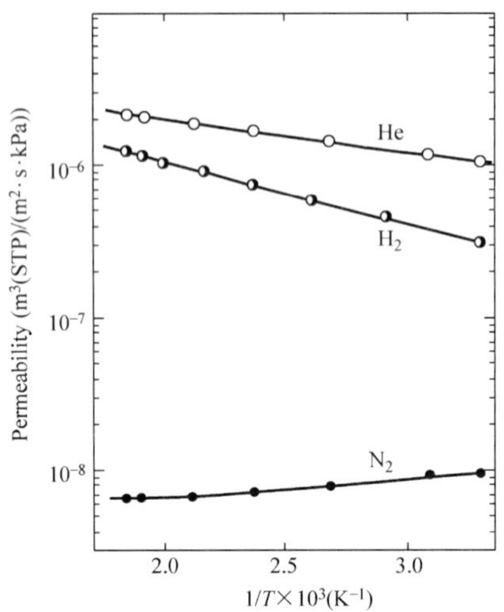

Fig.1.35 Arrhenius plots of the observed permeabilities of He, H_2 and N_2 in the range 30–270℃ (Kitao et al., 1990)

desired for separation based on the molecular sieving principle. However, some successful results are obtained in the highly selective separation of alcohol from an alcohol-water mixture by means of pervaporation, as schematically depicted in Fig.1.36. The alcohol-water system has an azeotropic point in the vapor-liquid equilibrium, so that a special method like the azeotropic distillation must be applied to attain a high concentration level. Separation by pervaporation using a silica-rich zeolite membrane is not limited by such a phenomena and shows a high alcohol selectivity as demonstrated in Fig.1.37, thereby making a single-stage operation possible (Sano et al., 1994). In this case, a strong affinity between the alcohol and the zeolite layer, not the sieving effect, contributes to the high selectivity and the permeation path is believed to be small openings remaining between fine crystals.

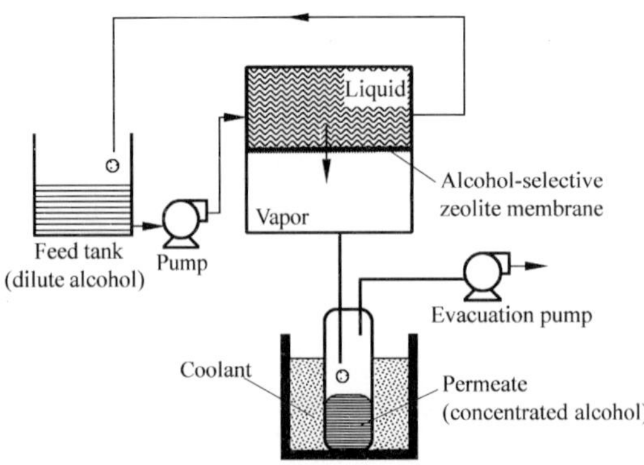

Fig.1.36 A pervaporation process for obtaining a solution of highly concentrated alcohol

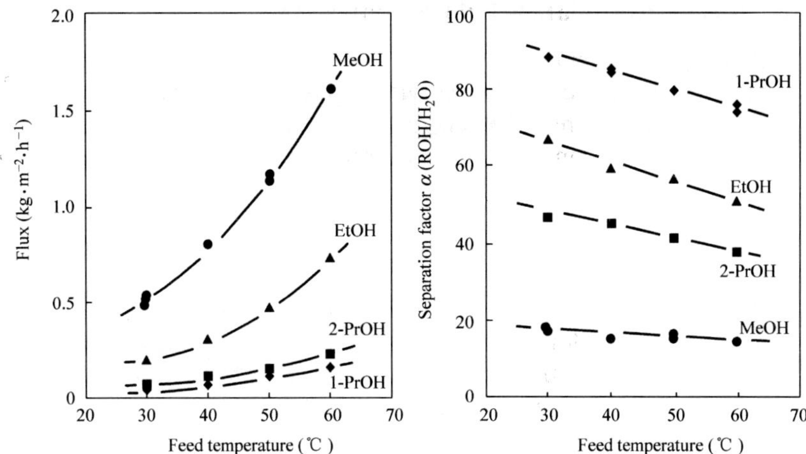

Fig.1.37 Temperature dependency of the flux and the separation factor in pervapora-tion of various alcohol-water mixtures where alcohol concentration in feed is fixed to be 1 mol% (Sano et al., 1994)

Regarding gas separation, although remaining vague points, some results representing a molecular sieving effect are reported. A typical instance is shown in Fig.1.38, in which gas permeabilities through a ZSM-5 membrane are correlated with the kinetic diameters of gases (Bakker et al., 1996). From the fact that ZSM-5 has a 0.55 nm straight channel close to the kinetic diameters of i-butane and i-octane, it can be partly understood that the permeabilities of those are extremely low.

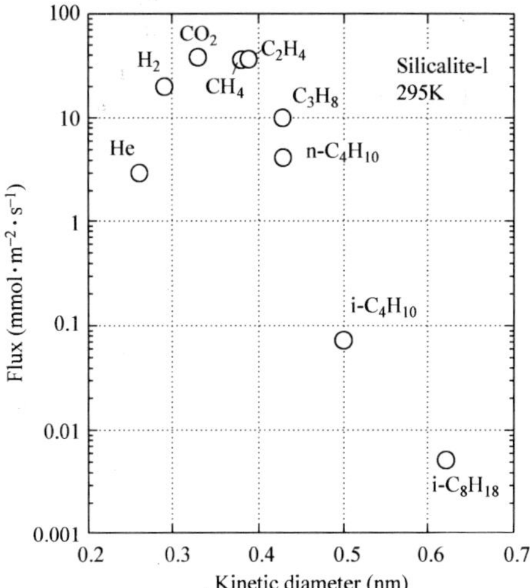

Fig.1.38 Relation between the flux through silicalite membrane and kinetic diameter (Bakker et al., 1996)

1.4.2.6 Dissociative Solution and Diffusion Regime

1) Solution-diffusion model

The solution-diffusion mechanism is very common in gas permeation through polymer membranes, where the permeation flux, $J(\mathrm{mol}/(\mathrm{m}^2 \cdot \mathrm{s}))$, can be represented by the following equation based on Fick's law.

$$J = \frac{D(c_{\mathrm{h}} - c_{\mathrm{l}})}{t_{\mathrm{m}}} \tag{1.10}$$

At steady state, the concentration of permeation species in the polymer membrane, c_{h} and $c_{\mathrm{l}}(\mathrm{mol}/\mathrm{m}^3)$, is in equilibrium with the corresponding partial pressure, p_{h} and $p_{\mathrm{l}}(\mathrm{Pa})$, in the gas phase. In the simple case, where c_i is proportional to p_i (Henry's law), these can be written as,

$$c_{\mathrm{h}} = Sp_{\mathrm{h}}, \; c_{\mathrm{l}} = Sp_{\mathrm{l}} \tag{1.11}$$

where S is the solubility constant. Then, (1.10) becomes,

$$J = \frac{DS(p_{\mathrm{h}} - p_{\mathrm{l}})}{t_{\mathrm{m}}} \tag{1.12}$$

So, if the pressure-composition (p-c) relation is given, the flux can be formulated similarly as presented here.

2) Dissociative solution of hydrogen in the amorphous metal

According to Kirchheim et al. (1982), the hydrogen dissolution in the disordered structure can be successfully described using a site energy function, $n(G)$, and a probability function, $o(G)$: the former represents a distribution of the energy levels of the interstitial sites available for hydrogen atoms, and the latter characterizes the thermal occupancy of hydrogen atoms. The total amount of hydrogen atom dissolved in an amorphous alloy, c, can be given by the following expression.

$$c = \int_{-\infty}^{\infty} n(G) \cdot o(G)\mathrm{d}G \tag{1.13}$$

It has also been shown by Kirchheim et al. that the Gaussian formula for the distribution of the site energy levels, $n(G)$, is appropriate for amorphous alloys and is able to describe either solubility or diffusivity of hydrogen.

$$n(G) = \frac{1}{\sigma\sqrt{\pi}}\exp\left[-\left(\frac{G - G^o}{\sigma}\right)^2\right] \tag{1.14}$$

where σ is the width of the Gaussian distribution, G^o is the mean free energy of dissolution of gaseous hydrogen at 1 atm.

Further, on an assumption that interstices in the amorphous lattice can be occupied by only one hydrogen atom, Fermi-Dirac statistics ($o(G)=\{1+\exp[(G-\mu)/RT]\}^{-1}$) are applied to give the probability $o(G)$ for the free energy, G, of a site to be occupied at a given temperature T. However, it can be approximated by the step function, that is,

$$o(G) = 1 \text{ for } \mu > G$$
$$o(G) = 0 \text{ for } \mu < G \tag{1.15}$$

where μ is the chemical potential of hydrogen and can be estimated from the hydrogen pressure in the gas phase, p_{H_2} (Pa), and the reference pressure p_o(=101325 Pa) as follows.

$$\mu = \frac{RT}{2} \ln\left(\frac{p_{H_2}}{p_o}\right) \tag{1.16}$$

Thus, (1.13) can be integrated using (1.14) and (1.15) as follows.

$$\mu = G^o - \sigma \cdot \mathrm{erf}^{-1}(1-2c) \tag{1.17}$$

Here, the composition, c, is defined as the ratio of the number of hydrogen atoms to the number of available interstices, which, however, is unknown in amorphous alloys, and therefore the atomic H/Pd ratio will be used. It can be seen from (1.17) that there is a linear relation between μ and $\mathrm{erf}^{-1}(1-2c)$, and that from the slope and the intercept of the straight line σ and G^o can be obtained respectively.

However, (1.17) including the inverse error function is too complicated to use in practice. Therefore, as an appropriate approximation for the inverse error function, $\ln 2c$ shows a good linearity with the $\mathrm{erf}^{-1}(1-2c)$. By putting this relation and (1.16) into (1.17), a concise formula is given as follows (Itoh et al., 1997a).

$$c = Sp_{H_2}^n \tag{1.18}$$

Equation (1.18) means that the relationship between the solubility and the pressure of hydrogen in the amorphous material can be described by the n-th power law. This is proven by the log-log plots, as shown in Fig.1.39, in which the n is dependent on temperature. Therefore, the hydrogen flux can be given by substituting (1.18) into (1.10).

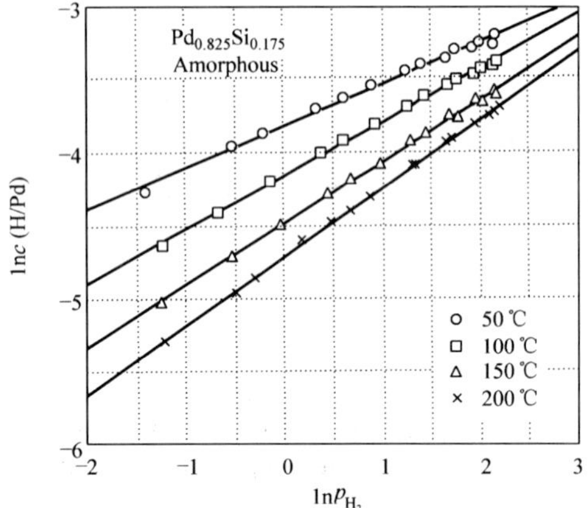

Fig.1.39 Double logarithmic plots of *p-c* isotherms for the amorphous Pd$_{0.825}$ Si$_{0.175}$ (Itoh et al., 1997)

$$J = \frac{DS(p_h^n - p_l^n)}{t_m} \tag{1.19}$$

According to the above relation, the flux, J, and ($p_h^n - p_l^n$) are correlated in Fig.1.40. A good linearity obtained indicates the validity of (1.19).

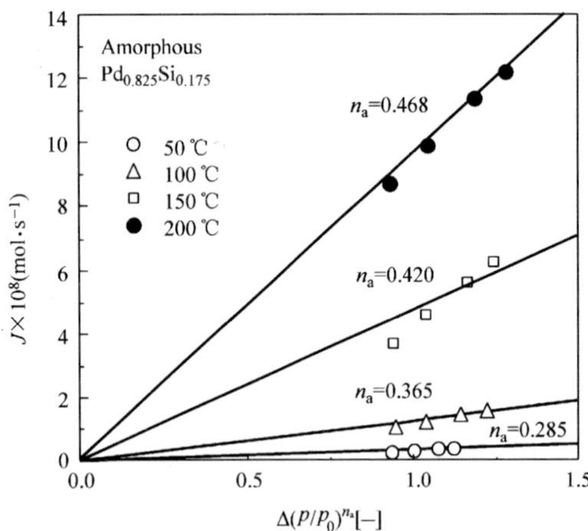

Fig.1.40 Correlation between the permeation rate of hydrogen and the pressure difference (Itoh et al., 1998b)

3) Crystalline metal

In the case of crystalline metals, $n(G)$ can be replaced by the Dirac delta function, $\delta(G - G^o)$, instead of (1.14). Thus the integration of (1.13) gives the following expression,

$$\mu = G^o + RT \ln\left(\frac{c}{1-c}\right) \tag{1.20}$$

When an absorption equilibrium holds, (1.20) is equated with (1.16) and is rearranged using a solubility constant, S, as follows.

$$\frac{c}{1-c} = S\sqrt{p_{H_2}} \tag{1.21}$$

If the solubility of hydrogen is quite low ($c \ll 1$), the following equation, well-known as Sievert's law (the half-power law), is obtained.

$$c = S\sqrt{p_{H_2}} \qquad (\text{for } c \ll 1) \tag{1.22}$$

Figure 1.41 is a plot representing a linear relationship between c and $\sqrt{p_{H_2}}$, where a crystallized $Pd_{0.825}Si_{0.175}$ is used (Itoh et al., 1998b). In this case, therefore, the hydrogen flux is given by,

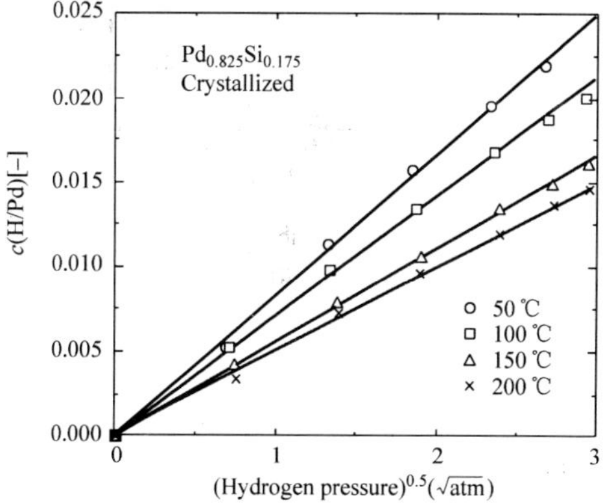

Fig.1.41 Relationship between the hydrogen solubility in the crystallized $Pd_{0.825}Si_{0.175}$ alloy and the half-power of hydrogen pressure (Itoh et al., 1997)

$$J = \frac{DS\left(\sqrt{p_h} - \sqrt{p_l}\right)}{t_m} \tag{1.23}$$

This equation is also valid as can be seen from Fig.1.42. Finally, a comparison of hydrogen permeabilities between the amorphous and crystal Pd-Si alloys is made in Fig.1.43. The amorphous form always outperforms the crystallized one

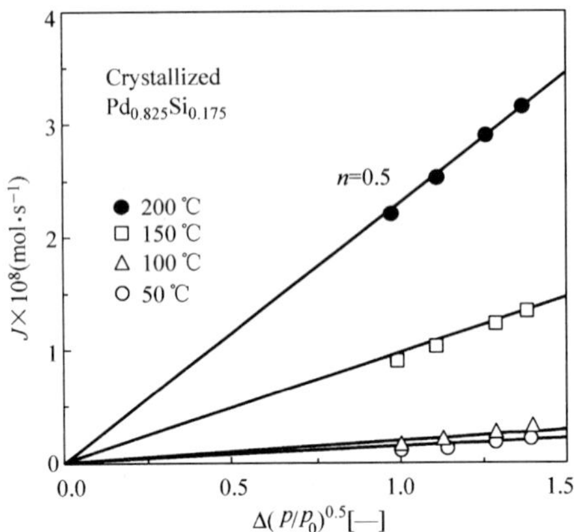

Fig.1.42 Correlation between the permeation rate of hydrogen and the pressure difference (Itoh et al., 1998b)

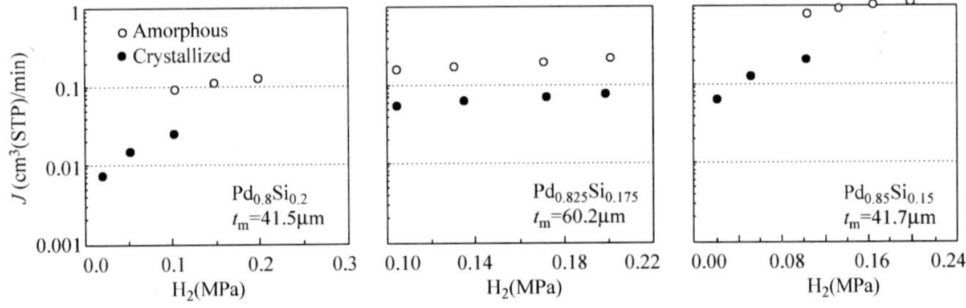

Fig.1.43 Comparison of the hydrogen permeabilities between the amorphous alloys and the corresponding crystallized ones at 250℃, where the permeate hydrogen was swept with argon gas of 15 ml/min and analyzed by a gas chromatograph to determine the flux, J (Itoh et al., 1998b)

in the hydrogen flux. This is, as mentioned above, because more sites for hydrogen atom can be available in the amorphous form. The whole mechanism of hydrogen permeation through the metallic membrane is represented by a schematic in Fig.1.44. When the membrane thickness is large and the diffusion process is the rate-determining step, (1.23) can be used to describe the flux. It should be noted that as the thickness becomes small the flux is affected by the surface processes like adsorption, dissociation and solution; J is not simply proportional to $1/t_m$.

The hydrogen fluxes and selectivities (H_2/N_2) of palladium-based membranes developed by various techniques are compared in Table 1.3, in which the flux of hydrogen is normalized by giving 1013hPa for the pressure on the feed side and 0hPa for the permeate side. So far, the electroless deposition technique seems to be the most successful and reliable.

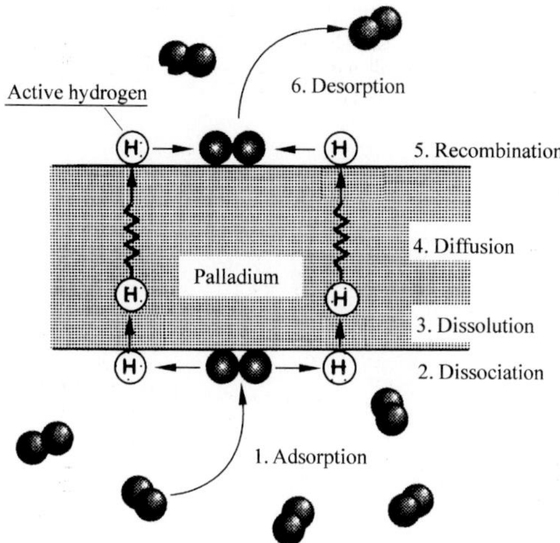

Fig.1.44　Mechanism of hydrogen permeation across palladium membrane

1.4.2.7　Ionic Conduction Regime

Solid electrolytes like yttria-stabilized zirconia have an oxygen ionic conductivity but no electronic one. Therefore, electrodes attached on both sides of the electrolyte and a lead connecting them are necessary to make oxygen anions flow. To obtain the driving forces for oxygen transport, there are two possible ways: producing a difference in partial pressures of oxygen on the both sides (concentration cell mode in Fig.1.45(a)), and applying a voltage externally (electrolytic mode in Fig.1.45(b)).

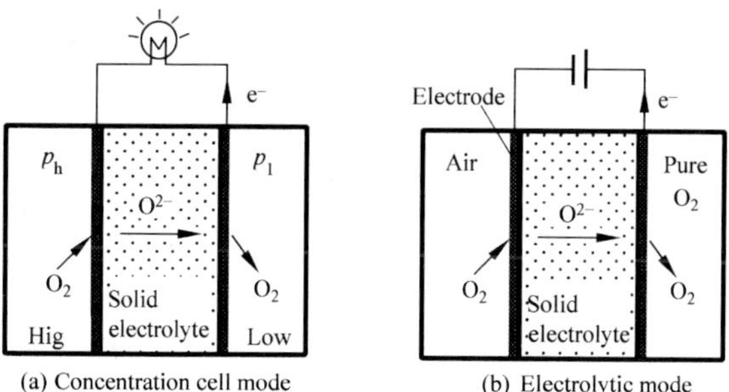

Fig.1.45　Oxygen anion conduction in the solid electrolyte

In the concentration cell mode, the oxygen flux, J, can be given by the following equation.

$$J = \frac{\sigma_i RT}{4n^2 F^2 t_m} \ln\left(\frac{P_h}{P_l}\right) \tag{1.24}$$

Table 1.3 Hydrogen fluxes and selectivities of palladium composite membranes

Investigator	Support		Method	Mambrane		Temperature	H_2flux[a]	Selectivity
	Material	Pore dia.		Material	Thickness			H_2/N_2
		(nm)			(μm)	(℃)	(cm^3/(cm^2 · min))	(−)
1. Uemiya et al.	Alumina tube	200	Electroless	Pd	4.5	400	61.6	>10000
(1991)			plating			450	70.0	>10000
						500	85.4	>10000
				$Pd_{77}Ag_{23}$	5.8	400	84.0	>10000
						450	95.2	>10000
						500	112	>10000
					13	400	43.5	>10000
					23	400	23.4	>10000
2. Collins and Way.	Alumina	200	Electroless	Pd	11.4	550	17	1180
(1993)	tube		plating			600	18	1180
3. Yan et al.	Alumina	150	CVD	Pd	2	100	1.4	100
(1994)	tube					300	13.6	1000
4. Li et al.	Alumina	4	Sputtering	$Pd_{86}Ag_{14}$	0.5	150	0.2	33[c]
(1999)	disk[b]					200	0.48	71[c]
						250	1.09	151[c]
						300	1.76	196[c]
5. Nam et al.	Stainless	500	Electroplating	$Pd_{78}Ni_{22}$	0.8	350	26	450
(1999)	disk					450	77	1100
						550	260	4500

| Investigator | Support | | Method | Mambrane | | Temperature | H_2flux[a] | Selectivity |
| | Material | Pore dia. | | Material | Thickness | | | H_2/N_2 |
		(nm)			(µm)	(℃)	($cm^3/(cm^2 \cdot min)$)	(−)
6. Itoh et al.	Anodic alumina tube	60	Electroplating	Pd	4	200	12.5	600
(2000)						250	17.7	780
						300	22.3	1260
						350	27.5	1640
7. Itoh and Xu			Rolling	$Pd_{77}Ag_{23}$ plate	100	200	4.4	∞
(1993a)						250	6.08	∞
						300	6.12	∞
						350	6.42	∞
				Pd Plate	100	200	2.1	∞
						250	2.98	∞
						300	3.66	∞
						350	3.88	∞
				$Pd_{93}Ni_7$ Plate	100	200	0.752	∞
						250	1.09	∞
						300	1.50	∞
						350	1.72	∞

(a) Δp=1013 hPa(0 hPa for the permeate side)   (b) H_2 permeance of substrate=$27 cm^3/(cm^2 \cdot bar)$ (c) H_2/He

42

where σ_i(S/m) is the ion conductivity, n is the ionic valence (n=2 in the oxygen ionic conduction), F(C/mol) is the Faraday constant, and p_h and p_l(Pa) are partial pressures of oxygen upstream and downstream, respectively. In this case, since a current spontaneously flows, an electric power can be generated in parallel. In particular, when hydrogen and oxygen are separately supplied to both sides, the system works as a fuel cell and can generate electric power. Fuel cells are expected to be one of the future power generators because the direct conversion method from chemical energy to electric power leads to a high energy efficiency compared with the conventional thermal power generation one.

In the electrolytic cell mode, J becomes

$$J = \frac{i}{2nF} \tag{1.25}$$

where i(A/m^2 or C/(m$^2 \cdot$ s)) is the current density observed. Equation(1.25) can be verified practically as can be seen in Fig.1.46. This mode will be applied to purification of oxygen from its mixture although electric power equivalent to the oxygen flow is consumed.

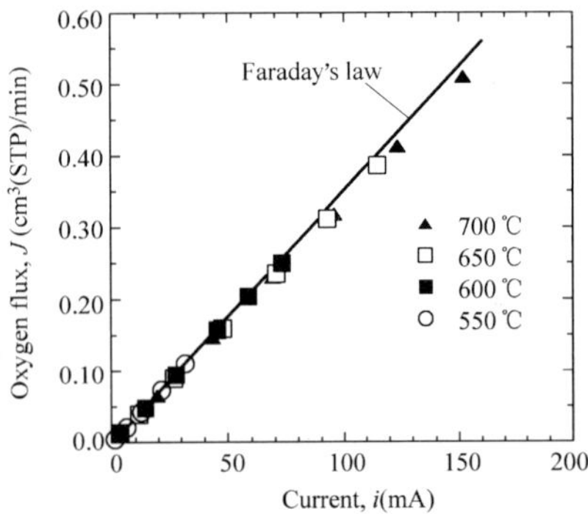

Fig.1.46　Linear relationship between current and oxygen flux (YSZ tube, Enomoto,1998)

1.4.2.8　Mixed Ionic Conduction Regime

The equation for oxygen transport across a mixed conductor has been derived based upon the concept that oxygen ions and electrons diffuse in a way shown in Fig.1.47, as follows.

$$J = \frac{RT}{4n^2 F^2 t_m} \frac{\sigma_e \sigma_i}{(\sigma_e + \sigma_i)} \ln\left(\frac{p_h}{p_l}\right) \tag{1.26}$$

where σ_e(S/cm) is the electronic conductivity. When the electronic conductivity, σ_e, is sufficiently larger than the ionic conductivity, σ_i, i.e., $\sigma_e \gg \sigma_i$, (1. 26) can be approximated by the following equation.

$$J = \frac{\sigma_i RT}{4n^2 F^2 t_m} \ln\left(\frac{p_h}{p_l}\right)$$

(1.27)

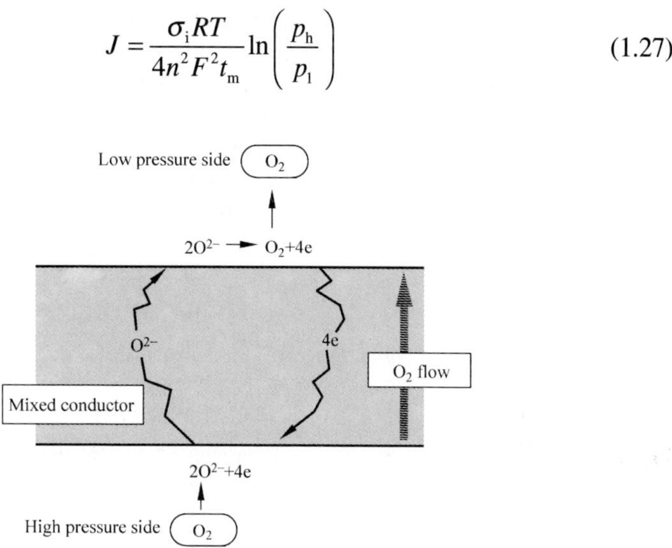

Fig.1.47 Ionic and electronic conduction in the mixed conductor

Figure 1.48 indicates the temperature dependence of oxygen permeation rate for the different Sr-content samples. Oxygen fluxes become substantial above 800 K and increase with increasing temperature, this is simply due to increasing diffusion rate, and is the highest when Sr content, X, is 0.6. In Fig.1.48, Vö$_x$ represents the concentration of oxygen hole, calculated on the assumption that the electric neutrality condition is satisfied. For instance, in the case of $La_{0.4}^{3+}Sr_{0.6}^{2+}Co^{3+}O_{3-\Delta}^{2-}$, it must be required that $(+3)\times0.4 + (+2)\times0.6+(+3)\times 1 + (-2)\times(3-\Delta) = 0$, i.e., $\Delta=0.3$, where Δ denotes the atomic ratio of oxygen vacancies. The amount of thus calculated oxygen vacancies is found to increase with increasing Sr content. Therefore, it is conceivable that the increment of concentration of oxygen vacancy, caused by replacing La^{3+} by Sr^{2+}, promotes the oxygen transport. As can be understood from (1.27), the oxygen permeation rate is inversely proportional to the thickness, t_m, so that developing thinner membrane is of great importance.

Oxygen-conductive perovskite oxides show a catalytic activity for oxidation of hydrocarbons. When this is combined with their oxygen separation function, simultaneous air separation and partial oxidation of methane become possible. The oxygen anion, passed through the membrane, can react with methane on the active membrane surface of other side to produce a syngas composed of CO and hydrogen (Balachandran, 1995, 1997; Tsai et al., 1997) or C_2 compounds (Elshof et al., 1995; Lin and Zeng, 1996; Zeng et al., 1998). Such a combination is promising as an innovative chemical reaction system.

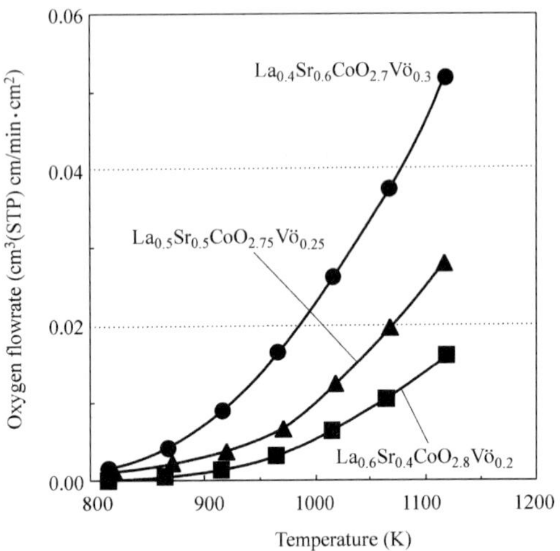

Fig.1.48 Oxygen flowrate observed with varying Sr content, where increasing Sr gives a higher flowrate of oxygen (Itoh et al., 1994)

1.4.3 Membrane Reactors

It is possible to prevent a reaction mixture from attaining equilibrium composition by using a reactor containing a membrane separation unit, which is used for continuous and selective removal of products from the reaction mixture. This composite type of reactor called a "membrane reactor", of course, will increase product yield and decrease the energy requirement for separation. There are three principal reactor configurations, a catalytic membrane reactor, a membrane reactor without catalyst and a catalyst-packed membrane reactor, as shown in Fig.1.49. More detailed reviews and discussions with inorganic membrane reactors appear in the recent book (Hsieh, 1996)

1.4.3.1 Catalytic Membrane Reactor

Let us call a catalytically active membrane a "catalytic membrane". When a reaction takes place on the surface of catalytic membrane, there are two possible ways as the reaction scheme. One is that a membrane-permeable species of products can pass to the other side of the membrane, that is, a product-removal mode. An earlier study on this mode was made for a partial dehydrogenation of cyclohexane to cyclohexene (Wood and Wise, 1966). The other is that an active species is supplied to the surface from the opposite side, that is, an-active-species-supply mode. Most studies have been concentrated on the latter mode. For instance, in the hydrogenation of 2-butynediol-1,4 into butenediol using a palladium- ruthenium alloy membrane, on which active hydrogen atoms emerged as shown in Fig.1.44, a selectivity of 98%–99% could be obtained

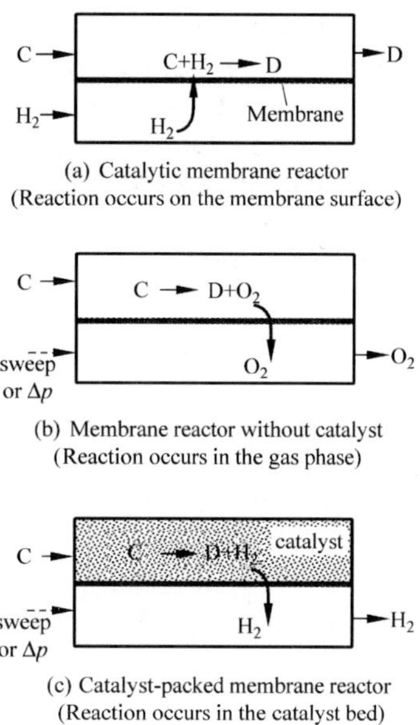

(a) Catalytic membrane reactor
(Reaction occurs on the membrane surface)

(b) Membrane reactor without catalyst
(Reaction occurs in the gas phase)

(c) Catalyst-packed membrane reactor
(Reaction occurs in the catalyst bed)

Fig.1.49 Membrane-reactor configurations, where the permeation should be driven by sweeping with inert gas or applying a pressure difference

while butandiol was produced in the hydrogen-bubbling system with the same membrane catalyst (Gryaznov, 1986). Similar results were obtained for dehydrolinalool hydrogenation to linalool using a $Pd_{94}Ru_6$ membrane (Gryaznov et al., 1993), for hydrogenation of ethylene, propylene, 1-butene and 1,3-butadiene (Nagamoto and Inoue, 1981, 1985), and for cyclohexene hydrogenation (Wood, 1968). Phenol could be selectively hydrogenated to cyclohexanone (an intermediate of nylon) by using palladium and its alloys as catalytic membranes. Palladium, of all those tested (Pd, $Pd_{93}Ni_7$, $Pd_{93}Ru_7$, $Pd_{77}Ag_{23}$) showed the best performance, 60% in conversion and 77% in selectivity (Itoh and Xu, 1993a).

Many perovskite-type oxides are also known to be useful as catalysts for the reduction of NO_x (Yasuda et al., 1990), so that its decomposition followed by in-situ separation of oxygen on the catalytic membrane composed of the oxide is possible. In addition, partial oxidation and coupling of methane using perovskite-oxide membranes, already mentioned in sect. 1.4.2.8, are of great interest, where oxygen separation from air is successfully incorporated.

1.4.3.2 Membrane Reactor without Catalyst

In the case of special reactions, which can go forward in a gas phase without catalyst, the membrane used is only for dividing a reactor into two compartments. There are only a few examples in the literature. Direct thermolysis of water vapor using a calcia-stabilized zirconia (CSZ) membrane tube at 1400–1800℃ (Carles and Baumard, 1982) and of carbon dioxide using either a CSZ membrane tube at 1400–1800℃ (Nigara and Carles, 1986) or an yttria-stabilized zirconia (YSZ) membrane tube at 1300–1500℃ (Itoh et al., 1993b) were attempted. In those high-temperature ranges, both CSZ and YSZ can show a little electronic conductivity in addition to an ionic one. Therefore the oxygen produced can be transferred in the form of O^{2-} ions without any electrodes or lead. An example of increasing the decomposition rate is shown in Fig.1.50 (Itoh et al., 1993b).

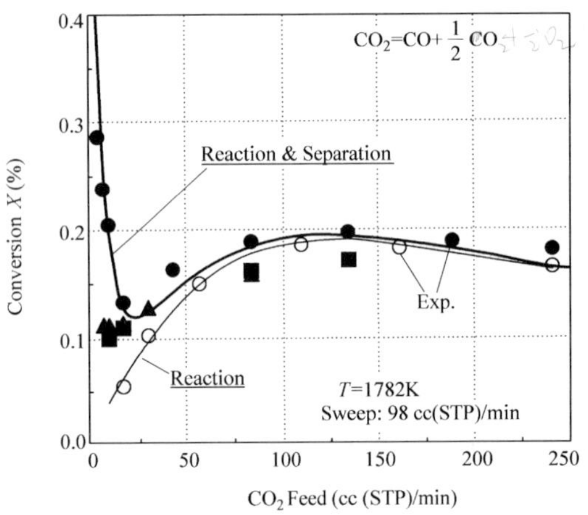

Fig.1.50 Increase in decomposition of carbon dioxide by selective separation of oxygen in a YSZ membrane reactor (Itoh et al., 1993)

1.4.3.3 Catalyst-packed Membrane Reactor

The most disadvantageous point in the catalytic membrane reactor is that the area for reaction is limited. To make a reaction progress at a sufficient rate, thereby increasing the quantity of the reaction per unit volume increase, a catalyst should be packed in the membrane reactor. Regarding such a membrane reactor, it has been shown that a selective separation of hydrogen from a reacting mixture using a microporous glass membrane can enhance H_2S decomposition (Dokiya et al., 1977, Kameyama et al., 1981) and C_6H_{12} dehydrogenation (Shinji et al., 1982) experimentally and do C_6H_{12} dehydrogenation (Itoh et al., 1983) and HI decomposition predictively. In fact, a higher conversion than that at equilibrium could be obtained. This kind of effect could be very much emphasized by using a palladium membrane for the dehydrogenation of cyclohexane (Itoh, 1987) because differently from the porous membrane the palladium membrane was impervious to other gases except hydrogen. In fact, a conversion of nearly 100% could be obtained for the dehydrogenation of cyclohexane to benzene, as shown in Fig.1.51. A more detail analytical and experimental study with respect to the palladium membrane reactors can be seen

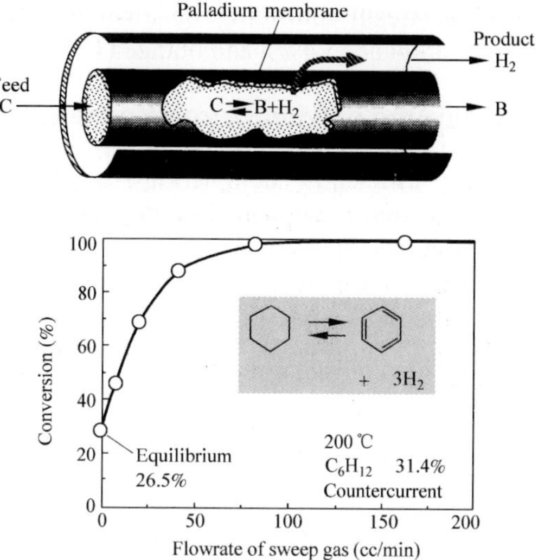

Fig.1.51   Chemical equilibrium shifting in a catalyst-packed palladium membrane reactor (Itoh, 1987)

in the literature (Itoh et al., 1990a, Itoh, 1995). Among numerous applications of palladium-membrane reactors to various hydrogen-generating reactions, which are equilibrium-limited, the most sophisticated and largest system is a membrane-reformer developed by a joint study between Tokyo Gas Corp. and Mitsubishi Heavy Ind., Japan. A 4 Nm³-H₂/hr bench-scale plant (Fig.1.52), in which 24 tubes composed of 20 μm thick palladium membrane supported on a stainless tube (20 mm in diameter, 600 mm long) are placed in the Ni catalyst-packed layer, can demonstrate that a conversion over 80% near 550℃ at 0.6

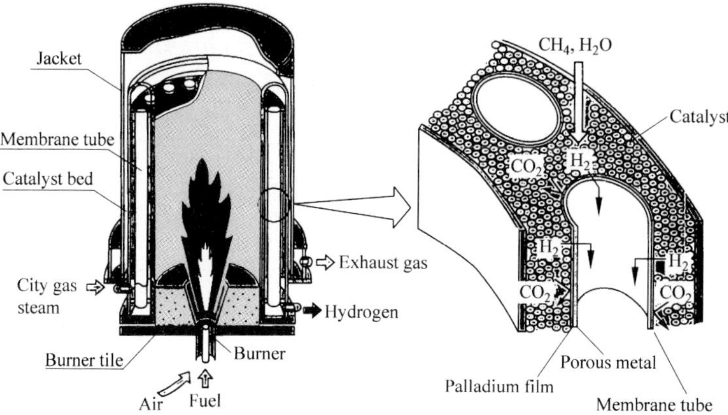

Fig.1.52   Membrane reformer for producing high-puririty hydrogen efficiently from steamreforming of methane (Kuroda et al., 1996)

MPa is attained for the steam reforming of town gas containing methane (88.5%), ethane (4.6%), propane (5.4%), and butane (1.5%).

1.4.4   Reaction Coupling in Membrane Reactors

As a means of more efficiently completing the reaction, coupling of dehydrogenation and oxidation in the palladium membrane reactor as illustrated in Fig.1.53 can offer an outstanding advantage. The hydrogen produced during the dehydrogenation easily passes through the membrane, and then can react readily with oxygen on the palladium surface. An example of the experimental results is shown in Fig.1.54 (Itoh et al., 1989, Itoh and Wu, 1997). A higher conversion can be obtained when 15.1% O_2-Ar is used compared to when an inert gas (Ar) is used. The coupling between dehydrogenation and hydrogenation also becomes true as demonstrated by Gryaznov et al. (1993).

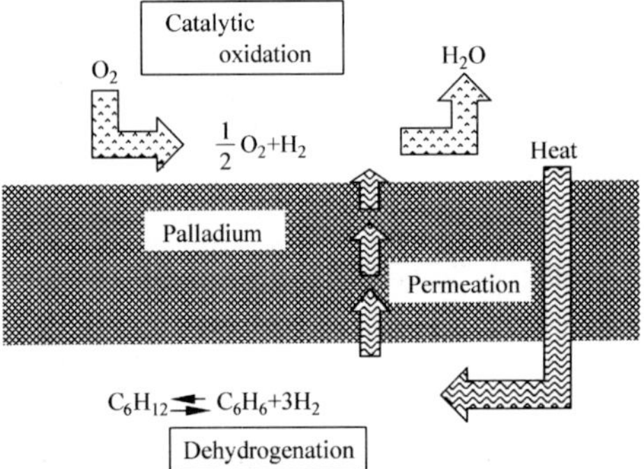

Fig.1.53   Reaction coupling in a membrane reactor, accompanied by heat reflux (Itoh et al., 1989)

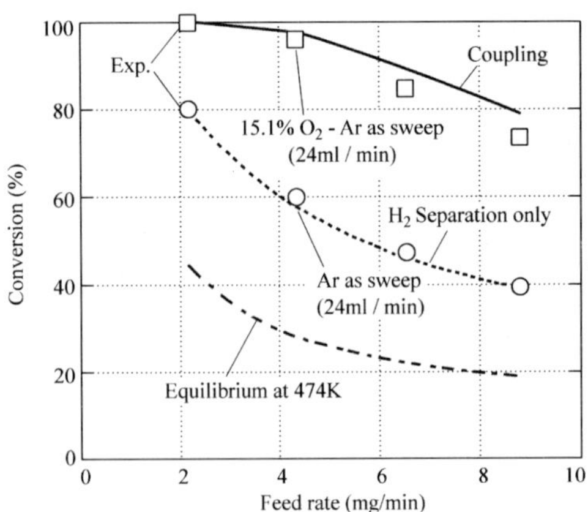

Fig.1.54   Increase in conversion of dehydrogenation of cyclohexane which is coupled with oxidation of hydrgen permeated by 15.1% O_2-Ar in a palladium membrane reactor (Itoh and Wu, 1997b).

Furthermore, if the reactor is thermally isolated from the exterior (adiabatic condition), the large amount of heat evolved by the oxidation can flow backward to the dehydrogenation side immediately, so that the endothermic heat required for the dehydrogenation is unnecessary to be supplied from the outside of the reactor, and simultaneously the dehydrogenation will go to completion (Itoh and Wu, 1997b). This bifunctional membrane reactor will lead to a "fuel-less reaction system" because the hydrogen combustion heat of -729 kJ/mol-cyclohexane ($3H_2$) is much larger than the dehydrogenation heat of 215 kJ/mol-cyclohexane.

1.4.5 Fuel Cells

One of the most attractive applications of solid electrolyte membrane is a fuel cell. For this purpose, stabilized zirconia membranes are used. In the concentration cell mode (Fig.1.46), when hydrogen is made to flow on the low oxygen pressure side, the system begins to work as a fuel cell with 1.23 V of standard potential. Since the mobility of oxygen anions in the zirconia becomes much larger above $800°C$, organic fuel gases like methane can be used directly whereas one is limited to hydrogen in polymer electrolyte fuel cells. To spread its practical application, oxygen-anion conductive inorganic materials, which can work at lower temperatures, are desired.

1.5 Summary

About half of the total amount of energy consumed in the petrochemical and chemical industry is estimated to be for separations. It is noted that nearly 90% of the separation is implemented by distillation, which, unfortunately, is an energy-intensive process with a phase change from liquid to gas by heating. In this sense, membrane technology without any phase changes is very significant and promising in terms of energy-efficient separation processing and is very competitive with other separation processes, as long as a mid-scale plant or less is required.

From another point of view, in the future industrial production system, key phrases such as shortage of natural resource and energy, environmental pollution, global warming, etc., which are directly related to increasing human population over the world, will be more significant. Therefore, minimizing use of materials and energy, and waste will be most important to establish a sustainable human society. Chemical reaction systems taking advantage of inorganic membranes and catalytic membranes, as mentioned in this chapter, may have the possibility of constructing a chemical process without further separation and integrating two or more reactions within a reactor (membrane reactor), thereby greatly improving the energy efficiency of the chemical process and being able to design totally compact chemical plant. It is, however, emphasized that such

50

noble expectations can not be realized before inorganic membranes with specific selectivity as well as extremely high permeability are developed.

References

Asaeda, M. and L.D. Du. J. Chem. Eng. Jpn. **19**, 72 (1986)

Asaeda, M. Proceedings of the Fifth International Conference on Inorganic Membranes. PL-1, Nagoya (1998)

Bakker, W.J.W, F. Kapteijin, J. Proppe, J.A. Moulijin. J. Membrane Sci. **117**, 57 (1996)

Balachandran, U., J.T. Dusek, R.L. Mievilee, R.B. Poeppel, M.S. Kleefisch, S. Pei, T.P. Kobylinski, C.A. Udovich and A.C. Bose. Appl. Catal. A. **133**, 19 (1995)

Balachandran, U., J.T. Dusek, P.S. Maiya, B. Ma, R.L. Mievilee, M.S. Kleefisch, C. A. Udovich. Catalysis Today. **36**, 265 (1997)

Brinker, C.J., T.L. Ward, R. Sehgal, N.K. Raman, S.L. Hietala, D.M. Smith, D.W. Hua and T.J. Headley. J. Membrane Sci. **77**, 165 (1993)

Cales, B. and J.F. Baumard. J. Mater. Sci. **17**, 3243 (1982)

Collins, J.P. and J.D.Way. Ind. Eng. Chem. Res. **32**, 3006(1993)

Defrance, L. and M.Y. Jaffrin. J. Membrane Sci. **152**, 203 (1999)

Dokiya, M., T. Kameyama and K. Fukuda. Denki Kagaku. **45**, 701 (1977)

Dong, J., T. Dou, X. Zhao and L. Gao. J. Chem. Soc., Chem. Commun. 1056 (1992)

Elshof, J.E., H.J.M. Bouwmeester and H. Verweij. Appl. Catal. A: General. **130**, 195 (1995)

Enomoto, T., Study on the catalytic membrane reaction systems using oxygen conductors. MS thesis, Science University of Tokyo (1998)

Gryaznov, V.M. Platinum Metals Rev. **30**, 68 (1986)

Gryaznov, V.M. Platinum Metals Rev. **36**, 70 (1992)

Gryaznov, V.M., O.S. Serebryannikova, Y.M. Serov, M.M. Ermilova, A.N. Karavanov, A.P. Mischenko and N.V. Orekhove. Appl. Catal. A: General. **96**, 15 (1993)

Hsieh, H.P. Inorganic Membranes for Separation and Reaction. (Elsevier, Tokyo, 1996)

Ishikawa, A., T.H. Chiang and F. Toda. J. Chem. Soc., Chem. Commun.. 764 (1989)

Itaya, K., S. Sugawara, K. Arai and S. Saito. J. Chem. Eng. Jpn., **17**, 514 (1984)

Itoh, N., Y. Shindo, K. Haraya, K. Obata, T. Hakuta and H. Yoshitome. Kagaku Kogaku Ronbunshu. **9**, 572 (1983)

Itoh, N., Y.Shindo, T.Hakuta and H.Yoshitome. Int. J.Hydrogen Energy. **9**, 835 (1984)

Itoh, N. AIChE J. **33**, 1576 (1987)

Itoh, N., K. Miura, Y. Shindo, K. Haraya, K. Obata and K. Wakabayashi. Sekiyu Gakkaishi. **32**, 47 (1989)

Itoh, N., Y. Shindo and K. Haraya. J. Chem. Eng. Jpn. **23**, 420 (1990)

Itoh, N. J. Chem. Eng. Jpn. **23**, 81 (1990)

Itoh, N. and W.C. Xu. Appl. Catal. A. **107**, 83 (1993)

Itoh, N., M.A. Sanchez C., W.C. Xu, K. Haraya and M. Hongo. J. Membrane Sci. **77**, 245 (1993)

Itoh, N., T. Kato, K. Uchida and K. Haraya. J. Membrane Sci. **92**, 239 (1994)

Itoh, N. Catalysis Today. **25**, 351 (1995)

Itoh, N., K.Kato, T.Tsuji and M.Hongo. J. Membrane Sci. **117**, 189 (1996)

Itoh, N., W.C. Xu, H.M. Kimura and T. Masumoto. J. Membrane Sci. **126**, 41 (1997)

Itoh, N. and T.H. Wu. J. Membrane Sci. **124**, 213 (1997)

Itoh, N., N. Tomura, T. Tsuji and M. Hongo. Microporous and Mesoporous Materials. **20**, 333 (1998a)

Itoh, N., W.C. Xu, S. Hara, H.M. Kimura and T. Masumoto. J. Membrane Sci. **139**, 29 (1998b)

Itoh, N., N.Tomura, T.Tsuji and M.Hongo. Micropor. Mesopor. Mater. **39**, 103(2000)

Jia, M.D., K.V. Peinemann and R.D. Behling. J. Membrane Sci. **82**, 15 (1993)

Kameyama, T., M. Dokiya, M. Fujishige, Y. Yokokawa and K. Fukuda. Ind. Eng. Chem. Fundam. **20**, 97 (1981)

Kanna, A., K. Kusakabe and S. Morooka. J. Membrane Sci. **141**, 197 (1998)

Kato, T. Study on inorganic membranes and its applications. MS thesis, University of Nihon (1994)

Kennard, E.H., Kinetic Theory of Gases. Chapter III, (McGraw-Hill, New York, 1938)

Kikuchi, E. and S. Uemiya. Gas Sep. Purif. **5**, 261 (1991)

Kikuchi, E. Sekiyu Gakkaishi. **39**, 301 (1996)

Kirchheim, R., F.Sommer and G.Schluckebier. Acta metall. **30**, 1059(1982)

Kita, H., K. Horii, Y. Ohtoshi, K. Tanaka and K. Okamoto. J. Mater. Sci. Lett. **14**, 206 (1995)

Kitao, S., H. Kameda and M. Asaeda. Membrane. **15**, 222 (1990)

Knapton, A.G. Platinum Metals Review. **6**, 44 (1962)

Kuroda, K., K. Kobayashi, Oouchida, Y. Ohta and T. Shirasaki. Mitsubishi Juko Giho. **33**, 346 (1996)

Kusakabe, K., S. Yoneshige, A. Murata and S. Morooka. J. Membrane Sci. **116**, 39 (1996)

Lewis, F.A. The Palladium Hydrogen System. (Academic Press, New York, 1967)

Li, A., W. Liang and R. Hughes. Sep. Purif. Technol. **15**, 113 (1999)

Lin, Y.S. and Y. Zeng. J. Catal. **164**, 220 (1996)

Loeb, S. and S. Sourirajan. Adv. Chem. Ser. **38**, 117 (1962)

Matsukata, M., N. Nishiyama and K. Ueyama. Microporous Materials. **1**, 219 (1993)

Michaels, A.S. Chem. Eng. Progr. **64**, 31 (1968)

Nagamoto, H. and H. Inoue. J. Chem. Eng. Jpn. **14**, 377 (1981)

Nagamoto, H. and H. Inoue, Chem. Eng. Commun. **34**, 315 (1985)

Nam, S.E., S.H. Lee and L.H. Lee. J. Membrane Sci. **153**, 163 (1999)

Nigara, Y. and B. Cales. Bull. Chem. Soc. Jpn. **59**, 1997 (1986)

Ohbayashi, H., T.Kudo and T. Gejo. Japanese J. Appl. Phys. **13**, 1 (1974)

Sakamoto, Y., F.L. Chen, M. Furukawa and K. Mine. J. Less-Common Met. **166**, 45 (1990)

Sano, T., Y. Kiyozumi, M. Kawamura, F. Mizukami, H. Takaya, T. Mouri, W. Inaoka, Y. Toida, M. Watanabe and K. Toyoda. Zeolites. **11**, 842 (1991)

Sano, T., M. Hasegawa, Y. Kawakami, Y. Kiyozumi, H. Yanagishita, D. Kitamoto and F. Mizukami. Stud. Surf. Sci. Catal. **84**, 1175 (1994)

Shimizu, S., Y. Kiyozumi and F. Mizukami. Chemistry Lett. 403 (1996)

Shinji, O., M. Misono and Y. Yoneda. Bull. Chem. Soc.. Jpn. **55**, 2760 (1982)

Shu, J., B.P.A. Grandjean, E. Ghali and S. Kaliaguine. J. Membrane Sci. **77**, 181 (1993)

Smith, A.W. J. Electrochem. Soc. **120**, 1068 (1973)

Stolz, U. and R. Kirchheim. J. Less-common Metals. **103**, 81 (1984)

Suzuki, K., Y. Kiyozumi, T. Sekine, K. Obata, Y. Shindo and S. Shin. Chemistry Express. **5**, 793 (1990)

Szokefalvi-Nagy, A., S. Filipek and R. Kirchheim. J. Phys. Chem. Solids. **48**, 613 (1987)

Teraoka, Y., T. Nobunaga and N. Yamazoe. Chem. Lett. 503 (1988)

Tsai, C.Y., A.G. Dixon, W.R. Moser and Y.H. Ma. AIChE J. **43**, 2741 (1997)

Tsuru, T., S. Wada, S. Izumi and M. Asaeda. J. Membrane Sci. **149**, 127 (1998)

Uemiya, S., Y. Kude, K. Sugino, N. Sato, T. Matsuda and E. Kikuchi. Chem. Lett. 1687 (1988)

Uemiya, S., T. Matsuda and E. Kikuchi. J. Membrane Sci. **56**, 315 (1991)

Uhlhorn, R.J.R, K. Keizer and A.J. Burggraaf. J. Membrane Sci. **66**, 271 (1992)

Wood, B.J. and H. Wise. J. Catalysis. **5**, 135 (1966)

Yan, S., H. Maeda, K. Kusakabe and S. Morooka. Ind. Eng. Chem. Res. **33**, 616 (1994)

Yan, Y., M.E. Davis and G.R. Gavalas. Ind. Eng. Chem. Res. **34**, 1652 (1995)

Yasuda, H., N. Mizuno and M. Misono. J.Chem. Soc., Chem. Commun. 1094 (1990)

Yoldas, B.E. Am. Cerm. Soc. Bull. **54**, 289 (1975)

Yoshinari, O. and R. Kirchheim. J. Less-common Metals. **172–174**, 890 (1991)

Zeng, Y., Y.S. Lin and S.L. Swartz. J. Membrane Sci. **150**, 87 (1998)

2 Sol-Gel Thin Films Synthesis and Properties of Sorbents and Catalysts

Y.S.Lin

2.1 Introduction

2.1.1 Catalysts and Adsorbents

Chemical reactions and separations are key processes in chemical and related industries. Among various chemical reaction processes heterogeneous reactions are most commonly found for chemical conversion. Heterogeneous reactions require use of solid catalysts to enhance the chemical reaction rates. On the other hand, adsorption has become, in the past two decades, a widely accepted industrial process for separation of gas and liquid mixtures. The adsorbent is a porous solid that can preferentially adsorb certain gases or liquids over others on its surface.

The most common mode of operation for a heterogeneous reaction or adsorption process is the fixed-bed process (Fogler, 1992; Ruthven, 1984), as shown in Fig.2.1. The fixed-bed is packed with solid catalyst or adsorbent. In the reaction process, the reactants are fed into the fixed-bed and the products, side-products and unrecalled reactants exit the fixed-bed as the effluents. At the steady state the production rate, yield and selectivity for the products depend on the catalytic properties of the catalyst and the reaction conditions (Fogler, 1992; Smith, 1981).

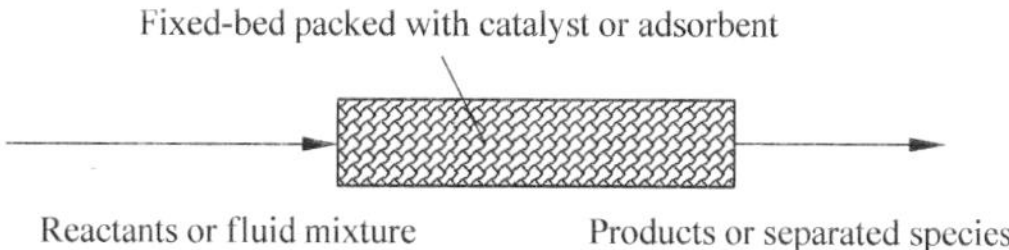

Fig.2.1 Fixed-bed process for heterogeneous chemical reaction or adsorption separation

The fixed-bed adsorption process operates in the unsteady-state mode. Fluid mixture is fed into the bed packed with a fresh adsorbent. Species preferentially adsorbed on the adsorbent are retained in the fixed-bed, and are the non- or less adsorbable species pass through the fixed-bed and are collected as pure species. The preferentially adsorbed species will eventually emerge from the fixed-bed and this point is called the breakthrough point. Once this happens the process has to be terminated and the fixed-bed needs to be regenerated by either lowering the pressure (pressure swing adsorption) or increasing the temperature (temperature swing adsorption). The more preferentially adsorbed species is collected in the regeneration step. Obviously, the purity of the separated fluid products, production rate and energy consumption depend on the properties of the adsorbents and operation conditions (Ruthven, 1984; Yang, 1997).

A solid catalyst usually consists of a porous support and active species dispersed on the support surface. Although the active species is critical to the selectivity and activity of the catalyst for a specific reaction, the support body often has strong effects on the catalytic properties of the catalysts in the following two ways. The support and the chemical/physical interaction between the support and active species affect the chemical properties of the active species. The pore structure of the supports is also important. Support with a large surface area is required to enhance the volume-based reaction rate. Pore size, which usually decreases with increasing surface area, may affect the transport of the reactants or products and therefore the reactivity and selectivity of the catalysts (Ruthven, 1984; Smith, 1981).

As for solid catalysts, the adsorbents are made of porous support with or without an active species. The latter is more common for physical adsorption and for these adsorbents certain gases or liquids can be preferentially adsorbed on the support surface via the van der Waals forces. Again, larger surface area is required for the adsorbent in order to obtain a higher adsorption capacity of the adsorbent based on the adsorbent unit volume. The adsorption rate is determined not only by the chemical nature of the adsorbing species and support surface but also by the pore size of the adsorbents. Adsorbents with a large pore size (low diffusional resistance), usually have small surface area. Very often a compromise has to be made between the surface area and pore size.

Practically the catalysts or adsorbents are used in the form of granules (pellets) of spherical or cylindrical shape with size in the range from 1 to 5 mm. This is required in order to reduce the pressure drop of the fluid through the fixed-bed and for easier handling of the catalysts or adsorbents during their transportation and packing. When the support body is made of microporous material with large surface area and small pore size (<2 nm, such as zeolites), the support in the form of fine powder (a few micrometers in size) should be used in order to reduce the diffusional resistance in the pore of the support. The granules or pellets of this type of catalysts or adsorbents consist of the fine support powder and inorganic binder with much larger pore size. As a result, a bimodal pore size distribution is found on this type of catalyst and adsorbent.

In addition to the catalytic properties, adsorption capacities, selectivity and kinetic rates, other major requirements for the granular catalysts or adsorbents are mechanical strength and attrition resistance. In the fixed-bed catalysts and adsorbents with good attrition resistance are easier to handle, and have a long application time. This also minimizes the formation of attrition powder that has to be separated from the downstream. The catalysts or adsorbents with good mechanical properties are especially important for applications in fluidized-bed or moving-bed chemical reaction and separation processes. The mechanical properties depend mainly on the binder used and the method and conditions of the granulation processes (Fayed and Otten, 1984).

2.1.2　Sol-Gel Process

Sol-gel processing refers to the process of fabrication of ceramic materials by preparation of a sol, gelation of the sol, and removal of the solvent (Brinker and Scherer, 1990). Sols are dispersions of colloidal particles in a liquid solvent. A gel is a solid matrix encapsulating a solvent. In a sol-gel process the sol can be formed from a solution of colloidal powders or hydrolysis and condensation of alkoxides or salt precursors. In the latter approach, which is much more popular, primary particles of uniform size are formed and grow in a sol and connect to each other to from aggregates during gelation. These aggregates forming the network of gel are broken apart into the primary particles in the drying step. Upon calcination and sintering, these primary particles are bound together strongly to form a very stiff solid network, and large interparticle spaces with uniform nanoscale pores are formed.

A typical sol-gel process using an alkoxide precursor involves the following steps: (1) formation of stable sols; (2) casting or shape formation; (3) gelation of the sols; (4) aging of the gel; (5) drying of the gel; (6) calcination; (7) sintering, if necessary. It should be pointed out that sol-gel processing can also be done with a wide variety of precursors in addition to alkoxides. With alkoxide ($M(OR)_n$) as a precursor, sol-gel chemistry can be described in terms of two classes of reactions.

$$\text{Hydrolysis:} \qquad -MOR + H_2O \Leftrightarrow -MOH + ROH$$
$$\text{Condensation:} \qquad -MOH + ROM \Leftrightarrow -MOM- + ROH$$
$$\text{or} \qquad -MOH + HOM- \Leftrightarrow -MOM- + H_2O$$

The above simple description of sol-gel chemistry identifies two key ideas. First, a gel forms because of the condensation of partially hydrolyzed species into a three-dimensional polymeric network. With time the colloidal particles and condensed species link together to become a 3-D network. The physical characteristics of the gel network depend greatly upon the size of the particles and extent of cross-linking prior to gelation. At gelation, viscosity increases sharply, and a solid object results in the shape of the mold.

Second, any factors that affect either or both of these reactions are likely to impact on the properties of the gel. In fact it is the control of many of these factors, generally referred to as sol-gel parameters, that separates sol-gel preparation from other methods. A representative but not exhaustive list of these parameters includes type of precursors, type of solvent, water content, acid or base content, precursor concentration, and temperature. These parameters affect the structure of the initial gel and, in turn, the properties of the material at all subsequent processing steps.

The time between the formation of a gel and its drying, known as aging, is also an important parameter. Aging of the gel, also called syneresis, involves maintaining the cast object or wet gel granules for a period of time, hours to days, completely immersed in liquid solvent. During aging, polycondensation continues along with localized solution and reprecipitation of the gel network, which increases the thickness of interparticle necks and decreases the porosity. The strength of the gel thereby increases with aging. An aged gel must develop sufficient strength to resist cracking during drying.

Another parameter that affects a sol-gel product is the drying condition. During drying the liquid is removed from the interconnected pore network, and large capillary stresses can develop during drying when the pores are small (<20 nm). These stresses will cause the gels to crack catastrophically unless the drying process is controlled by decreasing the liquid surface energy by addition of surfactants or elimination of very small pores, by supercritical evaporation, which avoids the solid/liquid interface, or by obtaining monodisperse pore sizes by controlling the rates of hydrolysis and condensation. Xerogel is obtained under conventional evaporative drying conditions. The resultant materials of supercritical drying of wet gels, known as aerogels, have high surface area, porous structure, and low density. In a way drying can be viewed as part of the overall aging process because the material can, and often does, undergo physical and chemical changes during this stage.

Calcination of gels at temperatures higher than drying temperatures causes further removal of solvents and dehydration of the gels. Those physically adsorbed water molecules are removed at the initial stage of calcination, and thermal dehydroxylation subsequently occurs. Porosity develops when, due to additional cross-linking or neck formation, the gel network becomes sufficiently strengthened to resist the compressive forces of the surface tension created during the calcination stage. Moderate densification of the gel skeleton also occurs due to the condensation reactions and structural relaxation. Sintering of the calcined gels at even higher temperatures will further consolidate the gel skeleton by eliminating the majority of the porosity of the gels. The primary particles in sintered gels are crosslinked together so closely that they generate the strongest binding for the material.

Ceramic thin films, sensors, nanoscale materials, multifunctional ceramic composites, optical fibers, ceramic membranes and many other products can be manufactured by the sol-gel process (Brinker and Scherer, 1990; Klein, 1988). The major applications of sol-gel processing are in the ceramic industry for

fabrication of oxide ceramics and glasses. Several studies have been reported on the preparation of supported catalysts and zeolite granular particles using the sol-gel technique (van der Grift et al., 1991; Wolff et al., 1993; Fanelli et al., 1991; Duisterwinkel and Frens, 1995; Li et al., 1990; Spek and van Beem, 1982; Mirsky et al., 1978). Sol-gel derived inorganic thin films and membranes have recently attracted attention from both academic and industry (Bhave, 1991; Hsieh, 1996; Burggraaf and Cot, 1996). Only limited studies have been carried out on the sol-gel fabrication of adsorbents for industrial separation or purification purposes.

Sol-gel prepared alumina adsorbents have also been used as flue-gas cleaning adsorbents and proved to be superior to the conventional adsorbents with respect to their chemical and mechanical stability (Duisterwinkel and Frens, 1995; Hakvoort et al., 1987). Hydrophobic alumina and silica aerogels with large specific surface area (~600 m^2/g) and large pore volume (~17 cm^3/g) were synthesized by supercritical drying (Fanelli et al., 1991; Armor and Carlson, 1987; Teichner et al., 1976), but no report on their adsorption property was found. Most recently, aerogel-like mesoporous siliceous molecular sieve material with large specific surface area (1075 m^2/g), large pore volume (1.36 cm^3/g) and promising adsorption properties was also successfully synthesized by a sol-gel method at conventional drying conditions (Guo, 1995; Guo et al., 1996).

Modified silica or diatomaceous earth oxides with impregnated active species have been applied as packings or supports in gas or liquid chromatography (Grassini-Strazza et al., 1989; Kohli and Badaisha, 1985). These materials generally have much lower surface area and were not studied for large-scale separation. Zirconia and titania were prepared by the sol-gel method to replace silica as packing materials for high-pressure liquid chromatography (HPLC) (Trudinger et al., 1990; Rigney et al., 1989; Weber et al., 1990; Blackwell and Carr, 1991). These adsorbents were used to separate structural isomers, amino acids and proteins by HPLC (Weber and Carr, 1990; Blackwell and Carr, 1992) and to separate peridines by ligand-exchange GC (Fujimara and Ando, 1977).

As a result of the unique characteristics of the sol-gel processing, the sol-gel-derived single component materials have high purity because of the quality of the available precursors. Furthermore, the textural properties of the product, most notably surface area and pore size distribution, can be tailored. These materials usually have large surface area, uniform pore size distribution, controlled average pore size, and good mechanical strength, which are particularly important to the materials for separation applications. Besides, the microstructure of the sol-gel-derived material can be controlled and tailored together with its bulk form (i.e., spherical particles, fibers and thin films).

However, it is believed that the area in which sol-gel preparation is going to make the most significant impact is multicomponent systems because of the following specific advantages of the sol-gel approach: (i) the ability to control structure and composition at a molecular level; (ii) the ability to introduce several components in a single step; (iii) the ability to impose kinetic constrains on a system and thereby stabilize metastable phases; and (iv) the ability to fine-tune the activation behavior of a sample and thereby trace the genesis of active species (Ward and Ko, 1995).

Our research conducted at Cincinnati has been primarily focused on sol-gel synthesis of alumina, zirconia, titania and silica. These metal oxides not only are commonly used as adsorbent or catalyst support but also have recently emerged as excellent materials for ceramic membranes. The objective of this chapter is to report synthesis and properties of these sol-gel-derived adsorbent materials with emphasis on development of a sol-gel granulation method and the properties of the sol-gel-derived granular adsorbents.

2.2   Sol-Gel-Derived Materials for Adsorbents and Catalysts

The adsorptive and catalytic properties of an adsorbent/catalysts depend mainly on the pore structure (surface area, pore size and pore volume) and surface chemical properties of the adsorbent. The sol-gel-derived ceramic adsorbents/catalysts possess unique pore structure defined by the microstructure of the sol-gel-derived materials. This section describes the synthesis and microstructure of several crystalline and amorphous adsorbents/catalysts prepared by the sol-gel method in our laboratory.

2.2.1   Crystalline Materials

Common crystalline adsorbent/catalyst materials include γ-alumina, zirconia and titania. These porous ceramic bodies consist of small crystallites of alumina, zirconia or titania. The alumina, zirconia or titania sols can be directly prepared by dispersing the fine solid particles of these oxides in aqueous solution. They can also be prepared from the inorganic or metalorganic precursors of these oxides.

In our laboratory, alumina sol was prepared by the Yoldas process (Yoldas, 1975). Typically, stable 1 M alumina (boehmite) sol was synthesized by dissolving 260 ml of aluminum tri-sec-butoxide in 1 liter of water at 70–90°C. The boehmite precipitate formed from the hydrolysis and condensation was peptized by adding 70 ml 1 M HNO_3 solution at 90–100°C under refluxing condition. Stable 0.25 M titania sol was prepared by dissolving 74 ml titanium tetra-isopropoxide (with 500 ml isopropanol) in 450 ml water in a nitrogen box. The titania precipitate was washed with water to remove alcohol and diluted with 1 l of water. The product was then peptized by adding 72 ml of 1 M HNO_3 at 75°C under refluxing conditions.

Stable 0.25 M zirconia sol was synthesized by modifying a procedure reported earlier from our laboratory (Chang et al., 1994). It was prepared by hydrolysis and condensation of 0.25 mole zirconium n-propoxide in a water (900 ml)/isopropanol (500 ml) solution. The white zirconia precipitates were filtered with vacuum suction and washed in water several times to remove the isopropanol. In this process, a small amount of water was added to the zirconia precipitates to help filtering, and the washed water was filtered again to prevent the loss of zirconia precipitates. The filtered zirconia cake was diluted in 1 l of water and peptized with 125 ml of 1 M HNO_3 solution at 90–100°C overnight.

[Xerogels of alumina, titania or zirconia in the form of thin sheet (20–200 μm thick) were prepared by pouring respective sols in given quantities in to petri dishes. The sols in the petri dishes were dried at 40°C and 40%–50% relative humidity. Gelation occurred after a sufficient amount of solvent (water) had evaporated. The xerogel samples were calcined in a temperature programmable box furnace at 450°C for 3 h, with carefully controlled heating and cooling rates.]

XRD data reveal that alumina particles in the sol are of boehmite crystalline structure and the particles in zircornia and titania sols are of amorphous structure (Chang et al., 1994). The alumina, titania and zirconia samples obtained from the sols after gelation and calcination at 450°C are respectively, in the phases of γ-alumina, tetragonal zirconia and anatase. These are thermodynamically meta stable phases, and may transform to the thermodynamically stable phases, which are α-alumina, monoclinic zirconia and rutile. The crystallite structure and lattice parameters of these phases are listed in Table 2.1.

Table 2.1 Characteristics of phase transformation of sol-gel-derived alumina, zirconia and titania

Material	Initial phase and lattice parameter(Å)	Final phase and lattice parameter(Å)
Alumina	γ-alumina (cubic)	α-alumina (hexagonal)
	a=7.8	a=4.8, b=13.0
Zirconia	Tetragonal	Monoclinic
	a=b=5.1, c=5.3	a=5.14, b=5.2, c=5.3, β=99.2°
Titania	Anatase (tetragonal)	Rutile (tetragonal)
	a=b=3.8, c=9.5	a=b=4.9, c=3.0

Figure 2.2 shows the pore size distributions of γ-alumina, titania (anatase) and tetragonal zirconia (after calcination at 450°C for 3 h). The pore structure data of these three adsorbent samples are compared in Table 2.2. The pore structure data were obtained from nitrogen adsorption isotherms measured by a Micromeritics ASAP-2000 adsorption porosimeter. As shown in Fig.2.2, the pore size distributions of these materials are rather narrow, with an average pore diameter of about 3 nm. Such a narrow size distribution and nanoscale average pore size are determined by the primary crystallite particles. The particles of the sol-gel-derived alumina, titania and zirconia, due to the Ostwald ripening

mechanism, are usually of nanoscale size, with a uniform particle size distribution (Brinker and Scherer, 1990). γ-alumina crystallites are of plate-shape (Leenaars and Burggraaf, 1984; Lin et al., 1991) with a size in the range from about 5 to 20 nm. The sol-gel-derived γ-alumina consists of such plate-shaped crystallite particles, which give rise to a relatively large surface area. Crystallites of tetragonal zirconia and rutile are of a more spherical shape, with a crystallite size of about 15 nm and 11 nm, respectively (Christian, 1975).

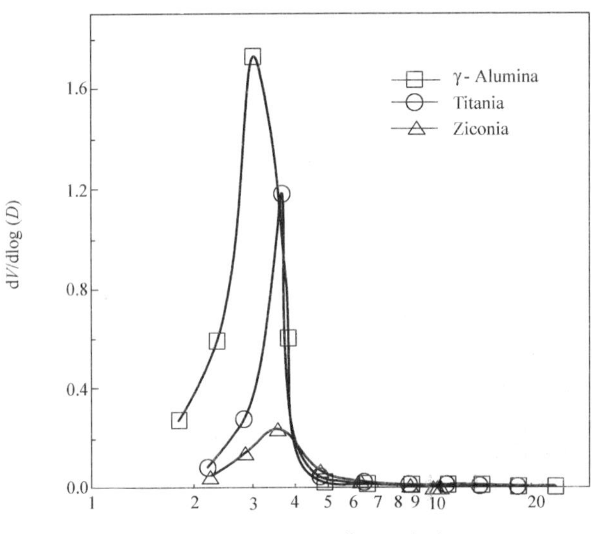

Fig.2.2 Pore size distributions of sol-gel-derived alumina, zirconia and titania

Table 2.2 Pore structure of γ-Al$_2$O$_3$,titania and zirconia (calcined at 450℃ for 3 h)

Materials	Average pore size (nm)	Pore volume (ml/g)	Surface area (m²/g)
γ-Al$_2$O$_3$	2.8	0.33	373
TiO$_2$	3.4	0.21	147
ZrO$_2$	3.8	0.11	57.2

Alumina, titania and zirconia in their metastable phases will transform to their stable phases. Such phase transformation usually occurs via a nucleation and crystal growth process. Kinetically, however, the phase transformation will not occur or be observed at low temperatures. For alumina, the γ-Al$_2$O$_3$ to α-Al$_2$O$_3$ (via δ-and θ-aluminas) phase transformation is found to start at temperatures around 900℃. This is accompanied by a sharp decrease in the surface area and increase in the pore size of the alumina adsorbent/catalyst. The surface area of γ-Al$_2$O$_3$ adsorbent will decrease with time at temperatures lower than 900℃ due primarily to sintering (Chang et al., 1994).

From the metastable tetragonal phase, zirconia starts transforming to the more stable monoclinic phase around 700°C. At temperatures lower than 700°C, sintering causes a change of the pore structure of zirconia (decrease in surface area and increase in pore size). At temperatures higher than 700°C, phase transformation plays a major role in affecting the pore structure of the zirconia. Phase transformation also results in a decrease in surface area and an increase in pore size (Chang et al., 1994; Gopalan and Lin, 1995). For titania the phase transformation from anatase to rutile, occurring around 450°C, also accompanies a substantial change in the pore structure of titania (Gopalan et al., 1995).

The activation energy for phase transformation is respectively about 600, 570 and 213 kJ/mol for the sol-gel-derived alumina, zirconia and titania (Chang et al., 1994). These data were obtained when these materials were exposed to air. The presence of steam in the atmosphere appears to reduce the activation energy for phase transformation, thus enhancing the rate of the pore structure change of the adsorbent/catalyst materials at a given temperature. For practical application, the pore structure of the materials can be kinetically stabilized by heat-treating the materials at a temperature a few hundred degrees higher than the application temperature.

2.2.2 Noncrystalline Materials

Although silica has/or catalyst several crystalline forms, only amorphous silica finds uses as adsorbents for separation and purification applications. Both particulate silica sol (with dense particles larger than 1 nm) and polymeric silica sol (without dense particles larger than 1 nm) can be prepared, depending on the precursor, solvent and, most importantly, the catalyst used. Silica prepared from particulate silica sol has mesopore size with surface area in the range comparable to γ-alumina. Microporous silica adsorbents with pore diameter smaller than 2 nm can be prepared from polymeric silica sol. This microporous silica has a surface area much larger than the crystalline adsorbents discussed above.

Silica sol is prepared by hydrolysis of a metal alkoxide and subsequent or simultaneous polycondensation (polymerization), which can be either acid-or base-catalyzed. The acid-catalyzed hydrolysis proceeds through an electrophilic attack of the H^+ ions. This means that the reactivity decreases as the number of OR groups decreases with the progression of hydrolysis (Brinker et al., 1984). The probability of formation of fully hydrolyzed silicon, $Si(OH)_4$, is thus very small. Since the condensation reaction starts before the silicon alkoxide is completely hydrolyzed, silicon alkoxide molecules will polymerize with nonhydrolyzed alkoxyl groups, in which the degree of crosslinking is low. The gyration radius of these small molecules is typically of the order of 1.5–1.7 nm (Brinker et al., 1982).

The base-catalyzed hydrolysis reaction proceeds through the nucleophilic substitution of OH^- ions. The reactivity increases as the number of OR groups decreases. The silicon alkoxide will tend to be completely hydrolyzed. The condensation reaction is the rate-determining step and the more crosslinked species will grow at the expense of the smaller one (Brinker et al., 1982). Therefore the base-catalyzed hydrolysis condensation reactions yield large, highly crosslinked polymers. Gels formed from these large polymers normally contain large pore (Iler, 1979). Therefore, the base-catalyzed synthesis will not result in microporous materials due to the large highly crosslinked particles. Only the acid-catalyzed reactions can result in microporous materials.

In our laboratory, microporous silica sol was experimentally prepared by the nitric-acid-catalyzed hydrolysis and condensation of tetraethoxysilane (TEOS). A typical sol composition (in molar ratio) was 3.8 for ethanol/TEOS, 6.5 for deionized water/TEOS and 0.09 for nitric acid/TEOS. This composition was adopted from Uhlhon et al. (1992). Solutions of TEOS/ethanol and nitric acid/water were prepared separately and then mixed in a spherical flask equipped with a reflux condenser. The mixture was rapidly heated to and eld at $90\,°C$ for 3 h with vigorous stirring. Finally, the resulted sol was cooled naturally to room temperature.

1.5 mol% Al_2O_3-doped SiO_2 was also prepared by the sol-gel method. In this case, a solution of 10% wt/vol $Al(NO_3)_3 \cdot 9H_2O$ in ethanol was prepared by heating $Al(NO_3)_3$ at moderate temperature for 15 min to ensure complete dissolution of the salt in ethanol. Ethanol was chosen as the solvent for the salt because it was also used in preparing the silica sol. Doping the silica sol with alumina was performed by mixing the sol with a controlled amount of the ethanol solution of aluminum nitrate.

Xerogel samples of silica- or alumina-doped silica were obtained by drying the corresponding silica sols in petri-dishes at $40\,°C$ and 60% relative humidity for one to two days. Silica (in the form of thin sheet) was obtained by calcining the xerogel samples at $400\,°C$ for 3 h with heating and cooling rates of $25\,°C/h$. Pore structure and surface properties of microporous silica were analyzed by nitrogen adsorption using an adsorption porosimeter (Micromeritics ASAP 2000 with micropore capability). The sample was first degassed at $300\,°C$ under vacuum for more than 10 h until the sample passed the degassing check up test. The sample was then weighed and transferred to the measurement port. The nitrogen adsorption isotherm was measured at liquid nitrogen temperature (78 K) and nitrogen pressure ranging from 10^{-7} to 1.0 p/p_0 (relative pressure of nitrogen). The Horvath-Kawazoe model (Horvath and Kawazoe, 1983), which was preinstalled in the micropore analysis program of the ASAP 2000 adsorption porosimeter, was applied to calculate the pore size distribution and micropore volume of the silica sample.

The pore size distribution of a typical sol-gel derived silica sample is given in Fig.2.3. As shown, the sol-gel-derived silica sample contains microporous (<2 nm) pores, with an average pore diameter of about 0.6 nm. The pore structure of the pure and alumina-doped silica samples is summarized in Table

2.3. The surface area of sol-gel-derived silica is much larger than the sol-gel derived crystalline samples.

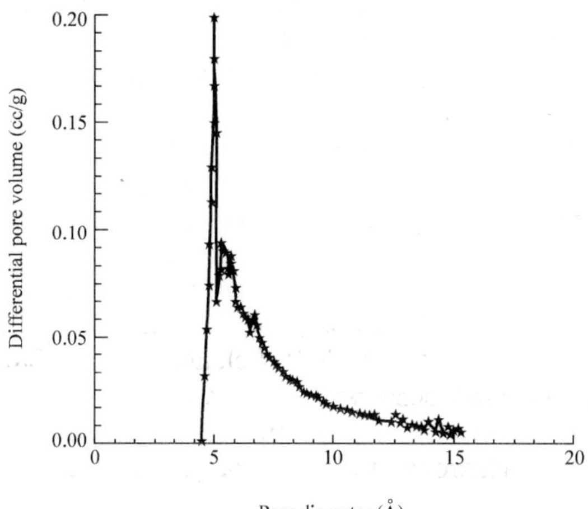

Fig.2.3 Pore size distribution of sol-gel-derived silica sample

Table 2.3 Pore structure of pure microporous silica and alumina doped silica

Materials	BET surface area (m²/g)	Average pore diameter (Å)	Pore volume (cm³/g)
Pure SiO$_2$	588	6.4	0.24
1.5%Al$_2$O$_3$-SiO$_2$	660	6.5	0.26

As with silica gels prepared by other methods, the sol-gel-derived microporous silica is not thermally stable at high temperatures, especially under humid conditions. Table 2.4 shows the pore structure data of both silica samples after heat treatment at 600℃ for 30 h. Figure 2.4 compares the pore size distributions of a silica sample before and after the heat-treatment. For pure silica, the heat treatment results in an 89% reduction in the surface area and a loss of 87% micropore volume. These results agree with the previous study that indicated similar effects of steam and heat on the stability of silica. Such a structural change of silica at high temperature is believed to be due to continuous condensation of surface silano groups and sintering.

Table 2.4 Pore structure of pure microporous silica and alumina-doped silica after heat-treatment at 600℃ for 30 h (data in parenthesis are the changes with respect to data in Table 2.3)

Materials	BET surface area(m²/g)	Average pore diameter (Å)	Pore volume (cm³/g)
Pure SiO$_2$	64.4(–89%)	8.4(31%)	0.03(–87%)
1.5%Al$_2$O$_3$-SiO$_2$	447.1(–32%)	6.9(6%)	0.18(–30%)

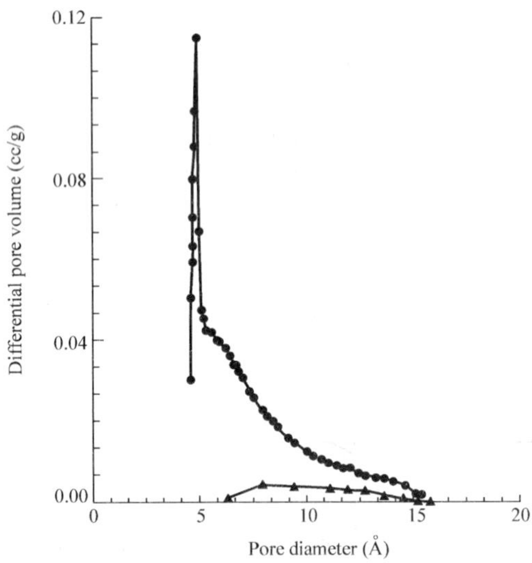

Fig.2.4　Comparison of the pore size distribution of a silica sample before doped silica(●)
and after(▲)heating treatment at 600℃ for 30 h

As to the alumina-doped silica sample, the reductions of surface area and
pore volume after the heat-treatment was approximately 32% and 30%, and the
increase of average pore diameter were just 6%. This improvement can be more
clearly illustrated by comparing the pore size distribution of the alumina-doped
silica sample before and after the heat treatment, as shown in Fig.2.5. The alumina-
doped silica clearly has a better thermal stability than the pure silica adsorbent.

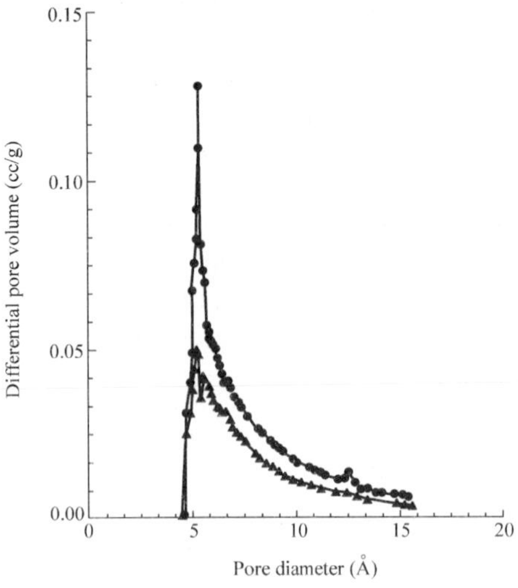

Fig.2.5　Comparison of the pore size distribution of a 1.5% alumina-doped silica sample
before (●) and after (▲) heating treatment at 600℃ for 30 h

2.3 Sol-Gel Granulation Process

Adsorbents/catalysts are practically used in the granular form. The performance of an adsorption/reaction process critically depends on the size, shape, pore texture and mechanical strength of the support body used. The sol-gel process has commonly been employed for preparation of ceramic films, fibers and powders. This section reports our work on synthesis and properties of the sol-gel-derived alumina granular particles. The principle of the sol-gel-derived granulation process should be applicable to preparation of granular particles of other ceramic materials discussed in the preceding section.

2.3.1 Granulation Processes

Conventional granulation processes can be classified as agitation methods (tumbling agglomeration and mixer agglomeration), pressure methods (compaction, extrusion and rolling), thermal methods (sintering), spray and dispersion methods and agglomeration from liquids (Capes and Fouda, 1984). With these established granulation methods, relatively large particles are usually formed from small particles through different intraparticle forces of cohesion. After granulation processes the microstructures of granules generally do not change significantly, but the bulk properties of the granules improve substantially. Granules can be prepared to have useful structural forms and shapes, better definition of quantity units, larger bulk density, less dusting losses, to be easier to handle, and to have better heat transfer and controllable porosity, which are very important for adsorbents and catalysts.

The process to make spherical granular supports, catalysts and adsorbents is mainly based on the methods that rely upon the buildup of the smaller particles into spheres by means of a rolling, or "snow-ball" technique. The production costs for this type of processes are low, but the spheres are irregular and the particle sizes are randomly distributed. A few studies were reported on the sol-gel routes for preparing spherical gel particles of various sizes ranging from 0.5 µm to 2 mm to meet different applications (Trudinger et al., 1990; Komameni and Roy, 1985; Marella et al., 1995; Ramsay 1994). Large spherical granules (typically in the range of 100 µm to 5 mm) are generally prepared by gelation of individual droplets generated from the starting sol by the so called "oil-drop" method. The spheres manufactured by individual shaping are more expensive. However, their shape and calibration are quite regular.

Among these reported research studies on sol-gel granulation or "oil-drop" process, most were on the synthesis of spherical silica (SiO_2) supports or zeolite granules (Barten et al., 1987; Kiel et al., 1992). For example, Spek and van Beem (1982) reported a sol-gel granulation process for fabrication of spherical silica supports. In their process, an aqueous sodium silicate (waterglass) solution comprising 12 wt% of SiO_2 was mixed continuously in a mixing chamber with another aqueous 1.2 N sulfuric acid solution in a volume ratio acid/waterglass of 0.65. The hydrosol formed in the mixing chamber was converted into droplets that were allowed to fall through a vertical disposed cylindrical tube with a length of 1.8 m filled with paraffinic hydrocarbon oil at 25°C. The spherical hydrogels particles formed were separated, washed with water and dried. The spherical silica particles can be used directly as catalyst supports. Li et al. (1990) used an aqueous solution of sulfuric acid, ammonium sulfate and waterglass as the starting materials to fabricate spherical SiO_2 particles, and synthesized binderless zeolite type A from SiO_2 in an aqueous solution of alkaline sodium aluminate at a given condition. Binderless faujasite zeolites were also synthesized this way by Mirsky et al. (1978).

Granular alumina supports were also fabricated by the "oil-drop" sol-gel method (Duisterwinkel and Frens, 1995; Cahen et al., 1979; Meyer and Noweck, 1982; Shepeleva et al., 1991; Svoboda et al., 1994). All these reported "oil-drop" processes for preparation of granular alumina supports start with a boehmite sol using pseudo-boehmite powder as precursor, followed by introduction of the sol dropwise through an orifice into a column of heated paraffin oil. Each drop of oil-insoluble mixture forms a sphere and converts to spherical gel in the hot oil. The microstructure, especially the pore texture, of the granular alumina supports prepared by these methods are basically determined by the properties of the primary particles of the starting pseudo-boehmite powder. It is difficult to control and tailor the microstructure of the granular alumina supports to meet various needs by these "oil-drop" processes since the properties of the starting pseudo-boehmite powder are source dependent and generally not controllable in the granulation process.

Sol-gel granulation processes were also used to fabricate high-density microspheres as nuclear fuels (Haas and Clinton, 1966; Ganguly et al., 1985; Ganguly et al., 1994). In the sol-gel process of fabrication of ThO_2-2% UO_2 fuel pellets, the starting sols were prepared by passing a controlled amount of ammonia gas through the thorium and uranyl nitrate solution. Droplets of the sol were then introduced through an electromechanical vibrator with a horizontal jetting nozzle inside a containment box that houses two horizontal ammonia gas pipes and the gelation bath. Thus, the droplets passed through a curtain of ammonia gas and quickly coated themselves with a gel skin before falling into the gelation bath filled with 1% NH_3OH and 4 M NH_4NO_3 with pH about 8. The suitable NH^{4+}/NH_3 buffer system established the conditions essential to achieve the desirable gelation rate. The gels were then washed, dried and calcined before they were pressed into pellets and sintered at high temperature. It was reported that this sol-gel approach could achieve a pellet density of higher than 94% of the theoretical density, and a uniform dispersion of the uranium in the fuel, because the sol-gel process has a much better control of the properties of the primary particles and microspheres (Ganguly et al., 1985).

2.3.2   Sol-Gel Preparation of Alumina Granular Particles

γ-alumina is perhaps the most common crystalline material used as support for catalysts or adsorbents. Preparation of porous γ-alumina granules with excellent mechanical properties and desirable pore structure is of great importance to the development of novel catalysts and adsorbents for various applications. The superior mechanical properties can be derived from the unique microstructure of the granules, which is defined by compacting small γ-Al_2O_3 crystallite particles bound together by the bridges of the same material formed through coarsening or sintering.   To obtain high surface area of the granules the γ-Al_2O_3 crystallites should be preferably within a few nanometers. Such nanostructured γ-Al_2O_3 can be prepared by the Yoldas process, as described earlier. In what follows we report preparation of sol-gel derived nanostructured γ-Al_2O_3 granules based on a granulation process that combines the Yoldas process and the "oil-drop" method.

The sol-gel granulation process includes generating sol droplets, shaping and partially gelating the droplets into spherical wet-gel granules, and gelating and aging the wet-gel granules into solid-gel granules. The wet-gel granules were then washed with water, dried and calcined. In our laboratory, the starting material for preparing the γ-Al_2O_3 granules was high concentration (>1.0 M) boehmite sol. Two methods were used to prepare the high-concentration boehmite sol. The first method is based on the exact Yoldas process, as described earlier. The resulting sol at 1 M aluminum concentration was concentrated by evaporation on a hot plate. The second method, which avoids the evaporation step and is referred to here as the modified Yoldas process, was direct preparation of the high concentration boehmite sol. This method is described in more detail next using the preparation of 2 M boehmite sol as an example.

The synthesis process started with slow addition of a total of 520 ml of aluminum tri-sec-butoxide (ALTSB) in 1000 ml distilled water, which was well stirred and controlled at 80℃. Large irregularly shaped clusters started to appear, and gradually precipitated in the flask during the hydrolysis and condensation process when the amount of ALTSB added exceeded 400 ml (equivalent to about 1.56 M). The slurry was then dispersed with addition of 140 ml of 1 M HNO_3 (that gave a H^+/Al^{3+} ratio of 0.07) and/or dilution with about 200 ml deionized water. Continuous addition of 520 ml of ALTSB in water took about 5 h. After that, the ALTSB/water mixture was vigorously stirred for half an hour and refluxed overnight at 90–100℃.

The 2 M boehmite sol was first mixed with a small amount of 1 M HNO_3 (acid/sol volume ratio of 1 ∶ 5). The modified sol was stirred at 70–80℃ with a magnetic stirrer. A dramatic decrease in the pH with the addition of HNO_3 accelerated the gelation process of the sol. After aging for 30 min at 60–70℃, the 2 M boehmite became so viscous that it could not be stirred with the magnetic stirrer set at maximum power. It was then transferred to droppers as the starting material for the granulation process.

The granulation process included generating sol droplets by the droppers, shaping and partially gelating the droplets into spherical wet-gel granules in an paraffin oil layer, and consolidating the structure of the wet-gel granules in a 8 wt % ammonia solution layer. The device used to prepare the gel particles is schematically shown in Fig.2.6. The temperature of the paraffin oil layer (white color, density: 0.7864 g/cm^3, kinematic viscosity: 34.5 centistokes at 40℃, from Fisher Scientific) was varied from 25℃ to 100℃, and the ammonia solution layer was kept at room temperature. The interface between the oil and ammonia-solution was slowly stirred with a stirrer (at 18–50 rpm) to facilitate transport of the wet-granules across the interface. After aging in the ammonia solution for at least 45 min, the spherical wet-gel particles were removed from the ammonia solution, and carefully washed sequentially with water and alcohol, dried at 40℃ for 48 h, and finally calcined in air at 450℃ for about 4 h.

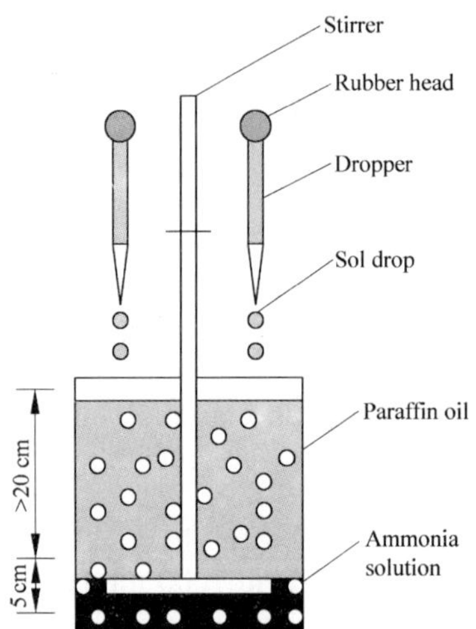

Fig.2.6 Schematic of the batch sol-gel granulation process

One of the key steps in this sol-gel granulation process is the formation of the sol droplets. There are basically two ways for droplet formation: gravity-force assistant method (Barten et al., 1987) and shear-force assistant method (Hu and Reeves, 1997). In both cases, the orifice of the dropper is immersed in the paraffin oil. In the second case, the shear-force is created by either moving the paraffin oil (orifice is fixed) or moving the orifice in the paraffin oil. When the sol is continuously pushed out through the orifice, the sol in the oil experiences the gravity force or drag-force (in the shear-force assistant method) that acts to separate the drop from the orifice. When the gravity force (minus buoyancy force) or the drag-force exceeds the forces acting to keep the drop in the orifice (e.g., interfacial tension), the drop breaks away from the orifice.

It was found (Ramsay, 1994) that the volume of the sol drop is proportional to the interfacial tension (between the sol and orifice material) and the diameter of the orifice, and inversely proportional to the density difference between the sol and oil. In general, larger spheres are obtained by increasing the diameter of the orifice. The sol concentration or density and interfacial tension are other important parameters. Thus, increases in the interfacial tension (γ) yield larger droplets and therefore larger spheres. If the kinetic and drag forces are considered, it can be shown that the drop size increases with increasing flow rate of the sol through the orifice and increasing viscosity of the continuous phase (oil phase).

Continuous generation of the droplets by the gravity-force assistant method may be facilitated by the use of a drop-generation device that generates sol droplets by intermittently forcing the sol through the orifice (Barten et al., 1987). The orifice of the dropper can be placed in the air above the oil layer. In this case the droplets experience more gravity because the buoyancy force in minimized. The shear-force assistant method (Hu and Reeves, 1997) appears to be more convenient for continuous generation of the sol droplets. In this method the size of the droplets is determined by the diameter of the orifice, the relative velocity of the orifice with respect to the oil phase, and the flow rate of the sol being pushed out of the orifice.

The rheological properties (viscosity and viscoelasticity) of the sol flowing through the orifice are very important in determining the size, uniformity and break-up of the droplet. Such properties are generally complex, since concentrated sol and liquid containing dissolved polymers, which are frequently employed in these processes, exhibit non-Newtonian behavior. With boehmite sol, the important parameters that control the rheological properties of the sol are the solid concentration and pH, and aging temperature and time of the sol before the droplet-generation step. These effects have been discussed in detail elsewhere (Deng and Lin, 1997; Wang and Lin, 1998).

The droplets generated from the dropper usually were not perfectly spherical, but they gradually become good spheres while falling through the hot oil layer. This is basically because the immiscibility between the boehmite sol and the paraffin oil makes the spherical shape thermodynamically most stable. At the time when the droplets changed to spheres, partial gelation of the thin outer layer of the droplets occurred due to the increase of temperature. The droplets become rigid enough to keep the spherical shape in the oil layer and then fall into the ammonia solution layer for further aging and conditioning.

During aging in the ammonia layer, ammonia will penetrate through the oil film to neutralize the acid in the partially gelled sol in the wet-gel granules. Thus, the wet-gel granules were further gelated, aged and became solid granules in the ammonia-solution layer. After being aged in the ammonia solution and separated from the liquid media the alumina wet-gel granules became rigid enough to be transported, washed and dried. The washing step may be required to remove those impurities such as ammonium nitrate, nitric acid or ammonia and hydrocarbons in the wet-gel granules before the drying process. In this case, a careful washing procedure should be followed in order to avoid breakage of the gel particles (Deng and Lin, 1997; Wang and Lin, 1998). Washing could also result in a loss of about 10%–15% of aluminum into the washing solvents. Thus, if possible, the washing step should be omitted. In these cases, the impurities in the wet-gel granules can be removed by directly drying and calcining the wet-gel granules. In the drying step, the aggregates in the gels broke apart into primary particles of pseudo-boehmite or boehmite, which bind together to form the skeleton of the final granules. Pores and porosity gradually developed as the calcination step progressed.

2.3.3 Physical Properties

A microscopic view of the sol-gel-derived alumina granules is shown in Fig.2.7. The multiple granules prepared under the same experimental conditions are nonaggregated spherical particles without lumps. The particle sizes of these granules are uniform. This smooth spherical shape and narrow particle size distribution provide the necessary homogeneity of γ-alumina support for catalysts or adsorbents for use in either fixed-bed, moving-bed or fluidized-bed reactors. Experiments performed in our laboratory showed that it was easy to make good spherical and smooth γ-alumina granules with diameters less than 3 mm using the sol-gel granulation method. Larger particles had the tendency to deform during the washing and drying stages.

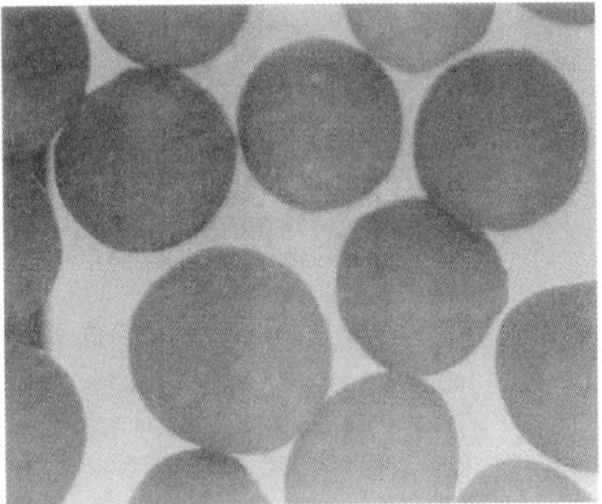

Fig.2.7 Appearance of the sol-gel-derived γ-alumina granules (particle diameter 2 mm)

The pore size distribution (calculated from the desorption isotherm) of the granule sample is shown in Fig.2.8. Similar to the nongranular sample, the alumina granules have a narrow pore size distribution (20–60 Å) with an average pore diameter of around 35 Å. The uniform pore size distribution indicates the presence of monodisperse pores in the granular particles defined by the intercrystalline space of the primary particles. The BET surface area and pore volume of the sol-gel-derived alumina granules are about 390 m^2/g and 0.5 mL/g, larger than the nongranular alumina samples (see Table 2.2). It should be noted that preparation procedure and purity level of the aluminum butoxide precursor may affect somewhat the pore structure of the sol-gel-derived alumina.

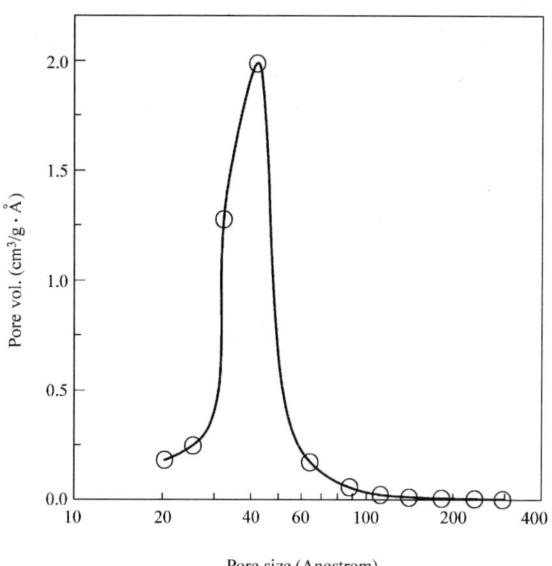

Fig.2.8   Pore size distribution of a sol-gel-derived alumina granular sample

The mechanical properties such as crush strength and attrition resistance are important properties for industrial adsorbents/catalysts, especially for uses in moving-bed or fluidized-bed processes. Adsorbents/catalysts with poor mechanical properties are usually not appropriate for industrial applications in fixed-bed processes due to the problems of dust formation in the adsorber/reactor vessels, which will change the particle size distribution of the adsorbents/catalysts and increase in the pressure drop of the adsorber/reactor system. Furthermore, the loss of the adsorbent weight due to dust formation will also degrade the performance of the adsorber/reactor system.

The crush strength of individual granular γ-alumina supports and adsorbents prepared by the sol-gel method and some commercial supports and adsorbents were experimentally determined by a universal testing instrument. During these experiments, a single granule was placed between two smooth and parallel compression surfaces made of steel. One of the flat surfaces was mounted on the base of loading frame of the instrument and the other flat plate attached to a crosshead moving towards the granule at a controlled speed of 2 mm/min. A compression load cell (force sensor, 0–400 lb) mounted on the moving crosshead was used to measure the force acting on the particle. Both the

72

load force and displacement of the sphere were recorded by a computer. At the end of the test, the granule crushed or collapsed when the force applied on the granule was large enough, and an abrupt decrease of the force signal was detected. The maximum force load applied to break the granule was taken as the side crushing strength.

Figure 2.9 shows the typical load-displacement curves obtained in the crushing strength test for a sol-gel-derived spherical γ-alumina granule and a commercial alumina granule sample. The alumina particles are elastic and the maximum displacements before crushed are about 40%–75% of the diameter of the spheres. The experimental side crushing strength of the sol-gel-derived alumina granules are compared with several commercial granular samples in Table 2.5. The crush strength of the sol-gel-derived γ-alumina granules (diameter is ~2 mm) prepared in this work is 160 N, several times larger than the commercial alumina and zeolite granular samples.

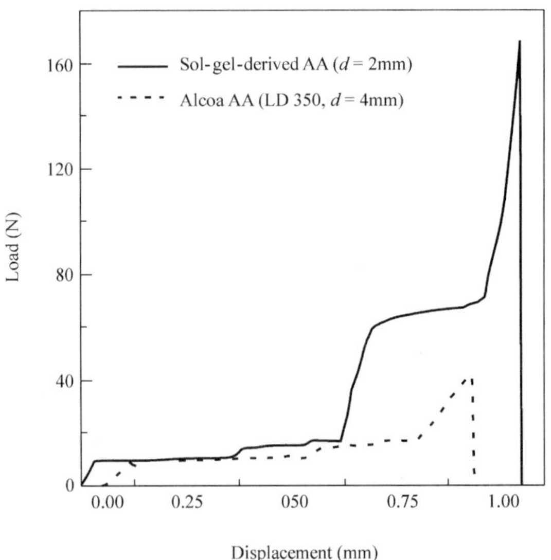

Fig.2.9   A comparison of load-displacement curves for the sol-gel-derived granular alumina particle and a commercial alumina particle (from Alcoa)

Two types of particle degradation are usually observed in the mechanical attrition of solid particles: "fracture" or deep disintegration of the adsorbent or catalyst particles; and "grinding" or the abrasive removal of the particle skin. Particle fracture yields an assortment of fragments from small size to large size (Doolin et al., 1993). The grinding mechanism leaves behind a particle reduced in size and quantity of fine particles. The forces to which the adsorbent or catalyst particles are exposed in the moving-bed or fluidized-bed unit involve mechanical motion between any given particle-particle or particle-wall interactions and high-speed-velocity impingement during movement of particles from vessel to vessel. These forces will result in both particle fracture and grinding in the moving beds and fluidized beds.

Attrition rates of these granules were measured by using an apparatus consisting of a geared motor and a stainless steel attrition testing cylinder (315 mm long, 10 mm in inter diameter) mounted at a point about 80 mm from the center. The testing cylinder rotated about an axis normal to the length at a controlled speed ranging from 20 rpm to 200 rpm. In the attrition test performed in this work, about 5 g of granules were placed in the testing cylinder and tested at a rotation speed of 60 rpm for 24 h. Thereafter the powder material generated during the attrition experiments was sieved off over a No. 35 sieve with mesh opening of 500 μm. The weight loss percentage rate (wt %/h) calculated by the following equation was used as the attrition index:

$$\text{Attrition index} = (\text{Initial weight–Remaining weight}) / \text{Initial weight} / \text{Time} \times 100\%$$

In attrition experiments, each granule sample after the attrition run was examined for its breakage and formation of particle fragments and fine powder. No breakage and particle fragments were found in all samples listed in Table 2.5 except for silicalite. Most of the silicalite extrudates were found to be broken into small particles and a certain amount of fine powder was generated after the attrition test. This may indicate that the attrition of silicalite pellets followed both particle fracture and abrasion mechanisms, and the attrition of other granules follows the abrasion mechanism only. The attrition index of the sol-gel-derived spherical γ-alumina granules along with the commercial supports and adsorbents determined by the Peter Spence method are summarized in Table 2.5. Among all samples listed in Table 2.5, the sol-gel-derived spherical γ-alumina granules prepared in this work have the smallest physical attrition index: 0.033 wt %/h, about 1/5 of that of the commercial alumina granule (LD-350 of Alcoa). This clearly demonstrates that the sol-gel-derived spherical γ-alumina granules have exceptional attrition resistance. These results also indicate that the attrition of the sol-gel-derived γ-alumina granules prepared in this work follows the "fracture" mechanism. In this case the attrition of the sol-gel-derived alumina

Table 2.5 Comparison of mechanical properties of sol-gel-derived alumina granules with commercial adsorbents

Adsorbents	Particle diameter (mm)	Crush strength (N)	Attrition rate (wt %/h)
Sol-gel alumina	2.0–2.5	160	0.033
Sol-gel alumina	2.6–2.8	190	–
Alcoa alumina (LD-350)	4.0–4.6	42	0.177
UOP silicalite	1.4–1.6	16	0.575
Degussa DAY zeolite	3.5–3.7	40	0.073

granules is extremely low if the mechanical stress is not large enough to break the particles in the attrition test.

The excellent mechanical strength, high attrition resistance, and desirable pore structure of the spherical γ-alumina granules prepared by the sol-gel method are unique properties of sol-gel derived ceramics. In synthesizing these granules, plate-shaped boehmite primary particles were formed and grew via the hydrolysis and condensation of aluminum alkoxide in aqueous solution (Leenaars and Burggraaf, 1984). The particles reached a final uniform nanoscale size (about 5–10 nm) due to the Ostwald ripening mechanism (Brinker and Scherer, 1990). The boehmite particles were connected during the partial gelation process while the sol droplets were falling through the oil layer. Complete gelation occurred when aged in the ammonia solution. After removal of solvent, the structure of the solid boehmite gel granules was consolidated by calcination in air at 450℃. The boehmite particles also transformed to γ-alumina crystallites during the calcination step. The final granules consist of nanoscale primary γ-alumina particles of rather uniform size. These uniform particles are strongly bound together by the alumina bridges formed in the calcination step, and, as a result, the granules become mechanically very strong.

2.4   Sol-Gel-Derived Granular Adsorbents/Catalysts

The sol-gel-derived ceramic granular particles described above may be directly used as the adsorbents for certain applications. More often, however, these granular particles are used as the support body on which an active species is coated. The active species coated on the internal pore surface of the support body provides active sites for preferential adsorption of selected gas and liquid species. Because of the unique coating process and microstructure of the sol-gel-derived support body, the sol-gel-derived supported adsorbents are expected to exhibit better properties than similar adsorbents prepared by the conventional methods. This section describes synthesis and adsorption properties of the sol-gel-derived granular adsorbents.

2.4.1   Synthesis of Adsorbents/Catalysts

Coating a metal or oxide species on the surface of adsorbent/catalyst supports have been traditionally accomplished by the solid dispersion method (Xie and Tang, 1990) and the wet-impregnation method (Deng and Lin, 1995, 1997). The first method is done by physically mixing the support body and solid active specie followed by heat-treatment to disperse the active species on the internal surface of the support. In the second method, the support body is brought into contact with a liquid solution containing the precursor of the active species. The precursor is transported into the pores of the support by capillary force. Solvent is removed by drying. The adsorbent is then calcined under specific atmosphere to convert the precursor to the desired active species.

The major disadvantage of the solid-dispersion coating method is its difficulty in dispersing oxide precursor with a high melting point on the internal pore surface of the support body. With the wet-impregnation method there is a problem of controlling the amount of the active species coated on the support body. In the case of the sol-gel-derived support body, the wet-impregnation method requires the drying and calcination steps in addition to those used in preparing the support body.

In our laboratory, we have developed a so-called solution-sol mixing method to prepare supported adsorbents by the sol-gel method (Wang and Lin, 1998). In this method, the stable particulate sols of the alumina, titania and zirconia are first prepared from corresponding metalorganic precursors. An aqueous solution containing appropriate precursor (usually a salt) is mixed with the stable sol, as shown in Fig.2.6. After a controlled aging period, the salt-doped sol is used as the starting material for preparing the coated granular adsorbents. In this method, the precursor is coated on the surface of primary particles in the sol, either during the mixing or subsequent drying step. This method allows coating an active species in a precisely controlled amount on the support. Other advantages include: (1) the active precursor can be homogeneously coated on the grain surface of the support as a result of the homogeneous mixing taking place in the liquid phase; (2) the method does not require additional drying/calcination steps in comparison with the wet-impregnation coating procedure.

Table 2.6 summarizes the pore structure of sol-gel-derived alumina coated with various active species by the solution-sol mixing method. Synthesis of the boehmite sol was described above. Nitric acid was used to peptize the particulate sol. The sol was mixed with the aqueous solution of corresponding salt solution in a predetermined ratio to give the loading listed on the second column in Table 2.6. The coated samples were dried and calcined in air at 450°C for 3 h.

Table 2.6 Pore structure of alumina coated with an active species by solution-sol mixing method using the boehmite sol as the starting material

Doping salt	Loading(M/A1)	$S_{BET}(m^2/g)$	$V_p(ml/g)$	$d_{av}(nm)$
Pure	0	323	0.36	3.1
$Ca(NO_3)_2$	0.20	298	0.25	2.9
$Cu(NO_3)_2$	0.14	302	0.28	2.9
$AgNO_3$	0.10	262	0.25	2.8
$La(NO_3)_3$	0.03	290	0.28	3.1
$CuCl_2$	0.20	207	0.18	2.9*

* Sol becomes unstable after mixing, with large aggregates precipitated.

For most cases shown in Table 2.6, the sols after doping still remained stable. The pore structures of the coated samples after calcination do not differ that much from the corresponding uncoated samples, as shown in Table 2.6. However, mixing a salt containing an anion different from the anion of the acid used in pepetization could destabilize the sols. The pore structure of the samples derived from the unstable sols is very different from those derived from the stable sols. Similar results were observed for zirconia and titania sols (Gopalan et al., 1995; Gopalan and Lin, 1995).

The pore structure of the samples after calcination is determined by the shape and size of the primary particles, and, to a less extent, by the manner in which these particles are packed. The slight differences in the pore structure between the uncoated and coated samples are due to the effects of coating material on the size and shape of the primary particles in the sol. The particulate sols contain fine particles of zirconia, boehmite or titania. Mixing a salt solution in the sol might slightly change the cation charge distribution, affecting the size and shape of the primary particles. The following were found to be critical to ensuring the stability of a sol after doping: (1) the precursor is preferred to be the salt containing the same anion as that of the acid used in pepetization of the sol; (2) the pH of the doped sol should remain essentially the same as the undoped sol and (3) the sol should be vigorously mixed while the precursor solution is added into the sol.

The exact location of the dopant ions in the doped sol is not clear. This requires further fundamental study on the solution-sol mixing method. However, it can be assumed that the doped salt is coated on the grain surface of the primary particles in the drying process. After calcination at $450°C$, the coated salt is converted to the corresponding oxide that may still remain on the grain surface of the primary particles.

A more detailed discussion on the solution-sol mixing coating method is given next with the preparation of CuO coated γ-Al_2O_3 granular particles as an example. The adsorbent was prepared from the 2 M boehmite sol described earlier. In experiments, a given amount of $Cu(NO_3)_2$ solution (with concentration varied from 0.52 M to 9 M, depending on the required amount of coating) was mixed with the 2 M boehmite sol in the volume ratio of 1:6. After aging for 30 min at $60–70°C$, the Cu doped 2 M boehmite was transferred to droppers as the starting material for the granulation process. Following the same granulation process for preparation of the pure γ-Al_2O_3 granular particles, the Cu-doped alumina particles were aged in ammonia solution, washed sequentially with water and alcohol, dried at $40°C$ for 48 h, and finally calcined in air at $450°C$ for about 4 h.

CuO-coated γ-Al_2O_3 granular adsorbents were also prepared by the wet-impregnation method for comparison. A given amount of the sol-gel-derived γ-Al_2O_3 granular particles was brought in contact with $Cu(NO_3)_2$ solution for over 16 h, followed by the same drying and calcination procedures described above. The amount of active species coated was estimated from the weights of support and $Cu(NO_3)_2$ that were brought into contact with each other (assuming that all $Cu(NO_3)_2$ was coated on the support). The $CuO/\gamma\text{-}Al_2O_3$ samples, prepared either by the solution-sol mixing method or the wet-impregnation method were further calcined in air at 550°C for 6 h to convert $Cu(NO_3)_2$ to CuO.

Table 2.7 compares the pore structure of pure $\gamma\text{-}Al_2O_3$ granular particles and $CuO/\gamma\text{-}Al_2O_3$ granular particles prepared by the wet-impregnation method and by the solution-sol mixing method. For the wet-impregnated $CuO/\gamma\text{-}Al_2O_3$ granular particles with CuO loading less than about 10 wt%, the particles exibit gray color with smooth surface. The particles of the samples with CuO loading exceeding 10 wt %, however, show gray color with a rough surface. XRD peaks of CuO crystallites appear in the samples with CuO loading of about 10 wt %. According to the monolayer loading theory (Xie and Tang, 1990), this result indicates that only limited amounts of CuO (<10 wt %) could be well dispersed on the internal pore surface of the alumina support by the wet-impregnation method used here. Excessive CuO may crystallize and is present on the external surface of the particles. For the $CuO/\gamma\text{-}Al_2O_3$ granular particles, the color of the wet samples (before drying) was blue. The color became darker as CuO loading increased. The color of the final calcined samples was green. XRD data indicate that CuO in a loading up to at least 18 wt % could be dispersed on the grain surface of these particles by the solution-sol mixing coating method.

Table 2.7 Pore structure of pure and CuO-coated $\gamma\text{-}Al_2O_3$ granular particles (particle diameter 2 mm) prepared by different methods

Coating method	CuO(wt %)	Surface area (m²/g)	Average pore diameter (nm)	Pore volume (ml/g)
Pure alumina	0	325	6.2	0.5
WI method	9.3	239	8.0	0.48
WI method	13.4	237	8.3	0.45
SSM method	9.0	244	7.5	0.45
SSM method	13.5	243	7.4	0.46

It should be noted that in a previous study (Deng and Lin, 1997) CuO at a much higher loading (>15 wt %) could be coated on the sol-gel-derived $\gamma\text{-}Al_2O_3$ granular particles by the wet-impregnation method. The only difference between these two studies is that the concentration of the $Cu(NO_3)_2$ solution used in the study described in the above paragraph is about 5 times that of the solution used in the earlier study (Deng and Lin, 1997) for wet-impregnation. Apparently, it is more difficult to disperse the precursor on the support surface with an impregnation solution of higher concentration. These results indicate the difficulty in controlling the coating process in the wet-impregnation method.

The surface area of the CuO-coated $\gamma\text{-}Al_2O_3$ samples prepared by both coating methods is about 25% smaller than the uncoated alumina samples. Coating CuO on $\gamma\text{-}Al_2O_3$ results in about 25% increase in the average pore size. The pore volume of the coated samples is the same as the uncoated ones. The pore size distribution data for the granular samples of pure $\gamma\text{-}Al_2O_3$, $CuO/\gamma\text{-}$

78

Al_2O_3 prepared by the wet-impregnation method, and CuO/γ-Al_2O_3 alumina by the solution-sol mixing method, are shown in Fig.2.10. The two coated samples have similar pore size distribution. Compared with the pure alumina sample, the pore size distribution data suggest that the coated CuO fills the smaller pores of γ-Al_2O_3, resulting in an appreciable decrease in the number of smaller pores, and, consequently, an increase in the average pore size. The results in Table 2.7 and Fig.2.10 also show that CuO-coated γ-Al_2O_3 samples prepared by the two coating methods have essentially the same pore structure.

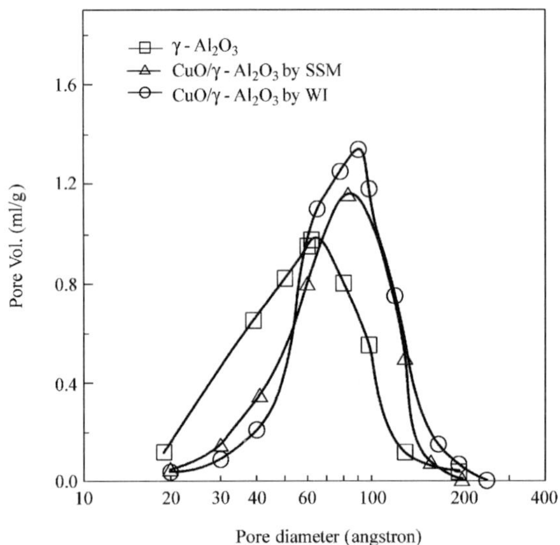

Fig.2.10 Pore size distribution of pure γ-Al_2O_3 granule, 9.3 wt % CuO coated γ-Al_2O_3 granule prepared by wet-impregnation method, and 9.0 wt % CuO coated γ-Al_2O_3 granule prepared by the solution-sol mixing method

2.4.2 Sorption Properties

Adsorption of SO_2 on a 30 wt % CuO/γ-Al_2O_3 granular samples prepared by the sol-gel method was measured by using a Cahn electronic microbalance, as described in detail in reference (Deng and Lin, 1997). The pore structure of the sorbent (Sample 1) is given in Table 2.8. The experimental results on the sulfation, regeneration and oxidation of the sorbent at 500°C and total pressure of 1 atm are presented in Fig. 2.11. Curve A is the weight gain of the sorbent when exposed to an air stream containing 0.8% SO_2. In this step SO_2 and O_2 react with CuO to form $CuSO_4$. The sulfation results in about 30% weight gain of the sorbent, equivalent to 3.75 mmol/g sorption capacity of SO_2. This sulfation capacity is much larger than similar sorbents prepared by the conventional method (Centi et al., 1990).

Table 2.8 Pore structure of sol-gel-derived adsorbents used in adsorption study

Sample #	Method	Active species	Loading (wt %)	S_{BET} (m²/g)	V_p(ml/g)	d_p(nm)
1	Sol-gel	CuO	30	228	0.5	5.4
2	Sol-gel	CuCl	30	283	0.43	4.9
3	Sol-gel	CuO	9	243	0.45	7.3
4	From UOP	CuO	7	158	0.60	15.2

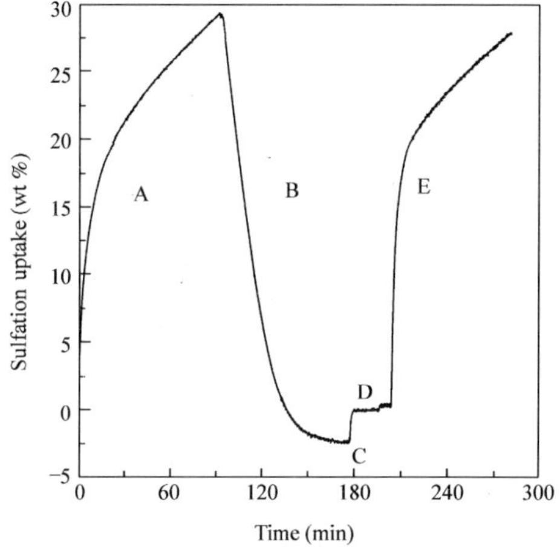

Fig.2.11 Sulfation, regeneration and oxidation curves of CuO/γ-alumina granular sorbent

The sulfated sorbent can be completely regenerated using 10 v.% methane in nitrogen, as shown in Curve B. Curve C-D shows weight gain of the sorbent when exposed to air due to oxidation of copper formed in the regeneration step. Curve E corresponds to weight gain in the second sulfation step, which is essentially identical to curve A. This indicates that the sorbent is highly recyclable.

Ethylene adsorption was measured on CuCl/γ-alumina granular particles prepared by sol-gel/wet-impregnation method. A cuprous solution was synthesized by mixing cuprous chloride, ammonium hydroxide, ammonium citrate and deionized water with the composition of 1.98 g:8.8 ml:0.4 g:20 ml. About 180 mg of alumina sample was used in each adsorption measurement run by Cahn balance. Before each run, the sample was activated at 250°C for 4 h in ethylene atmosphere to ensure that the Cu^{2+} ions, if any, were reduced to Cu^+ ions. Following that, the sample was exposed to nitrogen at 250°C for two more hours to completely remove the residual ethylene adsorbed on the sample. Finally, the sample was cooled down under vacuum (1.0 Pa) to the room temperature for adsorption measurement.

Figure 2.12 shows the adsorption isotherm of ethylene on the sol-gel-derived CuCl/γ-Al₂O₃ sorbent (Sample # 2 in Table 2.8). The isotherm is of type I. The equilibrium adsorption data of C_2H_4 at 1 atm pressure and 14–18°C on

CuCl/γ-Al$_2$O$_3$ (prepared by the solid-state mixing method) was previously reported in the literature (Xie and Tang, 1990). The amount of ethylene adsorbed on CuCl/γ-Al$_2$O$_3$ was only about 0.29 mmol/g. Yang and Kikkinides (1995) reported very good data of adsorption of 0.73 mmol/g of ethylene on CuCl/γ-Al$_2$O$_3$ adsorbent with support surface area of 340 m^2/g. The result of our sorbent is similar to the best result reported.

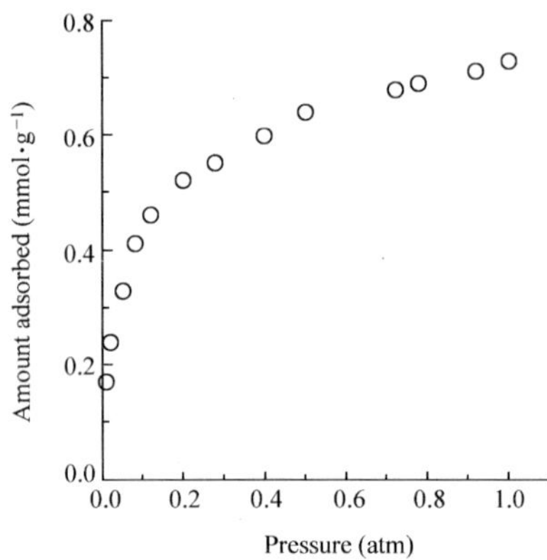

Fig.2.12 Ethylene adsorption isotherm on sol-gel-derived CuCl/γ-Al$_2$O$_3$ granule (1 atm= 101 325 Pa)

The availability of the sol-gel CuCl/γ-Al$_2$O$_3$ samples in granular form (particle diameter 2 mm) and in thin sheet form (0.1 mm thickness) allows us to examine the rate-limiting step of ethylene adsorption on the adsorbents (Wang and Lin, 1998). As shown in Fig.2.13, the time for reaching 95% uptake for both

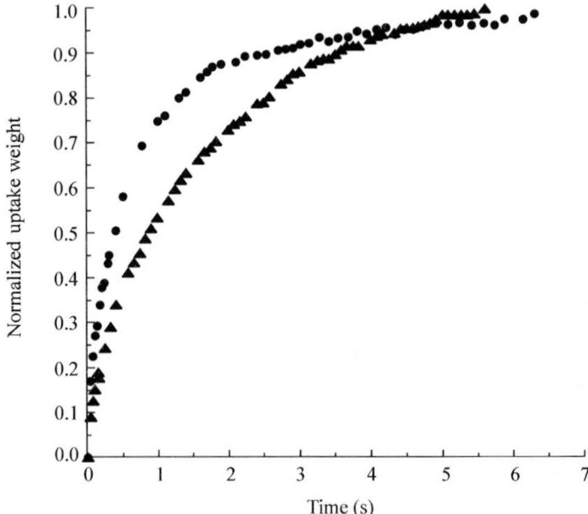

Fig.2.13 Ethylene adsorption uptakes in sol-gel-derived CuCl/γ-Al$_2$O$_3$ granule (2 mm) (▲)and thin sheet (0.1 mm)(●)

adsorbents subjected to an increase in the ethylene pressure (corresponding to a change from 0 to 0.2 mmol/g) is essentially the same (about 5 s) although they are different in the physical dimension. This indicates that adsorption of ethylene on the internal pore surface of adsorbent is the rate-limiting step.

The SO_2 removal capacity of the sol-gel derived sorbents was also studied in a fixed-bed adsorber system, which includes an adsorber column packed with sorbent pellets, a feeding flow-control system, and a sulfur analyzer (HORIBA, PIR-2000). A computer data-acquisition system was incorporated into this adsorber system in order to obtain continuous breakthrough curves of SO_2. The fixed-bed was made from dense α-Al_2O_3 tube of 6 mm ID and 8 mm OD. The central portion of the bed was packed with the sorbent, with both ends of the bed filled with quartz particles (0.5 mm in diameter). The feed was air containing 2000 ppm SO_2, and the fixed-bed temperature was 400℃.

Figure 2.14 shows the SO_2 breakthrough curve from the fixed-bed packed with 0.5 g of one of sol-gel-derived γ-Al_2O_3/CuO sorbent prepared by the solution-sol mixing method (Sample #3 in Table 2.8). The feed flow rate is 8.7 ml/min. The temperature for the experiment is 400℃. As shown, the concentration of sulfur dioxide in the effluent stream is zero in the first 3 h. After 3 h, the concentration of SO_2 in the effluent stream begins to increase, but it takes more than 50 h to reach the input concentration. These results indicate that the sorbent used in the experiment has very good capacity for removal of SO_2.

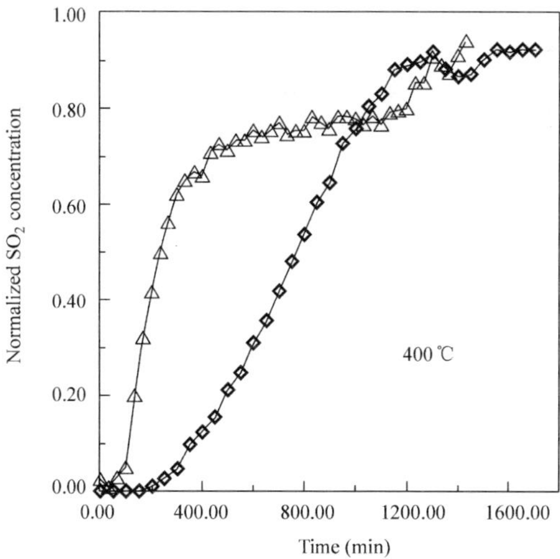

Fig.2.14 Comparison of SO_2 breakthrough curves at 450℃ from fixed-bed packed with CuO/γ-alumina sorbents prepared by solution sol mixing sol-gel method and conventional method

SO_2 breakthrough curves from fixed-bed packed with CuO/γ-alumina prepared by the conventional process (UOP SOX-3 sorbent (Harriott and Markussen, 1992; Yeh et al., 1985), 7wt % CuO on γ-alumina, sample #4 in Table 2.8) were also measured for comparison purpose under the same conditions as for the sol-gel-derived sorbent. The result is compared in Fig.2.14 with that for the sol-gel-derived sorbent (both at 400℃). At 100% SO_2 removal efficiency the sol-gel-derived sorbent could treat about 1800 ml flue gas (corresponding to 200 min breakthrough point), about three times that of the conventional sorbent (80 min). Considering the total area between the C/Co axis and the breakthrough curve, the total sorption capacity of the sol-gel-derived sorbent is also about three times that of the conventional sorbent. Obviously, the sol-gel-derived sorbent has much higher sulfation capacity than the conventional sorbent. The slightly higher amount of CuO coated on the sol-gel-derived sorbent alone would not contribute to the significantly improved SO_2 removal capacity of sol-gel-derived sorbent. The excellent sulfation reactivity of the sol-gel-derived sorbents is a result of more uniform dispersion of CuO on the support surface.

2.5 Conclusions

γ-alumina, zirconia and titania adsorbents/catalysts with uniform pore size distribution and an average pore diameter of about 3 nm are prepared by hydrolysis and condensation of their metalorganic precursors. These materials consist of nanoscale (5–20 nm) crystallites of which the intercrystalline space gives rise to the pore space. The as-synthesized adsorbents are in their metastable phases, and will transform to their stable phases upon heat treatment at high temperatures. The phase transformation accompanies a decrease in surface area and increase in pore size. The sol-gel γ-alumina, zirconia and titania adsorbents should be stable below 400℃, 300℃, 200℃, respectively.

Microporous (<2 nm) silica adsorbents are prepared by the sol-gel method following a polymeric sol route. The adsorbents have a large surface area (600–900 m^2/g). Doping using alumina in silica adsorbents improves their thermal and hydrothermal stability. These silica adsorbents can be safely used at temperatures lower than 200℃.

A sol-gel granulation process based on the oil-drop principle is described for the preparation of γ-alumina granular particles. The sol-gel-derived granular sorbents exhibit high crush strength, good attrition resistance and the desired pore structure. A solution-sol mixing coating method is used to coat active species on the pore surface of the granular γ-alumina particles. This coating method is particularly useful in preparation of the sol-gel-derived supported adsorbents. The method avoids additional drying and calcination steps as otherwise required in the wet-impregnation method. The supported adsorbents prepared by this method exhibit better dispersion of the active species on the support surface. Experimental results show that the sol-gel derived CuO/γ-alumina and CuCl/γ-alumina granular adsorbents have better properties for sorption of SO_2 or ethylene than similar adsorbents prepared by the conventional methods.

References[1]

Armor, J.N. and E.J. Carlson. J. Mater. Sci.. **22**, 2549 (1987)

Barten, B., F. Janssen, F. V. D. Kerkhof, R. Leferink, E. T. C. Vogt, A. J. van Dillen and J.W. Geus. In: Preparation of Catalysts IV, (Eds). B. Delmon, P. Grange, P. A. Jacobs and G. Poncelet. (Elsevier, Amsterdam, 1987) 103–111

Bhave, R.R. Inorganic Membranes: Synthesis, Characterization and Applications. (Van Nostrand Reinhold, New York, 1991)

Blackwell, J.A. and P. W. Carr. J. Liquid Chromatog. **15**, 1487 (1992)

Blackwell, J.A. and P.W. Carr. J. Chromatog. **549**, 59 (1991)

Brinker, C.J. and G.W. Scherer. Sol-Gel Science: The Physics and Chemistry of Sol-Gel Processing. (Academic Press, San Diego, CA 1990)

Brinker, C.J., K. D. Keefer, D. W. Schaeffer and C. S. Ashley. J. Non-Cryst. Solids. **48**, 47 (1982)

Brinker, C.J., K.D. Keefer, D.W. Schaeffer, T.A. Assink, B.D. Kay and C.S. Ashley. J. Non-Cryst. Solids. **63**, 45 (1984)

Burggraaf, A.J. and L. Cot (Eds.). Fundamentals of Inorganic Membrane Science and Technology. (Elsevier, Amsterdam, 1996)

Cahen, R.M., J.M. Ander and H.R. Debus. in Preparation of Catalysts 11, (Eds). B. Dehnon, P. Grange, P. A. Jacobs, and G. Poncelet. (Elsevier, Amsterdam, 1979) 585–594

Capes, C. E. and A.E. Fouda, in Handbook of Powder Science and Technology, (Ed).

Capes, C.E. in Particle Size Enlargement in Handbook of Powder Technology. (Eds). J. C. Williams and T. Allen, Vol. 1. (Elsevier, Amsterdam, 1986)

Centi, G., A. Riva, N. Passarini, G. Brambilla, B.K. Hodentt, B. Delmon, M. Ruwet. Chem. Eng. Sci. **45**, 2769 (1990)

Chang, C.-H., R. Gopalan and Y. S. Lin. J. Membr. Sci. **91**, 27 (1994)

Christian, J.W. The Theory of Transformation in Metals and Alloys. 2nd Edn. (Pregaimon Press, New York, 1975) Chap. 10, 12

Deng, S.G. and Y.S. Lin. AIChE J. **43**, 505 (1997)

Deng, S.G. and Y.S. Lin. AIChE J. **41**, 559 (1995)

Doolin, P.K., D.M. Gainer and J.F. Hofftnan. J. Testing & Evaluation. **21**, 481 (1993)

Duisterwinkel, A.E. and G. Frens. in Preparation of Catalysts IV, (Eds). G. Poncelet et al. (Elsevier Science Publishers, Amsterdam, The Netherlands, 1995) 1051–1058

Fanelli, A.J., S. Verma, T. Engelmann and J.V. Burlew. Ind. Eng. Chem. Res. **39**, 126 (1991)

Fayed, M.E. and L. Otten (Eds). Handbook of Powder Science and Technology. Chapter 7, (van Nostrand Reinhold, New York, 1984)

---

[1] Acknowledgement

The author would like to thank the former students in his laboratory for their contributions to the research summarized in this chapter.

Fogler, H.S. Elements of Chemical Reaction Engineering. (Prentice-Hall, Englewood Cliffs, New Jersey, 1992) Chap. 6

Fujimara, K. and T. Ando. Anal. Chem. **49**, 1179 (1977)

Ganguly, C., P.V. Hegde and G.C. Jain. Nuclear Technology. **105**, 346 (1994)

Ganguly, C., H. Langen, E. Zimmer and E.R. Merz. Nuclear Technology. **73**, 84 (1985)

Gopalan, R. and Y.S. Lin. Ind. Eng. Chem. Res. **34**, 1189 (1995)

Gopalan, R., C.H. Chang and Y.S. Lin. J. Mater. Sci. **30**, 3075 (1995)

Grassini-Strazza, G., V. Carunchio and A.M. Girelli. J. Chromatog. **466**, 1(1989)

Guo, C.J., C.W. Fairbridge and J.P. Charland. Synthesis of Mesoporous Catalytic Materials. U.S. Patent 5538,710 (1996)

Guo, C.J., in Zeolites: A Refined Tool for Designing Catalytic Sites. (Eds). L. Bonneviot and S. Kaliaguine, (Elsevier, Amsterdam, 1995) 165–171

Haas, P.A. and S.D. Clinton. Ind. Eng. Chem. Product Res. Dev. **5**, 236 (1966)

Hakvoort, G.C.M., J.C. van der Bleek, J.C. Schouten and P.J.M.Valkenburg. Thermochim Acta. **114**, 103 (1987)

Harriott, P. and J.M. Markussen. Ind. Eng. Chem. Res. **31**, 373 (1992)

Horvath, C. and K. Kawazoe. J. Chem. Eng. Japan, **16**, 470 (1983)

Hsieh, H.P. Inorganic Membranes for Separation and Reaction. (Elsevier, Amsterdam, 1996)

Hu, M.Z.-C. and M. Reeves. Biotechnol. Prog. **13**, 60 (1997)

Iler, R.K. The Chemistry of Silica. (Wiley, New York, NY, 1979)

Kiel, J.H.A., W. Prins and W. P. M. van Swaaij. Appl. Catal. B. Environ. **1**, 13 (1992)

Klein, L.C. (Ed). Sol-Gel Technology for Thin Films, Fibers, Performs, Electronics and Specialty Shapes. (Noyes Publications, Park Ridge, NJ, 1988)

Kohli, J.C. and K.K. Badaisha. J. Chromatog. **320**, 455 (1985)

Komameni, S. and R. Roy. Mater. Lett.. **3**, 165 (1985)

Leenaars, A.F.M. and A. J. Burggraaf. J. Mater. Sci. **19**, 1077(1984)

Li, S.A., Y. Ke, C. Tang and E. Zhao. The Preparation of Zeolite Molecular Sieve Type A in the Form of Bideless Spherical Granules. Chinese Patent CNIO10301B (1990)

Lin, Y.S., K.J. de Veries and A.J. Burggraaf. J. Mater. Sci.. **26**, 715 (1991)

Marella, M., M. Tomaselli, L. Meregalli, M. Battagliarin, P. Geronopoulosa, F. Pinna, M. Signoretto and G. Strukul. in Preparation of Catalysts IV, (Eds). G. Poncelet et al. (Elsevier, Amsterdam, 1995) 327–335

Meyer, A. and K. Noweck. Verfahren zur Herstllung von kugelfomiger Tonerde. Europaische Patentanmeldung, No. 0,090,994 A2 (1982)

Mirsky, Y.V., A.Z. Dorogochinskyy, N.F. Meged and A. P. Kosolapova. Process for the Production of Zeolites in the Form of Binderless Spherical Granules. U.S. Patent 4 113, 843 (1978)

Ramsay, J.D.F. in Controlled Particle, Droplet and Bubble Formation. (Ed.) D.J. Wedlock, (Butterworth Heinemann, Oxford, 1994) 1–38

Rigney, M.P., T.P. Werber and P.W. Carr. J. Chromatog. **484**, 273 (1989)

Ruthven, D.M. Principles of Adsorption and Adsorption Processes. (Wiley, New York, 1984)

Shepeleva, M.N., R.A. Shkrabina, Z.R. Ismagilov and V.B. Fenelonov. in Preparation of Catalysts V, (Eds.) G. Poncelet, P. A. Jacobs, P. Grange and B. Delmon. (Elsevier,

Amsterdam, 1991) 583–590

Smith, J.M. Chemical Engineering Kinetics. Chaps 7 and 8, (MsGraw-Hill, New York, 1981)

Spek, T.G., and M.J.L. van Beem. Silica Particles and Method for Their Preparation. European Patent EP 0067459 Al (1982)

Svoboda, K., W. Lin, J. Hannes, R. Korbee and C. M. van den Bleek. Fuel. **73**, 1144 (1994)

Teichner, S.J., G.A. Nicolaon, M.A. Vicarini and G.E.E. Grades. Adv. Colloid Interface Sci. **5**, 245 (1976)

Trudinger, U., G. Muler and K. K. Unger. J. Chromatog. **535**, 111(1990)

Uhlhom, R.J., R. J. R., K. Keizer and A. J. Burggraaf. J. Membrane Sci. **66**, 271 (1992)

van der Grift, C.J.G., A. Mulder and J. W. Geus. Colloids and Surfaces. **53**, 223 (1991)

Wang, Y. and Y.S. Lin. J. Sol-Gel Sci. Tech. **11**, 185 (1998)

Wang, Z. and Y. S. Lin. J. Catalysis. **174**, 43 (1998)

Ward, D.A. and E.I. Ko. Ind. Eng. Chem. Res. **34**, 421 (1995)

Webb, P.A. and C. Orr. Analytical Methods in Fine Particle Technology. (Micromeritics, Norcross, GA, 1997) Chap.3

Weber, T.P. and P.W. Carr. Anal. Chem.. **62**, 2620 (1990)

Weber, T.P., P.W. Carr and E. F. Funkenbusch. J. Chromatog. **519**, 31 (1990)

Wolff, E.H.P., A.W. Gerritsen and P.J.T. Verheijen. Powder Technol. **76**, 47 (1993)

Xie, Y.-C. and Y.Q. Tang. Adv. Catal. **37**, 1 (1990)

Yang, R.T. and E.S. Kikkinides. AIChE J. **41**, 509 (1995)

Yang, R.T. Gas separation by Adsorption Processes. (Imperial College Press; River Edge, New Jersey (Distributed by World Scientific, 1997)

Yeh, J.T., R.J. Demski, J.P. Strakey and J.I. Joubert. Environ. Prog. **4**, 232 (1985)

Yoldas, B.E. Amer. Ceram. Soc. Bull. **54**, 289 (1975)

3   Ferroelectric Thin Films and Applications

Cuozhong Cao

3.1   Introduction

Ferroelectrics are a special group of compound crystals that have noncentrosymmetric (or polar) crystal structure and possess reversible spontaneous electric polarization (electric dipoles). These dipoles can switch directions under the influence of an externally applied electric field or an internal electric field from the adjacent unit cells. This physical property is called ferroelectricity. The denominations of ferroelectrics and ferroelectricity are historical and due to the fact that there are many similarities between ferroelectrics and ferromagnetics such as spontaneous polarization, domain structure, hysteresis, and the Curie temperature. However, the fundamentals for ferroelectricity are totally different from ferromagnetism. For example, at the Curie temperature ferroelectrics undergo a phase transition that involves rearrangement of ions and change of crystal symmetry and the materials change between the paraelectric and ferroelectric states. In the case of ferromagnetic materials, the transition between the spontaneously magnetized and magnetically disordered states that occurs at the Curie point does not involve a change in crystal structure. Spontaneous electric polarization in ferroelectrics is due to the noncentrosymmetric crystal structure, while spontaneous magnetization is due to unpaired and unsaturated electron spins. Above the Curie point, spontaneous electric polarization disappears since the crystal structure changes to a centrosymmetric system, while individual magnetic moments or unpaired electron spins do exist, but are randomized and cancel one another completely due to thermal motion.

Ferroelectricity has been known since the 1920s and ferroelectricity in ceramic, $BaTiO_3$, was first discovered in the 1940s (Jona and Shirane, 1962; Megaw, 1957). Both ferroelectricity and ferroelectric materials have been subjects of extensive study for over half a century, and a huge amount of literature has been published on both fundamentals and applications (Buchanan, 1991; Cohen, 1992; Jaffe et al., 1971; Jona and Shirane, 1962; Lines and Glass, 1977; Megaw, 1957; Ramesh, 1998; Strukov and Levanyuk, 1998; Treece et al., 1998). A wide range of commercial and technological applications have been realized, since the reversible spontaneous polarization in a ferroelectric makes it capable of piezoelectric, pyroelectric and electro-optic behavior, in addition to its application in highdielectric constant capacitors. Although it is not necessary that piezoelectrics, pyroelectrics and electro-optic ceramics are ferroelectrics, ferroelectrics always possess very good piezoelectricity, pyroelectricity, and electro-optic properties. It is not surprising, therefore, that many devices based on piezoelectricity, pyroelectricity, and electro-optic activity are made of ferroelectric materials.

The current technology advancement imposes an ever-increasing demand of development of new ferroelectric materials and/or improved material performance. Ferroelectric materials with high dielectric constants, ε_r, promise possibilities for manufacturing high-density planar dynamic random access memories (DRAM) (Scott, 1998a; Summerfelt, 1998); while ferroelectrics with high remanent polarization, P_r, and low coercive field, E_c, have an important application in fabricating non-volatile random access memories (NvRAM) (Bondurant and Gnadinger, 1989; Jones et al., 1995; Scott and Paz de Araujo, 1989; Suzuki, 1995) for information storage. Piezoelectricity of ferroelectrics has been explored for applications such as accelerometers, displacement transducers, actuators, such as those required for inkjet printers and video-recording head positioning, and, recently in particular, micro-electro-mechanical systems (MEMS) (Kim et al., 1993; Polla and Francis, 1996; Polla and Francis, 1998). Pyroelectricity of ferroelectrics can be utilized in the fabrication of high-sensitivity, room temperature infrared detectors. Electro-optic activity of ferroelectrics can be used in color-filter devices, displays, image-storage systems, and optical switches for integrated optical systems. Although various applications of ferroelectric materials require a different combination of physical properties, dielectric constant, remanent polarization and coercive field are the key properties. Both P_r and E_c are directly related to reversible spontaneous electric polarization.

3.2   Charge Displacement and Spontaneous Polarization

When an electric field is applied to a nonconductive material (dielectric), there will be no long-range transport of charge and, thus, no electrical conduction. However, there will be a short-range transport of charge or charge displacement with positive charge shifting along the direction of electric field and negative charge moving in the opposite direction. This short-range transport of charge or charge displacement is called electric polarization, which occurs in four distinct mechanisms: electronic polarization, ionic polarization, dipolar polarization, and space charge polarization (Hench and West, 1990; Moulson and Herbert, 1990). Figure 3.1 schematically depicts the four types of polarization mechanisms (Moulson and Herbert, 1990). The overall polarization is additive of all four types of polarization.

88

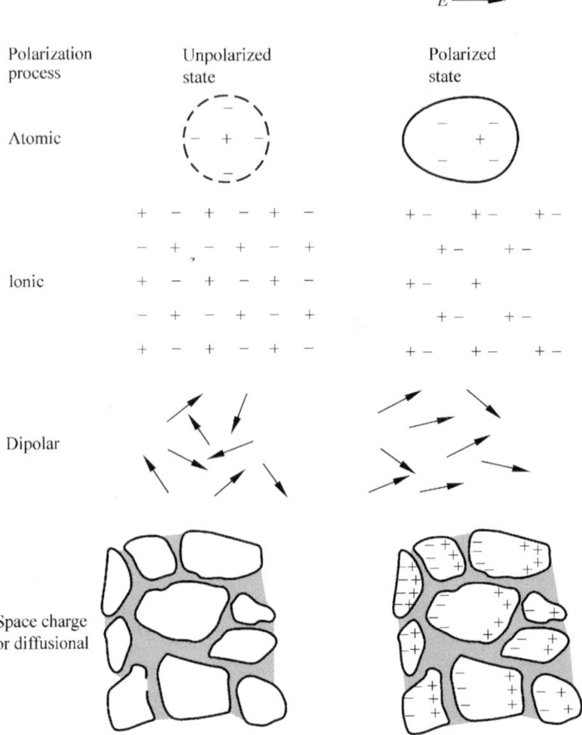

Fig.3.1 Schematic representation of four different electric polarization mechanisms (Moulson and Herbert, 1990)

Electronic polarization (also called atomic polarization) exists in all materials and is the response of the electrons and the atomic nuclei that shift their relative positions under an applied electric field and form an electric dipole (each per atom or ion). If an AC field is applied, these atomic dipoles resonate back and forth in step with the frequency of the alternating field. This polarization occurs instantaneously, in response to light electromagnetic field frequencies ($\sim 10^{15}$ Hz), since the electrons have a very high natural frequency ($\sim 10^{16}$ Hz). Electronic polarization, being resonant in nature, is independent of temperature. Perfect elemental single crystals, such as diamond and silicon, possess only electronic polarization.

Ionic polarization is the displacement of negative and positive ions toward the positive and negative electrodes, respectively. Since they are much more massive than electrons, the ions cannot become polarized as rapidly. Ionic polarization is limited to a maximum frequency of approximately 10^{13} Hz. This value is below the frequency of visible light. Therefore, the electromagnetic field of visible light cannot produce ionic polarization, as it does electronic polarization. Like electronic polarization, ionic polarization is resonant in nature and is independent of temperature. Ionic polarization occurs in compounds in which chemical bonds are not 100% covalent.

Molecular polarization is also called dipolar polarization, which occurs in materials consisting of polar molecules (or unit cells) only. In polar molecules, the center of positive charges is not coincident with that of negative charges and, thus, an electric dipole is present in each molecule (or unit cell). An example of a polar molecule is water. Molecular polarization is permanent, since it is inherent in the molecular structure. These dipoles can be oriented with an electric field. Furthermore, the polar molecules can flip back and forth in response to an AC field. The maximum frequency of response varies significantly from material to material depending on the size of molecule, it is, however, always less than that for electronic and ionic polarization and is typically less than 10^{10} Hz. Unlike the electronic and ionic polarization, dipolar polarization is temperature dependent.

Space charge polarization is also called interfacial polarization, which is a short-range electrical conduction. A space charge develops when there is a local conduction within a dielectric. Such a local conduction stops at a barrier or an energy potential, such as grain boundaries. Due to the nature of diffusion, space charge polarization occurs rather slowly and the typical frequency of response is of approximately 10^2 Hz. Like dipolar polarization, space-charge polarization is temperature dependent.

Figure 3.2 summarizes the frequency dependence of various polarization mechanisms and shows the polarization, represented by dielectric constant, ε_r, is additive (Moulson and Herbert, 1990). The combined value of electronic, ionic and space charge polarization in normal dielectric materials is relatively small, and thus exerts very small effects of electric polarization on a number of physical properties. In contrast, dipolar polarization may have a significant effect on many physical properties. One good example is ferroelectrics that, with reversible spontaneous polarization, possess a significantly large influence on many physical properties of the materials, such as the elastic, optical and thermal behavior.

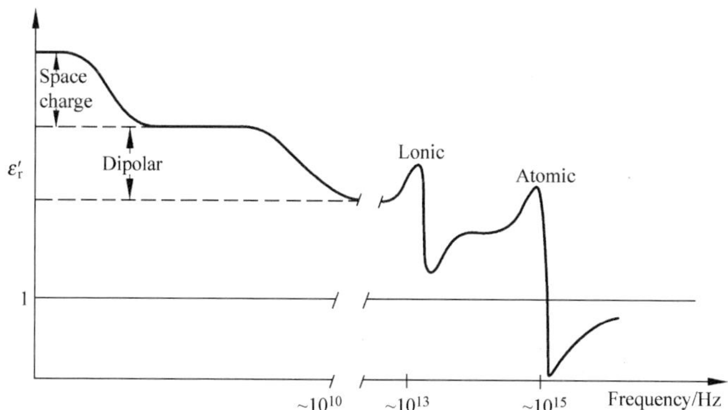

Fig.3.2 Frequency dependence of four polarization mechanisms (Moulson and Herbert , 1990)

Ferroelectrics are a subgroup of polar crystals. They contain a permanent electric dipole at the unit cell level as a result of the local atomic arrangement, and this electric dipole spontaneously aligns with those in adjacent unit cells to yield a net polarization over many unit cell dimensions. These many unit cells with the same polarization direction form a domain and the region that separates two domains is called a domain boundary. Domain boundaries may migrate, resulting in growth or shrink of domains. Although individual domains have net electric polarization, domains are oriented in such a way that the electric polarization of individual domains cancels completely and the system exhibits no net polarization. Barium titanate, $BaTiO_3$, the first ceramic material from which ferroelectric behavior was observed, is an ideal model for a discussion of the ferroelectricity and spontaneous polarization from the point of view of crystal structure. $BaTiO_3$ has a typical perovskite (ABO_3) structure. Unlike many other crystals, oxygen anions in perovskite do not form a close packing structure. Above the Curie temperature (approximately 130℃), the unit cell is cubic with the ions arranged in the way outlined in Fig.3.3 (Chiang et al., 1996, Kingery et al., 1976). Barium ions (A ions), which are large in size (~158 pm), occupy the corner sites, titanium ions (B ions), which are small in size (~60 pm), locate in the centers of the cubes (the oxygen octahedra) and oxygen anions are on the face-centers. At the Curie point, the crystal undergoes a phase transition which is also called displacement phase transition. Below the Curie point, the crystal structure is slightly distorted to the tetragonal form with cations shifting in one direction and anions in the opposite direction and consequently an electric dipole moment along the c-direction is formed (as shown in Fig.3.4a) (Jona and Shirane, 1962). In addition to the shift of cations relative to anions and the change of lattice constants, there is a distortion of oxygen octahedron as well. However, the oxygen octahedron distortion is not always necessary to accompany the formation of electric dipoles. In another perovskite, $PbTiO_3$, the

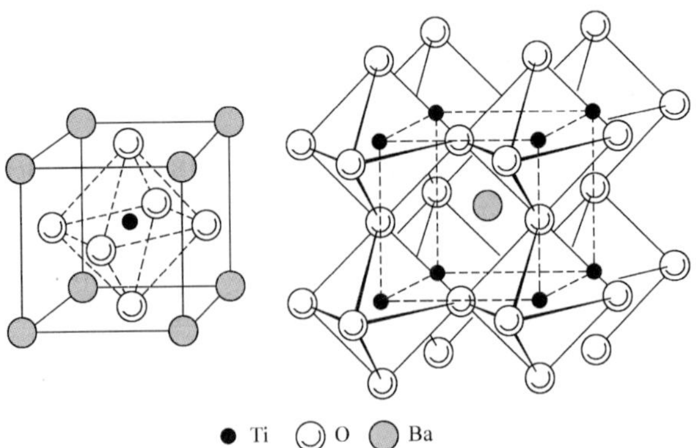

Fig.3.3   Crystal structure of $BaTiO_3$, an typical isotropic perovskite (ABO_3), above the ferroelectric-paraelectric transition temperature (Kingery et al., 1976)

phase transition at the Curie point involves only position shift of cations relative to anions, but no distortion in oxygen octahedron as shown in Fig.3.4b.

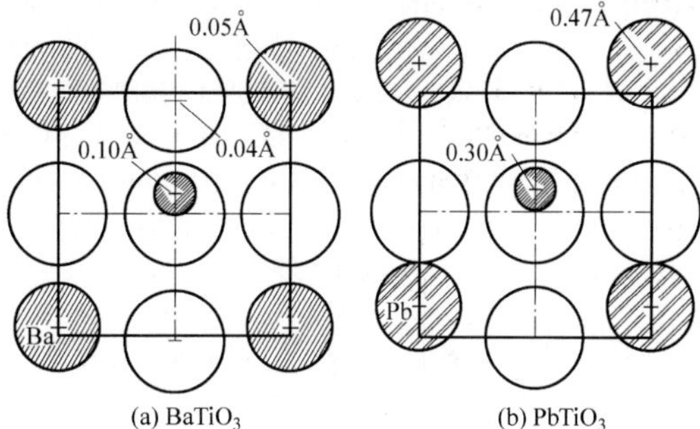

Fig.3.4 Spontaneous polarization in isotropic perovskites, $BaTiO_3$ and $PbTiO_3$. The polarizability is dependent on the sizes of A and B cations as well as the deformability of A cations (Jona and Shirane, 1962)

The spontaneous polarization value of each unit cell in perovskites is largely determined by the sizes of A and B cations (Jona and Shirane, 1962; Megaw, 1957) as well as the deformability of A cations. In general, a size decrease of B cations (located inside an oxygen octahedron) results in an increase in polarization due to an increased "rattling" space available for the B cations, provided that the perovskite structure (oxygen octahedron) is preserved. The size increase of the A cations would result in an increased size in the unit cell and thus an increased size of the oxygen octahedron. The influence of the sizes of A and B cations in perovskite ferroelectrics on the dielectric properties such as the Curie point will be discussed further later in this chapter. The deformability of A cations determines how much the oxygen octahedron can shift relative to A cations. Pb cation possesses a very good deformability and thus permits a large extent shift of oxygen octahedron, resulting in a large polarization value. This partly explains the fact that many important commercial and technological ferroelectrics and piezoelectrics are lead-containing perovskite materials, although lead oxide has long been known as a hazardous material and presents some technical challenges in materials synthesis and processing due to its low melting point and high vapor pressure.

3.3 Hysteresis

When an electric field is applied to dielectric materials, electric polarization will occur accordingly, involving the four polarization mechanisms described previously. For ferroelectrics, domains, consisting of many spontaneously polarized unit cells aligned in the same direction, will start switching directions in response to the electric field and eventually aligned with the electric field as closely as the crystal structure permits, when the externally applied electric field is sufficiently large. This processing is called poling. Ferroelectrics prior to poling are called virgin ferroelectrics, while ferroelectrics after poling are called poled ferroelectrics.

The hysteresis, or *P-E* loop, is the most important characteristic of a ferroelectric material (Hench and West, 1990; Moulson and Herbert, 1990). Figure 3.5 shows a typical hysteresis loop for a ferroelectric ceramic.

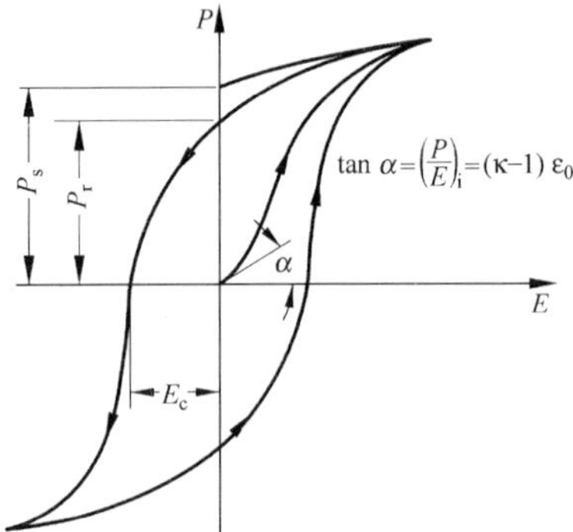

Fig.3.5 Typical ferroelectric hysteresis loops for a $BaTiO_3$ ceramic (Kingery et al., 1976)

Because a ceramic is composed of a large number of randomly oriented crystallites, it would normally be expected to be isotropic in its properties. Therefore, no preferred orientation effects should be evident in the loop and it should be symmetric about a point of origin. Starting at point O, as the electric field gradually increases, a little polarization is induced, being linearly proportional to the applied electric field. This polarization is reversible and due to electronic, ionic, and space charge polarization, similar to normal or linear dielectrics (or paraelectrics). As the electric field increases further, the polarization starts to increase abruptly, due to domain switching within the material in response to the externally applied electric field. After most domains switched direction to achieve the closest alignment with the externally applied electric field, polarization will slowly increase and eventually reach the saturation state as the electric field rises further. Reducing the electric field to zero leaves the material in a polarized condition with a remanent polarization, P_r, since not all domains have enough energy to switch back. Increasing the electric field in the opposite direction reduces the polarization by reversing the direction of the domains. The polarization first reaches zero at E_c, the coercive field of the material, and then further to the saturation state in the reverse direction. The polarization will reduce with a reducing electric field and reach a remanent polarization as the field reaches zero. If the opposite electric field is applied again, the domains reverse their polarization direction in response to the electric field and a complete hysteresis loop is then obtained when a saturation state is reached. Although it is possible to back switch the material to a state of zero polarization, the material is not the same as the starting condition, which can only be reinstalled by thermally depoling, i.e., heating the material to temperatures above the Curie point. Ferroelectric ceramics have relatively lower coercive force and remanent polarization than that of ferroelectric single crystals, which exhibit square-loop hysteresis. Hence ferroelectric ceramics are generally referred to as soft ferroelectrics with a smaller coercive field, E_c, since the domains, randomly oriented, are easier to switch. For the same reason, ferroelectric ceramics have relatively lower remanent polarization, P_r.

For many applications of ferroelectrics as piezoelectrics, pyroelectrics and electro-optics, a large spontaneous polarization is one of the most important requirements. A large spontaneous polarization can be achieved by a combination of a large spontaneous polarization in each individual unit cell and a close alignment of spontaneous polarization directions with the externally applied electric field. Like many physical properties of single crystals, spontaneous polarization in ferroelectrics is dependent on crystallographic directions, since the change in direction of the spontaneous polarization requires small ionic shifts in specific crystallographic directions. One good example is layered perovskite ferroelectrics such as $SrBi_2Ta_2O_9$ (SBT), in which spontaneous polarization occurs in both a-and b-directions, but little or no spontaneous polarization in the c-direction. Since a ceramic is composed of a large number of randomly oriented crystallites, spontaneous polarization in ferroelectric ceramics will be isotropic. A greater polarization would be achieved if there are a greater number of possible crystallographic directions in which spontaneous polarization can occur and thus a closer alignment of spontaneous polarization with the externally applied electric field can be achieved. The tetragonal structure allows six directions, while the rhombohedral

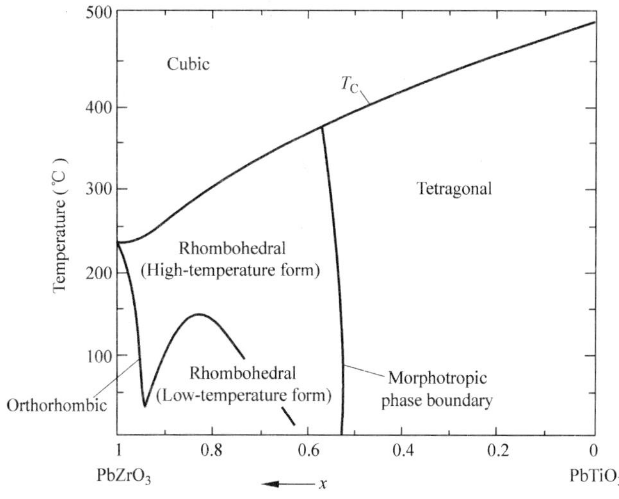

Fig.3.6 Phase diagram of the system $Pb(Ti_{1-x}Zr_x)O_3$ (Jaffe et al., 1971)

allows eight and so should permit greater alignment. If both tetragonal and rhombohedral crystallites are present at a transition point, where they can be transformed from one to the other by an electric field, the number of alternative crystallographic directions rises to 14 and the extra alignment attained will become of practical significance. Figure 3.6 shows the phase diagram of the lead zirconate titanate (PZT) system where it can be seen that the morphotropic phase boundary (MPB) is a significant feature (Jaffe, 1971). A MPB denotes an abrupt structural change with composition at constant temperature in a solid solution range. In the PZT system, it occurs close to the composition where the $PbZrO_3$: $PbTiO_3$ ratio is 1 : 1. At compositions near the MPB the piezoelectric coupling coefficient and the dielectric constant peak as shown in Fig.3.7 (Jaffe, 1971). This feature is explored in commercial compositions and also explains, in part, why PZT ceramics are now the most widely studied and commercialized of all piezoelectric ceramics.

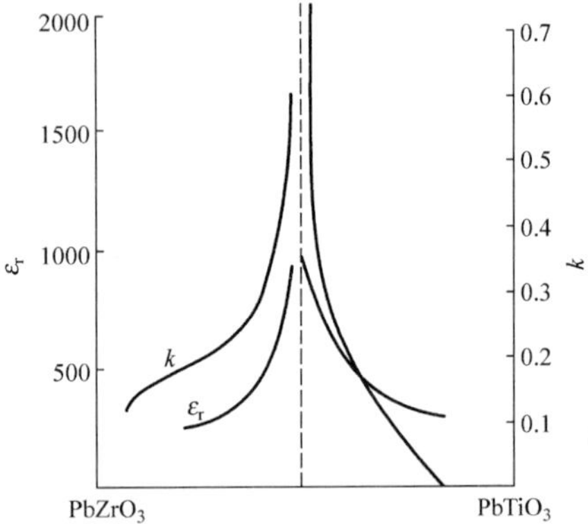

Fig.3.7   Dielectric constants and piezoelectric coupling coefficients across the PZT compositional range (Jaffe et al., 1971)

3.4   The Curie Point and the Phase Transition

Ferroelectrics become paraelectrics at temperatures above the Curie point and the spontaneous polarization disappears due to the change of crystal structure from noncentrosymmetric to centrosymmetric. Figure 3.8 shows the dielectric constant of a barium titanate ceramic as a function of temperatures (von Hippel, 1954). The dielectric constant peaks at the Curie point, because the energy barrier or the activation energy for the B ions to shift along certain crystallographic directions reduces to zero at the Curie point. Figure 3.9 shows a typical energy diagram for a ferroelectric (Jona and Shirane, 1962). Below the Curie point, the body-center ion (B ion) has to shift away from the center point to a position where the lowest energy is, forming an electric dipole (spontaneous

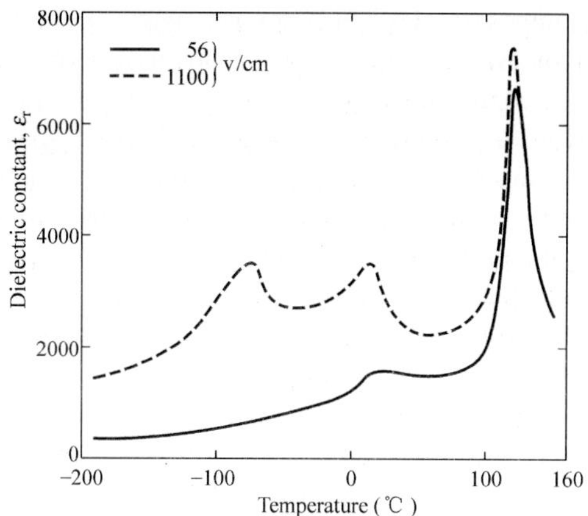

Fig.3.8 Dielectric constants of BaTiO₃ ceramic as a function of temperature (von Hippel, 1954)

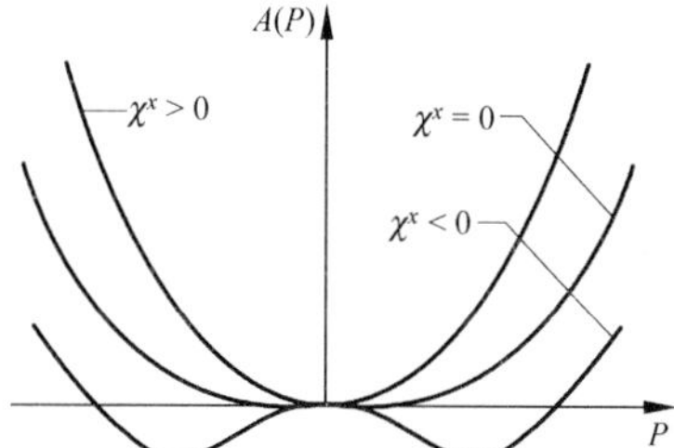

Fig.3.9 Variation in the potential energy of B ion along the c-axis, at temperatures higher than, equal to and lower than the Curie point (Jona and Shirane, 1962)

polarization). An activation energy is required to switch the direction of this electric dipole by shifting this B ion from one side to the other. Above the Curie point, the B ion possesses the minimum energy at the center position and therefore there would be no spontaneous polarization or electric dipoles. Energy is required to shift the B ion away from the center position. At the Curie point, the energy curve becomes rather flat near the center and the B ions can easily move away from the center point and shift back and forth. Consequently, the dielectric constant peaks at the Curie point for all ferroelectrics. However, the high dielectric constants of ferroelectrics at the Curie point are of little practical use due to its temperature sensitivity. For a given system (or material), the Curie point is determined by the chemical composition and varies as the chemical composition changes. Figure 3.10 shows the relationship between the Curie temperatures and chemical composition in the SrTiO₃-BaTiO₃-PbTiO₃ ternary system (Moulson and Herbert, 1990). The Curie point decreases linearly as Ba ions are substituted by Sr ions; while the Curie point increases as Ba ions are replaced by Pb ions. The decrease in the Curie point can be easily understood by

96

considering the change of size and deformability of A ions. Although the substitution of Ba by Sr does not result in a change of crystal structure, the lattice constants decreases linearly as the substitution proceeds. A decreased lattice constant leads to a reduced oxygen octahedron space in which Ti ion "rattles". This decrease in the "rattling" space has two consequences: a decrease in the Curie point and a reduced polarization. The substitution of Pb by Ba is a different situation. Ba ions are greater than Pb ions in size; however, Pb ions possess a much greater deformability and thus allow a greater extent of polarization (as depicted in Fig.3.4).

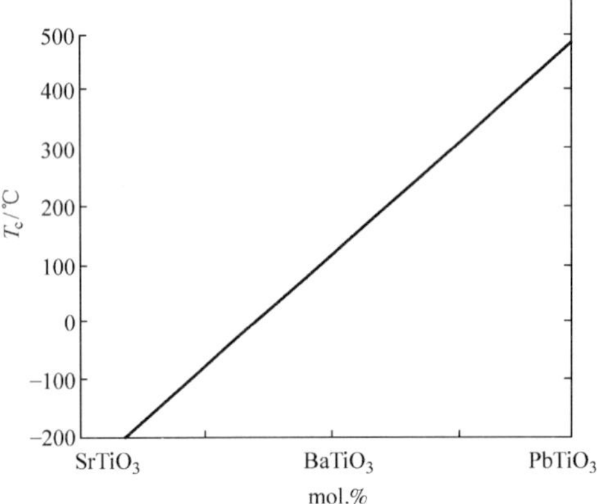

Fig.3.10 The effect on the Curie of the substitution of either strontium or lead for barium in BaTiO$_3$ (Moulson and Herbert, 1990)

Such a relationship between the Curie point and the lattice parameters or the size of cations has also been observed in layered perovskite ferroelectrics. The change in the Curie point may also result from the vacancy-induced variation of lattice constants. For example, recently Torii et al. (1998) reported that the ferroelectric properties of SBT layered perovskite ceramics are strongly dependent on the stoichiometry. The Curie temperature increased from approximately 300°C for a ceramic with a stoichiometric composition to 450°C with a composition of 30 at% Sr deficiency as shown in Fig.3.11a (Torii et al., 1998). This figure also indicates that below the Curie point, in general, the dielectric constants are seen to decrease as the Sr deficiency decreases. Although the deficiency of Sr in the layered perovskites is compensated by the increased amount of Bi, an appreciable linear decrease in parameters a and c was observed as given in Fig.3.11b (Torii et al., 1998). Although Torii et al. did not explicitly attribute the decrease in dielectric constant and the increase in the Curie point to the decrease in lattice parameters, a decrease in lattice parameters would certainly lead to a decrease in the "rattling" space for the B ions inside the oxygen octahedra. This conclusion is in a good agreement with other studies (Isupov, 1997a, b; Subbarao, 1996).

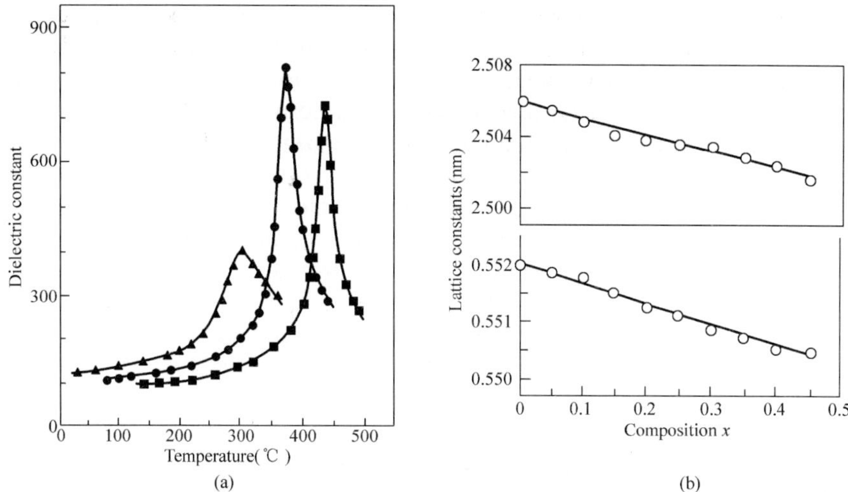

Fig.3.11 Effects of nonstoichiometry in SBT on (**a**) the dielectric constants and the Curie point and (**b**) the lattice constants (Torii et al., 1998)

Partial substitution of cations in perovskite ferroelectrics may result in broadening of the maximum peak at the Curie point, if the substitution constituent ions are randomly distributed and form micro-inhomogeneity in crystal structure (Bokov and Myl, nikov, 1961; Burfoot and Clarke, 1974; Cross et al., 1980; Jang et al., 1981). The ferroelectric-to-paraelectric phase transition in the material will no longer occur at a single temperature. Instead, the various regions with slightly different crystal structures will undergo the phase transition in a very broad range of temperatures. Consequently, a very high dielectric constant (typically over 10000) can be obtained in a broad range of temperatures. Such ferroelectrics are known as relaxor or relaxator-type ferroelectrics and have gained significant attention recently for many applications, particularly as piezoelectrics in MEMS.

3.5 Ferroelectric Thin Films for DRAM and NvRAM

Ferroelectric thin films for DRAM and NvRAM application have received considerable attention in both industrial and research communities and significant progress has been made in advancing the fundamental understanding of processing, properties, performance, and fabrication of ferroelectric based DRAM and NvRAM (Auciello et al., 1996a; Auciello et al., 1996b; Ramesh, 1998; Scott, 1998a; Treece et al., 1998; Waser and Lohse, 1998). Ferroelectrics are excellent candidates for the applications in information storage in digital memory systems. Interest in ferroelectrics for information storage systems stems from high dielectric constants and remanent polarization. For the new

98

generation of DRAM, high dielectric constant and low leakage current are among the most important requirements. Random access memories (RAMs) based on semiconductor integrated technology have been a great success; however, these semiconductor RAMs can retain information only when power is on. A serious drawback is that when power is interrupted, all information is lost (volatile memory). Furthermore, these RAMs are very sensitive to radiation and this is detrimental for military and space applications. The intrinsic nonvolatility (retention of memory when power is interrupted) and radiation hardness of ferroelectric RAMs brought about a move to substitute a thin film ferroelectric capacitor, which can be permanently polarized to store digital data (Auciello et al., 1996a; Auciello et al., 1996b; Bondurant and Gnadinger, 1989; Jones et al., 1995; Ramesh, 1998; Scott, 1998a; Scott and Paz de Araujo, 1989; Treece et al., 1998; Waser and Lohse, 1998; Desu et al., 1997). In recent years it has become possible to fabricate ferroelectric thin film memories onto standard silicon integrated circuits that combine very high speed (30 ns read/erase/rewrite operation), 5-V standard silicon logic levels, very high density (2×2 µm cell size), complete nonvolatility and extreme radiation hardness (Scott et al., 1996; Scott, 1998b). Many excellent comprehensive review articles have been published (e.g., Scott, 1998a; Scott, 1998b). In this section, we only briefly summarize the work done in this exciting field in the hope of providing readers with a simple clear overall picture of the current accomplishment and remaining challenges. Three families of ferroelectric materials have been studied for the DRAM application: barium strontium titanate (BST), PZT and strontium bismuth niobate tantalate (SBNT). The early work was focused on BST ferroelectrics; while the recent attention has been shifted to SBNT layered perovskite ferroelectrics.

At present, BST has been chosen to make DRAM by many companies (Hwang, 1998; Koyama et al., 1991; Mikami, 1998; Summerfelt, 1998; Yuuki et al., 1995). Although BST material has the Curie point below the room temperature, if Sr : Ba ratio exceeds 25 : 75, and thus is not a ferroelectric, the dielectric constant of BST films is relatively high (~300). The paraelectric characteristic of BST makes it free of fatigue, since there is no reversible spontaneous polarization. In addition, BST consists of no hazardous material such as lead and is relatively easy to make in film deposition. Extensive research on BST thin films for DRAM application has revealed two interesting issues. One is that the BST thin films have a rather different dielectric property from that of bulk BST materials. The other is the dependence of dielectric constant on the film thickness and the processing temperatures. Bulk BST has a very high dielectric constant above 1000 with a maximum of over 10000 at the Curie temperature. The Curie temperature for bulk BST varies between 130℃ for $BaTiO_3$ and approximately 50 K for $SrTiO_3$ (Fig.3.10). Like many other ferroelectrics, the dielectric constant of bulk BST ceramics varies significantly as temperature changes. Mikami (1998) showed that BST thin films with a thickness less than 320 nm behaved very differently from bulk BST. The dielectric constant of BST films has a minor variation as temperature changes and there is no abrupt change of dielectric constant at the Curie point. In addition, the characteristic *P-E* hysteresis loop for ferroelectrics was not observed in BST (Ba ∶ Sr = 75 ∶ 25) thin films at 77 K, below the Curie point of the corresponding bulk BST ceramic (~20 ℃). Currently, a thorough understanding of the above ferroelectric properties of BST thin films has not been achieved.

It is a well-known phenomenon that the dielectric constant of BST thin films decreases significantly as the film thickness reduces below 100 nm. In addition, the leakage current density increases rapidly as the film thickness decreases under a standard operation voltage (5 V). This strong dependence of ferroelectric properties on thickness imposes a great challenge to the further development of BST-based DRAM, as the devices miniaturize rapidly as technologies advance ever further. Several possible mechanisms, all based on microscopic considerations, have been suggested to explain the decrease in dielectric properties and absence of the ferroelectric characteristic hysteresis loop below the Curie point of BST thin films as film thickness is below 100 nm. One is that the highly disordered surface layer (or amorphous layer) in very thin films has a low dielectric constant and plays a significant role in the overall performance of the BST thin films. Small grain size, less than the film thickness, may change the dielectric constant and ferroelectric characteristic as well. Stress in BST thin films may also have a significant influence on the dielectric and ferroelectric properties. It is known that a low dielectric constant surface layer, a smaller grain size and stress in BST thin films all result in a reduction in dielectric constant. The increase of leakage current with a decrease in film thickness below 100 nm might also be explained by the above mechanisms. The film thickness dependence of dielectric and ferroelectric properties is not observed in all paraelectrics and ferroelectrics, however. As will be discussed below, a decrease of dielectric constant with a decrease in film thickness is not present in layered perovskite ferroelectric thin films. In addition, the above mechanisms could not satisfactorily explain the absence of ferroelectric characteristic, i.e., hysteresis loop, in BST thin films at temperatures below the Curie point. X-ray diffraction (XRD) analysis indicated that BST thin films do consist of polycrystalline BST perovskite with a relatively good crystallinity and no significant deviation in crystal structure has been established. At present, it remains an ongoing effort to achieve a better understanding of the mechanism of the dependence of dielectric properties on the thickness of BST thin films. A relatively high leakage current density in BST thin films is another challenge for achieving high density DRAM devices.

DRAM based on BST thin film is volatile, since BST is a paraelectric at the room temperature. In addition, BST thin films do not show the ferroelectric characteristic, i.e., hysteresis loop, even at temperatures below the Curie point.

For ferroelectric DRAM and NvRAM applications, the recent focus has been on PZT and SBNT thin films. PZT has an isotropic perovskite crystal structure with a high remanent polarization (30–50 $\mu C/cm^2$) and a coercive field of 70 kV/cm (Bondurant and Gnadinger, 1989; Grill et al., 1997; Shimizu et al., 1995; Tuttle, 1998; Vorotilov et al., 1995). Unfortunately, PZT films tend to degrade most of the initial amount of switching charge (so-called "fatigue") after 10^6–10^8 cycles of full polarization switching (Dimos et al., 1997; Haertling, 1997; Joo et al., 1997; Majumder et al., 1997; Ramesh et al., 1992; Warren et al., 1996). To be competitive with electrically erasable read-only memories (EEPROM), ferroelectric memories must be improved to withstand at least 10^{12} erase/rewrite operations or they must have qualitatively different nondestructive read operations. Several fatigue mechanisms for the PZT films have been proposed. One of the mechanisms is the 90° domain walls. Being an isotropic perovskite ferroelectric, PZT has two types of domain walls: 180° and 90°. At the Curie point, the cubic changes to tetragonal structure, which results in the formation of an electric dipole in each unit cell. The polarization direction (c-axis in tetragonal) can be any of the original a-axis directions in the cubic structure. Since the lattice constant, c, is longer than the lattice constant, a, there will be a distortion at the 90° domain walls, which induces a mechanical stress. Another mechanism is the "domain pinning". Domain pinning is due to oxygen vacancies. Oxygen vacancies in PZT thin films are due mainly to the volatile nature of PbO that results in a Pb deficiency and leads to the formation of oxygen vacancies. The reduction of Ti ions from four-plus-valence to three-plus valence may also result in the formation of oxygen vacancies. These relatively mobile oxygen vacancies and other substitutional impurities may accumulate at grain boundaries, domain boundaries and other defect sites and pin the domains resulting in the degradation of ferroelectric properties. Currently, the approach to reduce the degradation of ferroelectric properties or fatigue of PZT thin films is to use complicated electrodes, such as electronic conducting oxides, e.g., RuO_2, $(La,Sr)CoO_3$ and $SrRuO_3$ (Aggarwal et al., 1997; Araujo et al., 1995; Kim et al., 1998; Lee et al., 1995; Lin et al., 1998). Like BST thin films, the ferroelectric properties of PZT thin films also change drastically as the film thickness reduces below 200 nm (Mihara et al., 1995). Furthermore, PZT thin films suffer from a relatively high leakage current density.

Recently, bismuth oxide layered perovskite materials such as $SrBi_2Nb_2O_9$ (SBN), $SrBi_2Ta_2O_9$ (SBT), and $SrBi_2(Nb,Ta)_2O_9$ (SBTN) have become increasingly important due to their high fatigue resistance and ferroelectric properties comparable to PZT (Cross et al., 1980; Scott, 1998a; Takenaka et al., 1995). Bismuth layered ferroelectric oxides have large spontaneous polarization along the a- and b- axes, but little or no spontaneous polarization along the c- axis. Thus, most of the domain configuration in bismuth layered perovskite oxides is the 180° domain, leading to excellent fatigue resistance. Bismuth layered perovskite structure and its ferroelectricity were systematically studied by Aurivillius (Aurivillius, 1949; Aurivillius, 1950). Bismuth layered perovskite structure is sketched in Fig.3.12 and the compounds are generally represented by the following formula (Isupov, 1997a, b; Jona and Shirane, 1962; Subbrarao, 1996):

$$(Bi_2O_2)^{2+}(A_{m-1}B_mO_{3m+1})^{2-}$$

where A is Bi^{3+}, Ba^{2+}, Sr^{2+}, Ca^{2+}, Pb^{2+}, K^+ or Na^+; B is Ti^{4+}, Nb^{5+}, Ta^{5+}, Mo^{6+}, W^{6+} or Fe^{3+}; and m is an integer from 1 to 8 or a fraction such as 2(1/2) or 3(1/2). SBT system thin films have many outstanding characteristics for nonvolatile memory applications as compared with PZT thin films, being briefly summarized below (Takenaka et al., 1995):
1. fatigue-free characteristics without any complicated electrodes,
2. high signal/noise ratio larger than 8 at 1.2 V,
3. low-voltage operation at less than 2 V,
4. less surface effect owing to less space charge,
5. long data retention time and very stable imprint characteristics, and
6. low leakage current due to less space charge.

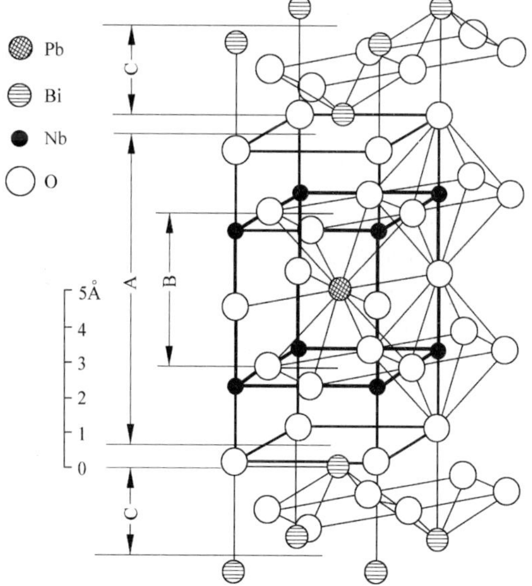

Fig.3.12　Crystal structure of layered perovskite, $SrBi_2Nb_2O_9$ (Jona and Shirane, 1962)

In addition, the remanent polarization of SBTN layered perovskite does not vary with the thickness of the films. This indicates that the SBTN system has the ability to maintain good ferroelectric properties even when the layer is less than 100 nm thick. The independence of ferroelectric properties of SBTN thin films on thickness is not understood yet. One possible reason　may be the surface effect of SBTN layered perovskite that is much smaller than that of PZT perovskite due to less space charge.

In comparison to PZT perovskite, however, the remanent polarization, P_r, of SBTN layered perovskite ceramic films is smaller, typically about 20 μC/cm^2 (Sedlar and Sayer, 1996). The small remanent polarization of layered perovskite ferroelectric ceramics may arise from two fundamental reasons (Jona and Shirane, 1962). One is that the change of crystal structure or the rearrangement of ions is greatly constrained by the $(Bi_2O_2)^{2+}$ layers in layered perovskites, so that there is a smaller polarization in comparison to the isotropic perovskites. Another reason is that there is very little or no spontaneous polarization along the c-axis, since continuous B-O-B-O chains are required for spontaneous polarization. Therefore, the spontaneous polarization can occur only along four possible crystallographic directions and would be relatively difficult to align with the externally applied electric field. The relatively small remanent polarization of the layered perovskites imposes a challenge for the applications in fabricating high-density DRAM and NvRAM, in spite of other superb properties of the layered perovskite ferroelectrics. In operation of a DRAM or NvRAM, the parameter of greatest importance for the capacitor is the charge density that is directly related to the dielectric constant (volatile DRAM) and the remanent polarization (NvRAM). The remanent polarization of SBNT ferroelectrics may have been enhanced using the following approaches. Microscopically, similar to isotropic perovskites, if substitution of A and B ions with appropriate ion sizes results in an enlarged "rattling" space for B ions, the polarizability or the saturation polarization, P_s, would certainly increase. In addition, a higher valence of the B ions would result in a higher polarization. To achieve a high remanent polarization, it is also necessary to create a large energy barrier so as to prevent the spontaneous back -switch of domains when the externally applied electric field is removed. Incorporation of an appropriate solute would result in an increased energy barrier so as to prevent the motion of domain wall. The difficulties lie in the fact that the crystal structure of layered perovskites is constrained by the $(Bi_2O_2)^{2+}$ layers. Consequently, unlike in isotropic perovskites, the size change of A ions will most probably not result in a significant change in lattice parameters. Difficulty also arises from the lack of pentavalent cations with ionic size close to Nb^{5+} and Ta^{5+}. In spite of the above difficulties, it is possible to manipulate the ion substitution and stoichiometry so as to change the lattice constants and the dielectric properties of layered perovskites as demonstrated by Torii et al. (1998).

Another approach to overcome the relatively small remanent polarization of layered perovskite ferroelectrics in DRAM and NvRAM applications is to grow epitaxial thin films. As discussed previously, ceramic films consist of randomly oriented crystallites, and thus it is not possible to align all the domains with the externally applied electric field. Consequently, relatively smaller remanent polarization and coercive field are obtained from ceramic films (as well as bulk ceramics) than from single crystals. For layered perovskite ferroelectrics, epitaxial films are much more important, since, as discussed above, there is little or no spontaneous polarization along the c-direction in layered perovskites; instead, the spontaneous polarization occurs only along a-and b-directions, i.e., there are only four possible crystallographic directions with which spontaneous polarization or domains could align. A heteroepitaxial structure can be grown with two models that were proposed by Suzuki (Suzuki, 1995) and are depicted in Fig.3.13. The heteroepitaxial structure requires that the electrode material has a good lattice match with the ferroelectrics. Such an electrode material is readily available, such as $(Ca,Sr)RuO_3$, (CSR), which was under an extensive study recently. Compared to conventional electrodes such as novel metals, CSR offers several additional advantages including: etchability, chemical stability (less inter-diffusion between the ferroelectrics and electrodes), and a better match of thermal expansion coefficients (Bensch et al., 1990; Eom et al., 1993; Lee et al., 1996; Szymanik and Edgar, 1991). The heteroepitaxial structure offers the elimination of the thin amorphous layer that is always present at the interface between dielectric/ferroelectric thin films and novel metal electrodes. This amorphous interfacial layer has a low dielectric constant, contains many defects, and results in property degradation of ferroelectric films, and thus is detrimental to DRAM and NvRAM device performance. With a heteroepitaxial structure, this amorphous layer is eliminated.

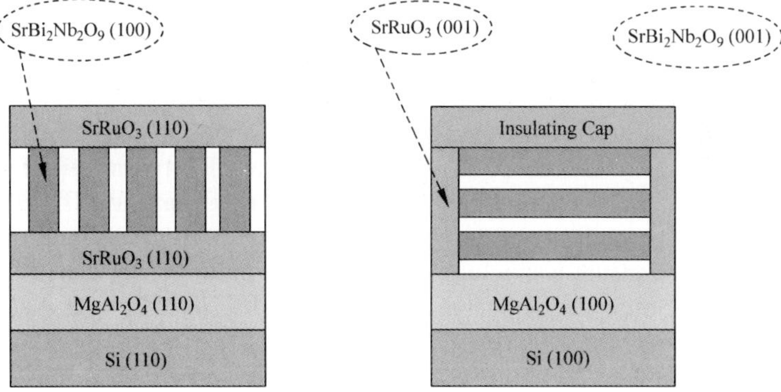

Fig.3.13 Two ideal models for the growth of textured or epitaxial films of layered perovskite SBNT and electronic conducting oxides (Ca, Sr) RuO$_3$ (Suzuki, 1995)

3.6 Deposition of Ferroelectric Thin Films

Many processing techniques have been applied successfully in ferroelectric thin film growth and device fabrication and there have been many excellent review articles on this subject published recently. A brief list of these recent review articles is give below so that the readers can easily find more detailed information if needed. A book titled "Thin Film Ferroelectric Materials and Devices" edited by Ramesh (Ramesh, 1998) provided a rather comprehensive summary of the recent advancement in the processing and growth of ferroelectric thin films and device fabrication. Auciello et al. (1996a; 1996b) edited two special issues of 1996 Materials Research Society Bulletin on topics: "Electroceramic Thin Films, Part I: Processing" and "Electroceramic Thin Films, Part II: Device Applications". Most recently, Scott (1998a) published an excellent review article on ferroelectric films for DRAM with emphasis "on breakdown mechanisms and limits, leakage current, electrodes and electrode interfaces, scaling to submicrometer geometry, and deposition techniques".

A process for ferroelectric thin-film-based device fabrication needs to satisfy the following requirements: (1) applicable to device fabrication such as patterning, (2) relatively low processing temperature so as not to damage the device structure involved, (3) preserving the materials with desired physical properties and chemical composition, (4) good film quality and good reproducibility, and (5) low cost (simple and environment-friendly). There are many processes and techniques studied for ferroelectric thin films growth which include: sputtering, pulsed laser ablation deposition (PLAD), chemical vapor deposition (CVD), liquid injection or mist CVD, metal-organic decomposition (MOD), chemical solution deposition, and hybrid and modified deposition processes.

Sputter deposition is a very popular technique in film deposition and has been extensively studied for ferroelectric thin film growth (Auciello et al., 1996c). Examples are $BaTiO_3$, $PbTiO_3$, BST, PZT, $(Pb,La)(Zr,Ti)O_3$ (PLZT), and SBT (Amanuma et al., 1995; Hong et al., 1998; Lee et al., 1997; Sreenivas et al., 1989). The plasma needed for sputtering the target and thus depositing the films can be generated by various power sources. DC and RF plasma sputtering are commonly used and often combined with bias electric or magnetic field, so as to achieve a better crystallinity of deposited films. A reactive gas such as oxygen is also often used to ensure a stoichiometric composition of the deposited film. The advantages of sputter deposition include: availability of all target materials, a relatively low vacuum required (typically range from a few to 100 mtorr), possibility of gas incorporation so as to modify the film chemical composition, textured or oriented films, and good adhesion to substrates. The major limitation of the technique is the control of stoichiometric composition, the thickness uniformity, and the conformal coverage of the films. For multiple component materials, the chemical composition of the deposited film is commonly different from that of the target, due to the different yield efficiency of various components and dissociation of compounds. Use of multiple targets and enrichment of some specific component(s) in the targets are the common approaches to obtain a desired stoichiometric composition in deposited films. The thickness uniformity and conformal coverage can be improved by using multiple targets and rotating the substrate. Although intensive basic and applied research on sputter deposition of ferroelectric thin films has been done and a fundamental understanding of the process has been greatly deepened, the difficulty in achieving good conformal coverage and desired stoichiometric composition remains a challenge in this processing technique.

Pulsed laser ablation deposition (PLAD), also called pulsed laser deposition (PLD), is another physical vapor deposition (PVD) technique that has received ample attention from the research community (Auciello and Ramesh, 1996; Chrisey and Hubler, 1994; Dat et al., 1995a; Dat et al., 1995b; Lichtenwalner et al., 1994; Tabata et al., 1995; Thomas et al., 1997; Yang et al., 1997). Conventional evaporation techniques are difficult to apply to most ferroelectric film growth, partly because the elements such as PbO and Bi_2O_3 that are commonly present in ferroelectrics have low melting point and high vapor pressures. For many multicomponent materials, PLAD can better transfer the chemical composition of the targets to deposited films. Using multiple targets and rotating the substrate holder, uniform coverage and good quality of ferroelectric films can be obtained. A variety of ferroelectric films have been deposited by PLAD techniques including SBT (Dat et al., 1995a; Tabata et al., 1995; Thomas et al., 1997). PLAD, however, also suffers from the common limitations of other physical vapor deposition techniques. For example, it remains difficult to precisely transfer the target stoichiometry to the deposited films, due to the volatility of some elements such as lead and bismuth. Uniform coating of large areas and conformal step coverage have not been firmly achieved. Another problem is the presence of large particulates in the deposited films.

Chemical vapor deposition (CVD) is a very attractive deposition technique for growth of ferroelectric thin films and much work has been published (Dey and Alluri, 1996; Eguchi and Kiyotoshi, 1997; Hintermaier et al., 1998; Keijser and Dormans, 1996; Lee et al., 1998; Li et al., 1996; Neumayer et al., 1998). There are various CVD processes depending on the form in which energy is supplied. Metalorganic compounds are commonly used as precursors in depositing ferroelectric films by CVD technique, which is often referred to as MOCVD (metalorganic CVD) or OMCVD (organometallic CVD). Xu in the next chapter of this book has provided a comprehensive coverage of the subject. CVD offers some distinct advantages such as uniform deposition over large areas, good conformal step coverage, and relatively easy control of stoichiometry of deposited films. However, CVD suffers from a limitation in growth of ferroelectric thin films, which is the limited availability of suitable precursors for high-Z elements. The vapor pressure of most high-Z element precursors is too low (e.g., below 1 mtorr at RT) to deliver a sufficient amount of material to the deposition chamber. Consequently a large-area uniform film would be difficult to achieve due to the depletion of the precursors. Many efforts have been made to overcome this obstacle. For example, CVD process with liquid or solid as transport precursors has been developed and such CVD

106

processes are generally referred to as liquid-injection, aerosol-assisted, or "mist" CVD (Isobe et al., 1997; McMillan et al., 1992; Solayappan et al., 1997; Van Buskirk et al., 1995).

The aerosol-assisted CVD process is different from other CVD techniques in technical details. In the aerosol-assisted CVD process, the precursors are delivered in the form of either liquid or solid by a carrier gas, while in conventional CVD processes, precursors are in the form of gas. Solid or liquid precursors are first dissolved in a solvent to form a liquid precursor. Such a liquid precursor can be directly injected by a high-pressure carrier gas into a CVD reaction chamber as schematically depicted in Fig.3.14 (Yamaguchi et al., 1998). Once the liquid is injected into the deposition chamber, solvent in the tiny liquid droplets evaporates immediately and organic components in the precursor undergo decomposition. The direct liquid injection method offers the advantage of high rate of precursor delivery and thus a high film growth rate. An additional advantage is that there would be no liquid condensation and no blockage of the precursor delivery system. However, it is potentially difficult to control the size of liquid droplets and large liquid droplets may result in incomplete decomposition of organic components and thus degradation of the film crystallinity. Another approach is to mistify the liquid to form tiny liquid droplets through either ultrasonication or uncentrification by high-speed rotating disc or using a commercially available atomizer. Submicrometer (<1 μm in diameter) liquid droplets are mixed with a carrier gas (so forming an aerosol or a mist). The aerosol may be transported either directly to the CVD reaction chamber,

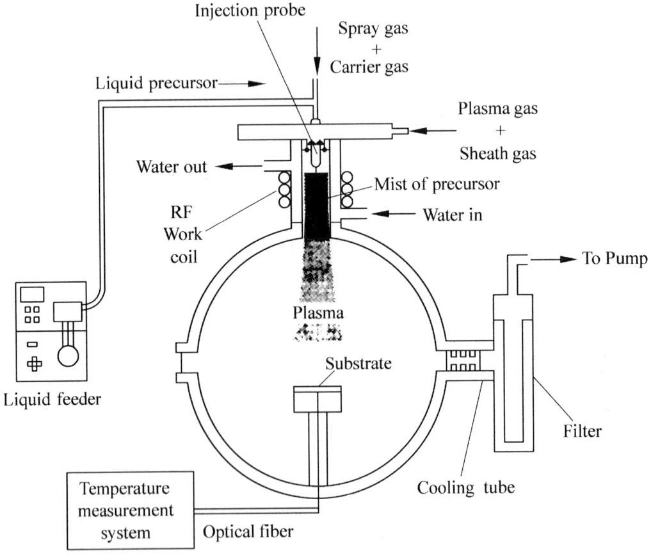

Fig.3.14 Schematic of thermal plasma spray CVD apparatus with a direct liquid injection probe (Yamaguchi et al., 1998)

or first to the so-called evaporizer before the aerosol enters the reaction chamber (as shown in Fig.3.15) (Xia et al., 1998). An evaporizer is used to evaporate the solvent from the liquid droplets so as to prevent liquid condensation and to avoid blocking the precursor delivery system. The temperature in the evaporizer can range from approximately 150 ℃ to several hundred degrees Celsius. Depending on the precursor and the temperature at the evaporizer, the precursor may be in the form of either solid or liquid when exiting the evaporizer. Aerosol-assisted CVD has demonstrated its great potential in the growth of ferroelectric thin film as well as other multicomponent films. With aerosol-assisted precursor delivery, ferroelectric films with good crystallinity and the desired stoichiometric composition have been reported.

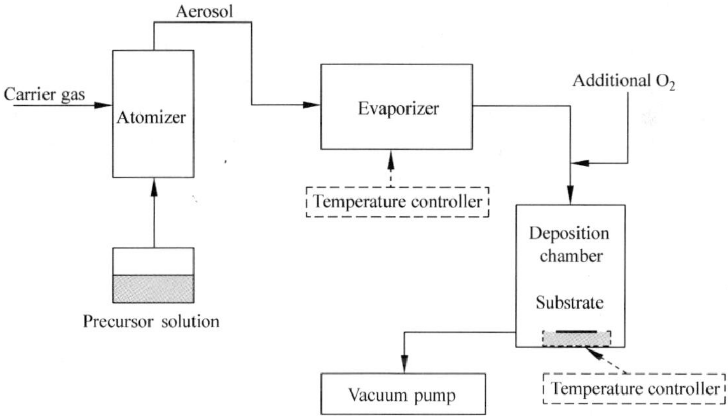

Fig.3.15 Schematic of aerosol-assisted chemical vapor deposition (CVD) apparatus with an evaporizer between reaction chamber and aerosol-generator (Xia et al., 1998)

The sol-gel process is very versatile and offers many advantages such as low capital investment, atomic-level homogeneity, mild processing conditions, applicability to multicomponent systems, and coatings on complex substrates (Brinker and Scherer, 1990; Pierre, 1998). The sol-gel process facilitates stoichiometric control of complex oxides better than other film growth techniques such as PVD and CVD. Furthermore, it is a process compatible with many semiconductor-fabrication technologies, and may be the deposition method of choice for applications that do not require conformal depositions and that have device dimensions of 2 μm or greater. Several excellent review articles on sol-gel processing of ferroelectric films have been published recently (Mackenzie and Xu, 1997; Schwartz, 1997; Tuttle and Schwartz, 1996; Xu et al., 1994) and provided a comprehensive summary of the recent advancement in the subject. Starting precursors for sol-gel processing can be either inorganic salts or metal alkoxides. The precursors are dissolved in a common solvent and undergo hydrolysis and condensation reactions. It is a common practice that the precursors with low reactivity are partially hydrolyzed prior to the addition of

108

more reactive precursors. During the hydrolysis and condensation, the starting precursors react to form clusters with a desired stoichiometric composition and such a stable solution with dispersed solid clusters is called a sol. When sol is deposited onto a substrate by either dip-, spin-, or spray-coating, solvent evaporates and a solid amorphous film forms. Such an amorphous film is typically porous and consists of many residual organic components. Post-deposition heat treatment is required to crystallize and densify the film as well as to remove the residual organic components. A great variety of ferroelectric films, such as PZT and SBT (Atsuki et al., 1995; Boyle et al., 1996; Ching-Prado et al., 1997; Ito et al., 1996; Kato et al., 1998; Klee and Mackens, 1995; Krupanidhi et al., 1992; Mackenzie and Xu, 1997; Schwartz, 1997; Tuttle and Schwartz, 1996; Xu et al., 1994), have been prepared by sol-gel processing. Oriented or textured films of layered perovskites grown on single crystal substrates have also been reported (Cho et al., 1998; Desu et al., 1996; Gu et al., 1996; Song et al., 1996; Wu and Cao, 1998). However, due to the nature of liquid processing, sol-gel processing suffers from drawbacks such as, size-limitation, film uniformity, and post-heating required for densification and crystallization. A typical procedure of sol-gel processing of SBN thin film was outlined in Fig.3.16 (Wu and Cao, 1998).

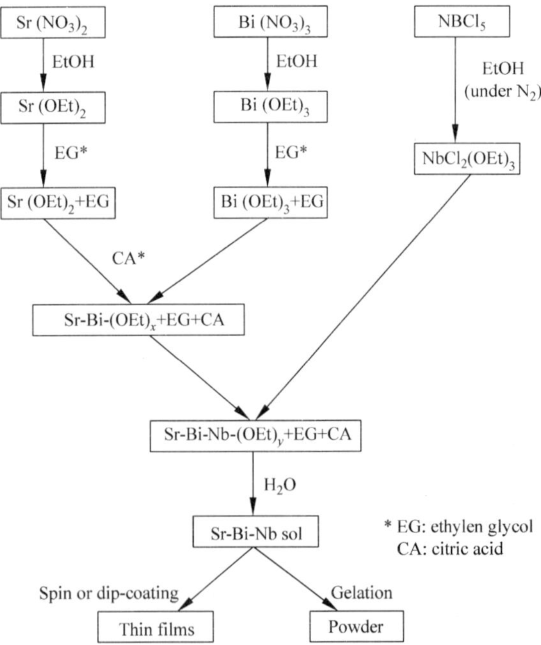

Fig.3.16 Schematic flow chart of a typical sol-gel processing of SBN layered perovskites (Wu and Cao, 1998)

Metalorganic deposition (MOD) is another wet-chemical technique currently under intensive study for the growth of ferroelectric thin films (Chu et al., 1996; Haertling, 1995; Joshi et al., 1997a; Joshi et al., 1997b; Joshi et al., 1998; Noma and Ueda, 1997; Solayappan et al., 1998; Schwartz, 1997; Xu et al., 1994). The precursors are first dissolved in either a common solvent or different solvents, and the resultant solutions are then combined together. The difference between the sol-gel processing and the MOD method is that for MOD, there is no chemical reaction necessary in the solutions. The choice of starting organic precursors is broad and the solution synthesis is straightforward. MOD also offers the advantage of easy tailoring of the chemical composition. MOD suffers several drawbacks as well, however. In addition to the limitations applied to sol-gel processing, it is not possible in the MOD method to precisely control the structural and phase evolution of the deposited films. Furthermore, due to the large organic ligands commonly used to minimize the reactivity of the precursors, the removal of these organic components often leads to crack formation. A proper control of solution concentration and thermal treatment is required to avoid the formation of cracks.

3.7 Summary

Significant progress has been made in understanding and exploring ferroelectric thin films in digital information storage applications. Ferroelectrics offer advantages such as high signal/noise ratio, chemical inertness, moisture insensitivity, and good thermal and mechanical properties. One of the significant advantages arises from the fact that the perovskite, unlike many other crystal structures, does not have a close-packing oxygen sublattice and, thus, provides a great flexibility in tailoring chemical composition and the crystal lattice through substitution with a great variety of cations. The substitution can take place in either A or B site, or both A and B sites. As a result, the chemical composition, lattice constants, and many physical properties of perovskites can be tailored significantly to satisfy various requirements depending on applications. Ferroelectrics have also been intensively studied and widely commercialized as piezoelectrics, pyroelectrics and electro-optic ceramics. The research on ferroelectric-based piezoelectrics has been greatly intensified recently, particularly driven by the emerging of micro-electro-mechanical systems (MEMS) technologies. The advances in computer science and engineering have generated an ever-increasing demand for a new generation of sensors and actuators. Many of these devices are based on electromechanical (piezoelectrics), electrothermal (pyroelectrics), or electro-optic (electro-optic ceramics) properties of ferroelectric materials. All these properties are closely associated with spontaneous polarization in ferroelectric materials and, thus, are dependent on the value of spontaneous polarization. For many applications, a large value of spontaneous polarization is critical, since a large signal/noise ratio can be obtained and a small size of devices can be achieved.

References

Aggarwal, S., B. Yang and R. Ramesh. in Thin Film Ferroelectric Materials and Devices. ed. R. Ramesh, (Kluwer, Norwell, MA, 1997) 221

Amanuma, K., T. Hase and Y. Miyasaka. Appl. Phys. Lett. **66**, 221 (1995)

Araujo, C.A. de, J.D. Cuchlaro, L.D. McMillan, M.C. Scott and J.F. Scott. Nature. **374**, 627 (1995)

Atsuki, T., N. Soyama, T. Yonezawa, and K. Ogi. Jpn. J. Appl. Phys. **34**, 5096 (1995)

Auciello, O. and R. Ramesh. Mater. Res. Soc. Bull. **21**(6), 31 (1996)

Auciello, O., R. Ramesh and F. Armani-Leplingard (guest editors). Mater. Res. Soc. Bull. **21(6)**, (1996a)

Auciello, O., R. Ramesh and F. Armani-Leplingard (guest editors). Mater. Res. Soc. Bull..**21(7)**, (1996b)

Auciello, O., A.I. Kingon and S.B. Krupanidhi. Mater. Res. Soc. Bull. **21(6)**, 25 (1996c)

Aurivillius, B. Arkiv Kemi. **1**, 463, 499 (1949)

Aurivillius, B. Arkiv Kemi. **2**, 519 (1950)

Bensch, W., H.W. Schmalle and A. Reller. Solid State Ionics. **43**, 171 (1990)

Bokov, V.A. and I.E. Myl'nikov. Soviet Phys.-Solid State (Engl. Transl.). **3**, 613 (1961)

Bondurant, D. and F. Gnadinger. IEEE Spectrum. **26(7)**, 30 (1989)

Boyle, T.J., C.D. Buchheit, M.A. Rodriguez, H.N. Al-Shareef, B.A. Hernandez, B. Scott and J.W. Zoller. J. Mater. Res. **11**, 2274 (1996)

Brinker, C.J. and G.W. Scherer. Sol-Gel Science: The Physics and Chemistry of Sol-Gel Processing. (Academic Press, San Diego, CA, 1990)

Buchanan, R.C.(ed). Ceramic Materials for Electronics: Processing, Properties, and Applications, 2nd. edition, (Marcel Dekker, New York, 1991)

Burfoot, J.C. and R. Clarke. Ferroelectrics. **8**, 505 (1974)

Chiang, Y.M., D.P. Birnie III and W.D. Kingery. Physical Ceramics: Principles for Ceramic Science and Engineering. (Wiley, New York, 1996)

Ching-Prado, E., W. Perez, A. Reynes-Figueroa, R.S. Katiyar, D. Ravichandran and A.S. Bhalla. Mater. Res. Soc. Symp. Proc. **474**, 49 (1997)

Cho, J.H., S.H. Bang, J.Y. Son and Q.X. Jia. Appl. Phys. Lett. **72**, 665 (1998)

Chrisey, D.B. and G.K. Hubler (ed.). Pulsed Laser Deposition of Thin Films. (John Wiley & Sons, New York, 1994)

Chu, P.Y., R.E. Jones, P. Zurcher, D.J. Taylor, B. Jiang, S.J. Gillespie, Y.T. Lii, M. Kottke, P. Fejes and W. Chen. J. Mater. Res. **11**, 1065 (1996)

Cohen, R.E. Nature. **358**, 136 (1992)

Cross, L.E., S.J. Jang, R.E. Newnham, K. Uchino and S. Nomura. Ferroelectrics. **23**, 187 (1980)

Dat, R., J.A. Greer and M.D. Tabat. J. Vac. Sci. Technol. **A13**, 1175 (1995a)

Dat, R., J.K. Lee, O. Auciello and A.I. Kingon. Appl. Phys. Lett. **67**, 572 (1995b)

Desu, S.B., D.P. Vijay, X. Zhang and B.P. He. Appl. Phys. Lett. **69**, 1719 (1996)

Desu, S.B., P.C. Joshi, X. Zhang and S.O. Ryu. Appl. Phys. Lett. **71**, 1041 (1997)

Dey, S.K. and P.V. Alluri. Mater. Res. Soc. Bull. **21(6)**, 44 (1996)

Dimos, D., W.L. Warren and H.N. Al-Shareef. in Thin Film Ferroelectric Materials and Devices. (ed.) R. Ramesh. (Kluwer, Norwell, MA, 1997) p. 199

Eguchi, K. and M. Kiyotoshi. Integ. Ferroelec. **14**, 33 (1997)

Eom, C.B., R.B. V. Dover, J.M. Phillips, D.J. Werder, J.H. Marshall, C.H. Chen, R.J. Cava and R.M. Fleming. Appl. Phys. Lett. **63**, 2570 (1993)

Grill, A., R. Laibowitz, D. Beach, D. Neumayer and P.R. Duncombe. Integ. Ferroelec.. **14**, 211 (1997)

Gu, H., W. Sun, S. Wang, T. Zhou, A. Kuang, J. Liu and X. Li. J. Mater. Sci. Lett.. **15**, 53 (1996)

Haertling, G.H. Integ. Ferroelec.. **10**, 257 (1995)

Haertling, G.H. Integ. Ferroelec.. **14**, 219 (1997)

Hench, L.L. and J.K. West. Principles of Electronic Ceramics. Wiley, New York, (1990)

Hintermaier, F., B. Hendrix, D. Desrochers, J. Roeder, T. Baum, P. Vanbuskirk, D. Bolten, M. Grossmann, O. Lohse, M. Schumacher, R. Waser, H. Cerva, C. Dehm, E. Fritsch, W. Honlein, C. Mazure, N. Nagel, P. Thwaite, H. Wendt. Integ, Ferroelec.. **21**, 367 (1998)

Hippel, A.R. von. Dielectrics and Waves. John Wiley & Sons, New York, (1954)

Hong, K., I.N. You, Y.S. Yu, S.K. Lee. Integ. Ferroelec.. **21**, 511 (1998)

Hwang, C.S. Mater. Sci. Engr.. **B56**, 178 (1998)

Isobe, C., T. Ami, K. Hironaka, K. Watanabe, M. Sugiyama, N. Nagel, K. Katori, Y. Ikeda, C.D. Gutleben, M. Tanaka, H. Yamoto and H. Yagi. Integ. Ferroelec.. **14**, 95 (1997)

Isupov, V.A. Inorganic Materials. **33**, 936 (1997a)

Isupov, V.A. Phys. Solid State. **39**, 116 (1997b)

Ito, Y., M. Ushikubo, S. Yokoyama and H. Matsunaga. Jpn. J. Appl. Phys.. **35**, 4925 (1996)

Jaffe, B., W.R. Cook and H. Jaffee. Piezoelectric Ceramics. Academic Press, New York, (1971)

Jang, S.J., L.E. Cross and K. Uchino. J. Am. Ceram. Soc.. **64**, 209 (1981)

Jona, F. and G. Shirane. Ferroelectric Crystals. Pergamon Press, New York, (1962)

Jones, Jr., R.E., P.D. Maniar, R. Moazzami, P. Zurcher, J.Z. Witowski, Y.T. Lii, P. Chu and S.J. Gillespie. Thin Solid Films. **270**, 584 (1995)

Joo, J.H., Y.J. Lee and S.K. Joo. Ferroelectrics. **196**, 321 (1997)

Joshi, P.C., S.O. Ryu, X. Zhang and S.B. Desu. Appl. Phys. Lett.. **71**, 1080 (1997a)

Joshi, V., C.P. DaCruz, J.C. Cuchiaro, C.A. Paz de Araujo. and R. Zuleeg. Integ. Ferroelec.. **14**, 133 (1997b)

Joshi, V., N, Solayappan, G, Derbenwick, C. Dehm and C. Mazure. Integ. Ferroelec.. **22**, 543 (1998)

Kato, K., C. Zheng, J.M. Finder and S.K. Dey. J. Amer. Ceram. Soc.. **81**, 1869 (1998)

Keijser, M. de and G.J.M. Dormans. Mater. Res. Soc. Bull.. **21(6)**, 37 (1996)

Kim, J.H., L. Wang, S.M. Zurn, L. Li, Y.S. Yoon and D.L. Polla. Integ. Ferroelec.. **3**, 21 (1993).

Kim, S.H., J.G. Hong, J.C. Gunter, H.Y. Lee, S.K. Streiffer and A.I. Kingon. in Ferroelectric Thin Films. VI, ed. by R.E. Treece, R.E. Jones, C.M. Foster, S.B. Desu and I.K. Yoo.

Mater. Res. Soc. Symp. Proc. **493**, (1998) 131

Kingery, W.D., H.W. Bowen and D.R. Uhlmann. Introduction to Ceramics, 2nd edn. Wiley, New York, (1976)

Klee, M. and U. Mackens. Microelec. Eng.. **29**, 185 (1995)

Koyama, K., T. Sakuma, S. Yamamichi, H. Watanabe, H. Aoki, S. Ohya, Y. Miyasaka and T. Kikkawa. Proc. IEDM. 823 (1991)

Krupanidhi, S.B., H. Hu and V. Kumar. J. Appl. Phys.. **71**, 376 (1992)

Lee, J.J., C.L. Thio and S.B. Desu. J. Appl. Phys.. **78**, 5073 (1995)

Lee, J.K., T.K. Song and H.Y. Jung. Integ. Ferroelec.. **15**, 115 (1997)

Lee, J.S., H.J. Kwon, Y.W. Jeong, H.H. Kim and C.Y. Kim. J. Mater. Res.. **11**, 2681 (1996)

Lee, J.S., K.S. Kim, E. Kim and K. No. Integ. Ferroelec.. **21**, 343 (1998)

Li, T., Y. Zhu, S.B. Desu, C. Peng and M. Nagata. Appl. Phys. Lett.. **68**, 616 (1996)

Lichtenwalner, D.J., O. Auciello and A.I. Kingon. J. Appl. Phys.. **74**, 7497 (1994)

Lin, C.H., B.M. Yen, H. Chen, T.B. Wu, H.C. Kuo and G.E. Stillman. in Ferroelectric Thin Films VI, ed. by R.E. Treece, R.E. Jones, C.M. Foster, S.B. Desu, and I.K. Yoo Mater. Res. Soc. Symp. Proc. **493**, (1998) 189

Lines, M.E. and A.M. Glass. Principles and Applications of Ferroelectrics and Related Materials. Clarendon Press, Oxford, (1977)

Mackenzie, J.D. and Y. Xu. J. Sol-Gel Sci. Tech.. **8**, 673 (1997)

Majumder, S.B., Y.N. Mohapatra and D. C. Agrawal. Appl. Phys. Lett.. **70**, 138 (1997)

McMillan, L.D., C.A. de Araujo, J.D. Cuchlaro, M.C. Scott and J.F. Scott. Integ. Ferroelec.. **2**, 351 (1992)

Megaw, H.D. Ferroelectricity in Crystals. Methuen London, (1957)

Mihara, T., H. Yoshimori, H. Watanabe and C.A. Paz de Araujo. Jpn. J. Appl. Phys.. **34**, 5233 (1995)

Mikami, N. in Thin Film Ferroelectric Materials and Devices. ed. by R. Ramesh. Kluwer, Norwell, MA, p.43(1998)

Moulson, A.J. and J.M. Herbert. Electroceramics: Materials, Properties. Applications. Chapman & Hall, London, (1990)

Neumayer, D.A., P.R. Duncombe, R.B. Laibowitz, T. Shaw, R. Purtell, A. Grill. Integ. Ferroelec.. **21**, 331 (1998)

Noma, A. and D. Ueda. Integ. Ferroelec.. **15**, 69 (1997)

Pierre, A.C. Introduction to Sol-Gel Processing. Kluwer, Norwell, MA, (1998)

Polla, D.L. and L.F. Francis. Mater. Res. Soc. Bull.. **21(6)**, 46 (1996)

Polla, D.L. and L.F. Francis. Annual Review of Materials Science. **28**, 563 (1998)

Ramesh, R. (ed.). Thin Film Ferroelectric Materials and Devices. Kluwer, Norwell, MA, (1998)

Ramesh, R., W.K. Chan, B. Wilens, T. Sands, J.M. Tarascon, V.G. Keramidas and J.T. Evans. Jr., Integ. Ferroelec.. **1**, 1 (1992)

Schwartz, R.W. Chem. Mater.. **9**, 2325 (1997)

Scott, J.F. and C.A. Paz de Araujo. Science. **246**, 1400 (1989)

Scott, J.F., F.M. Ross, C.A. Paz de Araujo, M.C. Scott and M. Huffman. Mater. Res. Soc. Bull.. **21(7)**, 33 (1996)

Scott, J.F. Annual Review of Materials Science. **28**, 79 (1998a)

Scott, J.F. in Thin Film Ferroelectric Materials and Devices. ed. by R. Ramesh. p.115 Kluwer, Norwell, MA, (1998b)

Sedlar, M. and M. Sayer. Ceram. Inter.. **22**, 241 (1996)

Shimizu, M., H. Fujisawa and T. Shiosaki. Microelec. Eng.. **29**, 173 (1995)

Solayappan, N., G.F. Derbenwick, L.D. McMillan, C.A. Paz de Araujo and S. Hayashi. Integ. Ferroelec.. **14**, 237 (1997)

Solayappan, N., V. Joshi, A. DeVilbiss, J. Bacon, J. Cuchiaro, L. McMillan and C.A. Paz de Araujo. Integ. Ferroelec.. **22**, 521 (1998)

Song, T.K., J.K. Lee and H.J. Jung. Appl. Phys. Lett.. **69**, 3839 (1996)

Sreenivas, K., M. Sayer and P. Garrett. Thin Solid Films. **172**, 251 (1989)

Strukov, B.A. and A.P. Levanyuk. Ferroelectric Phenomena in Crystals: Physical Foundations. Springer, Heidelberg, (1998)

Subbarao, E.C. Integ. Ferroelec.. **12**, 33 (1996)

Summerfelt, S. in Thin Film Ferroelectric Materials and Device. ed. by R. Ramesh. Kluwer, Norwell, MA, (1998) p.1

Suzuki, M., J. Ceram. Soc. Jpn, (Int. Edition). **103**, 1089 (1995)

Szymanik, B. and A. Edgar. Solid State Comm.. **79**, 355 (1991)

Tabata, H., H. Tanaka and T. Kawai. Jpn. J. Appl. Phys.. **34**, 5146 (1995)

Takenaka, T., T. Gotoh, S. Mutoh and T. Sasaki. Jpn. J. Appl. Phys.. **34(9B)**, 5384 (1995)

Thomas, D., A.I. Kingon, O. Auciello, R. Waser and M. Schumacher. Integ. Ferroelec.. **14**, 51 (1997)

Torii, Y., K. Tato, A. Tsuzuki, H.J. Hwang and S.K. Dey. J. Mater. Sci. Lett.. **17**, 827 (1998)

Treece, R.E., R.E. Jones, C.M. Foster, S.B. Desu and I.K. Yoo (eds.). Ferroelectric Thin Films VI. Mater. Res. Soc. Symp. Proc. **493**, (1998)

Tuttle, B.A. in Thin Film Ferroelectric Materials and Devices. ed. by R. Ramesh. Kluwer, Norwell, MA, (1998)p.145

Tuttle, B.A. and R.W. Schwartz. Mater. Res. Soc. Bull.. **21(6)**, 49(1996)

Van Buskirk, P.C., J.F. Roeder and S. Bilodeau. Integ. Ferroelec.. **10**, 9 (1995)

Vorotilov, K.A., M.I. Yanovskaya, L.I. Solovjeva, A.S. Valeev, V.I. Petrovsky, V.A. Vasiljev and I.E. Obvinzeva.Microelec.Eng.. **29**, 41(1995)

Warren, W.L., D. Dimos and R. M. Waser. Mater. Res. Soc. Bull.. **21(7)**,40 (1996)

Waser, R. and O. Lohse. Integ. Ferroelec.. **21**, 27 (1998)

Wu, Y. and G.Z. Cao. unpublished work. University of Washington, (1998)

Xia, C.F., T.L. Ward and P. Atanasova. J. Mater. Res.. **13**, 173 (1998)

Xu, Y., C.H. Cheng and J.D. Mackenzie. J. Non-Crystalline Solids. **176**, 1 (1994)

Yamaguchi, N., T. Hattori, K. Terashima and T. Yoshida. Thin Solid Films. **316**, 185 (1998)

Yang, P., N. Zhou, L. Zheng, H. Hu and C. Lin. J. Phys. D: Appl. Phys.. **30**, 527 (1997)

Yuuki, A., M. Yamamuka, T. Makita, T. Horikawa, T. Shibano, N. Hirano, H. Maeda, N. Mikami, K. Ono, H. Ogata and H. Abe. IEDM Tech. Digest. 115, (1995)

4 Metalorganic Chemical Vapor Deposition of Ferroelectric Thin Films

Ren Xu

4.1 Overview of Precursor Compounds

4.1.1 Background

The technology of metalorganic chemical vapor deposition (MOCVD) for oxides has experienced explosive advances in the past ten years due to current interest in oxide superconducting, ferroelectric and dielectric materials. These oxide compounds, with the exception of simple dielectrics such as SiO_2 and Ta_2O_5, tend to be complex in their composition and structure. They often involve metal elements with wide-ranging size, electronegativity and oxidation states that necessarily require complex organic ligands to form volatile and yet thermally stable precursor compounds. Consequently, the method of precursor delivery as well as the type of deposition reactions vary widely among components for any given multicomponent oxide. The characteristics of deposited oxide films hence show crucial dependence on the selection of precursor compound, deposition temperature, deposition environment, and kinetic factors such as the precursor partial pressure and flow rate. While the variety of precursor compounds known to us today is large, and the processes of film deposition complex, a simple classification of precursor compound and discussion on their physical and chemical characteristics within the context of vapor deposition can be useful to the practitioners of MOCVD.

Metalorganic compounds suitable as precursors for most main group and transition metal elements are now available from a number of commercial sources. When not available, or when stringent requirements are imposed, precursors are readily prepared in laboratories for the immediate use in vapor deposition processes. Table 4.1 is a partial list of some examples of commercially available or laboratory prepared precursor compounds used in MOCVD of oxides(Curtis and Brunner, 1975; Wernberg et al., 1993a; Wernberg et al., 1993b; Xie and Raj, 1993; Schieber et al., 1991; Hayashi et al., 1991; Ushida et al., 1992; Yamasaki et al., 1992; Zhang and Xu, 1994; Nishino et al., 1997; Sladek and Gibert, 1972; Depp et al., 1994; Peng and Dcsu, 1992; Adams

Table 4.1 Examples of metalorganic compounds for oxide deposition

Metal element		Diketonate	Alkoxide	Alkylate/ carboxylate	Oxide examples
Main Group					
I **A**	Li	thd	-OEth, n-Obut		$LiTaO_3, LiNbO_3$
II **A**	Mg,Sr,Ba	thd	-OEth		$Bi_2Sr_2Ba_2Cu_3O_x$ $YBa_2Cu_3O_{6+\delta}, MgAl_2O_4$
III **A**	Al,In	acac	-OEth, i-OPro	Benzoate	$In_2O_3, Al_2O_3, MgAl_2O_4$
IV **A**	Si,Sn,Pb	thd	-OEth	TEPb,eha	$SiO_2, SnO_2, PbTiO_3,$ $Pb(Zr,Ti)O_3$
V **A**	Bi			Phenyl	$Bi_2Sr_2Ba_2Cu_3O_x$
Transition Metal					
III **B**	Y,La,Eu	thd,tfa			$YBa_2Cu_3O_{6+\delta}, PLZT, Eu_2O_3$
IV **B**	Ti,Zr	thd,tfa,acac	n-OBut		$TiO_2, ZrO_2, YSZ, PbTiO_3$
V **B**	Nb,Ta		n-OBut, -OEth		$LiTaO_3, LiNbO_3, (Ta,Nb)_2O_5$
VI **B**	Cr	acac			Cr_2O_3
VII **B**	Mn	acac			$Mn_2O_3, La_x(Sr,Ca)_yMnO_3$
VIII **B**	Co	acac			$LaSrCoO_3$
I **B**	Cu	thd			$Bi_2Sr_2Ba_2Cu_3O_x$
II **B**	Zn	acac		Ethyl	ZnO

and Capio, 1979; Kwak et al., 1988; Mantese, 1989; West and Beeson, 1990; Shimizu and Shiosaki, 1995; Chour et al., 1997; Si et al., 1994; Hwang and Kim, 1993; Kim et al., 1995; Broorse and Burlitch, 1994; Studebaker et al., 1997; Gardiner et al., 1994; Kamata et al., 1994; Shimizu et al., 1990). There are four categories of metalorganic compound frequently used. These include: (i) β-diketonates such as the 2,2,6,6-tetramethyl-3,5-heptanedionate (thd), 2,4-pentanedionate (also known as acetoacetonate, acac); (ii) alkoxides, such as ethoxide (OEth), isopropoxide (i-OPr), and normal-butoxide (n-OBut); (iii) alkylmetal such as ethylzinc, and phenylbismuth; and (iv) carboxylates such as benzoate and ethylhexanoate (eha). The metal elements are usually in their correct oxidation state for oxides in β-diketonates, alkoxides and carboxylates, thus the presence of oxygen during deposition is not required. When alkylmetals are used in MOCVD, an oxidizing environment must be used. This is sometimes dangerous because of the combustability and toxicity of most alkylmetal compounds. The use of water vapor in MOCVD of oxides has a relatively long history dating back to the early 1970s. It was recognized that the catalytic effect of H_2O in alkoxide pyrolysis significantly enhances the deposition rate as well as the deposition quality of oxides. Al_2O_3, Nb_2O_5, Sb_2O_3,

TiO$_2$ and ZrO$_2$ films were deposited by Sladek et al.(1972) through vapor phase hydrolysis of alkoxides. An atmospheric apparatus was used to produce aerosol particles of TiO$_2$(Visca and Matijetic, 1979) and Al$_2$O$_3$(Ingebrethsen and Matijetic, 1980). Recently, the effect of H$_2$O vapor in enhanced quality at lower deposition temperatures for ZrO$_2$(Kim et al., 1995), YSZ(Chour et al., 1997) and LiTaO$_3$(Chour and Xu, 1995b) was noted. Laser-assisted pyrolytic deposition was used to induce oriented crystallization by surface nucleations(Ushida et al., 1992).

Due to the low volatility inherent to common precursors, most oxide deposition processes are achieved at a reduced pressure, ranging from a few torr to a few tens of torr. Occasionally, atmospheric vapor deposition becomes feasible when the precursor is sufficiently volatile. Such is the case for ZnO using Zn(acac)$_2$ as precursor(Kamata et al., 1994). The vapor pressure of precursor compounds at various temperatures has far-reaching consequences beyond the simple concern of operating conditions. One important aspect of multicomponent oxide deposition is in the control of the vapor phase composition. Since the majority of the oxide systems of present interest have melting points much higher than the typical deposition temperature, insufficient diffusion results in nonequilibrium deposition reactions. Consequently, the composition of the deposited film is highly dependent upon the individual precursor partial pressure immediately above the substrate surface. Presently, little effort has been devoted to the in-situ monitoring and control of the vapor phase composition in MOCVD experiments. The common practice is to use the composition analysis result obtained from the deposited films as a qualitative guide for the subsequent adjustment of relative carrier gas flow rate and the precursor evaporation temperature in order to reach an acceptable deposition condition. The lack of a viable analytical technique with sufficient accuracy has hindered the improvement of film qualities for practical device applications. On the other hand, there have been numerous reports on the sensitivity of film optical, electro-optical, and electrical properties on compositional(Carruthers et al., 1971; Fukukawa et al., 1992), structural(Nassau and Lines, 1970) and morphological(Fork et al., 1995) defects in oxides.

4.1.2 Precursor Preparations

Metal alkoxides are readily prepared by reactions of metal or metal chlorides with appropriate alcohols. For conversion from lower alkoxide to higher alkoxides, alcoholysis reactions are carried out through simple dry-atmosphere reflux and distillations(Bradley et al., 1978a). For example, alkali metal reacts directly with anhydrous alcohol in an inert and dry atmosphere to produce hydrogen gas and a transparent alkoxide solution in parent alcohol:

$$Li(Na,K) + HOC_2H_5 \xrightarrow{\text{exothermic}} Li(Na,K)OC_2H_5 + \frac{1}{2}H_2 \qquad (4.1)$$

Subsequent alcoholysis reactions convert the lower alkoxides to higher alkoxides

$$LiOC_2H_5 + HOC_4H_9 \xrightarrow{\text{reflux}} LiOC_4H_9 + HOC_2H_5 \qquad (4.2)$$

When subjected to distillation, the lower alcohol evaporates first, leaving behind the higher alkoxide solution in higher alcohol. Similar up-conversion can be performed to other metal alkoxides when a commercial source is not identified for the particular precursor.

β-diketonates are readily prepared by directly refluxing the respective β-diketones with lower alkoxides of the metal element in appropriate solvent. For example, $Al(OC_2H_5)_3$ and $Al(i\text{-}OC_3H_7)_3$ react with acetylacetone in benzene yielded tris-derivatives (Bradley et al., 1978a):

$$Al(OR)_3 + 3acacH \xrightarrow{\text{benzene,reflux}} Al(acac)_3 + 3HOR \qquad (4.3)$$

Carboxylates are generally prepared by directly refluxing the carboxylic acid with lower alkoxides and subsequently distilling out the alcohols. Alkylmetals are generally available from commercial sources and are too dangerous to prepare in laboratories otherwise.

4.1.3 Recent Advances

Liquid precursor delivery systems were recently developed, tested and recently became commercially available for MOCVD applications (Gardiner et al., 1994). Such apparatus is generally compatible with most MOCVD reactors and their operating conditions. Despite their low volatility, poor thermal stability metalorganic precursors can be handled collectively with some flexibility in component concentrations in the solutions if proper solvents are selected. The liquid injection method is similar to spray pyrolysis in that the composition of the vapor phase species can be prescribed to a certain degree. Further, the component concentration in the solution is preserved precisely until the liquid is quickly evaporated upon hitting a flash evaporator. A flash evaporator consists of a heated stainless steel surface that is situated adjacent the separately heated substrate (Zhang et al., 1994). In other cases the liquid is directly injected at repeated pulses at quantities equivalent to a single molecular layer deposition per pulse (Xie and Raj, 1993). These techniques were successfully employed in the deposition of $MgAl_2O_4$ and $LiTaO_3$ from nonvolatile and thermally unstable alkoxide complexes $MgAl_2(OC_2H_5)_8$ (Zhang et al., 1994) and $LiTa(OC_2H_5)_6$ (Xie and Raj, 1993) The general applicability of such liquid delivery evaporation systems was demonstrated for other oxide systems such as $BaTiO_3$, $YBa_2Cu_3O_{7-\delta}$, YSZ, $LaSrCoO_3$ and Cu metal, using mixtures of respective precursors in the solution(Gardiner et al., 1994).

Many heterometallic alkoxide complexes commonly referred to as double alkoxides are available at appropriate metal-to-metal ratios for a number of important multicomponent oxide systems. The thermal stability and volatility of these double alkoxides vary. In solutions using the parent alcohol as a solvent, these double alkoxides have sufficient stability to be recrystallized, or to show sharp transitions in potentiometric titrations using single alkoxides(Bradley et al., 1978a). These double alkoxides in their solution form, therefore, are ideal precursors for the deposition of the respective multicomponent oxides using liquid precursor delivery with flash evaporators.　Table 4.2 shows a partial list of some examples of double alkoxides matched with the respective multicomponent ferroelectric oxides.

Table 4.2　　Selected list of double alkoxides

Double alkoxide	Vapor pressure ($^\circ$C/mmHg)[*]	Oxide/application
LiTa(OEt)$_6$	230/0.2	LiTaO$_3$/pyroelectricity(PE)
LiTa(i-OPr)$_6$	160 –180/0.1	Electro-optics(EO)
LiTa(t-OBut)$_6$	110 –120/0.1	2nd harmonic generation (SHG)
LiNb(i-OPr)$_6$	<140/0.2	LiNbO$_3$/EO
LiNb(t-OBut)$_6$	110 –120/0.1	Surface acoustic device (SAW)
KNb(i-OPr)$_6$	<200/0.8	KNbO$_3$/EO
KNb(t-OBut)$_6$	<200/0.8	Nonlinear optics (NLO)
Ba[Nb(i-OPr)$_6$]$_2$	170 –180/0.1	Ba$_{4+x}$Na$_{2-2x}$Nb$_{10}$O$_{30}$/
NaNb(t-OBut)$_6$	110 –120/0.1	NLO,SHG
LiNb(t-OBut)$_6$	110 –120/0.1	K$_3$Li$_2$Nb$_5$O$_{15}$/
KNb(t-OBut)$_6$	<200/0.8	NLO,EO
KNb(t-OBut)$_6$	<200/0.8	K$_3$Li$_2$(Ta$_x$Nb$_{1-x}$)$_5$O$_{15}$/
KTa(t-OBut)$_6$	<200/0.8	EO
LiNb(t-OBut)$_6$	110 –120/0.1	
LiTa(i-Opr)$_6$	160 –180/0.1	
KNb(t-OBut)$_6$	<200/0.8	KTa$_{1-x}$Nb$_x$O$_3$/
KTa(t-OBut)$_6$	<200/0.8	SHG,EO,holographic storage (HS)
Sr[Nb(i-OPr)$_6$]$_2$	210 –220/0.1	Sr$_x$Ba$_{1-x}$Nb$_2$O$_6$/
Ba[Nb(i-OPr)$_6$]$_2$	170 –180/0.1	PE,EO,HS

[*] Data according to Bradley et al. (1978) 1mmHg=133.322Pa

For complex precursors such as LiTa(OR)$_6$ with limited stability, where R=C$_n$H$_{2n+1}$ (n=1– 6), the evaporation behavior can be considered, in general, as two competing processes—the simple evaporation of a complex molecule, and the decomposition-evaporation of the precursor. For the simple case of LiTa(OR)$_6$ system, the competing processes can be written as:

$$\text{LiTa(OR)}_6\,(\text{liquid}) \xrightarrow{\;\Delta H_V^{\text{LiTa(OR)}_6}\;} \text{LiTa(OR)}_6\,(\text{vapor}) \tag{4.4}$$

$$\text{LiTa(OR)}_6\,(\text{liquid}) \xrightarrow{\;K\;} \text{LiOR(solid)} + \text{Ta(OR)}_5\,(\text{liquid}) \tag{4.5a}$$

$$\text{Ta(OR)}_5\,(\text{liquid}) \xrightarrow{\;\Delta H_V^{\text{Ta(OR)}_5}\;} \text{Ta(OR)}_5\,(\text{vapor}) \tag{4.5b}$$

The composition of the vapor phase species is determined by the stability of the double alkoxide LiTa(OR)_6 and the relative volatility between LiTa(OR)_6 and Ta(OR)_5. For unstable LiTa(OR)_6, the liquid phase can be considered a simple mixture of LiOR and Ta(OR)_5. Since LiOR is mostly ionic with very low vapor pressure compared to that of Ta(OR)_5,(Bradley et al., 1978a; Chour et al., 1994) the vapor phase composition is simply determined by the relative volatility of LiTa(OR)_6 to that of Ta(OR)_5. A completely decomposed LiTa(OR)_6 has Ta(OR)_5 in a solution at 50 mol% concentration. It is thus informative to compare the vapor pressure versus temperature behavior of LiTa(OR)_6 with that of Ta(OR)_5 in a 50 mol% solution. Experimentally this can be performed for all varieties of double alkoxides and the respective volatile single alkoxides. Figure 4.1 shows the results of such measurements(Chour et al., 1998) for $\text{LiNb(n-C}_4\text{H}_9)_6$, $\text{Nb(n-C}_4\text{H}_9)_5$, $\text{LiTa(n-C}_3\text{H}_7)_6$ and $\text{Ta(n-C}_3\text{H}_7)_5$ as an example. For the $\text{LiNb(n-C}_4\text{H}_9)_6$, $\text{Nb(n-C}_4\text{H}_9)_5$ system, the vapor pressure-temperature bahaviors are sufficiently different between the double and single alkoxide, indicating the $\text{LiNb(n-C}_4\text{H}_9)_6$ is structurally different from a simple solution of $\text{Li(n-C}_4\text{H}_9)$ in $\text{Nb(n-C}_4\text{H}_9)_5$. The vapor pressure behaviors for the $\text{LiTa(n-C}_3\text{H}_7)_6$ and $\text{Ta(n-C}_3\text{H}_7)_5$ on the other hand, are identical within the errors of the experiment, suggesting that the structure of $\text{LiTa(n-C}_3\text{H}_7)_6$ is probably the same as a 50% Raultian solution of $\text{Ta(n-C}_3\text{H}_7)_5$.

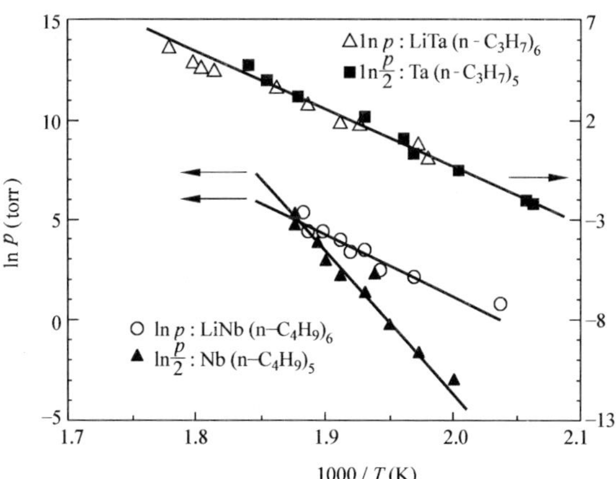

Fig.4.1 Vapor pressure-temperature behavior for LiNb(n-OBut)_6: Nb(n-OBut)_5 system (bottom,scale on left side)and LiTa(n-OPr)_6: Ta(n-OPr)_5 system(top,scale on right side). 1 torr=133.322Pa

120

The vapor pressure behaviors for various double alkoxide-single alkoxide pairs revealed Arrhenius relationship with temperature. The Arrhenius coefficients for each compound can be obtained through linear regression. It should be noted that the relative volatility between double and single alkoxides can be used as an indicator of composition of the vapor phase species. Consider a partially stable double alkoxide $LiTa(OR)_6$ with 50% decomposition at the evaporation temperature. The vapor pressure ratio between $Ta(OR)_5$ and $LiTa(OR)_6$ can be calculated on the basis of the Arrhenius equation:

$$\Delta \ln p(\text{torr}) = \Delta a - \frac{\Delta b \times 10^3}{T(\text{K})} \tag{4.6a}$$

Using the constants found from the linear fit of vapor pressure-temperature data, the fraction excess $Ta(OR)_5$ in the vapor phase, the nonstoichiometry factor K, with respect to $LiTa(OR)_6$ can be estimated as:

$$K = \frac{[Ta(OR)_5]}{[LiTa(OR)_6]} = \frac{p_{Ta(OR)_5}}{p_{LiTa(OR)_6}} \tag{4.6b}$$

$$\ln K = \Delta \ln p = \ln p_{Ta(OR)_5} - \ln p_{LiTa(OR)_6} \tag{4.6c}$$

Figure 4.2 shows the calculated nonstoichiometry factor K (The fraction deviation from stoichiometric composition proposed earlier) for five alkoxide systems studied. It is apparent that $LiNb(n-C_4H_9)_6$ and $LiTa(n-C_4H_9)_6$ are more likely to evaporate stoichiometrically.

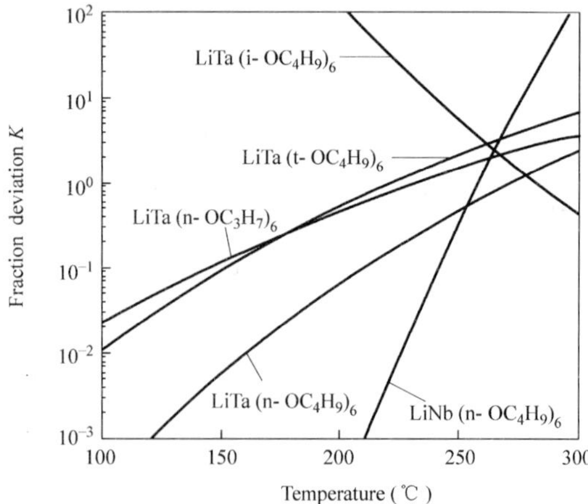

Fig.4.2 Nonstoichiometry factor in temperature range 100 – 300℃ for five double alkoxide systems

The stability of double alkoxide complexes is apparently maintained through a different mechanism from that for single alkoxides. For single alkoxides, generally a branched alkoxyl group favors thermal stability. Better charge screening is expected out of branched alkoxyl groups for alkaline and alkaline earth elements with strong ionic characters. Double alkoxides, on the other hand, appear to evaporate in the form of dimers as suggested by mass spectrometric study of double alkoxide vapor.(Chour et al., 1994) A linear, long chain alkoxyl group favors the formation of dimers with the long hydrocarbon chain wrapped around to provide charge screening to the alkaline elements. Such an argument appears to be in agreement with our observation in LiNb (n-C_4H_9)$_6$ and LiTa(n-C_4H_9)$_6$ systems.

The composition of sublimates for the LiTa(n-OC_4H_9)$_6$ system was analyzed by lattice parameter measurement. Figure 4.3 shows a series of standard samples prepared in this study. The lattice parameters were measured on annealed samples with varying compositions. Silicon powder was mixed prior to X-ray diffraction as an internal standard. The calibration curve was obtained by scaling the Li:Ta ratio 1:1 sample using JCPDS card No. 29-836. A similar standard prepared by Barns and carruthers. (1970) is also shown as reference. The individual lattice parameter of a stoichiometric sample by Abrahams and Bernstein. (1967) is indicated in the figure by an arrow. The lattice parameter of a Puratronic ® standard sample from Johnson Matthey, Co. is also shown in the figure. The calcined sublimate from LiTa(n-C_4H_9)$_6$ showed a lattice parameter of 5.1518Å, corresponding to $Li_{1.01\pm0.02}Ta_{0.99\pm0.02}O_{2.98\pm0.06}$. The evaporation of LiTa(n-OC_4H_9)$_6$ is thus stoichiometric within the composition detection limit by the X-ray lattice parameter method. Stoichiometric and epitaxial LiTaO$_3$ films were deposited from direct evaporation of LiTa(n-OC_4H_9)$_6$ precursor, and such a deposition process was referred to as autostoichiometric vapor deposition (Chour et al., 1998).

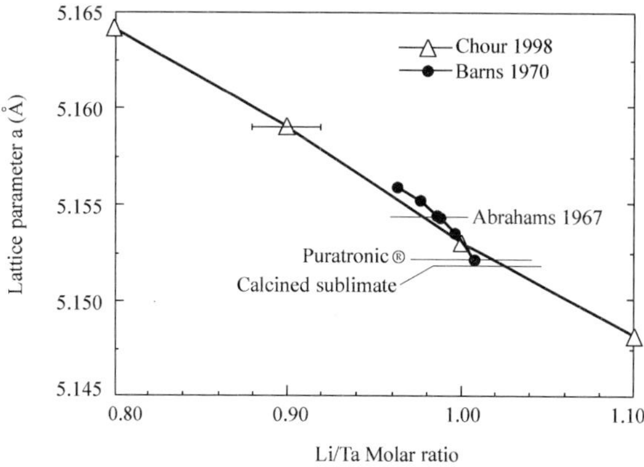

Fig.4.3 Stoichiometric evaporation of LiTa(n-OBut)$_6$ as evidenced by lattice parameter measurement

4.1.4 Ultimate Role of Precursor Compound

It should be recognized that between the preferred characteristics of high volatility and high thermal stability, difficulty remains in identifying appropriate metalorganic compounds as precursors.   Liquid injection and flash evaporators are important advances in recent years because a wider variety of metalorganic compounds can now be used in MOCVD of oxides. However, this technique is limited to the precursors that have sufficient solubility in some solvents. In addition, when liquid is directly injected on a substrate surface, the uniformity and the density of the layer, and ultimately the compositional and structural perfection of the deposited layer become suspect.   There have been no attempts at active monitoring and control of vapor phase precursor compositions. From a chemistry stand point, the ultimate role of a precursor compound should be one with the capability to deliver metal elements to the deposition site at a precisely controlled, constant dosage. The precision in dosage control should reach ppm to ppb levels, if not that of a single molecular level. The precursor should play an active role in regulating the relative ratio between metal elements, and the single precursor approach is an important step in that direction.   Finally, the precursor compound should permit at least one reaction mechanism through which a clean deposition reaction with little comtamination can be carried out. The MOCVD of oxides has advanced to the present stage when precursor compounds suitable for deposition can be identified, purchased or prepared on demand. When it comes to the quality of films deposited, much remains to be explored.

4.2 Theoretical Possibility of Autostoichiometric Vapor Deposition

4.2.1 Stoichiometry of MOCVD

Multicomponent oxide thin films have, in recent years, received wide interest for their attractive electrical and optical properties. Potential application of such materials as active media for optical signal processing prompted intensive investigations among materials scientists and engineers. Two fabrication techniques stand out as the most extensively studied for multicomponent oxides: Sol-gel processing of ferroelectric oxides(Xu and Mackenzie, 1992) and metalorganic chemical vapor deposition (MOCVD)(Desu et al., 1993). From the chemistry standpoint, deposition of multicomponent oxides involves essentially the delivery to the substrate surface metal elements of various forms in stoichiometric ratios.   Sol-gel processing achieves stoichiometry by means of a thin liquid solution film, on substrates, which contains soluble metalorganic compounds in prescribed proportions. Subsequent heating to cause crystallization generally does not disturb the film stoichiometry due to the relatively low temperatures required to effect crystallization of high melting point oxides. In other words, the stoichiometry is normally prescribed in sol-gel processing. The main difficulty associated with sol-gel is its discontinuous deposition of films. A single dip or spin coating generally results in films of a few thousand angstroms thickness. Multiple coating and firing cycles are necessary to obtain films with thicknesses in the micrometer range.

MOCVD provides a means of continuous deposition of oxides on the molecular level, thus rendering convenient control over film thicknesses and microstructure. However, when the oxide system contains more than one metal element, the stoichiometry can no longer be prescribed as is done in sol-gel processing. MOCVD of multicomponent oxide involves the use of separate metalorganic compounds for each components. In ordinary use, the precursor compounds are independently transported into the vapor phase by evaporation or other means. Whether a given vapor phase composition results in the desired film stoichiometry is determined only after a compositional analysis of the deposited film is performed. The information is then used as feedback to adjust individual precursor flow rates. Many cycles of operations are necessary before an acceptable approximation can be reached. In addition, carefully obtained operational parameters are not universal. These conditions are generally different from apparatus to apparatus, and from event to event.

The above situation can be analyzed within the framework of the stagnant-film model for vapor depositions (Grove, 1966; Treybal, 1955) ,shown in Fig.4.4. A precursor compound with concentration C_G in the vapor phase is brought to the vicinity of a heated substrate with surface concentration of precursor C_S, the flux of precursor molecules through the stagnant film is controlled by diffusion in the stagnant layer alone:

$$F_1 = D_G \frac{C_G - C_S}{\delta} \tag{4.7}$$

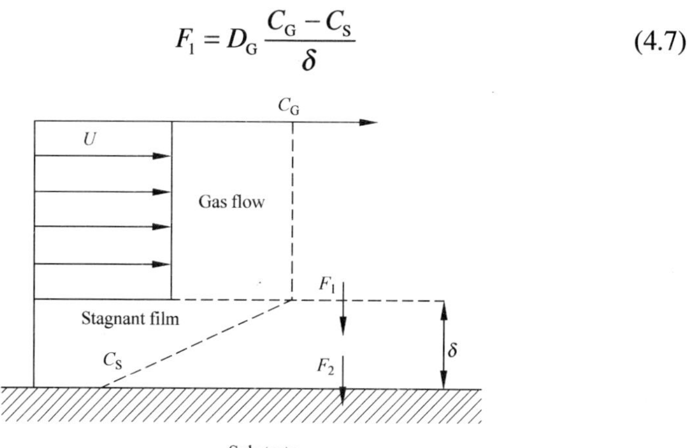

Fig.4.4　Schematic diagram of stagnant-film model for vapor deposition

where D_G is the gaseous phase diffusion coefficient, and δ the stagnant film thickness. The flux consumed by the surface reaction (pyrolysis, oxidation or other reactions) converting precursor compound to oxide is determined by the reaction kinetics. Assuming a first-order reaction at the surface, the flux is proportional to the surface concentration of the precursor and a rate constant k_S:

$$F_2 = k_S C_S \tag{4.8}$$

Under steady-state conditions, $F_1 = F_2$, and the precursor concentrations follow the relation:

$$C_S = \frac{C_G}{1 + \delta k_S / D_G} \tag{4.9}$$

The surface concentration C_S is uniquely determined by the gas-phase concentration C_G. If the deposition is mass-transfer controlled, $\dfrac{D_G}{\delta} \gg k_S$, and therefore $C_S = C_G$. If the deposition is surface-reaction controlled, $\dfrac{D_G}{\delta} \ll k_S$, and $C_S = 0$. For the vapor deposition of a multicomponent oxide AB_nO_m, normally at least two precursor molecules are involved, and the precursor concentration relationships for components A and B become respectively:

$$C_S^A = \frac{C_G^A}{1 + \delta k_S^A / D_G^A} \quad \text{and} \quad C_S^B = \frac{C_G^B}{1 + \delta k_S^B / D_G^B} \tag{4.10}$$

It can be shown that in order to grow stoichiometric compound AB_nO_m, the gas-phase concentrations C_G^A and C_G^B must be controlled to follow precisely:

$$\frac{C_G^A}{C_G^B} = \frac{k_S^B}{n k_S^A} \cdot \frac{1 + \delta k_S^A / D_G^A}{1 + \delta k_S^B / D_G^B} \tag{4.11}$$

In a practical deposition experiment, the initial levels of C_G^A and C_G^B are entirely empirical. It is only after the film is grown, characterized, and the composition information is used as feedback to correct the flow rate of each individual precursor, that an approximately stoichiometric film can be grown. Repeated operations for a number of cycles are needed to reach an acceptable film composition. A boundary-layer model based on fluid mechanics

(Schlichting, 1960) gives slightly modified expressions for the term $\dfrac{D_G}{\delta}$, the basic argument, however, remains the same for our purposes.

The preceding discussions suggest that there are inherent limitations and inconveniences in both sol-gel processing and MOCVD for the deposition of multicomponent oxides. It thus becomes of interest to investigate possibilities of devising a new deposition method or approach that possesses the combined advantages of both techniques. In the following we present one particular case through which such an objective can be met.

4.2.2 Autostoichiometric Vapor Deposition

Suppose a single molecular precursor $AB_n(OR)_{2m}$ exists, where OR stands for any alkoxy groups, and the following reaction can be used for the deposition of AB_nO_m:

$$AB_n(OR)_{2m} \xrightarrow{\text{vapor-phase reactions}} \xrightarrow{\text{surface reactions}} AB_nO_m + \text{organics} \qquad (4.12)$$

and further, the vapor-phase and surface reactions leading to the oxide deposition does not cause disturbance of the metal-metal ratio within the molecule, the molar relationship between A and B can be conserved throughout the deposition reactions, an inherently stoichiometric, autostoichiometric vapor deposition can be devised. In this case, the steady-state relation between gas-phase and surface species according to the stagnant film model becomes:

$$C_S^{A,nB} = \frac{C_G^{A,nB}}{1+\delta k_S^{A,nB}/D_G^{A,nB}} \qquad (4.13)$$

Hence the molar ratio between metals A and B is conserved throughout the deposition steps to give the correct stoichiometry in the final film.

4.2.3 Examples of Precursors

Double metal alkoxides are possible candidates for molecular precursors carrying heterometallic elements at the correct ratio for important multicomponent oxides. It has been established that a large number of such compounds exist in molecular form stable enough to be sublimed. Table 4.2 gives a list of a few such compounds after Bradley et al.(1978b). Also listed in the table are ferroelectric oxides with stoichiometry corresponding to the given double alkoxides. The fact that double metal alkoxides often carry heterometallic elements at the same metal-metal ratio as some multicomponent oxide ceramics may not be as incidental as it seems. The resemblance of stoichiometry as well as local structure of metal alkoxides to oxides have roots in the nature of the metal-oxygen bond. Chisholm (1983) has summarized the similarities between metal alkoxides and metal oxides based on their bond structures. As new heterometallic alkoxide complexes continue to be discovered (Beidel et al., 1992; Purdy and George, 1991; Campion 1991), the overall numbers of double alkoxides suitable for autostoichiometric vapor deposition are expected to grow.

Little information is known about the gas-phase reaction kinetics of double alkoxides. Early work on simple double alkoxides and recent work on complex systems concentrated on synthesis and structural characterizations. On the other hand, use of single metal alkoxides for the deposition oxides by MOCVD technique has steadily progressed in the past. For example, SiO_2 thin films have been made by MOCVD(Huppertz and Engl, 1979; Levin and Evans-Lutterodt, 1983; Becker et al., 1987; Desu, 1989) using tetraethylorthosilicate (TEOS, silicontetraethoxide) as precursor. However, utilization of double alkoxides as precursors in MOCVD is not by any means a straightforward practice. One of the key obstacles is the tendency for double alkoxides to prematurely decompose into single metal alkoxides of drastically different volatility. This tendency often leads to the deposition of an off-stoichiometry multicomponent oxide. In the following section, we analyze the gaseous phase reactions and discuss two reaction schemes through which an autostoichiometric vapor deposition can be realized.

4.2.4 Autostoichiometric Reactions

MOCVD of oxides in its conventional sense involves the pyrolysis of metalorganic precursors(Bradley, 1989). Often various auxiliary techniques are used to prompt pyrolytic reactions. Examples include introduction of free-radicals by plasma in plasma-assisted MOCVD (Chour et al., 1994) and laser-assisted MOCVD (Mehrotra et al., 1968). These techniques are aimed at prompting free-radical pyrolytic reactions, and generally induces fragmentation of precursor molecules. For the purpose of autostoichiometry, random fragmentation of precursor molecules is to be avoided. The hydrolysis-assisted pyrolysis, and hydrolysis-polycondensation are reaction schemes conforming to autostoichiometric vapor deposition. The first reaction is believed to be the dominating mechanism for the deposition of SiO_2 from TEOS. The second reaction is the basis for all sol-gel processing of ceramics. In the following we shall discuss briefly the implications of these reactions on autostoichiometry.

Pyrolytic reactions in general are inappropriate for autostoichiometry. For example, premature decomposition may occur for double alkoxide $AB_n(OR)_{2m}$ to give two alkoxides with drastically different volatility, $A(OR)$ and $B_n(OR)_{2m-1}$:

$$AB_n(OR)_{2m} \rightarrow A(OR) + B_n(OR)_{2m-1} \tag{4.14}$$

As a result, only $B_n(OR)_{2m-1}$ will be transferred into the vapor phase, and eventually participate in the deposition reactions. This is likely to result in a nonstoichiometric final film. Pyrolysis of alkoxides through free-radical processes were studied in detail (Desu, 1989) at 700°C. The decomposition reaction (4.14) occurs at lower temperature than that required for a free-radical reaction to commence. Consequently, one might not prefer to choose a pyrolysis reaction scheme, plasma-or laser-aided, for the purpose of autostoichiometric vapor deposition.

On the other hand, hydrolysis-assisted pyrolysis reactions are compatible with autostoichiometric vapor deposition. It is well known (Bradley, 1989) that metal alkoxides undergo pyrolysis in the presence of trace amount of H_2O according to the following steps:

$$M(OC_2H_5)_4 + H_2O \rightarrow MO(OC_2H_5)_2 + 2HOC_2H_5 \tag{4.15}$$

$$2HOC_2H_5 \xrightarrow{\text{Heat}} 2C_2H_4 + H_2O \tag{4.16}$$

It is not difficult now to see that reactions (4.15) and (4.16) are compatible with autostoichiometry when a precursor $AB_n(OR)_{2m}$ is involved:

$$AB_n(OR)_{2m} + H_2O \rightarrow AB_nO(OR)_{2m-2} + 2HOR \tag{4.17}$$

$$AB_nO(OR)_{2m-2} \xrightarrow{\text{on-substrate}} AB_nO_m + \text{organics} \tag{4.18}$$

$$2HOR \xrightarrow{\text{Heat}} \text{Olefin} + H_2O \tag{4.19}$$

where the reaction (4.17) generally does not cause fragmentation of $AB_n(OR)_{2m}$ due to steric hindrance at the bridging (tridentate)–OR groups ensures hydrolysis at terminal positions only. The subsequent pyrolysis reaction (4.18) of condensed, nonvolatile $AB_nO(OR)_{2m-2}$ occurs on the substrate surface, and no evaporation losses of either A or B component are expected. It is therefore possible to use a hydrolysis-assisted pyrolysis reaction to promote autostoichiometric deposition. In addition, the presence of H_2O will lower the deposition temperature and reduce the risk of random fragmentation.

The second reaction scheme, which is compatible with autostoichiometry is the hydrolysis-polycondensation, commonly used in sol-gel processing of oxide ceramics. In the presence of sufficient H_2O in the vapor phase, partial hydrolysis of the double alkoxide at low temperature allows the preservation of the A–(OR)–B link due to steric hindrance at tridentate oxygen sites. When performed properly, the following reactions can be implemented in a vapor deposition reactor to realize an autostoichiometric vapor deposition:

$$AB_n(OR)_{2m} + 2kH_2O \xrightarrow{\text{vapor-phase}} AB_n(OH)_{2k}(OR)_{2(m-k)} + 2kHOR \tag{4.20}$$

$$AB_n(OH)_{2k}(OR)_{2(m-k)} \xrightarrow{\text{on-substrate}} AB_nO_k(OR)_{2(m-k)} + kH_2O \tag{4.21}$$

$$AB_nO_k(OR)_{2(m-k)}+(m-k)H_2O \xrightarrow{\text{on-substrate}} AB_nO_m+2(m-k)HOR. \qquad (4.22)$$

Reactions (4.20) to (4.22) are essentially the same reactions involved in sol-gel processing with the exception that (4.20) is now facilitated in the vapor phase in an autostoichiometric vapor deposition. In practice, the hydrolysis-assisted pyrolytic reaction scheme differs from the hydrolysis-polycondensation scheme in that the former (being a thermally activated reaction) requires only a trace amount of water and happens at relatively higher temperature, while the latter requires more water and can occur at much lower temperatures. It should be noted that hydrolysis-polycondensation reactions of metal alkoxides generally occur at room temperature. If the objective is to convert alkoxide into oxide, in principle it is not necessary to raise the substrate temperature beyond that of the boiling point of the parent alcohol for the alkoxide.

4.2.5 Deposition Apparatus

Autostoichiometric vapor deposition is possible when double alkoxides are the choice of precursors and one of the above reaction schemes is used as the primary reactions leading to the oxide film depositions. This can be done through a simple low-pressure reactor capable of performing the following functions:

(1) Evaporation of double alkoxide precursors and transportation to the vicinity of a substrate;

(2) Controlled vapor phase hydrolysis of the precursors;

(3) Polycondensation of the partially hydrolyzed precursors onto substrate surfaces;

(4) In situ crystallization of the ferroelectrics by controlled heating of the substrates in the deposition chamber.

We show in Fig.4.5 the schematic block diagram of the essential features of the proposed reactor. First of all, the overall vacuum should be better than 10^{-2} torr, which is readily achievable via a liquid-nitrogen trapped mechanical pump. Metal alkoxides tend to have boiling points of about $200\,^\circ\mathrm{C}$ at 10^{-1} torr pressure. Therefore, the precursor evaporator consists of a simple heating mantle with temperature control, and regulated dry carrier gas flow. The water vapor generator can be a simple bubbler with regulated carrier gas. The deposition chamber needs to be equipped with a programmable heating unit to heat the substrate. In addition, the deposition chamber needs to be such that substrates can be easily removed. Water vapor and the precursor vapor should be allowed to mix and react in the vicinity of the substrate. Therefore, it is helpful to have the reaction chamber close to the deposition chamber, or simply designed to be the same unit.

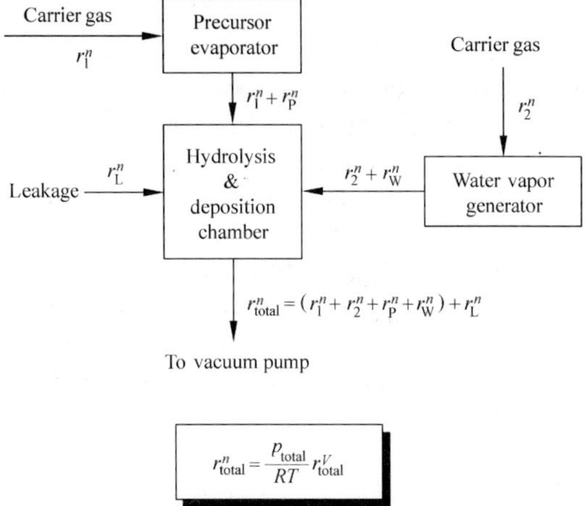

Fig.4.5   Schematic block diagram of deposition apparatus and mass flow analysis

4.2.6   Mass Flow and Rate Analysis

A simple flow analysis can now be conducted based on the block diagram in Fig.4.2. We define molar flow rate of gases in the unit of moles per minute "r^n". Carrier gas through the precursor is denoted by subscript 1. Carrier gas through the water generator is denoted by subscript 2. subscripts p, W and L denote the molar flow rate of precursor, water and the leakage, respectively. Mass flows in and out of each segment of the deposition apparatus are denoted in Fig.4.5. It follows that the total molar flow rate r_{total}^n is the sum of all molar rates:

$$r_{total}^n = (r_1^n + r_2^n + r_p^n + r_w^n) + r_L^n \tag{4.23}$$

The total molar rate is related to the mechanical pump displacement rate (liter per minute) r_{total}^V by the ideal gas law:

$$r_{total}^n = \frac{P_{total}}{RT} r_{total}^V \tag{4.24}$$

Equations (4.23) and (4.24) provide useful tools for the analysis and testing of the deposition apparatus. For example, when $r_1^n = r_2^n = r_p^n = r_w^n = 0$ i.e. all gas inlets are shut and the only source of gas into the deposition chamber is that from leakage, then the absolute chamber pressure can be used to calculate the leakage rate:

$$r_L^n = \frac{P_{abs}}{RT} r_{total}^V \tag{4.25}$$

A similar operation can be used to measure the water evaporation rate and precursor evaporation rate as functions of temperature. Further, these equations can be used to estimate deposition rate. The pressure range at which deposition will be conducted is 0.1 to 10 torr. The mean free path of alkoxide molecules is in the range of $10-100$ μm. Therefore, it can be assumed that under normal circumstances, metal alkoxides in the vapor phase will have at least undergone partial hydrolysis prior to their reaching the substrate surface. We further assume reactions (4.20) through (4.22) are sufficiently fast, and the rate-controlling step is mass transfer. The deposition rate can then be calculated from the molar rate of precursors:

$$\eta r_p^n = r_G^{th} \frac{A \cdot d}{MW} \tag{4.26}$$

where η is the deposition efficiency, which takes account of lost precursor materials due to deposition on the apparatus wall and other losses; A is the substrate area, d the density of the oxide being deposited, MW the molecular weight of the oxide, and r_G^{th} the deposition rate measured in the units of Å/min.

Under these conditions, in order to obtain a film deposition rate of 1 μm/h, i.e., $r_G^{th} = 167$ Å/min, at $A = 45.78$ cm^2 for a 3$''$ wafer, deposition efficiency of 10%, $d = 7.45$ g/cm^3 and $MW = 235.89$ g/mol for LiTaO$_3$, according to (4.26):

$$r_p^n = \frac{r_G^{th}}{\eta} \cdot \frac{A \cdot d}{MW}$$

$$= \frac{167 \times 10^{-8} (cm/min) \times 45.78 (cm^2) \times 7.45 (g/cm^3)}{0.10 \times 235.89 (g/mol)}$$

$$= 24 (\mu mol/min).$$

Experimentally, the evaporation rate and transfer rate of precursors can be measured according to (4.23) – (4.25). In the experimental demonstration(Chour et al., 1994), it will be shown that LiTa(OBut^n)$_6$, which is typical among double alkoxides, does show volatility and evaporation rates sufficient for a deposition rate of 34 μm/h.

4.2.7 Stoichiometry Factor

Vapor phase hydrolyses of double alkoxides are autostoichiometric due to steric hindrance at tridentate alkoxy sites. Therefore, one can assume that stoichiometry is determined at the moment of precursor evaporation. Consequently the stoichiometry of the final film can be determined and analyzed by the evaporated precursor, and for precursors not able to give stoichiometric vapor one must modify the evaporation method, for instance, lower the evaporator temperature by decreasing the chamber pressure.

It is useful to define a nonstoichiometry factor "K" for a simple double alkoxide $AB(OR)_n$: Assuming the reaction that produces non-stoichiometry is the thermal decomposition reaction:

$$AB(OR)_n(\text{vapor}) \xrightarrow{K} A(OR)(\text{solid}) + B(OR)_{n-1}(\text{vapor}) \qquad (4.27)$$

and equilibrium is established in the precursor evaporator:

$$\frac{p_{B(OR)_{n-1}}}{p_{AB(OR)_n}} = K \qquad (4.28)$$

The vapor phase composition (assuming $A(OR)$ is nonvolatile, such as $Li(OR)$) in the presence of both $AB(OR)_n$ and $B(OR)_{n-1}$ is:

$$\frac{[B]}{[A]} = \frac{p_{B(OR)_{n-1}} + p_{AB(OR)_n}}{p_{AB(OR)_n}} = K + 1 \qquad (4.29)$$

When $K=0$, the vapor-phase composition and consequently the deposited film will be precisely stoichiometric. When $K>0$ the per cent non-stoichiometry can be measured by the value of K. Therefore, the nonstoichiometry factor K represents the per cent nonstoichiometry of the double alkoxide precursor.

The nonstoichiometry factor can be experimentally determined for known double alkoxides by analyzing alkoxide composition before and after sublimation. For example, Mehrotra et al.(1968) have performed compositional analysis of double alkoxides of the type $AB(OR)_n$. Table 4.3 lists some of their results and the nonstoichiometric factor as computed from their data. For double iso-propoxides and tert-butoxides, the K factor ranges from 0.16 to 0.62, implying that in order to do autostoichiometric vapor deposition, precursor delivery must be achieved at lower temperature and higher vacuum. Alternatively, nonthermal precursor evaporation methods, such as solution nebulization by ultrasonic devices, may be used to avoid reaction (4.27). We have demonstrated a double alkoxide system that does have a negligible value for the nonstoichiometry factor K, and allows for autostoichiometric vapor deposition. For double alkoxides with small K factors thermal evaporation of precursor can be used for the autostoichiometric vapor deposition, otherwise milder means of precursor evaporation must be used. Clearly double alkoxides with different alkoxy groups have different K factors. Therefore, it becomes necessary to synthesize double alkoxides with different organic moieties to achieve low K values, such as fluorinated alkoxides with high volatility and thus low boiling points. More systematic work along this line will certainly help alleviate problems associated with high K factors.

Table 4.3 Nonstoichiometry factor calculated based on data from Mehrotra et al.(1968)

Double alkoxide	Calculated Ta- (wt%)	Sublimed Ta- (wt%)	K-factor
$LiTa(OPr^i)_6$	33.4	34.7	0.32
$NaTa(OPr^i)_6$	32.4	33.2	0.16
$KTa(OPr^i)_6$	31.5	35.2	0.62
$LiTa(OBut^t)_6$	28.9	30.1	0.32
$NaTa(OBut^t)_6$	28.2	29.8	0.37

4.3 Stoichiometry of Vaporization Processes

4.3.1 Background

Deposition of a multicomponent oxide $M'_x M''_y O_z$ can be carried out autostoichiometrically using complex precursors of the kind $M'_x M''_y (OC_n H_{2n+2})_{2z}$. When a single metalorganic precursor $M'_x M''_y (OC_n H_{2n+2})_{2z}$ with sufficient stability is used for the deposition, the ultimate stoichiometry of the deposited material can be made to follow that in the precursor compounds by two successive processes–the stoichiometric vaporization of precursor and the stoichiometric deposition reactions. The basic strategy of an autostoichiometric vapor deposition is to maintain the metal-to-metal ratio carried in the precursor complex throughout the vaporization and deposition reactions such that the multicomponent oxide composition is determined by the precursor's chemical stability and the characteristics of the deposition reactions. In such operations, the deposition stoichiometry is independent of the precursor flow rate. Given a sufficiently stable precursor compound and favorable deposition conditions, the stoichiometry of the multicomponent oxide can be controlled precisely on the molecular level.

MOCVD of oxides using complex precursors was originally developed to enhance the volatility for cuprous and alkaline earth metalorganic precursor compounds. Among those are $Ba\{Cu[OCCH_3(CF_3)_2]_3\}_2$ for high T_c superconductor YBCO,(Purdy and George, 1991) and $Ba_4Ti_{13}O_{18}[O(CH_2)_2OCH_3]_{24}$ for $Ba_4Ti_{13}O_{30}$(Campion et al., 1991). A technique that provided fast evaporation was demonstrated for the deposition of $MgAl_2O_4$ from $MgAl_2(OC_2H_5)_8$ (Zhang et al., 1994), $LiTa(OC_2H_5)_6$ (Wernberg, 1993a, b) and $LiTa(t-OC_4H_9)_6$ (Xie and

Raj, 1993) were used in a liquid-injection method for the deposition of $LiTaO_3$. It was found that $LiTa(n\text{-}OC_4H_9)_6$ is a robust precursor for the stoichiometric deposition of epitaxial $LiTaO_3$ with low defect density(Xu, 1995; Chour and Xu, 1995b; Chour et al., 1998). $LiTa(n\text{-}OC_4H_9)_6$ provided sufficient thermal stability to be used in a conventional MOCVD apparatus, while, $MgAl_2(OC_2H_5)_8$ $LiTa(OC_2H_5)_6$ and $LiTa(t\text{-}OC_4H_9)_6$, were delivered to the deposition chamber in liquid form due to a lack of volatility or thermal stability at vaporization temperatures. The fact that $LiTa(OC_2H_5)_6$, $LiTa(t\text{-}OC_4H_9)_6$ and $LiTa(n\text{-}OC_4H_9)_6$ must be used under different vaporization conditions suggests that various organic derivatives of double alkoxides present drastically varying thermal stability and volatility.

In a series of experimental studies(Chour et al., 1998; Zhang and Xu, 1997), a simple procedure that helped in identifying stable double alkoxides was reported. Vapor pressure measurements on double and single alkoxide precursors in the vaporization temperature range provided insights into the thermal stability of the double alkoxides studied. Such estimation is possible because the fully decomposed double alkoxides tend to exhibit vaporization activation energy identical to that of the more volatile single alkoxides in our measurements, while stable double alkoxides exhibit drastically different vapor pressure and activation energy. Further analysis of vapor pressure as a function of temperature was performed for the class of compounds: $LiM(x\text{-}OC_nH_{2n+1})_6$, where M=Ta,Nb; $n=3,4$ and $x=$n,i,t; corresponding to the normal-, iso- and tert-derivatives of the respective alkoxides where applicable. It was found that by comparing $\ln p$ vs. $1/T$ curves for all double and single alkoxides in the class of compounds $LiM(x\text{-}OC_nH_{2n+1})_6$, a simple strategy emerged for the identification as well as the use of double alkoxides for the autostoichiometric vapor deposition of $LiTaO_3$ and $LiNbO_3$: The precursor vaporization should be performed in the temperature and pressure range in which the maximum deviation in vapor pressure exists between the double and single alkoxides, and where the vapor pressure of the double alkoxide exceeds that of the single alkoxide. For those derivatives that have single alkoxides more volatile than double alkoxides throughout the temperature and pressure range, the prospect of such double alkoxides being a good precursor candidate is low.

The composition of the vapor phase alkoxides at a given vaporization condition depends on the thermal stability of the double alkoxide as well as the relative volatility of the double and single alkoxides. Consequently, it is necessary to characterize not only the relative volatility of double and single alkoxides, but also the thermal stability of double alkoxides. A readily available approach to studying the thermal stability is the analysis of sublimate or distillate compositions. In the following we discuss the thermal stability of double alkoxides of the type: $LiM(x\text{-}OC_nH_{2n+1})_6$, where M=Ta,Nb; $n=3,4$ and $x=$ n,i,t. We discuss some of the important features in their vaporization behavior. Sublimate and distillate compositions were analyzed by inductively coupled plasma emission spectroscopy (ICPES). The stable double alkoxides thus identified along with the optimum vaporization conditions were used to successfully deposit stoichiometric $LiNbO_3$ and $LiTaO_3$ from new precursors $LiNb(n\text{-}C_4H_9)_6$ and $LiTa(i\text{-}C_4H_9)_6$.

4.3.2 Physical Chemistry of Evaporation Processes

Double alkoxides of the form $LiM(x\text{-}OC_nH_{2n+1})_6$, when heated, may undergo two competing evaporative processes. The first is the most desired in autostoichiometric vapor deposition-the stoichiometric vaporization:

$$LiM(x\text{-}OC_nH_{2n+1})_6(\text{liquid}) \xrightarrow{\text{heating}} LiM(x\text{-}OC_nH_{2n+1})_6(\text{vapor}) \qquad (4.30)$$

If this evaporation channel dominates, the vapor-phase composition is determined exclusively by the $LiM(x\text{-}OC_nH_{2n+1})_6$ vapor. Vaporization through this channel is referred to as stoichiometric evaporation. The second possible channel is the decomposition-vaporization:

$$LiM(x\text{-}OC_nH_{2n+1})_6(\text{liquid}) \xrightarrow{\text{heating}} Li(x\text{-}OC_nH_{2n+1})(\text{liquid})$$
$$+M(x\text{-}OC_nH_{2n+1})_5(\text{liquid}) \qquad (4.31)$$
$$Li(x\text{-}OC_nH_{2n+1})(\text{liquid}) \xrightarrow{\text{heating}} Li(x\text{-}OC_nH_{2n+1})(\text{vapor}) \qquad (4.32)$$
$$M(x\text{-}OC_nH_{2n+1})_5(\text{liquid}) \xrightarrow{\text{heating}} M(x\text{-}OC_nH_{2n+1})_5(\text{vapor}) \qquad (4.33)$$

When the decomposition-vaporization and the stoichiometric vaporization occur simultaneously, the vapor-phase composition is determined by the combined effects of the double alkoxide stability constant and the relative volatility of all single and double alkoxides present. For the sake of simplicity we assume the decomposition reaction (4.31) reaches equilibrium in the precursor evaporator, and the equilibrium constant is K. We further assume the vapor pressure-temperature behaviors for the vaporization of all single and double alkoxides follow the Arrhenius relation:

$$\ln p = a - \frac{b \times 10^3}{T} \qquad (4.34)$$

where p and T are the vapor pressure and temperature. The constants a and b can be measured experimentally from the vapor pressure of single and double alkoxides at various temperatures. For the class of precursor compounds LiM $(x\text{-}OC_nH_{2n+1})_6$ in this study, the single alkoxide $Li(x\text{-}OC_nH_{2n+1})$ is known to be polymeric and nonvolatile (Bradley, 1978a, b). The vapor-phase composition is thus determined by the decomposition reaction (4.31) and the relative volatility of $LiM(x\text{-}OC_nH_{2n+1})_6$ and $M(x\text{-}OC_nH_{2n+1})_5$ following reactions (4.30) and (4.33).

If we assume a small fraction of double alkoxide δ has decomposed in the liquid at the vaporization temperature, the equilibrium of reaction (4.2) would require:

$$\frac{\delta^2}{1-\delta} = K^*$$

(4.35)

where K^* is the equilibrium constant for reaction (4.31). The partial pressures of vapor phase species $M(x\text{-}OC_nH_{2n+1})_5$ and $LiM(x\text{-}OC_nH_{2n+1})_6$ can then be expressed following the behavior of a Raoultian solution:

$$p_{M(x\text{-}OC_nH_{2n+1})_5} = p_M^0\delta$$

(4.36)

$$p_{LiM(x\text{-}OC_nH_{2n+1})_6} = p_{LiM}^0(1-\delta)$$

(4.37)

where p_M^0 and p_{LiM}^0 are the vapor pressures of pure $M(x\text{-}OC_nH_{2n+1})_5$ and $LiM(x\text{-}OC_nH_{2n+1})_6$, respectively. The vapor pressure measurement on the surface of $M(x\text{-}OC_nH_{2n+1})_5$ thus gives p_M^0 as a function of temperature, while the measurement on $LiM(x\text{-}OC_nH_{2n+1})_6$ gives $p_M^0\delta + p_{LiM}^0(1-\delta)$ as a function of temperature. Here the decomposition of a fraction of the double alkoxide is included in the consideration. Consequently, the experimentally measured constants a and b can be used to calculate the ratio between p_M^0 and $p_M^0\delta + p_{LiM}^0(1-\delta)$:

$$\Delta a - \frac{\Delta b \times 10^3}{T} = \ln \frac{p_M^0}{p_M^0\delta + p_{LiM}^0(1-\delta)}$$

(4.38)

where $\Delta a = a_{M(x\text{-}OC_nH_{2n+1})_5} - a_{LiM(x\text{-}OC_nH_{2n+1})_6}$ and $\Delta b = b_{M(x\text{-}OC_nH_{2n+1})_5} - b_{LiM(x\text{-}OC_nH_{2n+1})_6}$, both can be calculated from experimentally measurable parameters. Finally, the composition of sublimates or distillates from double alkoxides $LiM(x\text{-}OC_nH_{2n+1})_6$, if expressed in terms of the molar ratio between group V metal M and Li is:

$$y = \frac{M}{Li} = \frac{p_M^0\delta + p_{LiM}^0(1-\delta)}{p_{LiM}^0(1-\delta)}$$

(4.39)

By combining equations (4.38) and (4.39) we have:

$$\delta = \frac{1 - \dfrac{1}{y}}{\exp\left(\Delta a - \dfrac{\Delta b \times 10^3}{T}\right)} \tag{4.40}$$

Equations (4.35), (4.37) and (4.40) thus allow the estimation of the equilibrium constants for double alkoxide decomposition (4.31) as well as the resolved vapor pressure of double alkoxide p_{LiM}^0 as a function of temperature.

4.3.3 Experimental Assessment of Evaporation Stoichiometry

4.3.3.1 Vapor Pressure Measurement

The vapor pressures of double alkoxides are not directly measurable due to decomposition reactions accompanying vaporization. The low vapor pressure of both single and double alkoxides in addition to the highly condensable nature of metal alkoxide vapor render a direct vapor pressure measurement difficult. Presently, very little is known about the vapor pressure and thermal stability of double alkoxides. In most cases, our knowledge is limited to the temperature and pressure at which sublimation was performed. For example, both LiNb $(\text{t-OC}_4\text{H}_9)_6$ and $\text{LiTa(t-OC}_4\text{H}_9)_6$ were known to have a boiling point of 110–120 °C under a pressure of 0.1 torr(Zhang and Xu, 1997).Based on the discussions and equations presented above, it is now possible, however, that substantially more information can be derived from a systematic vapor pressure measurement and composition analyses of sublimates or distillates.

The apparatus used for the measurement of vapor pressure is a Schlenk vacuum line with controllable pressure in the range 10^{-2} to 100 torr (Chour et al., 1998). A sample container is used to observe the boiling point of alkoxides under a fixed pressure and a slow heating ramp at a constant rate. Experiments were performed for double and single alkoxides for the class of compounds $\text{LiM}(x\text{-OC}_n\text{H}_{2n+1})_6$(Chour et al., 1998). Based on our analysis as shown by (4.35), (4.38) and (4.39), it is apparent that the thermal stability and decomposition of double alkoxides play a significant role in determining the vapor pressure and sublimate composition for double alkoxides. In addition, the vapor pressure measured on a single alkoxide corresponds to that of a pure single alkoxide, p_{M}^0, while the vapor pressure on the surface of a double alkoxide consists of contributions from the vapor-phase double alkoxide as well as the more volatile single alkoxide as a result of decomposition, $p_{\text{M}}^0 \delta + p_{\text{LiM}}^0 (1 - \delta)$.

4.3.3.2 Sublimation/Distillation

We have now established that the sublimation in different temperature and pressure ranges yields drastically different sublimate compositions. The precursor evaporation conditions used during the vapor deposition determines the composition of the vapor phase species. Consequently, it is necessary to analyze compositions of sublimates obtained in various representative temperature and pressure ranges.

The sublimation or distillation of double alkoxides were conducted under controlled temperature and pressure to study thermal decomposition behaviors of double alkoxides. The same Schlenk vacuum line with controllable pressure is used to provide the ambient for sublimation. A sublimator was designed(Zhang and Xu, 1997) and constructed to perform the collecting of solid sublimates or liquid distillates. Figure 4.6 shows schematically the design of the sublimator used in this study. The sublimation or distillation were conducted at three different temperatures through the range of temperature in which the vapor pressure measurement was performed for the double alkoxide. The pressure was also controlled to be at the corresponding vapor pressure. For all samples, the total amount of sublimate/distillate collected ranged from 500 mg to 1 g for the composition analysis.

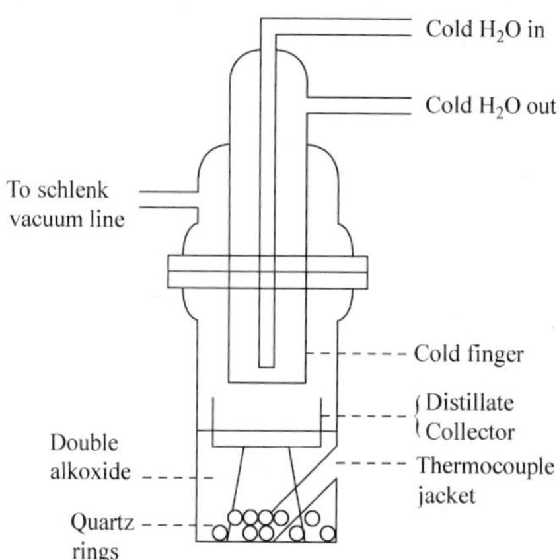

Fig.4.6   Schematic diagram showing the design of the sublimator

4.3.3.3   Composition Analysis and Decomposition Equilibrium

The solid sublimates and liquid distillates from double alkoxides in various temperature and pressure ranges were sent for composition analysis by Galbraith Laboratories, Inc. Knoxville, Tennessee. The analytical method employed was inductively coupled plasma emission spectroscopy using a Perkin-Elmer P2000 or an Optima 3000 instrument. The samples were charred by H_2SO_4, $HClO_4$ and HNO_3. Principal emissions used in the Galbraith Laboratories, Inc. analysis for lithium, tantalum and niobium were at 670.78 nm, 226.230 nm and 309.418 nm, respectively. The standards used were Li_2CO_3 of 99.999% purity, metal basis Ta and Nb of 99.99% purity. Puratronic® standard powder samples $LiNbO_3$ and $LiTaO_3$, from Johnson Matthey, Co., were sent with the sublimates as additional references. The accuracy of the composition analysis when expressed in Ta(Nb) : Li ratio, was $\pm$ 0.02 for Ta : Li and $\pm$0.09 for Nb : Li. These were determined by the analysis results for standard samples of Puratronic® $LiNbO_3$ and $LiTaO_3$.

4.3.3.4 Deposition of Multicomponent Oxide Thin Films

The thin film deposition procedures and the apparatus are similar to those used by Chour et al.(Chour and Xu, 1995b; Chour et al., 1998) with one exception: The depositions were carried out at three different precursor vaporization conditions. For example, deposition was attempted at precursor evaporator temperature settings corresponding to the low, median and high values found on the $\ln p$ (torr)vs. $1/T$(K) curve for the double alkoxide used for the deposition. The pressure settings in the evaporator were determined by the $\ln p$ (torr) vs. $1/T$(K) curve with the respective temperatures. The objective was to test our understanding of the double alkoxide vaporization stoichiometry in a real MOCVD environment. New precursor compounds $LiNb(n-C_4H_9)_6$ and $LiTa$ $(i-C_4H_9)_6$ were used to deposit stoichiometric $LiNbO_3$ and $LiTaO_3$ under varied vaporization temperature and pressure. X-ray diffraction patterns were used to identify the phases present in the films.

4.3.4 Experimental Results

The experimental constants in (4.34) were taken from Chour et al.(1998) Δa and Δb as defined by $\Delta a = a_{M(x-OC_nH_{2n+1})_5} - a_{LiM(x-OC_nH_{2n+1})_6}$ and $\Delta b = b_{M(x-OC_nH_{2n+1})_5} - b_{LiM(x-OC_nH_{2n+1})_6}$ are thus derived from these vapor pressure results. According to our analysis that gave rise to (4.38), (4.39) and (4.40), the fraction decomposition δ and the decomposition equilibrium constant K^* can be calculated using the composition analysis results. Table 4.4 lists the result of composition analysis y, defined as the molar ratio between group V metals (Nb or Ta) and lithium. The fraction decomposition of double alkoxides at the specified temperatures were obtained from (4.40) using Δa, Δb and δ. The decomposition equilibrium constants were obtained from (4.35) based on data for δ. The resolved vapor pressure of double alkoxides p^0_{LiM} are related to the measured vapor pressure $p_{LiM(x-OC_nH_{2n+1})_6}$ and the fraction decomposition δ by (4.37).

A special case arises in the $LiNb(n-OC_4H_9)_6$ system, where dimers $Li_2Nb_2(n-OC_4H_9)_{12}$ and $Nb_2(n-OC_4H_9)_{10}$ are the dominating vapor phase species. As we shall see in what follows, the dominating species in the vapor phase for double alkoxides are $Li_2Nb_2(n-OC_4H_9)_{12}$ and $Li_2Ta_2(n-OC_4H_9)_{12}$, while those for the single alkoxides are mostly monomeric such as $Ta(n-OC_4H_9)_5$ for all double alkoxide systems in this study with the exception of $Nb_2(n-OC_4H_9)_{10}$. The consequence of this is that the vapor pressure measurement on $Li_2Nb_2(n-OC_4H_9)_{12}$

Table 4.4 Vaporization stoichiometry of double alkoxides

Alkoxides	$T(°C)$	Δa	Δb	y	δ	K^*
$LiNb(n\text{-}OC_4H_9)_6$	195	70.3[a]	37.1[a]	1.34		
	230			9.92		
	265			high[b]		
$LiTa(n\text{-}OC_3H_7)_6$	202	11.3	5.5	0.98[c]	0.0	$<9.9\times10^{-5}$
	245			1.02	0.1	9.9×10^{-5}
	280			high[b]		
$LiTa(t\text{-}OC_4H_9)_6$	180	14.5	6.9	0.93[c]	0.0	$<1.6\times10^{-3}$
	225			1.07	3.9×10^{-2}	1.6×10^{-3}
	270			1.4	5.3×10^{-2}	3.0×10^{-3}
$LiTa(i\text{-}OC_4H_9)_6$	220	−27.7	−15.8	1.04	5.1×10^{-4}	2.6×10^{-7}
	251			1.06	5.0×10^{-3}	2.5×10^{-5}
	282			high[b]		
$LiTa(n\text{-}OC_4H_9)_6$	205	18.5	9.8	1.01	7.3×10^{-2}	5.8×10^{-3}
	245			1.09	1.3×10^{-1}	1.8×10^{-2}
	285			1.98	1.9×10^{-1}	4.7×10^{-2}

Notes:

a   Vapor pressure measurement unreliable due to formation of dimers: $Nb_2(n\text{-}OC_4H_9)_{10}$.

b   Lithium content too low to be detected.

c   Possible lithium-rich sublimates.

represents contribution from both $Li_2Nb_2(n\text{-}OC_4H_9)_{12}$ and $Nb(n\text{-}OC_4H_9)_5$, while similar measurement on single alkoxide is that of $Nb_2(n\text{-}OC_4H_9)_{10}$. If the double alkoxide decomposes extensively, the measured vapor pressure on the surface of $Li_2Nb_2(n\text{-}OC_4H_9)_{12}$ can exceed that on the surface of $Nb_2(n\text{-}OC_4H_9)_{10}$ since the $Nb(n\text{-}OC_4H_9)_5$ produced as a result of decomposition is considerably more volatile than $Nb_2(n\text{-}OC_4H_9)_{10}$. Therefore, for the $Li_2Nb_2(n\text{-}OC_4H_9)_{12}$ system, although the vapor pressure measurement shows higher volatility through out the entire temperature range, the vapor-phase species from the vaporization of $Li_2Nb_2(n\text{-}OC_4H_9)_{12}$ are predominantly $Nb(n\text{-}OC_4H_9)_5$. It turned out that $LiNb(n\text{-}OC_4H_9)_6$ can be used to deposit stoichiometric $LiNbO_3$ due to a different reason.   For the system of $Li_2Ta_2(n\text{-}OC_4H_9)_{12}$ and monomeric $Ta(n\text{-}OC_4H_9)_5$ the preceding theoretical discussions and equations remain applicable.

4.3.4.1   Thermal Stability of Double Alkoxides

The fraction decomposition of liquid precursors, d, can be used to calculate the Ta : Li ratio for the volatile components in the liquid prior to vaporization. Here, we again consider $Li(x\text{-}OC_nH_{2n+1})$ as the sole non-volatile component in the liquid at fraction decomposition δ. The volatile components thus have a Ta : Li ratio: $y_{liquid}=1/(1-\delta)$. It is rather informative to compare the sublimate composition to the volatile component ratio in the liquid. The difference is due to the vapor pressure difference between that of double and single alkoxides.

140

Figure 4.7 shows the results for LiTa(n-OC$_4$H$_9$)$_6$, LiTa(i-OC$_4$H$_9$)$_6$, LiTa(t-OC$_4$H$_9$)$_6$ and LiTa(n-OC$_3$H$_7$)$_6$. At temperatures below 250℃, all precursors showed Ta ∶ Li ratios within the stoichiometric range, since the accuracy of composition analysis is±0.02 for the Ta ∶ Li ratio and±0.09 for the Nb ∶ Li ratio. At temperatures above 250℃, the extent of decomposition in the liquid is apparently lower compared to the composition of the sublimates for LiTa (n-OC$_4$H$_9$)$_6$ and LiTa(t-OC$_4$H$_9$)$_6$, indicating more volatile single alkoxides compared to double alkoxides at that temperature. From the standpoint of a stoichiometric vaporization, it is advantageous to perform vaporization at lower temperatures if the volatility of the double alkoxide allows for a sufficient precursor delivery rate.

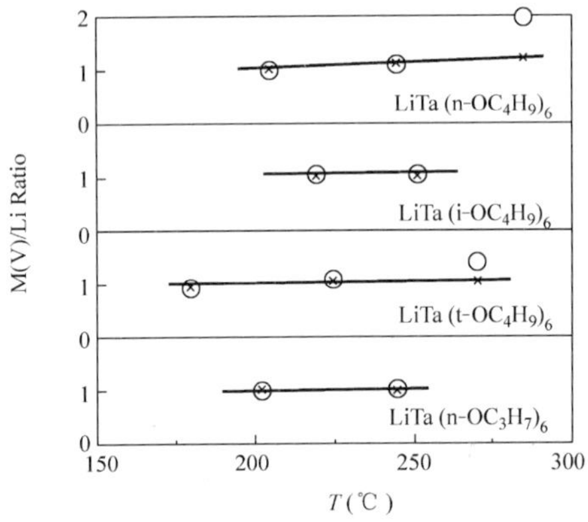

Fig.4.7 Compositions of the sublimates/distillates(open circles)as compared to the liquid composition (crosses) at sublimation temperature.The liquid composition at sublimation temperature is estimated from fraction decomposition δ. The lines are guides to the eye

4.3.4.2 Resolved Vapor Pressure of Double Alkoxides

It is now possible to separate the effect of decomposition from the double alkoxide vapor-pressure measurement result. The resolved vapor pressure for double alkoxides can be calculated based on (4.37). Figure 4.8 shows the resolved vapor pressure for double alkoxides LiTa(n-OC$_4$H$_9$)$_6$, LiTa(i-OC$_4$H$_9$)$_6$, LiTa(t-OC$_4$H$_9$)$_6$ and LiTa(n-OC$_3$H$_7$)$_6$. The relative volatilities of these double alkoxides are clearly presented in Fig. 4.8. Within the temperature range studied, the volatility of these double alkoxides follow the order LiTa(t-OC$_4$H$_9$)$_6$>LiTa (n-OC$_4$H$_9$)$_6$, LiTa(n-OC$_3$H$_7$)$_6$> LiTa(i-OC$_4$H$_9$)$_6$. The significance of resolved vapor pressure of double alkoxides can be further appreciated in (4.39). At low temperatures, when thermal decomposition $\delta \ll 1$, (4.39) becomes:

$$y = 1 + \frac{\delta\, p_M^0}{p_{LiM}^0} \tag{4.41}$$

The choice of precursors and vaporization temperature-pressure conditions should be such that $\delta p_M^0 \ll p_{LiM}^0$, in order to maintain stoichiometric vaporization. The fraction off-stoichiometry in the vapor phase species is in general:

$$K = \frac{\delta p_M^0}{(1-\delta)p_{LiM}^0} \qquad (4.42)$$

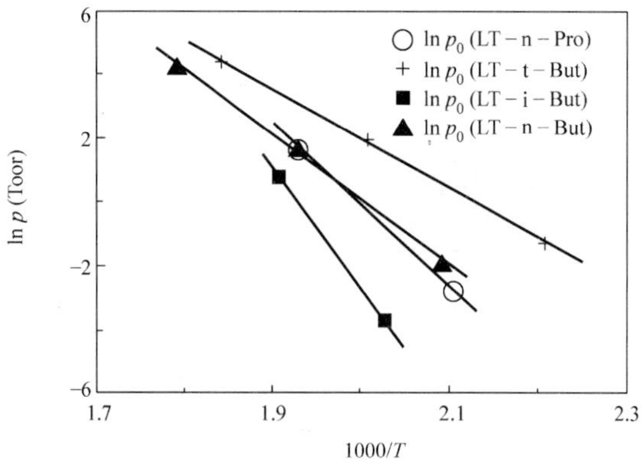

Fig.4.8 Resolved vapor pressure of double alkoxides LiTa(i-OC$_4$H$_9$)$_6$ (triangles),LiTa (i-OC$_4$H$_9$)$_6$ (squares), LiTa(t-OC$_4$H$_9$)$_6$ (crosses) and LiTa(n-OC$_3$H$_7$)$_6$ (open circles). The lines are guides to the eye

4.4 Autostoichiometric Vapor Deposition

4.4.1 Deposition of LiTaO$_3$ from LiTa(i-OC$_4$H$_9$)$_6$

The use of LiTa(i-OC$_4$H$_9$)$_6$ in the vapor deposition of stoichiometric LiTaO$_3$ has not been attempted before. According to the sublimate composition analysis and vapor pressure analysis,LiTa(i-OC$_4$H$_9$)$_6$ may be used at vaporization temperatures below 250℃. At 250℃, the vapor pressure of LiTa(i-OC$_4$H$_9$)$_6$ is about 2.2 torr, and the sublimate composition in terms of Ta : Li is 1.06±0.02. Three depositions were attempted, with vaporization conditions corresponding to the high, median and low temperatures covered in this study. Table 4.5 is a summary of typical deposition conditions. At higher precursor vaporization temperatures, the precursor carrier gas pressure is decreased to reduce the overall precursor flow and the deposition rate. An additional leak pressure is used to suppress precursor boiling since the vapor pressure is markedly higher at

142

high vaporization temperatures. The LiTaO$_3$ films were deposited on the (111) face of silicon substrates. Figure 4.9a – c are the SEM pictures of the LiTaO$_3$ films. All samples were annealed at 700℃ for 2 h to induce crystallization, since the as-deposited films were amorphous at the substrate temperature of 600 ℃. The crystallized films show typical polycrystalline morphology. It appears that a low vaporization temperature for the precursor favors a dense film. Figure 4.10a – c show the X-ray diffraction patterns for the LiTaO$_3$ films in Fig.4.9a – c.

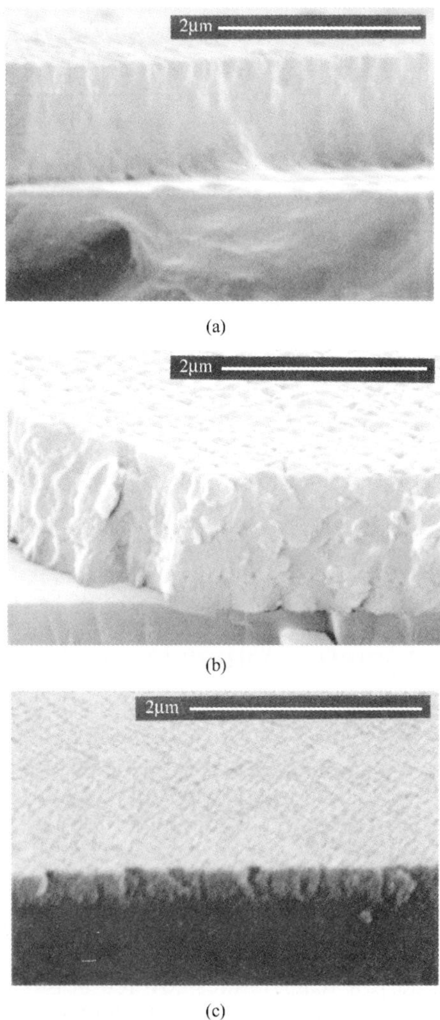

Fig.4.9 SEM micrographs of LiTaO$_3$ films deposited by vaporization of LiTa(i-OC$_4$H$_9$)$_6$ at temperatures of (**a**) 165℃, (**b**) 200℃ and (**c**) 135℃

At 165℃ vaporization temperature, shown in Fig.4.10a, the film appears to be single phase with no detectable second phases from non-stoichiometry. At 220 ℃ vaporization temperature, as shown in Fig.4.10b, a very small amount of second phase started to appear, indicating excess tantalum content in the vapor during deposition. At 315℃ vaporization temperature, Ta_2O_5 peaks were identified, indicating a further deviation from stoichiometry. The overall observation is consistent with the stoichiometry of vaporization behavior of $LiTa(i\text{-}OC_4H_9)_6$.

Table 4.5 Experimental conditions for the deposition of $LiTaO_3$ from $LiTa(i\text{-}OC_4H_9)_6$

	165	220	315
Precursor temperature (℃)	165	220	315
Substrate temperature (℃)	600	600	600
Annealing temperature (℃)	700	700	700
Initial pressure (torr)	8.0×10^{-2}	6.6×10^{-2}	6.5×10^{-2}
Water vapor+carrier gas pressure (torr)	1.0	1.0	2.0
Precursor+carrier gas pressure (torr)	5.0	4.0	2.0
Additional leak pressure (torr)	0	7.0	70
Deposition time (s)	180	255	120

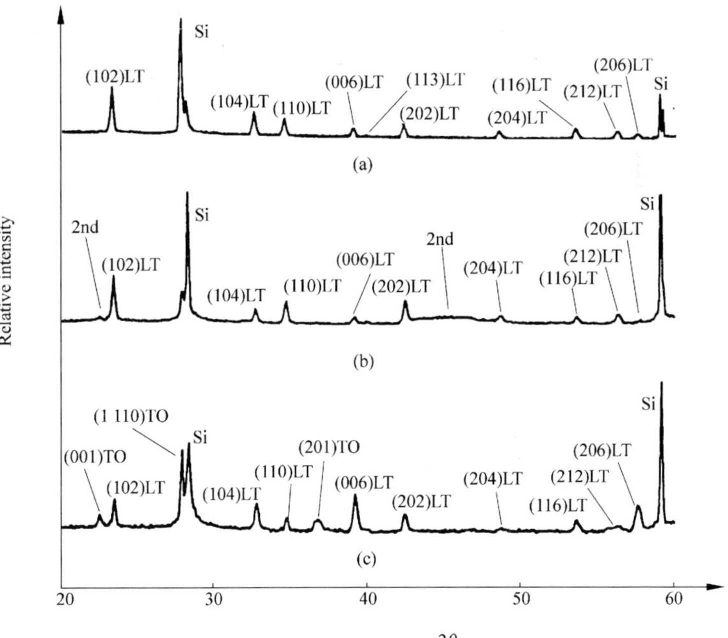

Fig.4.10 X-ray diffraction patterns of the $LiTaO_3$ films shown in Fig.4.9a – c.

4.4.2 Deposition of LiNbO$_3$ from LiNb(n-OC$_4$H$_9$)$_6$

Double alkoxide LiNb(n-OC$_4$H$_9$)$_6$ appears to decompose extensively at temperatures of 195℃, 230℃ and 265℃. The sublimate composition indicated high niobium content, as shown in Table 4.4. However, the vapor pressure measurement on LiNb(n-OC$_4$H$_9$)$_6$ and Nb(n-OC$_4$H$_9$)$_5$ indicated a more volatile LiNb(n-OC$_4$H$_9$)$_6$ than Nb(n-OC$_4$H$_9$)$_5$ (Chour and Xu, 1995b; Chour et al., 1998). Presently, this is attributed to the Nb$_2$(n-OC$_4$H$_9$)$_{10}$ dimers that dominate the vapor phase species above boiling Nb(n-OC$_4$H$_9$)$_5$ based on the knowledge that group V metal alkoxides tend to form dimers in the liquid form, as indicated by the molecular complexity from melting-point measurements(Zhang and Xu, 1997). This observation is also independently confirmed by a mass spectrometry study of the vapor phase species from the vaporization of LiNb(n-OC$_4$H$_9$)$_6$. Extensive formation of dimers Nb$_2$(n-OC$_4$H$_9$)$_{10}$ was found in the vapor phase, while no detectable Ta$_2$(n-OC$_4$H$_5$)$_{10}$ was found in the vapor above LiTa(n-OC$_4$H$_5$)$_6$. The formation of dimers in Nb(n-OC$_4$H$_9$)$_5$ apparently lowered its vapor pressure to the extent that the decomposed LiNb(n-OC$_4$H$_9$)$_6$, which produces monomeric Nb(n-OC$_4$H$_9$)$_5$, appeared to be more volatile in the boiling-point measurement.

As we have discussed before, the stoichiometry of the deposited films is ultimately determined successively by the stoichiometry of vaporization and the deposition reactions. In other words, without some knowledge of the nature of the deposition reactions, the ability to use LiNb(n-OC$_4$H$_9$)$_6$ for the deposition of stoichiometric LiNbO$_3$ has to be tested experimentally at present. We have made an attempt to deposit LiNbO$_3$ from LiNb(n-OC$_4$H$_9$)$_6$ under various vaporization conditions. Figure 4.11(a) and (b) show the SEM pictures of LiNbO$_3$ films deposited at vaporization temperatures of 230℃ and 270℃, respectively. The substrate temperature during deposition was 600℃, and the samples were annealed at 700℃ in air for 2 h. The films show typical polycrystalline forms due to the large lattice mismatch between silicon and that of LiNbO$_3$. X-ray diffraction patterns of the films in Fig.4.11a and b are shown in Fig.4.12a and b. In both cases, single-phase polycrystalline diffraction patterns of LiNbO$_3$ are observed, indicating that stoichiometric deposition of LiNbO$_3$ was achieved under given conditions. Such results, however, are unexpected. Presently we do not have a firm explanation for this. One possibility is that the Nb(n-OC$_4$H$_9$)$_5$ from thermal decomposition of LiNb(n-OC$_4$H$_9$)$_6$ does not participate in the deposition reaction in the temperature range covered in this study, possibly due to formation of dimers Nb$_2$(i-OC$_4$H$_9$)$_{10}$ prior to their arrival at the deposition chamber. Consequently, although a niobium-rich vapor is present, the component that did participate in deposition reactions under hydrothermal conditions is LiNb(n-OC$_4$H$_9$)$_6$, thus resulting in stoichiometric deposition of LiNbO$_3$. This point must be addressed in future studies of deposition reactions. A systematic study of deposition reactions of alkoxide precursors is currently underway in our laboratory.

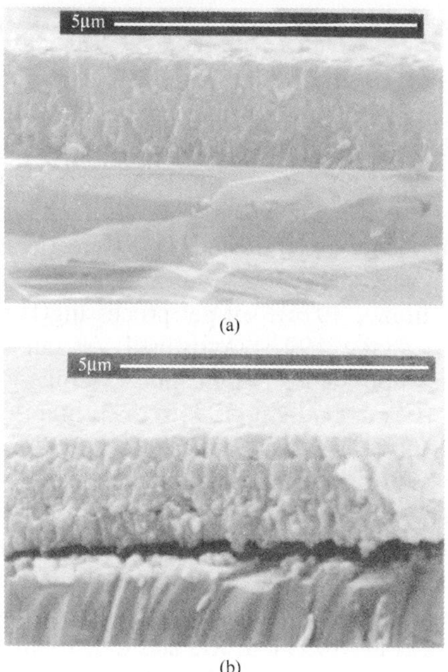

Fig.4.11 SEM micrographs of LiNbO$_3$ deposited by vaporization of LiNb(n-OC$_4$H$_9$)$_6$ at temperatures of (**a**) 230℃ and (**b**) 270℃

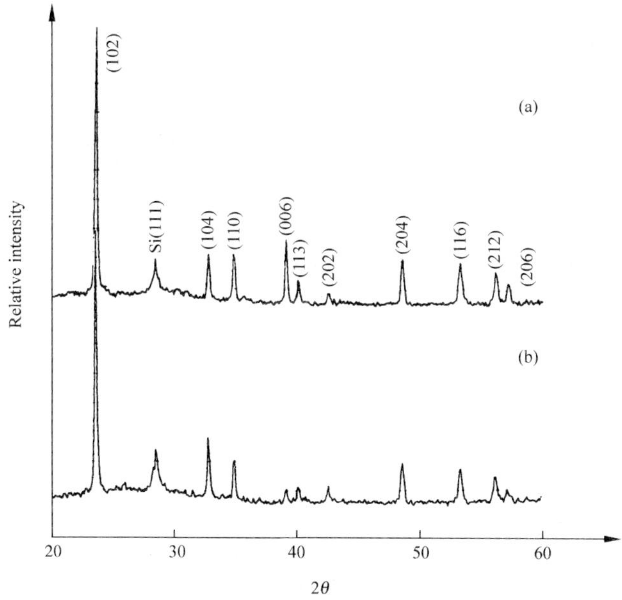

Fig.4.12 X-ray diffraction patterns of the LiNbO$_3$ films shown in Fig.4.11a and b

4.4.3   Deposition of LiTaO$_3$ from LiTa(n-OC$_4$H$_9$)$_6$

LiTaO$_3$ has long been considered one of the most promising materials for electro-optic device applications. Advantages inherent to LiTaO$_3$ include simple composition, high Curie point, low dielectric constant, high electro-optic and piezoelectric coefficients. Considerable effort has been devoted to the deposition of low-loss waveguides of LiTaO$_3$ on various single-crystal substrates for frequency-doubling applications. These are represented in attempts by sputtering techniques(Kanata et al., 1987; Saito and Shiosahi, 1991), liquid phase epitaxy(Fukuda and Hirano, 1976), sol-gel processing(Hirano and Raj, 1989), laser ablation(Hung et al., 1993; Agostinelli et al., 1993), as well as metalorganic chemical vapor deposition methods(Wernberg et al., 1993; Xie and Raj, 1993; Xu, 1995; Chour and Xu, 1995b). The high loss in optical signals presently remains a major obstacle to practical applications of waveguide devices. The optical loss was attributed to the combined effect of the scattering from the waveguide surface roughness, the internal absorption and the scattering due to various crystallographic and compositional defects(Fork et al., 1995).

Recently, emphasis was given to the precise control over the composition of LiTaO$_3$ in the given deposition processes as one approach to produce low-loss waveguides (Wernberg et al., 1993a, b; Xie and Raj, 1993; Xu, 1995; Chour and Xu, 1995b). Such a strategy was adopted because without precision in composition control, reduction of defects in an epitaxial film cannot be achieved on a consistent basis. In addition, the deposition temperature in most techniques, typically at 500 – 900 ℃, is far below that of the melting point of LiTaO$_3$ at 1650 ℃. Consequently, the thermodynamic driving force for stoichiometry, present for most compound semiconductors such as GaAs, is absent for LiTaO$_3$. The composition of the as-deposited amorphous films of LiTaO$_3$ is thus determined by the partial pressure of each precursor compound in the vapor phase and their reaction kinetics. Post-deposition annealing usually causes epitaxial crystallization in the solid phase for LiTaO$_3$. Therefore, the ideal scenario in vapor-phase composition control is to use associated precursors, such as a double alkoxide LiTa(n-OC$_4$H$_9$)$_6$ with the desired metal-to-metal ratio on the molecular level.

Little is known about the intrinsic thermal stability of double alkoxide precursors at their evaporation temperatures. For double alkoxides with low thermal stability, the precursor is usually delivered to the substrate in a liquid form. For example, Wernberg et al. used the liquid-injection method to deposit LiTaO$_3$ from LiTa(OC$_2$H$_5$)$_6$ solution in toluene(Wernberg, 1993a, b). Xie and Raj(1993) used a pulsed liquid-injection method to deposit LiTaO$_3$ from LiTa (t-OC$_4$H$_9$)$_6$ solution in toluene(Xie, 1993). We reported, on the other hand, in a vapor deposition environment the deposition of LiTaO$_3$ by stoichiometric evaporation of volatile double alkoxide LiTa(n-OC$_4$H$_9$)$_6$. (Xu, 1995; Chour and Xu, 1995b). It was found that the composition of the vapor phase as the result of vaporization of the double alkoxide is determined by its stability as well as the relative volatility between single and double akoxides. If the initial precursor is of a pure double alkoxide, the single alkoxide may arise from decomposition at the evaporation temperature. A nonstoichiometry factor K, the per cent off-stoichiometric composition in the vapor phase, was proposed to account for the decomposition and the subsequent co-evaporation of single and double alkoxides. $LiTa(n\text{-}OC_4H_9)_6$ was found to present a negligible K factor in the evaporation temperature range of $100-250\,^{\circ}C$. Stoichiometric, single-phase $LiTaO_3$ was deposited on a variety of substrates(Chour and Xu, 1995b). The fact that the decomposition and evaporation processes are both temperature dependent suggests that the nonstoichiometric factor K should also possess temperature dependences for all known double alkoxides. It is thus possible that all double alkoxides will evaporate with an acceptable K factor if the evaporation is conducted in an appropriate temperature and pressure range. A systematic investigation of decomposition and evaporation behavior of relevant precursors is thus warranted. Since the ultimate objective is to deposit stoichiometric, and defect-free epitaxial $LiTaO_3$ for device applications, a detailed characterization of the stoichiometry and defect structures of the epitaxial $LiTaO_3$ film is also necessary.

In this section, we attempt to re-enforce our previous discussions on the essential features for a stoichiometric vapor deposition of epitaxial $LiTaO_3$. We report a systematic investigation of the evaporation behavior of $Li(Ta,Nb)$ $(OC_nH_{2n+1})_6$, $n=3,4$ family double alkoxide precursors and present the temperature dependence of the nonstoichiometric factor for five double alkoxide systems. The stoichiometry of evaporation was studied by compositional analysis of sublimate. The crystallization behaviors under pyrolytic and hydrolytic deposition conditions were studied by the X-ray diffraction method. The film composition was analyzed by the Rutherford backscattering technique as well as precision lattice parameter measurement. The epitaxial layer was characterized by X-ray diffraction, double-crystal rocking curve measurement, as well as triple-crystal X-ray diffraction.

4.4.3.1 Experimental

The single alkoxide precursors were reagent grade $LiOCH_3$, $Ta(OC_2H_5)_5$ and $Nb(OC_2H_5)_5$ from Johnson Matthey Company, Ward Hill, Massachusetts. The solvents used to prepare double alkoxides were anhydrous 1 and 2-propanol, 1, 2 and 3-butanol. The alcoholysis procedures of single and double alkoxide preparation are similar to those reported previously(Fork et al., 1995). The double alkoxides were subjected to thermal analysis on a Seiko Instrument TG/DTA–220 for evaporation and pyrolysis behavior studies.

The typical alkoxide vapor pressures are between 0.1 torr to 500 torr in the temperature range between $200\,^{\circ}C$ to $300\,^{\circ}C$. Because these alkoxides are highly condensable, and their vapor pressure is strongly dependent on the local temperature, it is difficult to measure their vapor pressure directly. The boiling point of the alkoxides under controlled total pressure, however, can be measured rather precisely. In this experiment, a capillary valve was used to control the influx of nitrogen into the vacuum line in order to maintain a constant total pressure. The total pressure can be controlled to within $\pm 5\%$ in the range 7×10^{-3} torr to 500torr. The temperature of the alkoxide was monitored and controlled with a thermocouple inserted into a submerged glass capillary tube connected to a programmable infrared line heater model 5193 and controller model 8651, Research Inc., Minneapolis, Minnesota. The vacuum line was described in detail previously(Zhang and Xu, 1997). A small number of quartz rings was placed in the precursor compound to facilitate boiling surfaces. The boiling point thus observed are within inconsistency. The total pressure was used as the nominal vapor pressure of the alkoxide at the temperature. It should be noted that the vapor pressure-temperature dependences thus measured are not absolute, as calibration was not performed. It is the relative vapor pressure versus temperature behavior between double alkoxides and the respective single alkoxides that is of interest in this study. Under suitable total pressure, double alkoxides were sublimed and collected as condensates on the cold wall for composition analysis.

The films were deposited in two different environments where the precursor was subjected to hydrolytic or pyrolytic conditions. The double alkoxide precursors were typically heated to between $200\,°C$ and $250\,°C$. Nitrogen gas was used to carry the precursor to the deposition chamber. The substrate was placed on a graphite or stainless steel support heated to $500 - 800\,°C$ by a program-controlled infrared line heater. A separate stream of nitrogen bubbled through a water reservoir, controlled with a capillary leak valve was passed through the substrate surface for hydrolytic deposition. The deposition apparatus and conditions were similar to those used in previous experiments (Chour and Xu, 1995b; Fork et al., 1995; Zhang and Xu, 1997). The deposition rates were estimated by electron microscopy measured film thickness. X-ray diffraction was then conducted on the as deposited films. The films were subsequently annealed in a Lindberg 1700 tube furnace in air at various temperatures and again subjected to X-ray diffraction for crystallographic analysis.

The sublimed precursor powder was calcined in the Lindberg furnace at $750\,°C$ for 5 h. A silicon standard powder, from Johnson Matthey Company, was mixed with the calcined powder as an internal reference in the X-ray diffraction studies. The diffraction angles corresponding to (31 10), (30 12), (416), (328), (22 12), (505) and (422) from Cu $K\alpha_1$ ($\lambda = 1.5406$Å) were used in a refinement program to calculate the lattice parameter of the $LiTaO_3$. The lattice parameter was compared with those from various standards in the literature and a standard prepared by mixing, calcining and annealing Puratronic ® grade Li_2CO_3 and Ta_2O_5 from Johnson Matthey company.

The composition of the epitaxial $LiTaO_3$ on sapphire (006) substrate, was studied by the Rutherford backscattering method on a NEC MAS 1700 instrument at 1.522 MeV incident beam energy, and at 170° detector angle. The composition and film thickness were calculated based on the kinematic factors and the ^4He stopping crosssection data in the literature(Ziegler and Chu, 1974). Precision measurement of lattice parameter of the film by double-crystal rocking curve was also used to determine the composition of the epitaxial $LiTaO_3$.

Direct X-ray diffraction on the film surface was performed on a Siemens D5000 diffractometer. Double-crystal and triple-crystal diffractions were conducted on a Bede D_3 diffractometer with generator settings at 40 kV and 30 mA. An SEM cross-sectional picture was taken on a Cambridge-Stereoscan 240 electron microscope. A NanoScope III atomic force microscope, Digital Instruments, Santa Barbara, California, was used to study the surface roughness of the epitaxial layer.

4.4.3.2 Results

The evaporation behavior of the precursor with limited stability can be considered, in general, as two competing processes – the simple evaporation of a double alkoxide molecule, and the decomposition-evaporation of the precursor. For the case of the $LiTa(OR)_6$ system, the competing processes can be written following (4.4), (4.5a) and (4.5b). The composition of the vapor phase species is determined by the stability of the double alkoxide $LiTa(OR)_6$ and the relative volatility between $LiTa(OR)_6$ and $Ta(OR)_5$. For unstable $LiTa(OR)_6$, the liquid phase can be considered a simple mixture of LiOR and $Ta(OR)_5$. Since LiOR is mostly ionic with very low vapor pressure compared to that of $Ta(OR)_5$, (Bradley et al., 1978a; Chour et al., 1994) the vapor-phase composition is simply determined by the relative volatility of $LiTa(OR)_6$ to that of $Ta(OR)_5$. A completely decomposed $LiTa(OR)_6$ has $Ta(OR)_5$ in a solution at 50mol% concentration. It is thus informative to compare the vapor pressure versus temperature behavior of $LiTa(OR)_6$ with that of $Ta(OR)_5$ in a 50mol% solution. Experimentally this can be performed for all varieties of double alkoxides and the respective volatile single alkoxides. Figure 4.1 shows the results of such measurements for $LiNb(n\text{-}OC_4H_9)_6$, $Nb(n\text{-}OC_4H_9)_5$, $LiTa(n\text{-}OC_3H_7)_6$ and $Ta(n\text{-}OC_3H_7)_5$ as an illustration. For the $LiNb(n\text{-}OC_4H_9)_6$, $Nb(n\text{-}OC_4H_9)_5$ system, the vapor pressure-temperature behaviors are sufficiently different between the double and single alkoxide, indicating the $LiNb(n\text{-}OC_4H_9)_6$ is structurally different from a simple solution of $Li(n\text{-}OC_4H_9)$ in $Nb(n\text{-}OC_4H_9)_5$. The vapor pressure behaviors for the $LiTa(n\text{-}OC_3H_7)_6$ and $Ta(n\text{-}OC_3H_7)_5$ on the other hand, are identical within the errors of the experiment, suggesting that the structure of $LiTa(n\text{-}OC_3H_7)_6$ possibly is the same as a 50% Raultian solution of $Ta(n\text{-}OC_3H_7)_5$.

We have studied the vapor pressure befavior of five double alkoxide-single alkoxide pairs. In all cases, an Arrhenius relationship with temperature was found. Table 4.6 is a list of the Arrhenius coefficients for each compound obtained by linear regression. It should be noted that the relative volatility between double and single alkoxides can be used as an indicator of the vapor-phase composition. Consider a partially stable double alkoxide $LiTa(OR)_6$ with 50% decomposition at the evaporation temperature. The vapor pressure ratio between $Ta(OR)_5$ and $LiTa(OR)_6$ can be calculated on the basis of the Arrhenius equation following equation (4.6). Using the constants listed in Table 4.6, the fraction excess $Ta(OR)_5$ in the vapor phase with respect to $LiTa(OR)_6$, the nonstoichiometry factor K, can be estimated as shown in (4.7) and (4.8). Figure 4.2 shows the calculated K factor, the fraction deviation from stoichiometric composition for all five alkoxide systems studied. It is apparent that LiNb $(n-OC_4H_9)_6$ and $LiTa(n-OC_4H_9)_6$ are likely to evaporate stoichiometrically in the typical precursor vaporization temperature range of $100-200°C$.

Table 4.6 Vapor pressure constants according to equation (4.34)

Alkoxides	a	$b \times 10^3$
$LiNb(1-C_4H_9)_6$	56.7	27.6
$Nb(1-C_4H_9)_5$	127	64.7
$LiTa(1-C_3H_7)_6$	50.9	25.5
$Ta(1-C_3H_7)_5$	62.2	31.0
$LiTa(3-C_4H_9)_6$	34.5	16.2
$Ta(3-C_4H_9)_5$	49.0	23.1
$LiTa(2-C_4H_9)_6$	72.2	37.4
$Ta(2-C_4H_9)_5$	44.5	21.6
$LiTa(1-C_4H_9)_6$	44.3	22.1
$Ta(1-C_4H_9)_5$	62.8	31.9

The stability of double alkoxide complexes is apparently maintained through a different mechanism from that for single alkoxides. For single alkoxides, generally a branched alkoxyl group favors thermal stability. Better charge screening is expected out of branched alkoxyl groups for alkaline and alkaline earth elements with strong ionic characters. Double alkoxides, on the other hand, appear to evaporate in the form of dimers as suggested by mass spectrometric study of double alkoxide vapor(Chour et al., 1994) A linear, long chain alkoxyl group favors the formation of dimers with the long hydrocarbon chain wrapped around to provide charge screening to the alkaline elements. Such argument appears to be in agreement with our observation in LiNb $(n-OC_4H_9)_6$ and $LiTa(n-OC_4H_9)_6$ systems.

Thermogravimetric analysis of $LiTa(n-OC_4H_9)_6$, $LiTa(i-OC_4H_9)_6$ and $LiTa(t-OC_4H_9)_6$ were conducted in a nitrogen atmosphere. Figure 4.13 shows the relative weight loss at various temperatures for these double alkoxides. A differential thermal analysis of the precursors indicated that all major weight loss events were endothermic, which is consistent with their evaporation or decompositional evaporation. The largest weight loss for $LiTa(n-OC_4H_9)_6$ is at about 300℃, while that for LiTa(i-OC$_4$H$_9$)$_6$ and LiTa(t-OC$_4$H$_9$)$_6$ were at lower temperatures, indicating possible decomposition evaporation, since the evaporation of Ta(i-OC$_4$H$_9$)$_5$ and Ta(t-OC$_4$H$_9$)$_5$ is expected to be at lower temperature than that of LiTa(i-OC$_4$H$_9$)$_6$ and LiTa(t-OC$_4$H$_9$)$_6$. This finding is consistent with our previous observations of the deposition of LiTaO$_3$ from LiTa(n-OC$_4$H$_9$)$_6$, and Ta$_2$O$_5$ from LiTa(i-OC$_4$H$_9$)$_6$ and LiTa(t-OC$_4$H$_9$)$_6$ (Chour and Xu, 1995a). The final decomposition temperatures for all three systems were at 400℃. This is the temperature above which the substrate should be heated during deposition.

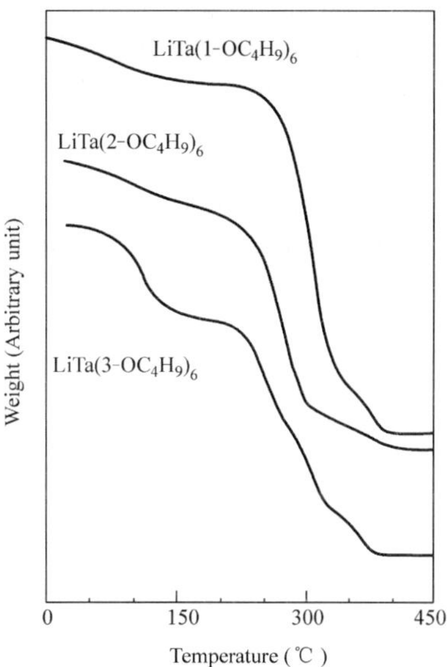

Fig.4.13 Thermogravimetric analysis of LiTa(n-C$_4$H$_9$)$_6$, LiTa(i-C$_4$H$_9$)$_6$ and LiTa(t-C$_4$H$_9$)$_6$

The composition of sublimate for the LiTa(n-OC$_4$H$_9$)$_6$ system was analyzed by lattice parameter measurement. Figure 4.3 shows the results along with that for a series of standard samples prepared in this study. Lattice parameters were measured on annealed samples with varying compositions. Silicon powder was mixed prior to X-ray diffraction as an internal standard. The calibration curve was obtained by scaling the Li : Ta ratio 1 : 1 sample using JCPDS card No. 29 – 836. A similar standard prepared by Barns and Carruthers (1970) is also shown for reference. Individual lattice parameters on a stoichiometric sample by Abrahams and Bernstein (1967) are indicated in the figure by an arrow. The lattice parameter measured on a Puratronic ® stoichiometric powder sample from Johnson Matthey, Co. is also shown in the figure. The calcined sublimate from LiTa(n-OC$_4$H$_9$)$_6$ showed a lattice parameter of 5.1518Å, corresponding to Li$_{1.01 \pm 0.02}$Ta$_{0.99 \pm 0.02}$O$_{2.98 \pm 0.06}$. The evaporation of LiTa(n-OC$_4$H$_9$)$_6$ is thus stoichiometric within the composition detection limit by the X-ray lattice parameter method at $\pm$0.02 for Li/Ta ratio.

152

Deposition of LiTaO$_3$ from double alkoxide LiTa(n-OC$_4$H$_9$)$_6$ was conducted under pyrolytic and hydrolytic conditions. Metal alkoxides are known to produce oxides through pyrolysis, oxidation, and hydrolysis-polycondensation. In the presence of water vapor, the pyrolytic reactions take on a water-catalyzed route with substantially lower activation energy. The presence of water also increases the deposition rate dramatically, possibly due to hydrolysis-polycondensation of alkoxide precursors. Figure 4.14 shows the Arrhenius plot of deposition rate-temperature dependence for pyrolytic and hydrolytic depositions of LiTaO$_3$ from LiTa(n-OC$_4$H$_9$)$_6$. At deposition temperatures below 700°C, the hydrolytic deposition shows a higher rate than pyrolytic deposition. As a result, the deposition can be realized at unusually low temperatures, as low as 200 °C under hydrolytic conditions, indicating the important role of hydrolysis-polycondensation in the deposition of oxides from hygroscopic alkoxide precursors. A common characteristic of the vapor deposition of LiTaO$_3$ from alkoxide precursors is that only amorphous films were deposited initially. Subsequent annealing is required to induce crystallization. Table 4.7 is a summary of the crystallization behavior for the LiTaO$_3$ system. All films in their as-deposited form are amorphous throughout the deposition temperature range covered by the present apparatus at 200–800°C. After annealing at temperatures 50–150°C above that for the deposition, the amorphous films crystallized. The exact cause of such behavior is not clear at this point although the observation is consistent with low-temperature and high deposition-rate experiments. In this study, we are unable to lower the deposition-rate sufficiently to observe in-situ crystallization. Xie and Raj (1993) using a piezoelectric pulser to inject the equivalent of one monolayer of precursor molecules at a time, with 5 s annealing period at deposition temperatures of 550–780°C, were able to obtain crystalline LiTaO$_3$ as deposited. This is indicative of the fact that given sufficient time for annealing, at sufficiently low deposition rate, in-situ crystallization may be realized.

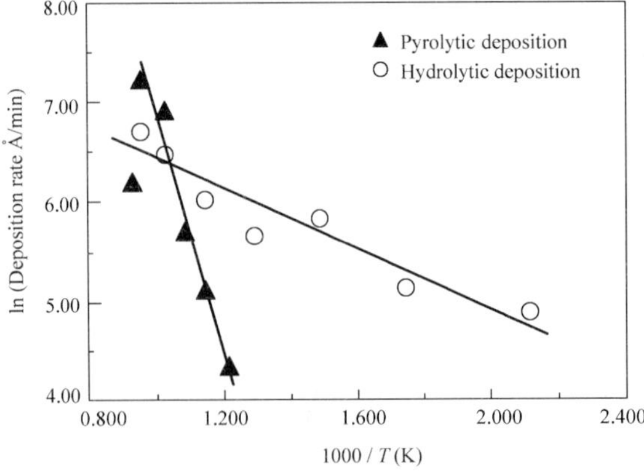

Fig.4.14　Deposition rate at various temperatures for pyrolytic (▲) and hydrolytic (○) depositions.The lines are guides to the eye.

Table 4.7   Crystallization behavior of LiTaO$_3$

Annealing Deposition	500℃	600℃	700℃	800℃
450℃	Amorphous	Crystalline	Crystalline	Crystalline
550℃	Amorphous	Amorphous	Crystalling	Crystalline
650℃	Amorphous	Amorphous	Crystalling	Crystalline
750℃	Amorphous	Amorphous	Amorphous	Amorphous

Of central concern in the present study is the stoichiometry and defect structures in the epitaxial LiTaO$_3$ films. Figure 4.15 shows the RBS result at 1.522 MeV, 170° detector angle for a film deposited on sapphire (012) substrate. The simulation data are summarized in Table 4.8. The simulated curve was found to be consistent with a film composition of $Li_{1.00}Ta_{1.00-0.05}O_{3.00+0.05}$. The accuracy of the composition analysis by RBS is limited. More accurate analysis may be provided by the lattice-parameter measurement. The film thickness estimated from RBS is in general agreement with SEM observation. Figure 4.16 is the high-resolution double axis rocking curve for the sample LiTaO$_3$(006) on LiNbO$_3$(006) substrate, with the inset an SEM picture of the fracture surface on the side of the film. The rocking curve shows an excellent quality layer on a substrate. The fringes stem primarily from an extremely clean, i.e. take over dislocation-"free" interface, as we shall see in the triple-axis contour map. The

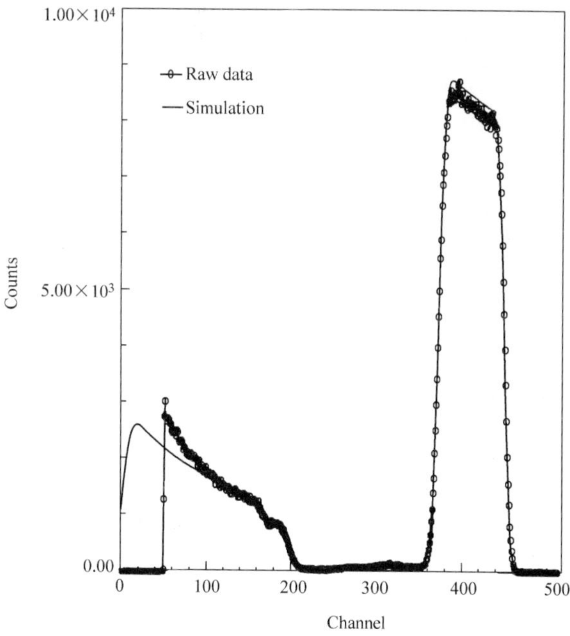

Fig.4.15   RBS analysis(－⊖－)and simulation(—)result for LiTaO$_3$ film on sapphire(012) substrate

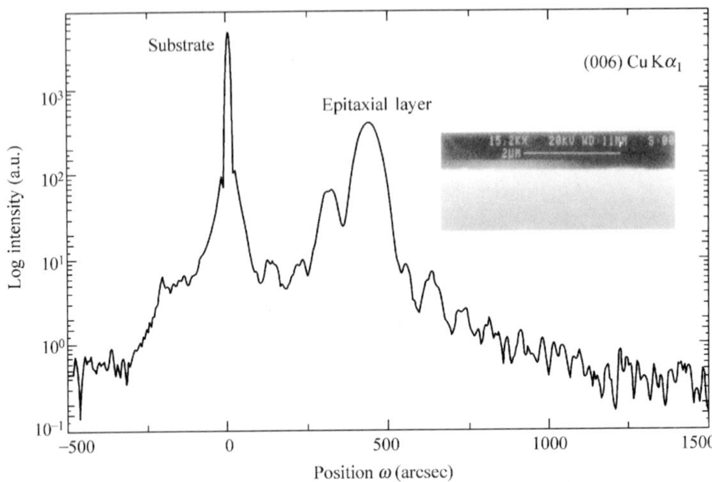

Fig.4.16　High-resolution double-crystal rocking curve of $LiTaO_3$ epitaxial layer on $LiNbO_3$ (006) substrate (inset,SEM view of the side of film)

Table 4.8　RBS simulation data summary

Parameters	Ta	O	Li
Kinematic factor	0.9159	0.3625	
Characteristic energy (MeV)	1.394	0.552	
Yield(counts)	7895	304	
Energy shift ΔE (MeV)	0.308		
^{4}H stopping cross section (10^{15}eVcm^2/atom)			
at 1.522MeV	114.5	41.26	17.72
at 1.394MeV	116.7	42.81	18.42
Film thickness(μm) RBS (SEM)	0.32(0.36)		
Lithium tantalate stoichiometry	$Li_{1.00}Ta_{1.00-0.05}O_{3.00+0.05}$		

degree of strain in the layer is rather low. The asymmetry to the fringes indicates that there is some strain grading in the layer, most likely at the interface. The sharp peak corresponding to the 0 arcsec position is that of the $LiNbO_3$(006) substrate. The single crystal substrate was of congruent melt composition at 48.5mol% in lithium, with lattice parameter a_H=5.148Å. c_H=13.863Å (1Å =0.1nm). The tall peak at 441 arcsec on the right side, corresponding to the epitaxial $LiTaO_3$, has a 2θ off-set of 0.245° from the substrate $LiNbO_3$(006) peak. The calculated lattice parameter for the epitaxial $LiTaO_3$ layer is thus c_H=13.780 Å. Using this value, combined with the density measurement of Barns and Carruthers, (1970) the lattice parameter a_H=5.152Å. This value is identical to that of the Puratronic ® $LiTaO_3$ standard, as shown in Fig.4.3, indicating a stoichiometric LiTaO₃ epitaxial layer. The fringes on either side of the LiTaO₃ peak are known to be film thickness interference fringes. The fringe period corresponds to 84.5 ± 8.4 arcsec. This can be used to estimate the epilayer thickness according to:

$$\Delta\theta = \frac{\lambda}{2t\cos\theta_{\mathrm{B}}} \tag{4.43}$$

Where $\Delta\theta$ is the average period of the fringes, λ is the wavelength of the CuKα_1 line, θ_{B} is the Bragg angle at the main peak, and t is the film thickness in Å. The film thickness was estimated at 0.20 ± 0.02 μm. The low thickness value obtained by the precision lattice measurement method as compared to those from SEM and RBS is currently attributed to the partial lattice relaxation near the interface. As a result, the measured lattice parameter on the epitaxial layer tends to shift toward that of the substrate, resulting in an underestimate of the actual LiTaO₃ film thickness. Figure 4.17 shows a triple-axis measurement, which gives an isointensity contour plot with the substrate at the left, and the layer and fringes shown at the right. The broadening along the tilt direction is induced by damage in the LiNbO₃ substrate commonly grown from an off-stoichiometry congruent melt. The LiTaO₃ epitaxial layer assumed this broadening. However, this broadening is of the same magnitude as in the substrate, indicating an essentially "defect-free" epitaxial LiTaO₃ layer. An atomic force microscope scan of the surface for LiTaO₃/LiNbO₃(006) was used to calculate the RMS roughness. Figure 4.18a and b show the AFM images of LiTaO₃/LiNbO₃(006) and LiTaO₃/Si(111), respectively. The RMS surface roughnesses are 39Å for LiTaO₃/LiNbO₃(006) and 45Å for LiTaO₃/Si(111). This

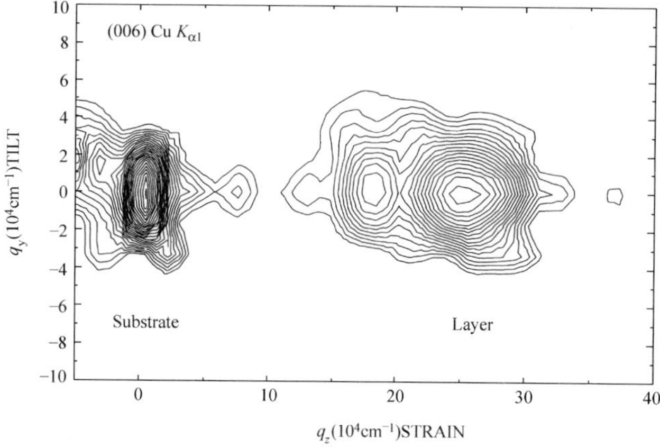

Fig.4.17 Triple-axis contour map of LiTaO₃ epitaxial layer on LiNbO₃(006)substrate. The fringes along the strain (q_z) direction are due to thickness interference. The tile angles are typical of oxide single crystals at $\pm 3 \times 10^{-4}\mathrm{cm}^{-1}$

observation is similar to those reported by Fork et al.(1995) The RMS roughness is aproximately 1% of the film thickness.

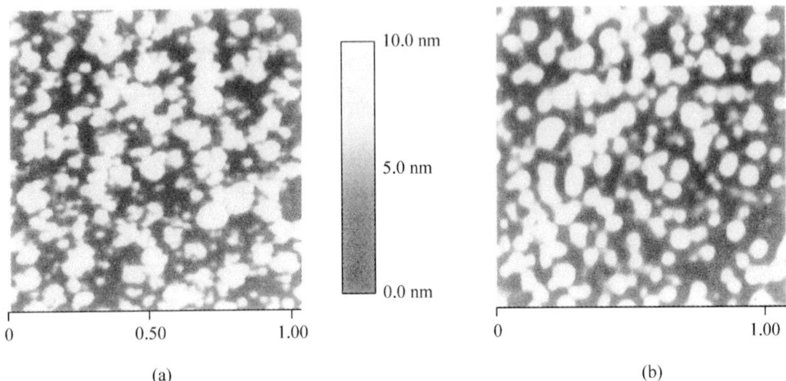

Fig.4.18　AFM scan of (**a**) LiTaO$_3$ on LiNbO$_3$(006) and (**b**) LiTaO$_3$ on silicon(111).The RMS roughness measured was 39 Å for (**a**) and 45 Å for (**b**)

References

Abrahams, S. C. and J. L. Bernstein. J. Phys. Chem. Solids. **28**, 1685(1967)

Adams, A. C. and C. D. Capio. J. Electrochem. Soc. **126**, 1042(1979)

Agostinelli, J., G. H. Braunstein and T. N. Blanton. Appl. Phys. Lett. **63**, 123(1993)

Barns, R. L. and J. R. Carruthers. J. Appl. Cryst. **3**, 395(1970)

Becker, F. S., D. Pawlik, H. Anzinger and A. Spitzer. J. Vac. Sci. Technol. **B5**, 1555 (1987)

Beidell, W., V. Shklover and H. Berke. Inorg. Chem. **31**, 5561 (1992)

Bradley, D. C., R. C. Mehrotra and D. P. Gaur. Metal Alkoxides. (Academic Press, London, 1978a), 46

Bradley, D. C., R. C. Mehrotra and D. P. Gaur, Metal Alkoxides. (Academic Press, London, 1978b), Chap. 5

Bradley, D. C. Chem. Rev. **89**, 1317 (1989)

Broorse, R. S. and J. M. Burlitch. Chem. Mat. **6**, 1509(1994)

Carruthers, J. R., G. E. Peterson, M. Grasso and P. M. Bridenbaugh. J. Appl. Phys. **42**, 1846(1971)

Campion, J. F., D.A. Payne, H. K. Chae, J. K. Maurin and S. R. Wilson. Inorg. Chem. **30**, 3245(1991)

Chisholm, M. H, Inorganic Chemistry: Toward the 21st Century. (American Chemical Society, Washington, DC., 1983), Chap. 16

Chour, K. W., G. D. Wang and R. Xu. Mat. Res. Soc. Symp. Proc. **335**, 65(1994)

Chour, K. W. and R. Xu. J. Integ. Ferroelectric. **7**, 459(1995a)

Chour, K. W. and R. Xu. J. Mater. Res. **10**, 2542(1995b)

Chour, K. W., J. Chen and R. Xu. Thin Solid Films. **304**, 106(1997)

Chour, K. W., R. C. Zhang, M. S. Goorsky, T. Takada, E. Akiba, T. Kumagai, M. L. Jensen, C. Eves and R. Xu. J. Cryst. Growth. **183**, 217(1998)

Curtis, B. J. and H. R. Brunner. Mat. Res. Bull. **10**, 515(1975)

Depp, A. F., M. T. Andreas, S. A. Duraj, E. B. Clark, D. G. Hehemann, D. A. Scheiman and P. E. Fanwick. Mat. Res. Soc. Symp. Proc. **335**, 227(1994)

Desu, S. B. J. Am. Ceram. Soc. **72**, 1615(1989)

Desu, S. B., D. B. Beach, B. W. Wessels, and S. Gokoglu (eds.) Metal-Organic Chemical Vapor Deposition of Electronic Ceramics. (Materials Research Society, Pittsburgh, 1993) Vol. 335

Fork, D. K., F. Armani-Leplingard and J. J. Kingston. Mat. Res. Soc. Symp. Proc. **361**, 155(1995)

Fukuda, J. and H. Hirano. Appl. Phys. Lett. **28**, 575(1976)

Fukukawa, Y., M. Sato, K. Kitamura, Y. Yajima and M. Minakata. J. Appl. Phys. **72**, 3250(1992)

Gardiner, R. A., P. C, Van Buskirk and P. S. Kirlin. Mat. Res. Soc. Symp. Proc. **335**, 221(1994)

Grove, A. S., Ind. & Eng. Chem. **58**, 48 (1966)

Hayashi, H., Y. Yamada, D. J. Baar, T. Sugimoto, K. Sugawara, Y. Shiohara and S. Tanaka. J. Cryst. Growth. **115**, 782(1991)

Hirano, S. and K. Kato. Mat. Res. Soc. Symp. Proc. **155**, 181(1989)

Hung, L. S., J. A. Agostinelli, J. M. Mir and L. R. Zheng. Appl. Phys. Lett. **62**, 3071(1993)

Huppertz, H. and W. L. Engl. IEEE Trans. Electr. Dev. **ED-26**, 658(1979)

Hwang, C. S. and H. J. Kim. J. Mater. Res. **8**, 1361(1993)

Ingebrethsen, B. J. and E. Matijetic. J. Aerosol Sci. **11**, 271(1980)

Kamata, K., J. Nishino, S. Ohshio and K. Maruyama. J. Am. Ceram. Soc. **77**, 505(1994)

Kim, J. S., H. A. Marzouk, P. J. Reucroft. Thin Solid Films. **254**, 33(1995)

Kwak, B. S., E. P. Boyd and A. Erbil. Appl. Phys. Lett. **53**, 1702(1988)

Kanata, T., Y. Kobayashi and K. Kubota. J. Appl. Phys. **62**, 2989(1987)

Levin, R. M. and K. Evans-Lutterodt. J. Vac. Sci. Technol. **B1**, 54(1983)

Mantese, J. Mat. Res. Soc. Bul. **14**, 48(1989)

Mehrotra, R. C., M. M. Agrawal and P. N. Kapoor. J. Chem. Soc. A, 2673(1968)

Nassau, K. and M. E. Lines. J. Appl. Phys. **41**, 533(1970)

Nishino, J., T. Kawarada, S. Ohshio and K. Kamata. J. Am. Ceram. Soc. **16**, 629-637 (1997)

Peng, C. H. and S. B. Desu. Appl. Phys. Lett. **61**, 16(1992)

Purdy, A. P. and C. F. George. Inorg. Chem. **30**, 1970(1991)

Saito, Y. and T. Shiosaki. Jpn. J. Appl. Phys. **30**, 2204(1991)

Schieber, M., S. C. Han, Y. Ariel, S. Chokron, T. Tsach, M. Maharizi, C. Deutcher, D. Racah, A. Raizman and S. Rotter. J. Cryst. Growth. **115**, 31(1991)

Schlichting, H. Boundary Layer Theory. (McGraw-Hill Book Co., New York, Chap. 1960), 7

Shimizu, D., T. Katayama, T. Shiosaki and A. Kawabata. J. Cryst. Growth. **99**, 399(1990)

Shimizu, M. and T. Shiosaki. Mat. Res. Symp. Proc. **361**, 295(1995)

Si, J., S. Desu and C. Y. Tsai. J. Mater. Res. **9**, 1721(1994)

Sladek, K. J. and W. W. Gibert. Proc. 3rd Intern. Conf. on Chem. Vap. Deposition. (Salt Lake

City, Utah, 24–17 April 1972) p. 215

Studebaker, D. B., G. Doubinina, C. Seegal, A. Rakovska, T. H. Baum. Mat. Res. Soc. Symp. Proc. **475** (1997) in print.

Treybal, R. E. Mass-Transfer Operations. (McGraw-Hill Book Co., New York, Chapter 1955), 3

Ushida, T., H. Higa, K. Higashiyama, I. Hirabayashi and S. Tanaka. Jpn. J. Appl. Phys. **31**, L608(1992)

Visca, M. and E. Matijetic. J. Colloid. Interface Sci. **68**, 308(1979)

Wernberg, A. A., H. J. Gysling, A. J. Filo and T. N. Blanton. Appl. Phys. Lett. **62**, 946(1993a)

Wernberg, A. A., G. Braunstein and H. Gysling. Appl. Phys. Lett. **63**, 2649(1993b)

West, G. and K. W. Beeson. J. Mat. Res. **5**, 1573(1990)

Xie, H. and R. Raj. Appl. Phys. Lett. **63**, 3146(1993)

Xu, Y. H. and J. D. Mackenzie. Integ. Ferroelectrics. **1**, 17 (1992)

Xu, R. J. Mater. Res. **10**, 2536(1995)

Yamasaki, H., E. Endo, Y. Nakagawa, M. Umeda, S. Kosaka, S. Misawa, S. Yoshida and K. Kajimura. J. Appl. Phys. **72**, 2951(1992)

Zhang, J., G. T. Stauf, R. Gardiner, P. Van Buskirk and J. Steinbeck. J. Mater. Res. **9**, 1333(1994)

Zhang, R. C. and R. Xu. Integrated Ferroelect. **18**, 197(1997)

Ziegler, J. F. and W. K. Chu. At. and Nucl. Data Tables. **13**, 481(1974)

5 Functional Nanocomposite Thin Films by Co-Sputtering

Feng Niu

5.1 Introduction

Nanocomposites consist of nanosized particles (several nanometers to tens of nanometers) embedded in different matrix materials. The particle phase could be metals, semiconductors, dielectric, etc. and the matrix materials could be metals, semiconductor, and dielectrics (including ceramics, glass, polymers, etc.).

Composite thin films possess many unusual physical properties that are not realizable with the pure single constituents. In recent years, ultrafine nanocrystalline materials have shown novel physical, chemical, magnetic, optical and electronic properties. Some interesting phenomena such as nonlinear optical behavior and quantum confinement effects of carriers have also been reported in nanocomposite materials. Nanocrystalline and nanocomposite materials are believed to play an important role in explaining some fundamental physical problems, such as microstructural and property transitions between the molecular and bulk solid state and also may have technological applications in the future in structural engineering, optoelectronic devices, catalysts for chemical reactions, magnetic storage and optical coatings.

The main objectives of this work are: (1) to develop novel nanocomposite thin film materials with potential in the optoelectronic device applications of the future, and to investigate their unusual optical and electrical properties and correlate these with the microstructures. The potential applications are expected to lie in the areas of optical selective filters, optical sensors and detectors, photoconductors, thermionic emitters, etc; (2) to study quantum confinement effects in metal particles and their influences on the optical and electrical properties of the composites; (3) to understand how we can control and design the dielectric permittivity of composite materials based on the knowledge of the optical and electronic properties of the constituents. In order to do these, we employ theoretical tools developed for the effective medium theories and model and predict the optical and electrical properties of composite materials.

The main focus has been on nanocomposite thin films with small metal particles finely dispersed in a Si matrix. The small metal particle system has been of great interest to scientists due to the apparent deviation of their physical and chemical properties from the bulk values. The surface plasma resonance absorption mode, for example, which is the most dominant optical characteristic of a small metal particle system does not exist in bulk materials. Generally, the optical and electrical properties of metal particle in a non-metal matrix nanocomposite will depend on the metal and matrix species, the volume fraction of the particles, the particles sizes and shapes and interparticle spacings.

In the current research a nanocomposite thin film material consisting of nanosized Ag metal particles embedded in an amorphous silicon matrix have been manufactured by the co-sputtering technique in a Nordiko RF sputtering system. The microstructures of the as-deposited films have been characterized by a combination of X-ray diffractometry (XRD), conventional and high-resolution transmission electron microscopies (CTEM and HREM), scanning electron microscopy (SEM) and differential scanning calorimetry (DSC). The optical absorption and reflection properties of the nanocomposite films have been investigated by near UV, visible and near IR spectrometry in the wavelength regions of 200–3000 nm. The electronic transport properties of the films as a function of metal content and temperature ranging from 120–373 K have also been studied by the four-probe and direct two-contact sheet resistance measurements.

An effective medium description of the nanocomposite materials was extended to predict the optical properties of the Si-Ag films where the amorphous or crystallized Si matrices are semiconducting and a strong absorber of light in the near ultraviolet and visible. It is realized that these nanocomposites offer model systems for testing the validity of the various effective medium theories in describing the optical and electrical properties of the composites beyond the classic metal/insulator system. It has been proved through the present simulation work that the effective medium description is qualitatively and, to some extent, quantitatively consistent with the measured optical properties. Furthermore, a possible electric transport mechanism in the metal/semiconductor composite is suggested based on the measured temperature dependence of the resistivity.

5.2 Classification of Nanocomposite Films, Their Properties and Applications

The term "nanocomposite" was first coined by Roy, Komarneni and their colleagues sometime during the period 1982–1983 to describe the major conceptual redirection of the sol-gel process, i.e., using the solution sol-gel process to create maximally heterogeneous rather than homogeneous materials (Hoffman et al., 1984). Nanocomposites should be clearly differentiated from nanocrystalline and nanophase materials, which refer to single phases in the nanometre range. Nanocomposites refers to composites of more than one Gibbsian solid phase where at least one dimension is in the 1–20 nm range (Komarneni, 1992). The solid phases can be metals, semiconductors, dielectrics or combinations of these. They can be amorphous or crystalline or combinations thereof. They can be inorganic or organic (e.g., polymers), or both, and essentially of any composition.

Although nanocomposites can be classified based on other criteria such as material function, processing, microstructures, etc., here we have identified several major families of nanocomposites mainly based on their physical properties as any combination of particle materials and matrix materials. These include the classic nanocomposite, i.e., metal / inorganic dielectric or insulator (nanocermet) or granular metal films and some emerging novel nanocomposite materials such as metal/polymer, metal/metal, metal/semiconductor, semiconductor/inorganic dielectric or insulator, etc. In the present research we refer especially to the nanocomposite thin films consisting of nanosized metal particle materials embedded in different nonmetal matrix materials.

The chemical and physical properties of inhomogeneous media nowadays constitute a wide area of research covering many fields of applications, i.e., astronomy, meteorology, petrology, biology, chemical engineering, energy conversion, new building materials etc.For example, among the metal particle/non-metallic matrix nanocomposites, the granular metal films or nanocermet films have been investigated for many decades (Abeles, 1976). A wide range of scientific and technological applications have been made in this, the largest family of nanocomposite films, such as cermet film resistors in microelectronics, granular superconductors and their application as superconducting quantum interference devices, granular ferromagnetism and their application as magnetic core materials, optical materials and selective absorbers in solar energy applications,and granular metals for television camera applications,etc. Metal/polymer nanocomposites are another active part of the metal/dielectric family, which can combine the advantages of organic polymers such as chemical inertness, processability, optical transparency in the visible region, excellent adhesion and the forming of continuous films even at low thicknesses (Heilmann et al., 1993) with optical and electrical properties of metal particles to obtain colored and conducting polymer films.

Since the end of the 1970s, as the interest in high-temperature energy conversion has risen, refractory coatings based on transition metal-refractory ceramic compounds have been developed, e.g., $Ni\text{-}Al_2O_3$, $Pt\text{-}Al_2O_3$, $Mo\text{-}Al_2O_3$, and $Co\text{-}Al_2O_3$. The cermets are promising because of their particular properties. First, their optical(high-frequency) and electrical (zero-frequency) properties show marked variations as a function of the volume fraction, the size and shape and the statistical distribution of the metallic inclusions. These variations are of particular importance near a critical metallic inclusion volume fraction f_c (the so-called electrical percolation threshold) where a transition of the electrical and optical properties between dielectric and metallic behavior occurs. At a given volume fraction (and over a wide fraction range) the dielectric function exhibits a resonance as a function of frequency;this so-called dielectric anomaly is attributed to the excitation of a metallic grain polarization mode. These two essential properties render the cermets of particular interest as selective absorbing coatings for the photothermal conversion of solar energy. To understand the basic selective optical properties of these materials it is necessary to know the dielectric function of the composite, especially in the visible and the near infrared (IR) ranges. Some novel optical properties have been also reported in a certain granular metal films such as surface enhanced resonant Raman scattering, enhanced IR absorption by small metal particles and optical nonlinearity. All these phenomena have promoted the development of the effective medium theories,percolation theory, and other theoretical approaches. There have been four international conferences specially dedicated to the increasing active area called "electrical transport and optical properties of inhomogeneous media" since the end of the 1970s (Garland and Tanner, 1978; Lafait and Tanner,1989; Mochan and Barrera, 1993; Dykhne et al., 1997) with gradually shortened intervals.

Since the 1980s increasing emphasis has been put on nanosized metal and semiconductor particles that show novel physical properties related to quantum size effects (Halperin, 1986; Alivisatos, 1996; Weller, 1993). Therefore, nanocomposite design can utilize combination effects of conventional composites with some tailored, improved or even superior electrical, optical or optoelectronic properties when compared with their monophase or microcomposite alternatives, and can also incorporate novel quantum-based properties associated only with nanophase materials.

While much research has been done on the metal-insulator systems, very few investigations have been made on the optical and electrical properties of metal particle/semiconductor matrix nanocomposite films. In terms of optical properties, early interest was shown in the optoelectronic properties of the S-1 photocathode, which consists of $Cs_{11}O_3$-coated Ag particles (5 nm) in vacuum, and embedded in a Cs_2O semiconductor matrix (Bates, 1984). Bates and his colleagues (Alexander and Bates, 1984; Chen and Bates, 1988) used an extended dynamic effective medium approximation theory (DEMA) to interpret successfully the photoelectronic quantum yield and absorption coefficient of the S-1 cathodes in the near IR and further investigated the influences of various factors on the optical properties of the metal-semiconductor composite thin films (Bates, 1993; Bates and Chen, 1997). They found double free electron plasma resonance absorption peaks associated with the isolated small Ag particles around 410 nm and for the aggregated Ag particles around 800 nm when surrounded by vacuum. In a composite consisting of small Ag particles (100 nm) in a Si matrix, Chen and Wang (1988) also observed double free electron plasma resonance absorption peaks in the near IR (located at 1×10^6 $\mu m/2 \times 10^6$ μm, respectively) associated with isolated and aggregated Ag particles. In addition they showed that the inclusion of the Ag particles in Si enhanced the absorption coefficient of the composite by three orders of magnitude, which offers an opportunity for long wavelength infrared detection application. Niklasson and Craighead (1983) reported a surface PRA in the visible in electron-beam deposited Si-Al composite films that consist of a mixture of Al nanocrystallites (5.5–11.5 nm) and probably even smaller Si particles or a-Si. The effective medium theory of the Bruggeman formula can give a good agreement with experimental reflectivity when the Si volume fraction is less than 33%. Co-sputtering of Si and Ag at very low Ag content(<6 at %) produced an amorphous $Si_{1-x}Ag_x$ alloy, and the optical bandgap of this alloy decreased systematically with increasing Ag concentration (Choudhury, 1987). In Ge-Ag graded-index composite thin films, different optical properties ranging from selective infrared reflectance to broad IR absorption were reported and the thin film coatings of this system were applied to an infrared tunnel sensor (micro-Golay cell) (Lamb and Nagendra, 1994). The same researchers (Nagendra and Lamb, 1995) also observed that when ultrafine Ag particles (2–30 nm) were embedded in a Ge amorphous matrix, the nanocomposite film showed a transition of optical properties from semiconducting to metallic, and also showed varying IR optical characteristics. These phenomena suggested potential applications as selective infrared reflectors and high-efficiency infrared absorbers.

In terms of electrical properties, Li and Wu (1991) observed a transition in a nanocomposite consisting of nanosized Ag particles (3–15 nm) embedded in surface layer of a Cs_nO polycrystalline matrix from polycrystalline semiconductor to metallic behavior as characterized by the conductivity-temperature variations (lg σ vs. $1/T$),and a five orders of magnitude increase in room temperature conductivity with surface Ag content increasing from an equivalent thickness of –2Å to –20Å. These results were difficult to explain with a model of direct electron tunneling through the Schottky barrier at the Ag-particle/Cs_nO interface, and an indirect hopping conduction mechanism through an impuritystate of the semiconducting matrix has been proposed to explain the electronic conduction behavior. Ndlela and Bates (1991) investigated the electrical transport properties in various metal-semiconductor composites, i.e., Ag-a-Si, Ag-poly-Si, Ag-CuInSe$_2$ and Au-CuInSe$_2$, and found a $T^{-1/4}$ temperature dependence for the resistivity. They also suggested a hopping conduction mechanism. Bates (1995) suggested a design for low-temperature, high-current thermionic emitters by burying small metal particles (10–100 nm) in semiconductor matrices, making use of the metal/semiconductor contact Schottky barrier, which could be modified by the QSEs in a Au-CuInSe$_2$ composite.

In terms of the fabrication and microstructure of metal-semiconductor composites, Bates (1995) found surface segregation of Ag in Si-Ag composites prepared by co-sputtering at high Ag content (50 vol%), but this could be prevented by depositing a thinner Si overlayer immediately after the Ag source was turned off. Lee (1994) reported some detailed results on the microstructures of Si-Ag nanocomposite thin films, and found that ultrafine Ag particles exhibited some preferred orientation and were aligned along the film growth direction.

5.3 Optical Properties of Small Particles Embedded in Different Matrixes

5.3.1 Definition of the Optical Constants of Matter

There are two sets of quantities that are often used to describe optical properties of matter: the real and imaginary parts of the complex refractive index $N= n+i\kappa=c\kappa/\omega$, where n is the real refractive index and κ the absorption index and the real and imaginary parts of the complex dielectric function (or relative permittivity) $\varepsilon=\varepsilon_1+i\varepsilon_2$. The two sets of quantities are not independent, they are related by the definition $\varepsilon = N^2=(n+i\kappa)^2$ or

$$\varepsilon_1 = n^2 - \kappa^2 \tag{5.1}$$

$$\varepsilon_2 = 2n\kappa = \frac{4\pi\sigma}{\omega} \tag{5.2}$$

$$n = \sqrt{\frac{\sqrt{\varepsilon_1^2 + \varepsilon_2^2} + \varepsilon_1}{2}} \tag{5.3}$$

$$\kappa = \sqrt{\frac{\sqrt{\varepsilon_1^2 + \varepsilon_2^2} - \varepsilon_1}{2}} \tag{5.4}$$

where σ is the conductivity of the matter and ω is the angular frequency of the light. Either of these two sets of quantities (n, κ) and $(\varepsilon_1, \varepsilon_2)$ may be thought of as describing the intrinsic optical properties of matter. For some purposes one set is preferable and for other purposes the other set is preferred. For example, in considering wave propagation, n and κ are preferred. However, in considerations of microscopic mechanisms that are responsible for optical effects, ε_1 and ε_2 are more appropriate (Papavassiliou, 1980; Bohren and Huffman, 1983; Kerker, 1969).

The ratio of intensity $I(x)$ at a distance x to the initial intensity I_0 is

$$\frac{I(x)}{I_0} = \exp(-2\omega\kappa x/c) = \exp(-ax)$$

where $\alpha = \dfrac{2\omega\kappa}{c} = \dfrac{4\pi\kappa}{\lambda}$ is the absorption coefficient and λ is the wavelength of the light. The complex optical constants n, κ and ε_1, ε_2 are not directly experimentally measurable, they must be derived from the measured quantities such as reflectance (R) and transmittance (T) by applying a suitable theory.

Of the total radiation energy incident on an object, a fraction R is reflected from the top surface and a fraction T is transmitted through the bottom surface. The remaining fraction is lost through absorption processes(A) and by scattering (S) at surface and volume imperfections (Ohring, 1992). Adding the various contributions gives $R+T+A+S=1$ according to the conservation of energy. In the simplest case of normal incidence on a material of complex refractive index N_1 and ignoring scattering

$$R = \frac{[(n_0 - n_1)^2 + \kappa_1^2]}{[(n_0 + n_1)^2 + \kappa_1^2]} \tag{5.5}$$

where $n_0=1$ is the refractive index of free space, $N_1=n_1+i\kappa_1$ is the complex refractive index of the material.

This formula is true for a very thick film of material. $T=I/I_0$ is defined as the transmittance while the absorbance A is defined as the negative logarithm of transmittance, i.e.,

$$A = -\ln T = -2.303 \lg T \tag{5.6}$$

and thus

$$\alpha = A / t \tag{5.7}$$

where t is the film thickness (cm).

5.3.2 Classic Mie Theory for Scattering and Absorption of Small Spherical Particles

The calculations of an effective optical dielectric function by self-consistent effective medium theories are based on the classic electromagnetic scattering theory that may be derived from the Maxwell equations in combination with "materials equations" (Palik, 1985, 1993). So firstly let us introduce the classic Mie theory for small spherical particles.

5.3.2.1 General Formula

In an effort to understand the varied colors in absorption and scattering exhibited by small colloidal particles of gold suspended in water, Mie (1908) obtained a rigorous solution for the absorption and scattering of light by a spherical particle. The theory of absorption and scattering by a single sphere can be directly applied to a collection of spheres based on the following conditions: (1) the distance between spheres must be larger than the wavelength so that the spheres scatter independently; (2) the spheres must be randomly located to avoid interference effects between waves scattered from different spheres; (3) the optical density must be small enough that essentially the same irradiance strikes each sphere. Under such conditions the energy removed from the incident beam of light by scattering and absorption can be expressed in terms of an extinction coefficient (γ) defined by the relation $T = \dfrac{I}{I_0} = \exp(-\gamma x)$. Starting from the Maxwell equations, Mie gave the phenomeonlogical description of the small spherical particles. The wave scattered by each particle is expanded in a series of partial waves caused by electric and magnetic partial oscillations of the electromagnetic field within the particle. The extinction coefficient γ (including absorption and scattering (γ') of a collection of spheres is given by (Papavassiliou, 1980; Bohren and Huffman, 1983; Kerker, 1969)

$$\frac{\gamma}{\Omega V} = \frac{3\pi \varepsilon_B^{3/2}}{\lambda a^3} \sum_{L=1}^{\infty} (2L+1)\,\mathrm{Re}(A_L + B_L) \tag{5.8}$$

where Ω is the number of spheres per unit volume, V is the volume of each sphere, ΩV is the volume concentration of the particles, λ is the wave length of light in vacuum, ε_B is the dielectric constant of the surrounding medium, $a = 2\pi R_A \varepsilon_B^{1/2} / \lambda (\varepsilon_B^{1/2} = N_B = n_B + i\kappa_B)$, and R_A is the radius of the spheres.

The coefficients A_L and B_L describe the contributions (amplitude) of the L-th electric and magnetic partial oscillations of the electrons in the particles, respectively.

$$A_L = \frac{\varepsilon_A j_L(b)[a j_L(a)]' - \varepsilon_B j_L(a)[b j_L(b)]'}{\varepsilon_A j_L(b)[a h_L^1(a)]' - \varepsilon_B h_L^1(a)[b j_L(b)]'} \tag{5.9}$$

These are given by

$$B_L = \frac{j_L(b)[a j_L(a)]' - j_L(a)[b j_L(b)]'}{j_L(b)[a h_L^1(a)]' - h_L^1[b j_L(b)]'} \tag{5.10}$$

where $b = 2\pi R_A \varepsilon_A^{1/2} / \lambda$, ε_A is the complex dielectric constant of the particles ($\varepsilon_A^{1/2} = N_A = n_A + i\kappa_A$), $j_L(z)$ is the spherical Bessel function of the first kind,

$h^1(z)=j_L(z)+iy_L(z)$ and $y_L(z)$ is the spherical Bessel function of the second kind. The scattering coefficient γ' is given by

$$\frac{\gamma'}{\Omega V} = \frac{3\pi\varepsilon_B^{1/2}}{\lambda a^3} \sum_{L=1}^{\infty} (2L+1)(|A_L|^2 + |B_L|^2) \tag{5.11}$$

5.3.2.2 Small Spherical Particles

Now we consider a special case where the radius R of the particle is much smaller than the wavelength of the incident light $(R \ll \lambda)$, The contributions from the first-order $(L=1)$ and second-order $(L=2)$ electric and magnetic oscillations are given by the following formula if we maintain the coefficients A_L and B_L of (5.9) and (5.10) to terms of order x^6,

$$A_1 = -\frac{i2x^3}{3}\frac{m^2-1}{m^2+2} - \frac{i2x^5}{5}\frac{(m^2-2)(m^2-1)}{(m^2+2)^2} + \frac{4x^6}{9}\frac{(m^2-1)^2}{(m^2+2)^2} + O(x^7) \tag{5.12}$$

$$A_2 = -i\frac{x^5}{15}\frac{m^2-1}{2m^2+3} + O(x^7) \tag{5.13}$$

$$B_1 = -\frac{ix^5}{45}(m^2-1) + O(x^7) \tag{5.14}$$

$$B_2 = O(x^7) \tag{5.15}$$

where $x = \dfrac{2\pi N_B R_A}{\lambda}$, and $m=N_A/N_B$, and N_B are the complex refractive indices of the particle and the surrounding medium, respectively. It will be sufficient to consider only the first term A_1, namely, the absorption is caused mainly by the electric dipole oscillation and ignore the size parameter terms higher than x^3, and then (5.8) and (5.11) can be simplified to the following equations independent of R_A, supposing that ε_B is real,

$$\frac{\gamma}{\Omega V} = \frac{18\pi\varepsilon_B^{3/2}\varepsilon_{A2}}{\lambda}\frac{I}{(\varepsilon_{A1}+2\varepsilon_B)^2 + \varepsilon_{A2}^2} \tag{5.16}$$

$$\frac{\gamma'}{\Omega V} = \frac{4\pi a^3\varepsilon_B^{1/2}}{\lambda}\frac{(\varepsilon_{A1}-\varepsilon_B)^2 + \varepsilon_{A2}^2}{(\varepsilon_{A1}+\varepsilon_B)^2 + \varepsilon_{A2}^2} \tag{5.17}$$

It is noted that only the electric dipole contribution is considered in this approach. Depending on the value of x, magnetic dipole contribution (B_1), electric quadrupole (A_2), magnetic quadrupole (B_2), etc. may also be excited upon an incident electromagnetic wave. Based on this simplification, Bruggeman's effective medium approach may be derived.

5.3.2.3 Concentric Particles (Coated Spheres or Composite Spheres)

For small concentric spheres with a core (phase A) of radius R_A and permittivity ε_A and a shell (phase B) with a radius R_B and permittivity ε_B embedded in a matrix material (phase C) of permittivity ε_C, when $R_B \ll \lambda$, the formulas (5.9) and (5.10) can be reduced to consider the first few terms as:

$$A_1 = -iI_1' v^3 + i\bar{I}_1' v^5 + \cdots \tag{5.18}$$

$$B_1 = i\bar{I}_1'' v^5 \tag{5.19}$$

$$A_2 = -iI_2' v^5 \tag{5.20}$$

where $v = \dfrac{2\pi R_B}{\lambda}$,

$$I_1' = -\frac{2}{3}\frac{(m_2^2-1)(m_1^2+2m_2^2)+q^3(2m_2^2+1)(m_1^2-m_2^2)}{(m_2^2+2)(m_1^2+2m_2^2)+q^3(2m_2^2-2)(m_1^2-m_2^2)} \tag{5.21}$$

where $m_1^2 = \dfrac{N_A^2}{N_C^2} = \dfrac{\varepsilon_A}{\varepsilon_C}$, $m_2^2 = \dfrac{N_B^2}{N_C^2} = \dfrac{\varepsilon_B}{\varepsilon_C}$.

Then Eq.(5.8) can be simplified to

$$\frac{\gamma}{\Omega V} = \frac{8\pi R_B}{\lambda}\,\mathrm{Im}\left(\frac{(\varepsilon_B-\varepsilon_C)(\varepsilon_A+2\varepsilon_B)+q^3(2\varepsilon_B+\varepsilon_C)(\varepsilon_A-\varepsilon_B)}{(\varepsilon_B+2\varepsilon_C)(\varepsilon_A+2\varepsilon_B)+q^3(2\varepsilon_B-3\varepsilon_C)(\varepsilon_A-\varepsilon_B)}\right) \tag{5.22}$$

where $q=R_A/R_B$.

5.3.3 Simulation of Optical Properties of Nanocomposite Materials by the Effective-Medium Theories (EMTs)

5.3.3.1 Introduction

In order to determine the optical properties of the heterogeneous or composite systems that are more accurately described as mixtures of separate regions of two or more materials, each of which retains its own dielectric identity, effective-medium theories must be introduced. The fundamental ideas behind the effective-medium theories are these: when a heterogeneous material is probed by an electromagnetic wave, the resolution limit of the probing wave is set by $\lambda/2$, where λ is the wavelength. In addition, the Rayleigh scattering strength is known to vary as λ^{-4}, so when the particles are orders of magnitude smaller than the wavelength of the radiation, $\lambda \gg L$, L is the scale of the inhomogeneities, the heterogeneous material would appear homogeneous to the probing wave, or lab probes can measure just an average of microscopic responses. The optical properties of the composite material can thus be treated in terms of an effective medium whose dielectric permittivity can by obtained by a suitable averaging over the dielectric permittivity of the two constituents. So the basic definition of an effective medium is that the random unit cell(RUC) when embedded in the effective medium, should not be detectable in an experiment using electromagnetic radiation confined to a specific wavelength range. In other words, the extinction of the RUC should be same as if it were replaced with a material characterized by ε_{eff}. An approach first proposed by Stroud and Pan (1978) and that has been used in recent years reduces to the requirement that ε_{eff} be chosen so that the forward scattering amplitude $S(0)$ of the particles embedded in the medium should vanish on the average as defined by:

$$S(0) = \frac{1}{2}\sum_{L=1}^{\infty}(2L+1)(A_{\text{L}} + B_{\text{L}}) = 0 \tag{5.23}$$

There are different models and further improvements in the framework of the EMTs, among which we will introduce the Maxwell-Garnett theory (MGT) (Garnett, 1904 and 1906), Bruggeman's effective medium approximation (EMA) (Bruggeman, 1935) and Ping Sheng theory (SPT) (Sheng, 1980).

5.3.3.2 Various Effective Medium Theories and Random Unit Cells (RUCs)

For generality, a two-phase inhomogeneous system is considered in which two materials A and B have dielectric permeability ε_{A} and ε_{B} and filling factors f_{A} and f_{B} ($=1-f_{\text{A}}$). The effective dielectric permittivity of the composite material is basically a function of ε_{A} and ε_{B} and the microstructure, as was first rigorously proved by Brown (1955). The microstructure is critically important in the EMTs. This approach regards a two-phase composite material as two types of typical structures or building units as shown in Fig.5.1, i.e., a separated grain structure with particles of A embedded in a continuous host of B as shown in Fig. 5.1a; or an aggregate structure in which A and B enter on an equal footing to form a space filling random mixture as shown in Fig. 5.1b. For the separated grain structure, the RUC or basic microstructure unit is a core of A surrounded by a concentric shell of B as depicted in Fig. 5.1c. The ratio of the core volume to the shell volume is equal to f_{A}. For the aggregate structure, the inherent structural equivalence of the components is ensured by letting the RUC have a probability f_{A} of being A and a probability f_{B} of being B, as shown in Fig. 5.1d. By using the effective-medium condition, we get the expression for the effective dielectric constant ε_{MG} of the dispersed inclusion microstructures:

$$\varepsilon_{MG} = \varepsilon_B \frac{\varepsilon_A + 2\varepsilon_B + 2f_A(\varepsilon_A - \varepsilon_B)}{\varepsilon_A + 2\varepsilon_B - f_A(\varepsilon_A - \varepsilon_B)} \tag{5.24}$$

This is well-known the Maxwell-Garnett theory(MGT) (Garnett, 1904 and 1906).

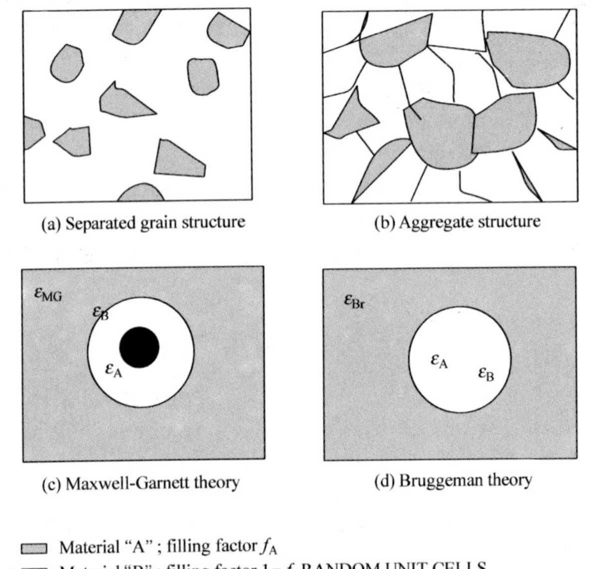

Fig.5.1　RUCs and their typical microstructures of the MGT and EMA

The same equation can be derived by using the condition (5.23) and the electric dipole coefficient of a coated sphere (5.18) and (5.21), suppose $\varepsilon_C = \varepsilon_{MG}$:

$$S(0) = -\frac{3}{2}i I_1' v^3$$

$$= v^3 \frac{(\varepsilon_B - \varepsilon_{MG})(\varepsilon_A + 2\varepsilon_B) + f_A(2\varepsilon_B + \varepsilon_{MG})(\varepsilon_A - \varepsilon_B)}{(\varepsilon_B + 2\varepsilon_{MG})(\varepsilon_A + 2\varepsilon_B) + f_A(2\varepsilon_B - 2\varepsilon_{MG})(\varepsilon_A - \varepsilon_B)} = 0$$

which requires

$$(\varepsilon_B - \varepsilon_{MG})(\varepsilon_A + 2\varepsilon_B) + f_A(2\varepsilon_B + \varepsilon_{MG})(\varepsilon_A - \varepsilon_B) = 0$$

This gives the same result as (5.24).

For the symmetric microgeometry, the resulting equation is

$$f_A \frac{\varepsilon_A - \varepsilon_{Br}}{\varepsilon_A + 2\varepsilon_{Br}} + f_B \frac{\varepsilon_B - \varepsilon_{Br}}{\varepsilon_B + 2\varepsilon_{Br}} = 0 \tag{5.25}$$

where ε_{Br} is the effective dielectric function of the composite. This is the well-known Bruggeman symmetric effective-medium approximation (EMA) (Bruggeman, 1935). This equation can be derived by using (5.23) and the series expansion for the small spherical particles (5.12).

Smith (1977) suggested that in considering the effective dielectric constants for metal-dielectric matrix systems, three classes of problem should be discussed: for the first two cases one constituent exists as particles embedded in the other, these two types can be described by the MGT(when f_A is low) or inverted MGT(when f_A is high, where the dielectric is broken up into islands).

$$\varepsilon_{MG} = \varepsilon_A \frac{\varepsilon_B + 2\varepsilon_A + 2f_B(\varepsilon_B - \varepsilon_A)}{\varepsilon_B + 2\varepsilon_A - f_B(\varepsilon_B - \varepsilon_A)} \tag{5.26}$$

For the third case where both are in particle form, an interpolation (or crossover) between (5.24) and (5.26) will have to be used and this depends on the topography of the given system. Smith suggested this might be determined directly using electron micrographs or indirectly by testing the possible results against optical observations.

Based on Smith's idea, Ping Sheng (1980) developed a theory in which a RUC consisting of a coated sphere whose core and shell can be either A or B is suggested, as depicted in Fig. 5.2. The relative occurrence of the two varieties of the RUC is determined by counting the number of equally possible configurations corresponding to different positions of the inner sphere in the RUC. When A is the core and B the shell, this number is $v_1 = (1 - f_A^{1/3})^3$. For the opposite situation $v_2 = (1 - (1 - f_A)^{1/3})$ is obtained.

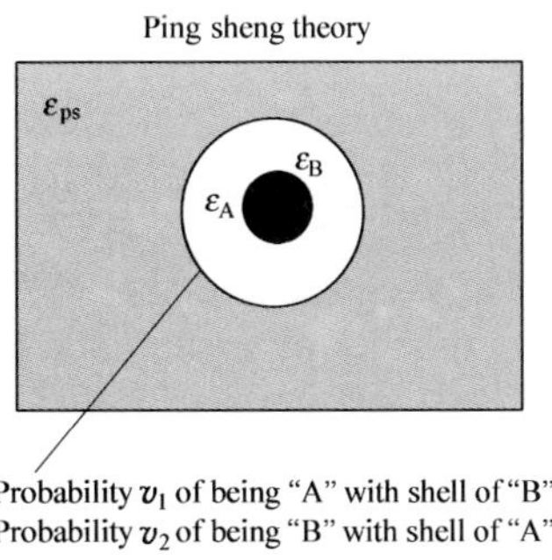

Probability v_1 of being "A" with shell of "B"
Probability v_2 of being "B" with shell of "A"

Fig.5.2   RUC of the SPT

The Ping Sheng theory (SPT) gives:

$$D_1 \frac{(\varepsilon_B - \varepsilon_{ps})(\varepsilon_A + 2\varepsilon_B) + f_A(2\varepsilon_B + \varepsilon_{ps})(\varepsilon_A - \varepsilon_B)}{(\varepsilon_B + 2\varepsilon_{ps})(\varepsilon_A + 2\varepsilon_{ps}) + 2f_A(\varepsilon_B - \varepsilon_{ps})(\varepsilon_A - \varepsilon_B)} +$$

$$D_2 \frac{(\varepsilon_A - \varepsilon_{ps})(\varepsilon_B + 2\varepsilon_A) + f_B(2\varepsilon_A + \varepsilon_{ps})(\varepsilon_B - \varepsilon_A)}{(\varepsilon_A + 2\varepsilon_{ps})(\varepsilon_B + 2\varepsilon_{ps}) + 2f_B(\varepsilon_A - \varepsilon_{ps})(\varepsilon_B - \varepsilon_A)} = 0$$

$$(5.27)$$

where ε_{ps} is the effective dielectric function by the SPT. This theory is sometimes called the symmetrical MGT. By using this model Sheng (1980) found success in explaining the conductivity of W-Al$_2$O$_3$ and Au-SiO$_2$ cermets as did Gibson and Buhrman (1983) in Au-Al$_2$O$_3$ near the percolation threshold.

The MGT predicts the existence of the optical dielectric anomaly, namely the plasma resonance absorption, which has been observed in granular metal films. However, because of the inherently asymmetrical treatment of the two constituents of the composite, the MGT predictions disagree with experimental optical and transport results in cermets when the volume fraction of the dispersed phase becomes comparable to or greater than that of the matrix phase. In particular, the theory does not produce the observed percolation threshold in granular metals. The EMA does give a percolation threshold (f_A=1/3), but it is low compared with the experimental result. Moreover, the EMA yields no dielectric anomaly (Gittleman and Abeles, 1977). The SPT, however,predicts both the optical dielectric anomaly and the percolation threshold.

The EMTs have been proved to be quite successful in describing the dielectric functions of composite or inhomogeneous materials at least qualitatively, because they are based on the classical theory of scattering and absorption of light, and further consider the local field enhancement of polarizable particles (Aspnes, 1982) and absorbing effects of the embedding medium. The various EMTs differ only in the choices of host matrices, and this choice does imply different microstructures. The MGT is suitable for the cases where f_A is much smaller than the percolation threshold volume fraction f_c, while the EMA is suitable for the cases like a spatial structure associated with an assembly of microcrystals or polycrystals.

5.3.3.3 Limitation of the Effective-Medium Theories

1. Particle Size Limit on the Validity of the Effective-Medium Theories

Under the EMTs, it is usually considered that individual particles are small enough to be insensitive to the dynamic characteristics of the incident electromagnetic wave. Thus they see a constant field region (static electric field). Such theories are classed as "quasistatic approximations". They can be quantitatively described as $R/\lambda \ll 1$, R is the particle radius. If this condition is taken then for metals at visible and near infrared wavelengths,sizes over 10 nm could not be considered.However Smith (1979) suggested that certain relaxed limits are tolerable based on some experimental results. He based the size and separation criteria on the magnetic dipole coefficient B_1 in the leading order (5.19) and (5.21), where

$$B_1 = \frac{(2\pi R_B)^2}{45\lambda^2} f(\varepsilon_A - \varepsilon_B) \times [f^{2/3} - 3[(\varepsilon_A + 2\varepsilon_B) - f(\varepsilon_A - \varepsilon_B)]^{-1}] \qquad (5.28)$$

where $f = (R_A/R_B)^3$, R_A and R_B are average radius of particles and shell. Smith considered a special case in chrome-black solar absorbers with fine chromium particles inside. Table 5.1 gives the upper limits for R_B ($2R_B$ is considered to be the average particle separation) and the particle diameter.

Table 5.1 The upper limits for R_B and d at different wavelengths assuming that the both the real and imaginary parts of B_1 are negligible ($B_1 \ll 0.01$)

λ (μm)	Limits on R_B (nm)		Limits on particle diameter (nm)	
	$f=0.5$	$f=0.2$	$f=0.5$	$f=0.2$
0.35	26	56	41	66
1.0	64	105	100	120

2. Magnetic Dipole Effects

When metal particle sizes are larger than 10 nm or beyond the Smith relaxed limits, contributions of higher-order spherical multipoles from the single particles to the scattering and absorption spectra may dominate, i.e., the quasistatic approximation no longer applies and the EMTs break down. Stroud and Pan (1978) extended the EMA to finite frequencies by considering the full multipole expansion(5.12–5.15), especially the magnetic dipole contribution to the scattering from small particles (5.14). In this approach the particles no longer see just a static electric field as in the spirit of the EMTs but a dynamic electromagnetic field when an electromagnetic wave propagates through the composite. So the extra absorption resulted from eddy currents in the metal particles (equivalently induced magnetic dipole effects) must be taken into account. This approach is called the dynamic effective medium approximation (DEMA). Stroud and Pan found that the far-IR absorption by metal particles in an insulating composite could be dramatically enhanced by considering the induced magnetic dipole effects in the metal spheres.

   1) DEMA

By considering the magnetic dipole coefficient (5.14), the extended DEMA can be derived (Stroud and Pan, 1978):

$$f_A\left[\frac{\varepsilon_A - \varepsilon_{eff}}{\varepsilon_A + 2\varepsilon_{eff}} + \zeta(\varepsilon_A - \varepsilon_{eff})\right] + f_B\left[\frac{\varepsilon_B - \varepsilon_{eff}}{\varepsilon_B + 2\varepsilon_{eff}} + \zeta(\varepsilon_B - \varepsilon_{eff})\right] = 0 \qquad (5.29)$$

where $\zeta = \dfrac{\omega^2 R_A^2}{30c^2}$ (ω is angular frequence, R_A is particle radius). The equation is an improvement of the EMA of Bruggeman when the particle size effects are considered. If the magnetic dipole terms (in the brackets following the ζ) are neglected, the resultant equation is in agreement with the "static" EMA of Bruggeman.

2) DMGT (dynamic MGT)

By considering the magnetic dipole coefficient (5.19), the extended DMGT can be also derived (Chylek and Srivastava, 1983),

$$\frac{(\varepsilon_B - \varepsilon_{\text{eff}})(\varepsilon_A + 2\varepsilon_B) + f_A(2\varepsilon_B + \varepsilon_{\text{eff}})(\varepsilon_A - \varepsilon_B)}{(\varepsilon_B + 2\varepsilon_{\text{eff}})(\varepsilon_A + 2\varepsilon_B) + 2f_A(\varepsilon_B - \varepsilon_{\text{eff}})(\varepsilon_A - \varepsilon_B)} +$$
$$\zeta f_A \left[\frac{\varepsilon_B - \varepsilon_{\text{eff}}}{f_A^{5/3}} + \varepsilon_A - \varepsilon_B \right] = 0 \tag{5.30}$$

where $\zeta = \dfrac{1}{30}\dfrac{\omega^2 R_A^2}{c^2}$ (ω is angular frequence, R_A is particle radius). The equation is an improvement of the MGT when the particle size effects are considered. If the magnetic dipole term (in the bracket following the ζ) is neglected, the resultant equation is in agreement with the "static" MGT of Maxwell-Garnett.

3) Alexander and Bates (1984)extended the formulation of Stroud and Pan to include two types of coated particles found in cermets, i.e., isolated particles and their clusters, the following summation about the forward scattering amplitudes of the two types of particles $S_1(0)$ and $S_2(0)$:

$$f_1 S_1(0) + f_2 S_2(0) = 0 \tag{5.31}$$

where weighting factors f_1 and f_2 are volume fractions of the two types of particles separately, which depend on, and are related to, the probability of each of the grains forming through different processing. They applied the approach to Au-SiO$_2$ cermet and found that for larger Au particle sizes up to 50 nm radius the minimum in the optical transmission was shifted to the ultraviolet with a corresponding increase in the magnitude when magnetic dipole contributions are included.

3. Effects of Multiple Scattering from Small Particles

The EMTs are based on the well-isolated particle or dilute solution approximation, in which case it is assumed that the incident electromagnetic wave just undergoes a single scattering from one particle, i.e., the scattering waves from neighboring particles are independent. For nondilute samples of larger particles (finite-sized particles) or many-particle aggregates, this approximation is no longer valid. The multiple scattering of light by the particles must be considered. A series of new approaches have been developed to deal with this problem beyond the EMT. These theories include explicitly the direct electromagnetic multipole coupling between the particles. The model of Clippe (Clippe, Evrard and Lucas, 1976; Ausloos et al., 1978) was the first to deal with the aggregation effect on the infrared absorption spectrum of small ionic crystals by solving a dipole-dipole interaction Hamiltonian equation of motion. The importance of the electromagnetic interaction between particles was clearly shown in a series of experiments by Kreibig on Au particles in gelatin, where a novel process was developed to control the states of aggregation and make possible the optical absorption measurement uniquely determined by different degrees of aggregation (Schonauer and Kreibig, 1985).The most striking features of the interaction are to cause the single particle surface plasma peak to split up into several peaks and shift towards the IR while the absorption from the surface plasma of non-interacted particle drops by about two orders of magnitude, and the splitting energies depend on symmetry, the size of aggregates and the particle distances (Kreibig and Genzel, 1985). Based on the classic Maxwell electromagnetic theory and taking into account interactions of the multipole modes of the particles (dipole-dipole, dipole-quadrupole, quadrupole-quadrupole), Quinten (Quinten et al., 1985; Kreibig, 1989) give a satisfactory explanation of extinction spectra of Au particles in gelatin as shown in Fig.5.3. Further discussion of the effects of different cluster topologies on the splitting of the plasma peaks can also be found in (Kreibig, 1989).

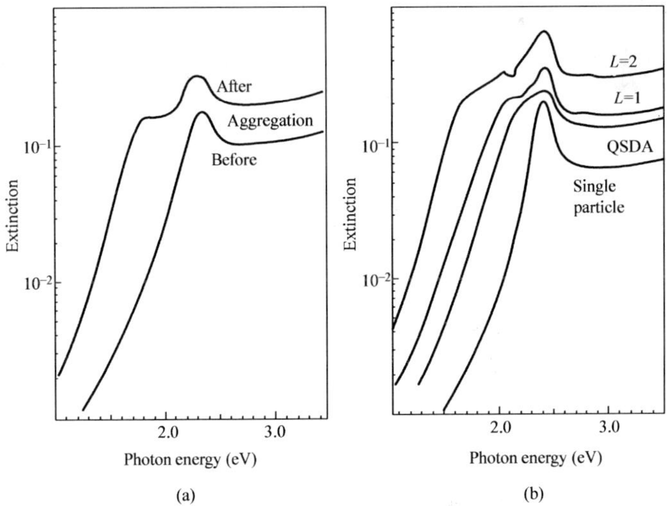

Fig.5.3 Extinction spectra of Au particles in gelation. (**a**) Measured spectra; (**b**) Spectra calculated with the quasistatic dipole approximation(QSDA), with dipole excitations and interactions ($L=1$), and including quadrupolar effects ($L=2$), respectively taken from Kreibig (1989)

5.3.4 The Bergman-Milton Theory of the "Bounds" on the Dielectric Constant of Inhomogeneous Media

The dielectric functions of inhomogeneous or two-phase composite materials are strongly microstructure dependent. The MGT and EMA gives only the ε of a special microstructural type, spherical particles. Supposing that not much information about the detailed microstructures is available, it is still possible to locate the regime where the ε falls, in other words, the allowed range of the ε is not arbitrary but has well-defined limits. We give here the most important formulas derived by Bergman-Milton (BM) to describe the bounds (Milton, 1980; Bergman, 1980). Their results are expressed as the bounds on the effective dielectric permeability (to a region of the complex plane) that depend on the amount of structural information that is at hand and which become narrower as more structural information is added.

Supposing the two phases are denoted as A and B, according to what is known about the composite, one of three cases is appropriate: (i) If we have no knowledge of the microgeometry except ε_A and ε_B of the two phases, then it can be stated that the ε is confined to a region $\Omega\,(\varepsilon_A, \varepsilon_B)$ of the complex plane; (ii) If we know the volume fractions f_A and $f_B=1-f_A$ of the components, then the ε is confined to a smaller region $\Omega'\,(\varepsilon_A, \varepsilon_B; f_A, f_B)$; (iii) If we know f_A and f_B and that the structure of the composite is statistically isotropic then the ε is confined to a still smaller region $\Omega''(\varepsilon_A, \varepsilon_B; f_A, f_B; d)$, here d is the dimensionality of the system (d=2 or 3). Figure 5.4 gives Milton's bound diagram taken from (Milton, 1980) in which

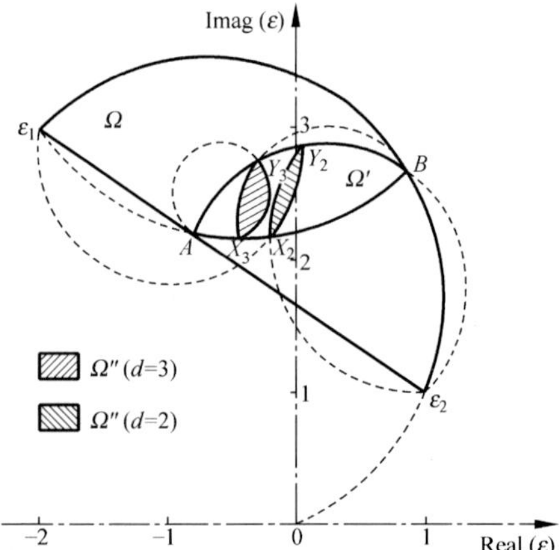

Fig.5.4 Construction of the bounds. Given ε_1, ε_2 and f_1, the points A, B, X_2, X_3, Y_2 and Y_3 are first plotted. Then a straight line is drawn joining ε_1 to ε_2, and arcs of circles are constructed as depicted above. The arc $\varepsilon_1 B \varepsilon_2$ and the straight line $\varepsilon_1 A \varepsilon_2$ bound Ω. The pair of arcs AY_3Y_2B and AX_3X_2B bound Ω''

$$A = f_A \varepsilon_A + f_B \varepsilon_B$$

$$B = \frac{1}{f_A / \varepsilon_A + f_B / \varepsilon_B}$$

$$X_d \equiv \varepsilon_A + \frac{d f_B \varepsilon_A (\varepsilon_B - \varepsilon_A)}{d\varepsilon_A + f_A (\varepsilon_B - \varepsilon_A)}$$

$$Y_d \equiv \varepsilon_B + \frac{d f_A \varepsilon_B (\varepsilon_A - \varepsilon_B)}{d\varepsilon_B + f_B (\varepsilon_A - \varepsilon_B)}$$

According to the definitions, the straight line A through ε_A, ε_B defines the lower bound of ε; the circular arc B define the upper bound of ε. It is noteworthy that when $d=3$, X_3, Y_3 actually represent the MGT and the EMA, respectively.

5.3.5 Optical Properties of Metals in the Near UV, Visible and Near IR Regions

5.3.5.1 Determination of the Dielectric Function of Small Metal Particles

Usually, as a simple approximation, the dielectric function of small metal particles can be presented by that of their bulk metal:

$$\varepsilon^{(f)} = 1 - \frac{\omega_{pf}^2}{\omega^2 + i\omega\gamma_b} \tag{5.32}$$

$$\varepsilon^{(b)} = \sum_j \frac{\omega_{pj}^2}{\omega_j^2 - \omega^2 - \omega\gamma_j} \tag{5.33}$$

$$\varepsilon = \varepsilon^{(f)} + \varepsilon^{(b)}$$

where $\varepsilon^{(f)}$ represents the dielectric function of the free-electron metal, i.e., the Drude model. $\varepsilon^{(b)}$ comes from the contribution of bound electrons which is deduced from the Lorentz oscillator model (Bohren and Huffman, 1983). $\gamma_b = 1/\gamma_b$ is the damping constant or collision frequency of the bulk conduction electrons, ε_b denotes an average time between collisions of free electrons. $\omega_{pf}^2 = \dfrac{Ne^2}{m\varepsilon_0}$ is square of the free electron plasma frequency, N is the free electron density, m is the effective electron mass, and ε_0 is the dielectric constant of free space.

The $\varepsilon^{(f)}$ is valid for free electron types of metal such as the alkali metals Li, Na, K, and Rb and multivalent metals such as Mg, Al, and Pb. For Ag and Cu, on the other hand, bound electrons strongly affect their optical properties, so the $\varepsilon^{(b)}$ is very important. Taking Ag for example, the $\varepsilon^{(b)}$ results from 4d→5s, 5p electron interband transition which determine its unique optical properties (Ehrenreich, 1962).

5.3.5.2 The Surface Plasma Resonance Absorption (PRA) on Small Metal Particles

Spectra of the absorption and the scattering of light by small particles of a solid have more than one resonance peak caused by several different mechanisms such as plasmas (free electrons), phonons (lattice vibrations) and excitons (electron-hole pairs). For small metal particles, the most interesting absorption peaks are those caused by plasma. The surface plasma resonance of small particles can be decided by the formula,

$$\varepsilon_{A1}(\omega) = -\left(\frac{m+1}{m}\right)\varepsilon_{B1}(\omega) \tag{5.34}$$

where ε_{A1} and ε are the real parts of dielectric functions of the metal particles and the surround dielectric. There are many possible PRA modes of different ordering(m). For the lowest-order mode, i.e., $m=1$,

$$\varepsilon_{A1}(\omega) = -2\varepsilon_{B1}(\omega) \tag{5.35}$$

is given. The frequency ε_F where this equation is satisfied is called the surface plasma resonance frequency or Frohlich frequency. For a metal particle /air system, this corresponds to $\varepsilon_{A1} = -2$ and for a free electron metal in air this give $\omega_F = \omega_{pf}/\sqrt{3}$. The position, the intensity and the width of the surface plasma depends on the shape, the size of the particles, and the dielectric constants of the surrounding medium. These effects were explained in detail in Mie theory (Bohren and Huffman, 1983; Papavassiliou, 1980). By controlling these factors, one may modify and even improve the optical properties of metal embedding composites. For example, it was found that the surface plasma of the Ag particles exhibited a significant "blueshift" when excess electrons were stored on the particles via free radicals produced by a microsecond pulse of highenergy radiation and the same researcher also found that the surface plasma was modified and even masked by coating a Cd metal overlay of different thickness on the Ag particles (Henglein, 1995).

The origin of the PRA comes from a collective dipole oscillation of induced charges on the spherical particle surfaces caused by a restoring force when an external electromagnetic field interacts with the metal particles. The restoring force originates from long-range correlation of conduction electrons caused by Coulomb forces (Kawabata and Kubo, 1966).

5.3.5.3 Limitation of the Mean Free Path(MFP) by Small Particle Boundary Scattering

When metal particles are smaller than the mean free path of conduction electrons in the bulk metal, the mean free path of conduction electrons in metal particles can be dominated by collisions with the particle boundary. Considering this effect, Kreibig (1974) suggested that the damping factor γ_e in the Drude model, which is the inverse of the collision time γ_e for conduction electrons, is increased because of additional collisions with the boundary of the particles. So γ_e can be modified as

$$\gamma_e = \gamma_{bulk} + 2\pi C \times \frac{v_F}{L} \tag{5.36}$$

where γ_{bulk} is the damping constant of the bulk metal, v_F is the electron velocity at the Fermi surface, and L is the effective mean free path for collisions with the boundary, Kreibig used $L=4R/3$ for a sphere of radius R. C is of the order of unity.

5.3.5.4 Optical Properties of Pure Metals

In order to select a metal/dielectric nanocomposite, it is first necessary to conduct a thorough appraisal of the dielectric permittivities of pure metals and dielectrics. A systematic literature survey was thus made of the optical properties of most elemental metals including alkali, alkaline earths, p-block metals, transition metals, noble metals, and group B metals. The optical constants of the most alkali, transition and noble metals can be found from Palik (1985 and 1993). The optical properties of some pure metals can also be found in Abeles (1966), but the data for other metals are scattered in the literature.

The optical properties of metals are determined primarily by the free electron density and fundamental absorption processes corresponding to the free electron intraband transition at low energies and the interband transition at a photon energy beyond a threshold. The interband transitions are of two types, as shown in Fig.5.5, either a transition from an occupied state on the Fermi surface of the conduction band to an unoccupied state (A), or a transition from a state in a fully occupied band to an unoccupied state on the Fermi surface (B). Above the critical photon energy threshold, transitions in which neither the initial nor the final states lie on the Fermi surface become possible (C) (Abeles, 1966).

1) Alkali metals (Na, K, Rb)

These are nearly free-electron metals and show behaviour predicted by the Drude model (5.32). However, they usually have low melting points(312-375 K) and are extremely reactive, so they are not suitable for co-sputtering experiments.

2) Alkaline earth metals (Mg, Ca, Sr, Ba)

These are also nearly free-electron metals and they show behavior predicted by the Drude theory. Figure 5.6a shows the imaginary and real parts of the dielectric function of Mg (Hagemanm, 1975). These metals are slightly less reactive and great care has to be taken to avoid oxidation in the sputtering and subsequent characterization.

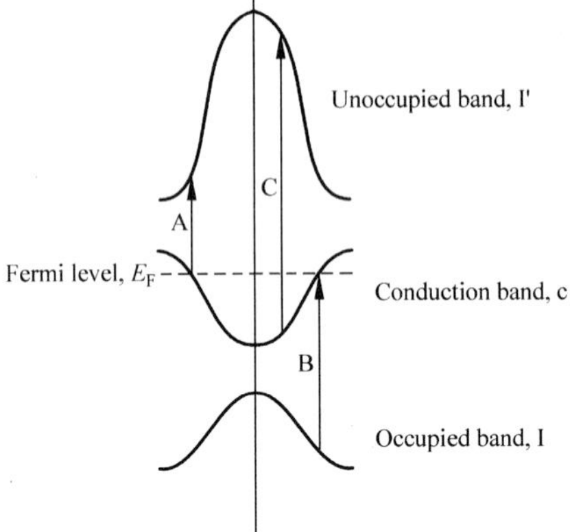

Fig.5.5 Various kinds of interband transition in metals

3) Group 3 metals (Al, In, Tl)

These trivalent metals also show pronounced nearly free-electron behavior. The only interband transitions occur at low energies (1–2 eV) that are associated with s-like and p-like bands. Figure 5.6b give the real (ε_1) and imaginary (ε_2) parts of the dielectric function of Al (Palik, 1985).The weak peak at 1.4 eV, which corresponds to an electron transition from an s-like band to p-like band, is only a slight perturbation to the free electron behavior because ε_1 is already quite negative at this energy (Ehrenreich, 1963a). In dium has optical properties similar to Al(Theye, 1969; Koyama, 1973), however, it has low melting point (430 K), thus it was proved by our experiments to be not suitable for sputtering.

4) Group 4 metals (Sn, Pb)

These polyvalent metals show broad or multiple absorption peaks in the visible region and near IR regions. The peaks probably originate from interband transitions associated with s-like band to p-like band transitions. It becomes difficult to divide the absorption spectra into a free electron part and an interband part as is customary for other metals, because interband transitions may occur at energies close to zero due to the possible crossing of two different bands at the Fermi level.Figure 5.7a gives the imaginary part of the dielectric function of Pb and optical absorption spectra, the broad peak (between 1.5 eV and 3 eV) comes from superimposed interband transitions. The existence of a pronounced interband transition at around 1 eV makes it impossible to separate the inter-and intraband contributions to the dielectric constant (Liljenvall, 1970; Hutter, 1995). Very little published data for Sn were found. Jezequel (1977) measured the reflectance spectra of white Sn, and several peaks in the reflectance were observed below 5 eV. Due to these complications Pb and Sn are not suitable for exploring the quantum-size effects in terms of interface plasma resonance absorption because no well-defined features are present.

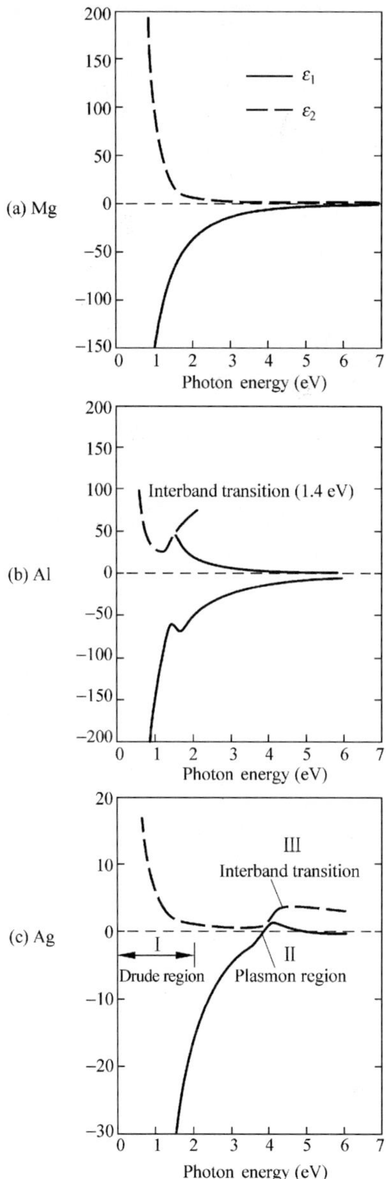

Fig.5.6 The dielectric functions of (**a**) Mg; (**b**) Al; (**c**) Ag

5) Noble metals (Cu, Ag, Au)

These metals show partially free-electron behavior. The bound electron or the interband transition associated with excitation from the d band to the Fermi surface makes a large contribution to the dielectric permittivity. The onset energies

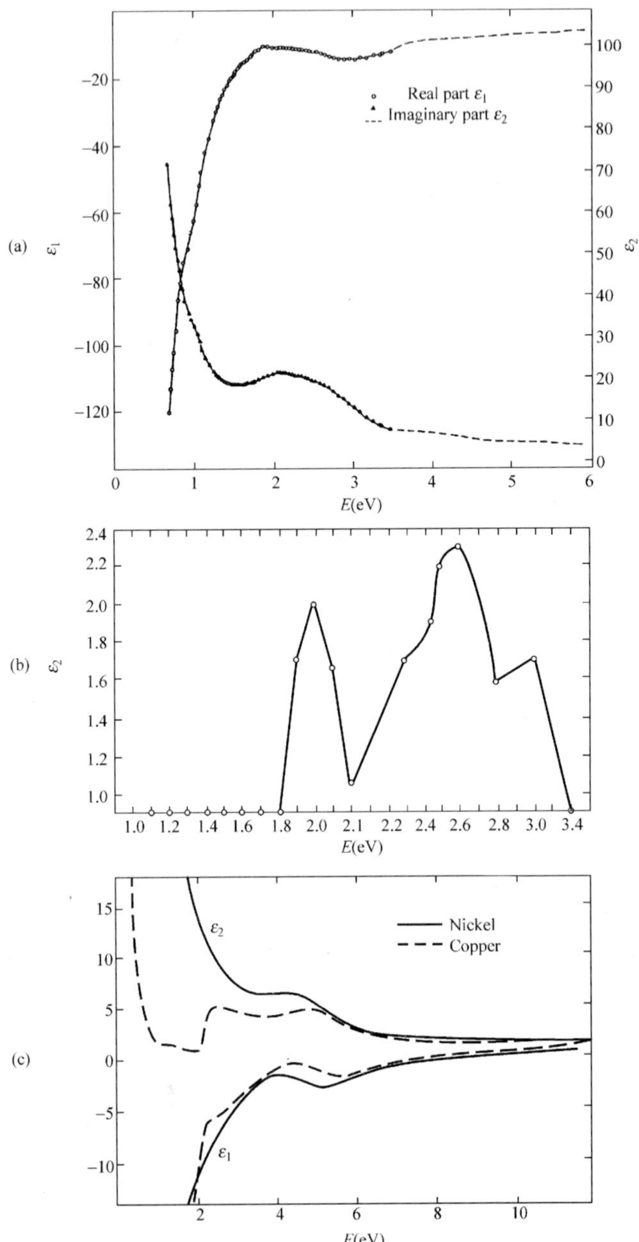

Fig.5.7 (**a**) The real part ε_1 and the imaginary part ε_2 of the dielectric function for Pb as a function of energy taken from (Liljenvall, 1970). (**b**) Relation between imaginary part ε_2 and photon energy for Zn film of thickness 75 nm. (**c**) Spectra dependence of real and imaginary parts of the dielectric function of Ni and Cu

for the interband transitions are in the range 2-4 eV (Ehrenreich and Philipp, 1962; Johnson and Christy, 1972; Cooper and Ehrenreich, 1965; Doremus, 1964; Doremus, 1965). Ag is a very interesting metal because of its unique $\varepsilon(\omega)$ behavior. Figure 5.6c shows the imaginary and real parts of the dielectric function of bulk Ag. The three different regions can be identified:

(I) Drude behavior region (energy <2 eV)

(II) Plasma region (3.8 eV) where the real part of the dielectric function ε_1 goes from negative values to zero and then increases. When $\varepsilon_1=0$ we have the condition for a volume plasma excitation. However, the strong 4d to Fermi interband transition also dominates the behavior just above the plasma resonance energy.

(III) Strong 4d-Fermi surface transition region. Because of this strong absorption, the "free electron plasma" is shifted from 9.2 eV to 3.8 eV. Also, because 4d electrons are already highly localized, it is difficult to observe quantum-confinement effects.

6) Group 2 B metals (Zn, Cd)

Their electronic structures are quite similar to those of the noble metals. The onset energies for the interband transitions seem to be lower that those of their neighbors Cu and Ag. Figure 5.7b shows the ε_2 of Zn (Abeles, 1966; El-Sahhar, 1987). So the bound electron behaviors seem to be dominant in the near UV and visible regions for Zn and Cd.

7) Transition metals

They have strong interband contributions in the near IR because the d band is close to the Fermi surfaces. Otherwise their optical properties are similar to those of noble metals (Abeles, 1966). Figure 5.7c shows the real and imaginary parts of the dielectric function of Ni in comparison with those for Cu (Ehrenreich, 1963a). The ε_2 for Ni is much higher than the one for Cu in the near IR. So they are not suitable for exploring the quantum size effects in terms of interface plasma resonance absorption because of the strong localization and not well-defined features from conduction electrons.

In this research, the focus has been put on Mg, which is an ideal free-electron metal, Al, which is a nearly-free-electron metal, and Ag which has a strong contribution from bound electrons and well-defined plasma features.

5.4   Electronic Transport Characteristics of Nanocomposites

5.4.1   Electronic Transport Properties in Amorphous Semiconductors

Amorphous materials process the intrinsic structure of the short-range order and the long-range disorder. Because of this, they remain in the fundamental energy bandgap and the concept of the density of states similar to crystallite materials, but there exist the localized states at the band edge or in the bandgap which are called the band tail states. An energy E_C separates the localized states at the band

184

edge from the nonlocalized ones (extended states), which is called a "mobility edge". Various models have been put forward for the density of states $N(E)$ for amorphous semiconductors in the bandgap as shown in Fig.5.8. Davis and Mott (1979) suggested that there are three mechanisms of conduction that may be expected to find in appropriate ranges of temperature in amorphous semiconductors.

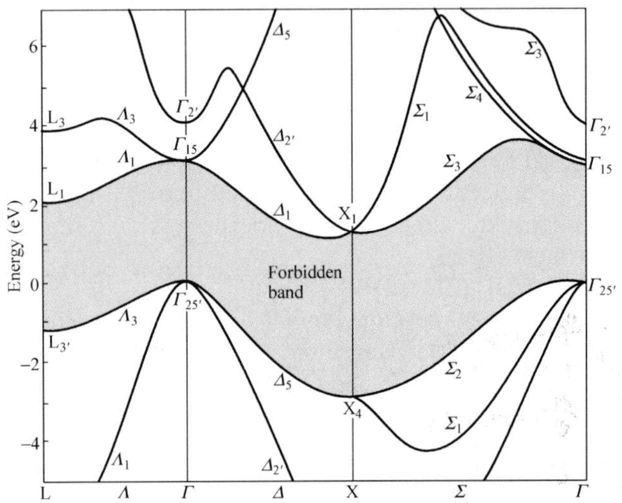

Fig.5.8   Electronic band structure of crystalline silicon taken from Davis and Mott (1979)

1) Transport by carriers excited beyond the mobility edges into non-localized (extended) states at E_C or E_V. The conductivity is (for electrons)

$$\sigma = \sigma_{min} \exp\left(-\frac{E_C - E_F}{k_B T}\right) \tag{5.37}$$

where σ_{min} is the minimum metallic conductivity and k_B is Boltzmann's constant.

2) Transport by carriers excited into localized states at the band edges and hopping at energies close to E_A or E_B. For this process, assuming again conduction by electrons,

$$\sigma = \sigma_1 \exp\left(-\frac{E_A - E_F + w_1}{k_B T}\right) \tag{5.38}$$

where w_1 is the activation energy for hopping.

3) If the density of states at E_F is finite, then there will be a contribution from carriers with energies near E_F that can hop between localized states.

$$\sigma = \sigma_2 \exp\left(-\frac{w_2}{k_B T}\right) \qquad (5.39)$$

where w_2 is the hopping energy, of the order of half the width of the band of states if the form of the density of states is as shown in Fig.5.8. At temperature such that $k_B T$ is less than the band gap, or if $N(E)$ is as shown in Fig.5.8, hopping will not be between nearest neighbors and variable-range hopping conductivity of the form,

$$\sigma = \sigma_2' \exp\left(-\frac{B}{T^{1/4}}\right) \qquad (5.40)$$

is to be expected at a temperature sufficiently low for $N(E_F)$ to be considered constant over an energy range k_B T. The various transport process are illustrated on a plot of $\ln \sigma$ versus $1/T$ in Fig.5.9. Davis and Mott indicated that if the density of defect states at E_F is high, then process (2) may not be dominant in any temperature range and a direct transition from process (1) to (3) will result.

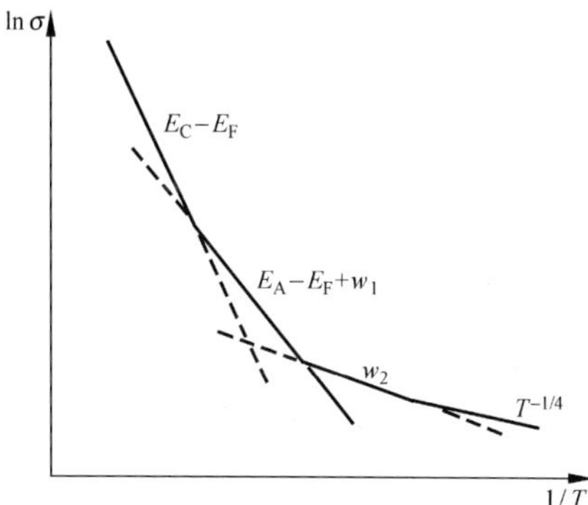

Fig.5.9 Illustration of the temperature dependence of conductivity and activation energies associated with various electronic transport process in amorphous semiconductors

5.4.2 Electronic Transport in Granular Metal Films(Cermet Films)

The interest in the electronic transport as well as optical properties in inhomogeneous media has greatly intensified in last two decades, in part because of basic scientific reasons but also because of the relevance of these materials to technology. As far as the electronic transport properties are concerned, most of the theoretical work has been concentrated on the improvement of the existing EMTs and percolation theory. Excellent reviews about the history and development of this subject can be found in Landauer (1978), Mclachlan (1990) and Sheng (1994).

186

Some common characteristics of the electronic transport properties of cermet films have been found over the last 2-3 decades (Abeles, 1976 Sheng, 1983 and 1994; Heinrich, 1992). Depending on the metal volume fraction f, there are three regimes of electronic conduction in cermet films as shown in Fig.5.10 measured in the Ni-SiO$_2$ cermet system. When f of Ni particles is large ($f > 0.54$), the metallic grains form a connected metallic network so that the electrons can percolate directly through connected metallic channels. This is denoted as the metallic regime. When f is small ($f < 0.54$), the metal grains form isolated dispersions in an insulating matrix. Electrical conduction in this dielectric regime is by the hopping mechanism in which the charge carriers are transported from one metal grain to another via thermally-activated tunneling. The transition between the metallic and the dielectric regimes is characterized by a percolation threshold f_c (0.54) at which the metallic network first becomes disconnected and the resistivity increases dramatically as f decreases further. The temperature coefficient of resistivity (θ) also changes from a positive value to a negative one as f goes down from above f_c to below f_c. In general the percolation threshold occurs at $0.4 < f_c < 0.6$.

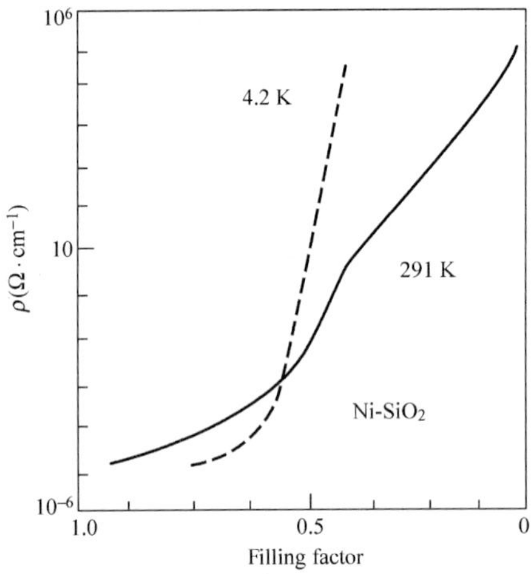

Fig.5.10   Resistivity as a function of filling factor of Ni in sputtered Ni-SiO$_2$

5.4.2.1   Conductivity in the Metallic Regime

Since the electrical conductivity in this regime is a direct result of electron transport through metallic channels, the microgeometry of the metallic channels plays an important role in the actual value of the conductivity. The conductance is determined by scattering by impurities, channel boundaries and grain boundaries. The effective-medium theory can be used to calculate such conductivity.

In principle, the effective electrical conductivity σ, i.e., DC conductivity and conductivity at high frequency can be derived by replacing the ε in the different EMTs as discussed in Sect. 5.3.3. If σ_A, σ_B are defined as the conductivities of particle (or high conductivity) phase and matrix (or low conductivity) phase, then the MGT is given by

$$\sigma = \sigma_B \frac{\sigma_A + 2\sigma_B + 2f_a(\sigma_A - \sigma_B)}{\sigma_A + 2\sigma_B - f(\sigma_A - \sigma_B)} \qquad (5.41)$$

and the EMA is given by

$$f_A \frac{\sigma_A - \sigma}{\sigma_A + 2\sigma} + f_B \frac{\sigma_B - \sigma}{\sigma_B + 2\sigma} = 0 \qquad (5.42)$$

In a cermet where σ_B=0, the MGT becomes

$$\frac{\sigma}{\sigma_A} = \frac{2f_A}{3 - f_A}$$

and the EMA becomes

$$\frac{\sigma}{\sigma_A} = \frac{3f_A - 1}{2}$$

which predicts a percolation threshold f_A=1/3.

Figure 5.11 shows the effective conductivity σ plotted as a function of metal volume fraction f_A by the MGT, EMA and SPT. Some discrepancies arise from these EMTs, e.g., the MGT gives no percolation threshold and the EMA does predict a percolation threshold (1/3) but somewhat different from experimental values. So further efforts has been made to improve the EMTs, i.e., the SPT (Sheng, 1980), the EMT with dipole-dipole interaction (Granqvist, 1978) and a generalized effective-medium theory (Mclachlan, 1988).

5.4.2.2 Conductivity in the Dielectric Regime

A general expression for ρ-T relationship may be written as

$$\rho(T) = \rho_0 \exp\left(\frac{A}{T^\alpha}\right) \qquad (5.43)$$

Measurements on a variety of cermets made by different workers have shown that the resistivity ρ in the dielectric regime has a characteristic temperature dependence given by α=1/2, which covers a large range of temperature

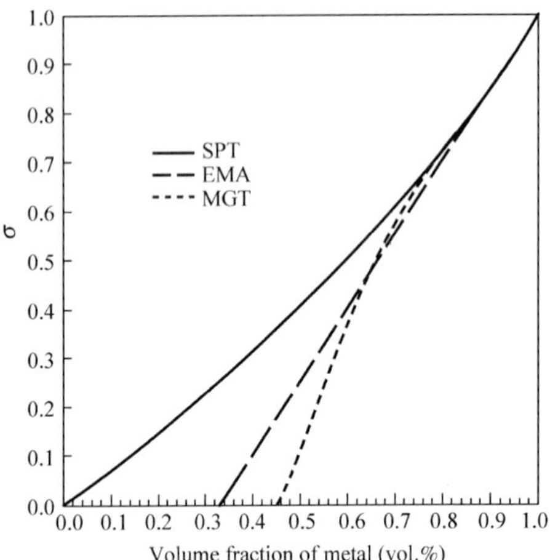

Fig.5.11 Normalized effective conductivity as a function of metal volume fraction calculated by the various EMTs

and resistivity (Abeles, 1976). Sheng and Klafter (1983) suggested that this is ascribed to an interpolation between the high-temperature thermally activated behavior ($\alpha > 1/2$) and the low-temperature behavior ($\alpha = 1/2$) which is the well-known variable range hopping between localized states in amorphous systems (Davis and Mott, 1979). The temperatures at which the crossovers occur depend on the distribution of the metal grain sizes.

The dominant electronic transport mechanism in the dielectric is believed to be the hopping of charge carriers between isolated metal particles. The hopping conduction is a term coined to denote the electrical conduction mechanism that is the result of two combined processes: thermal activation and tunneling. Mott (1968) was the first to show that these two nominally independently processes can actually couple to result in an apparently non-activated type of behavior.

In the Ping Sheng theory (Sheng, 1980; Sheng and Klafter, 1983), an important concept, the charging energy E_c was introduced. E_c is defined as the energy needed to remove or add a charge carrier from one neutral grain to another, $E_c = e^2 / KD$, where e is the electronic charge, D is the grain size, and K is the dielectric constant of the inhomogeneous system.

Consider the conductance G_{ij} between two grains, i and j. The G_{ij} may be expressed as the product of the probabilities for thermal activation and tunneling,

$$G_{ij} = G_0 \exp\left(-2\chi S_{ij} - \frac{E_{ij}}{k_B T} \right) \tag{5.44}$$

where χ is the tunneling parameter, S_{ij} is the tunneling distance between the two grains. $E_{ij}=1/2 \ (|E_i-E_j|+|E_i|+|E_j|)$ denotes a potential barrier between two charged grains that is produced by the charging energy. k_B denotes Boltzmann's constant. Physically, the fractional temperature dependence comes from the continuous shifting of the globally optimal conduction paths as the temperature is varied. At high temperatures, activation is easy, so the limiting factor is the tunneling distance. The optimal path is thus the one with the shortest tunneling distances between the successive sites, and the temperature dependence is expected to be $-\ln \ G_c \propto T^{-\alpha}$ where G_c is the percolation conductance (proportional to the macroscopic conductance of the sample), $1/2<\alpha<1$. As the temperature is lowered, activation becomes more of a limiting factor, and the tradeoff between a lower activation energy and a larger tunneling distance becomes favourable to the latter, and the temperature dependence is expected to be $-\ln G_c \propto T^{-1/4}$ which is the well-know "variable-range hopping".

5.4.3 Electronic Transport in Metal/Semiconductor Systems

In metal particle/insulating matrix systems, the matrix has a negligible contribution to the electrical conductivity, so when $f < f_c$, the dominant conduction mechanism is hopping between metal islands. However, in metal particle/semiconducting matrix systems, the matrix may contribute to the σ of the composite film. So different conduction mechanisms can co exist in these systems, such as amorphous semiconductor-like conductivity, polycrystalline semiconductor-like conductivity and even Schottky barrier-dominant conductivity.

5.4.3.1 Schottky Barrier (SB) Contact

When a metal is contacted with a semiconductor, a potential barrier arises from the separation of charges at the metal-semiconductor interface such that a high-resistance region devoid of mobile carriers is created in the semiconductor (Shoema, 1984). Figure 5.12a shows the electron energy band diagram of a metal of work function φ_m and an n-type semiconductor of work function φ_s, which is smaller than φ_m. E_F is the Fermi level. E_C is the edge of the semiconductor conduction band. χ_s is defined as the electron affinity. Figure 5.12b shows the energy band diagram after the contact is made and equilibrium has been reached. When the two substances are brought into intimate contact, electrons from the conduction band of the semiconductor which have higher energy than the metal electrons, flow into the metal until the Fermi level on the two sides is brought into coincidence. As a result, the conduction band edge E_C bends up, and an electric field is established from the semiconductor to the metal. From the semiconductor side, a contact potential difference, which an electron moving from the semiconductor into the metal has to surmount, is given by $qV_i=\varphi_m-\varphi_s$. However, the barrier looking from the metal side, is different and is given by $\varphi_B = \varphi_m-\varphi_s = qV_i+\varphi_n$. It is worth mentioning that the formulae are only suitable for an ideal SB junction. For a real BS junction some other factors should be considered such as the surface states.

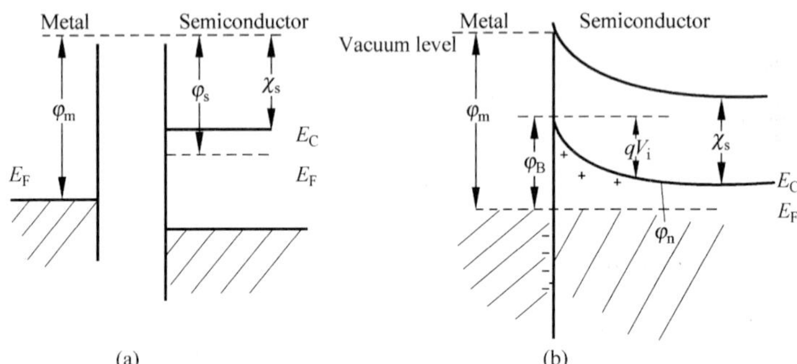

Fig.5.12 Schottky barrier contact formation. (**a**) before the contact is made; (**b**) after the contact is made

Once the Schottky barrier is established, the current flowing through the metal-semiconductor junction is determined by four different mechanisms: (a) thermionic emission over the barrier; (b) quantum tunneling through the barrier; (c) carrier recombination (or generation) in the depletion region; and (d) carrier recombination in the neutral region of the semiconductor, which is equivalent to the minority carrier injection (Sze, 1981).

It should be noted that all the above concepts and formula are established based on the bulk BS junctions, it is still uncertain if these well-established concepts can be extended to the nanosized junction at the interface between the nanosized metal particles and a semiconductor matrix.

5.5   Quantum Size Effects of Small Metal Particles

5.5.1   Introduction

It is well known that the conduction electron energy spectrum of a bulk metal is usually considered to be a continuum (the band theory). This is always an appropriate point of view at high temperatures and for an infinitely large system. However, for small metal particles having a small number of conduction electrons and a finite dimension, the electronic energy states are no longer continuous but are discrete. Consequently, all of the thermodynamic properties such as the specific heat and the magnetic susceptibility are significantly altered from their bulk values owing to the discrete character of the allowed electronic energy states. This is called the quantum size effect (QSE) or the quantum confinement effect of the conduction electrons (Perenboom and Wyder, 1981; Halperin, 1986). This idea was first discussed by Frohlich (1937), but it was not until Kubo and his collaborators (Kubo, 1962; Kawabata and Kubo, 1966; Kawabata, 1970; Kubo, 1984) independently addressed this problem that the subject developed widespread interest.

Kubo gave the average spacing of the successive quantum levels by

$$\Delta E = E_F/N \propto d^{-3} \text{(more precisely } 4E_F/3N)$$

where E_F is the Fermi energy and N is the number of valence electrons in the particles, which is proportional to the particle volume d^3.

Two conditions must be satisfied in order to observe possible QSEs: $k_B T$ must be less than ΔE and the lifetime τ of the electronic states must be much larger than $h/\Delta E$. For a silver particle with an electron density of $n=6 \times 10^{22}$ cm^{-3}, in order to have a level spacing of $\Delta E/k_B=1$ K, such a particle would have to have a diameter of $d=14$ nm. Figure 5.13 gives an average electron-level spacing as a function of particle diameter for a selection of metallic elements (Halperin, 1986). Kubo (1984) has pointed out that these discrete electron eigenstates will be broadened by the τ limitation, this broadening may possibly increase with decreasing particle size, QSEs should vanish in particles where the energy level broadening δE exceeds the mean spacing ΔE. Since ΔE depends on the particle size, a critical diameter d_c can be defined with:

$$d < d_c, \delta E < \Delta E \text{ (quantum size region);}$$
$$d < d_c, \delta E < \Delta E \text{ (quasicontinuous bands).}$$

The critical diameter will be a function of temperature and will depend on the particle preparation method because both δE and ΔE may be influenced by the form and the crystalline structure of the particles, by impurities, and also by the nature of the embedding matrix.

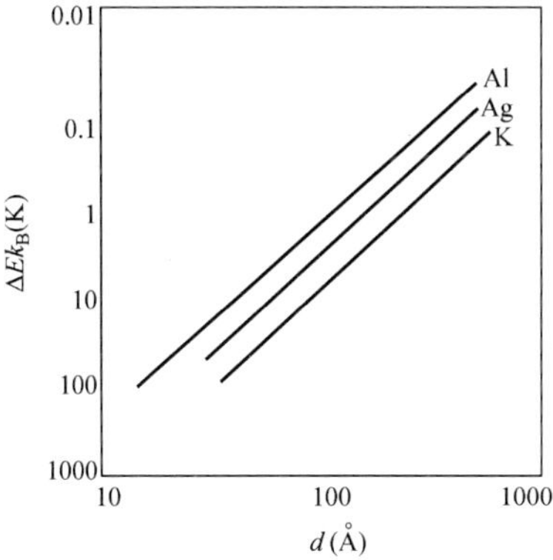

Fig.5.13 Average electron-level spacing as a function of particle diameter (d) for a selection of the metallic elements

If $d>d_c$, the conduction band is quasicontinuous as in the bulk materials, and a mean free path or a collision frequency may be used for calculating the optical and electrical properties as discussed in Section 5.3.5.3. Thus the electronic properties depend on particle size differently in the quantum size region and in the free-path limitation region (Kreibig, 1974).

For semiconductors and insulators, besides the discreteness of the energy band, reduction of dimensions also cause energy-gap widening. So the optical absorption edge shifts to shorter wavelengths when the particle size is reduced.

5.5.2 Quantum Size Effect (QSE) on the Optical Properties of Small Metal Particles

The Drude-like dielectric constant plus the modification of the particle boundary scattering of the valence electrons is usually called the classical size effect and it predicts a shift of the surface plasma resonance frequency ω_F towards longer wavelengths (redshift) and a broadening of the full width at half maximum(FWHM) when the particle size decreases. This prediction agrees with the experimental data for an average diameter of about 10 nm (Perenboom and Wyder, 1981). But apparent deviations from the experimental results appear when the particle size is smaller, especially smaller than 5 nm. (Genzel, 1975; Doremus, 1964; Doremus, 1965). This means that the classical interpretation of the surface plasma oscillation damping based on particle-like behavior is no longer valid. Kawabata and Kubo (1966) suggested a totally different quantum picture in which the particle boundary is not really a scatterer but does determine the eigenstates of electrons that are rather bound states in the finite volume of a sphere. Damping of the plasma oscillation is caused by transfer of its energy to excitation of the individual modes of electrons. Following this idea, many quantum-mechanical models were suggested aiming at calculating the dielectric constant of quantum size particles and predicting the size dependence of the PRA, such as Kawabata and Kubo's semiclassical theory (Kawabata and Kubo, 1966) based on current-current correlation functions, particle-in-a-cube

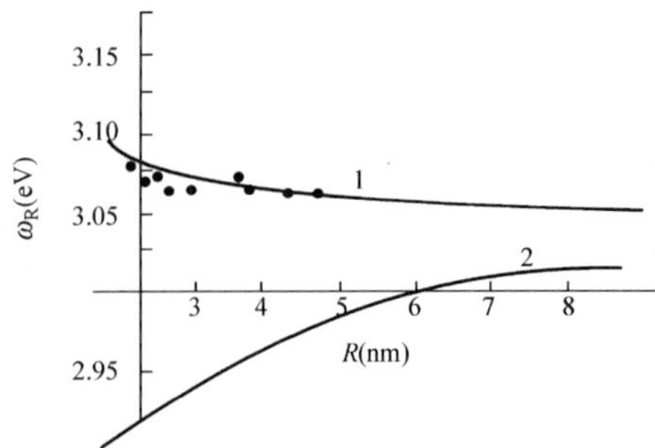

Fig.5.14 Surface PRA frequency from different calculations.Curve 1,theoretical simulation from the QSM;Curve 2, theoretical simulation from the classical size effect

models (Wood and Ashcroft, 1982), quantum sphere model (QSM) (Huang and Lue, 1994), etc. These approaches all predict a blueshift and FWHM sharpening of the PRA spectra. Of these models, the QSM seems to reveal a better agreement with experiments as shown in Fig.5.14. It is noteworthy that the shift of the surface plasma absorption wavelength towards both directions has been observed experimentally, i.e., Smithard (Smithard, 1974; Dupree and Smithard, 1972) reported an appreciable "redshift" on Ag particles in stained glass. Kreibig (1985) concluded that this confusing redshift and blueshift could originate from various other effects.

5.6 Preparation of Nanocomposite Thin Film by Co-Sputtering

Figure 5.15 shows a schematic diagram of the Nordiko RF sputtering system. The Ar ions produced by the low-pressure glow discharge (plasma) hit the target surface and cause some of the surface atoms to be knocked out of the surface and travel across the vacuum chamber and condense on the substrate. If the vacuum pressure is low, most of the atoms expelled from the target will collide with the gas molecules and return to the target or form a loose particulate deposit. The target is usually immersed in a plasma and a high negative voltage is applied to the target surface to attract positive ions.

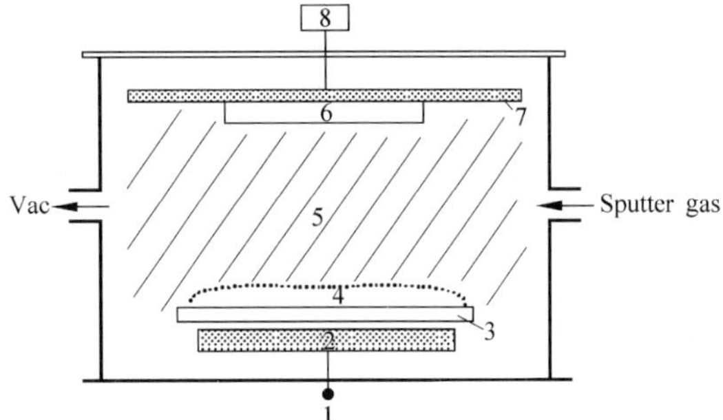

Fig.5.15 Schematic diagram of the Nordiko RF sputtering system
1-RF power supply, 2-Metal backing electrode, 3-Dielectric target, 4-RF plasma sheath, 5-Low-pressure gas discharge, 6-Substrate, 7-Work table, 8-Heater

In the DC sputtering system the target is composed of metal, since the glow discharge is maintained between the metallic electrodes. By simple substitution of an insulator for the metal target in the DC sputtering system, the sputtering discharge can not be sustained because of the immediate buildup of a surface charge of positive ions on the front side of the insulator. To sustain the glow discharge with the insulator target, an RF-voltage is supplied to the target. This is called RF-diode sputtering. In the Nordiko RF sputtering system, the target material may be either an electrical conductor or an insulator. An insulating target is normally bonded to a metal backing electrode that can be directly connected to an RF power supply as shown in Fig.5.15. In this case an RF voltage is induced on the front surface of the target by capacitive coupling through the target, so the thin films of the insulator can be sputtered directly from the insulator target (Wasa and Hayakawa, 1995).

When a voltage is applied to the target the luminous region of the plasma can be seen to be separated from the target surface to leave a "dark sheath". Most of the voltage applied to the target is dropped across this sheath, as shown in Fig.5.15.

Ar ions in the luminous parts part of the plasma move around with a random motion. Only those positive ions that enter the sheath region during their random motion are accelerated by the voltage across the sheath and bombard the target on each half cycle of the RF field.

Composite thin films and coatings of immiscible phases can be fabricated by conventional RF sputtering-deposition using co-sputtering from a single composite target or from different targets. However, it is difficult to avoid undesirable compound or alloy formation when the phases are thermodynamically miscible.

The Si-Ag composite thin films were manufactured in the Nordiko RF. co-sputtering apparatus. Composite targets were used consisting of 10 mm $\times$ 10 mm sheets of pure metal placed on top of a 203-mm diameter disc of 99.99% pure polycrystalline Si. Film compositions were roughly controlled by adjusting the area ratio of each element taking into account their relative sputtering yields. The sputtering yields of various elements with the variation of the incident ion energy can be found in (Maissel and Glang, 1970; Wasa and Hayakawa, 1995; Smith, 1995). The distribution of the metal sheets on the Si disc was made as uniform as possible in order to make the composition of the deposited films as homogeneous as possible and the film thickness as uniform as possible. The nominal metal content X_m (at %) in the cosputter-deposited composite film was roughly estimated by the following formula:

$$X_m = \frac{S_m Y_m}{S_m Y_m + (S - S_m) Y_T}$$

where S_m is the total area occupied by the metal sheets on top of the Si disk, S is the total area of the Si disk, Y_m (atoms/ion) and Y_T (atoms/ion) (T is Si or SiO_2) are the sputtering yields of the metal and polycrystalline Si or SiO_2 glass. The background pressure of the chamber was maintained below 10^{-6} torr, and sputtering was carried out in a 10^{-3} Torr Ar atmosphere at 250 W input power. Before commencing to deposit over a substrate it is usual to use the substrate shutter to cover the substrate and to sputter the target for 10–30 min. to remove unwanted surface layers.The substrate holder is water-cooled and the substrate electrode is also allowed to be heated to 200°C by proper electrical connections to a heater. The power source of the heater is a 6-A variable transformer connected to 240 V AC mains. The substrate temperatures are controlled by adjusting the transformer output. The substrate surface temperature-transformer output curve was precalibrated to give the correct temperature.

5.7 Microstructure, Optical and Electronic Transport Properties of Si-Ag Nanocomposite Thin Films

5.7.1 Introduction

To the author's knowledge, there have been no systematic studies on the optical and electronic transport properties and microstructure of Si-Ag composite thin films over the wide composition range from 0 to 60 at % Ag. The motives behind this research are: (1) It is well known that Ag particles exhibit a strong surface plasma resonance that covers the visible range of wavelengths depending on the size, shape and volume fraction of the particle and matrix materials (Aspnes, 1982; Bohren and Huffman, 1983). We expect to see interface-enhanced plasma resonance absorption (PRA) when the Ag particle size is small and also possible quantum-confinement effects associated with this PRA. Amorphous and crystalline Si are well-known and important semiconductors with many applications in semiconductor devices (INSPEC, 1985; INSPEC, 1988), e.g., they have dielectric permittivities higher than those of oxide insulators and this property may considerably shift the interfacial PRAs in metal-Si composites to lower energies and thus offer better chances to observe quantum-confinement effects. (2) These effects may be used to enhance the optical properties in the visible and IR of the matrix in which the Ag particles are embedded. (3) We are also interested in the possible existence of a large number of Schottky barrier interfaces due to high surface-volume ratio when the metal particle size is very small and this may give interesting and unusual visible and IR photoconduction behavior and nonlinear *I-V* characteristics (Sze, 1981). Si has a higher carrier mobility than oxide insulators and this could improve the efficiency of photoconduction and electronic transport. By measuring the relationships between (a) resistivity and metal volume fraction and (b) resistivity and temperature, it is hoped that the dominant electrical conduction mechanism in the metal particle/ semiconductor matrix composites may be clarified.

The Si-Ag phase diagram shows a simple eutectic system with a eutectic around 835°C and 89 at % Ag. The mutual solubilities of Si and Ag in the solid state are negligible. Two metastable phases are reported on the Ag-rich side of the phase diagram: Ag_2Si and Ag_3Si (β) obtained in rapidly solidified alloys at cooling rates above 10^5°C/s by melt spinning (Massalski, 1990). Recently, the

Ag$_3$Si metastable phase was also found in RF sputtered Ag$_{1-x}$Si$_x$ alloys when deposited on a liquid-cooled substrate (Leedy and Rigsbee, 1996).

5.7.2 Determination of the Film Composition, Thickness and Mean Ag Particle Size

Energy dispersive X-ray spectrometry (EDX) was used to determine the film compositions and surface profiling (SP), and cross-sectional TEM (CSTEM) techniques were used to measure the film thicknesses. The mean Ag particle size was estimated roughly by the broadening of the Ag(111) diffraction peak in X-ray diffraction(XRD). More accurate results were obtained as listed in Table 5.2 by counting and measuring many Ag particles on HREM micrographs and determining the mean particle sizes directly. The Ag particle volume fraction of each film was calculated from the measured film composition (EDX results) taking into account the mass densities of Ag and a-Si, assuming that Ag nanosized particles have the same density as bulk crystalline Ag. This assumption is reasonable because all the measured lattice parameters from Ag nanosized particles by XRD were close to those of the bulk crystalline Ag. The density of a-Si depends very much on the preparation conditions and it can be anywhere in the range 70%-100% of that of crystalline Si (INSPEC, 1988). Therefore, our estimates of the Ag volume fraction could be in error, but the trend with Ag composition is the more important issue in the present study. Tables 5.2 and 5.3 give the measured film composition, film thickness from the different methods, deduced Ag particle volume fraction and mean Ag particle size. All compositions cited in the later text refer to EDX results.

Table 5.2 Sample IDs, nominal and measured film compositions, film thicknesses, mean Ag particle sizes and Ag particle volume fractions from the co-sputtered Si-Ag nanocomposite thin films

Sample ID	Nominal composition (at % Ag)	Film thickness (nm)	Mean Ag particle size (nm)	Ag particle volume fraction (vol %+error %)
R0	0	60	0.0	0
R5	7	40	0	0
R7	15	36	0	0
R10	18	32	2.8	13$\pm$3
R15	19	45	3.0	15$\pm$3
R20	27	60	3.2	21$\pm$4
R30	30	60	3.4	23$\pm$6
R40	40	60	4.5	30$\pm$5
R60	61	45	6.0	51$\pm$6

Table 5.3 Sample IDs, measured film compositions and thicknesses from the co-sputtered Si-Ag nanocomposite thick films

Sample ID	Composition (at %)		Thickness (nm)	
	EDX	SP		CSTEM
S15	18	860		
S20	24	890		
S25	33	630		
S30	42	764		
S35	47	655		515

5.7.3 Microstructural Characterization

5.7.3.1 X-ray Diffraction (XRD)

A series of XRD spectra were collected from the as-deposited thick films containing a range of Ag contents from 18(S15), 24(S20), 33(S25), 42(S30) and 47(S35) at % as shown in Fig.5.16. Several important features were observed. First the broad peaks were found to belong to Ag(111) diffraction and the B values (the broadening of diffraction peak measured at half its maximum intensity (Cullity, 1956), became smaller with increasing Ag content, which suggests that the average Ag particle sizes from the different films increased upon increasing Ag content. Secondly, though the diffraction angles of the Ag(111) diffraction peaks from all the thick films are quite close to that of bulk Ag

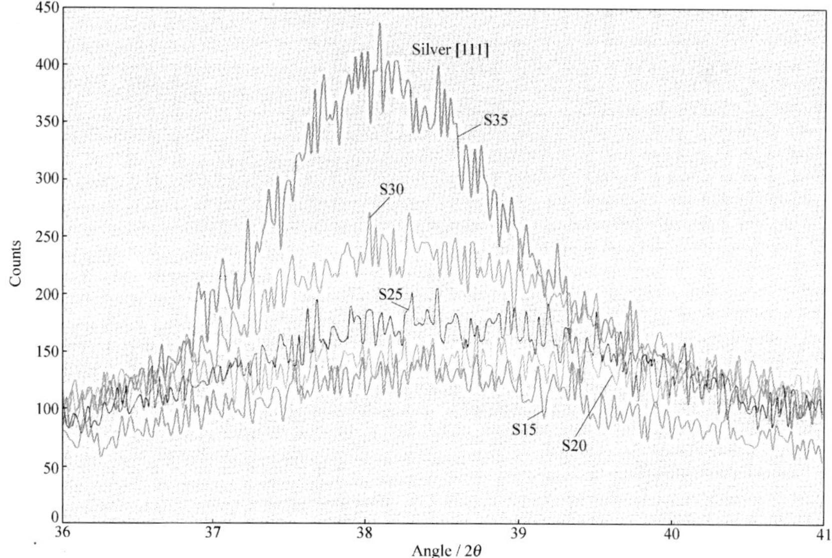

Fig.5.16 XRD spectra from the as-deposited Si-Ag nanocomposite thick films containing nominal Ag contents of 15(S15), 20(S20), 25(S25), 30(S30), and 35(S35) at %

198

$(2\theta = 38.12°)$ (JCPDS), they showed a trend to shift to the higher diffraction angles with respect to $2\theta = 38.12°$ with decreasing Ag content, which suggests a particle-size dependent lattice contraction possibly resulting from the elastic contracted stress inside Ag particles, i.e., the smaller the particle size, the greater the internal stress.

5.7.3.2 Conventional TEM Results

Each of the as-deposited thin films containing different Ag content from 0 at % to 61 at % was examined in the conventional TEM. Figure 5.17 shows the bright-field (BF) micrograph, centered dark-field (CDF) micrograph and corresponding selected area diffraction pattern (SADPs) taken from each of the films. It was found that depending on the Ag content in the films, rather different morphology was observed, i.e., Fig.5.17a and b, taken from a pure amorphous Si

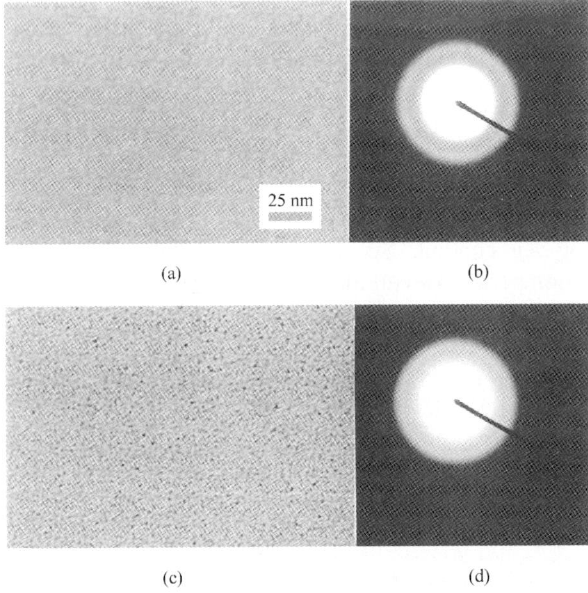

(a)　　　　　　　　　　(b)

(c)　　　　　　　　　　(d)

Fig.5.17　BF, CDF electron micrographs and corresponding SADPs from the as-deposited Si-Ag nanocomposite thin films containing a range of Ag contents from 0 to 61 at %. (**a**) and (**b**) BF micrograph and corresponding SADP from pure amorphous-Si; (**c**) and (**d**) BF micrograph and correspo ding SADP from the Si-7 at % Ag thin film; (**e**) and (**f**) BF micrograph and corresponding SADP from the Si-18 at %Ag thin film; (**g**) and (**h**) BF micrograph and corresponding SADP from the Si-27 at %Ag thin film; (**i**), (**j**) and (**k**) BF micrograph,CDF micrograph by using a portion of Ag(220)diffraction ring and corresponding SADP from the Si-30 at %Ag thin film; (**l**), (**m**) and (**n**) BF micrograph,CDF micrograph by using a portion of Ag (111) diffraction ring and corresponding SADP from the Si-40 at %Ag thin flim; (**o**) and (**p**) BF micrograph and corresponding SADP from the Si-61 at %Ag thin film

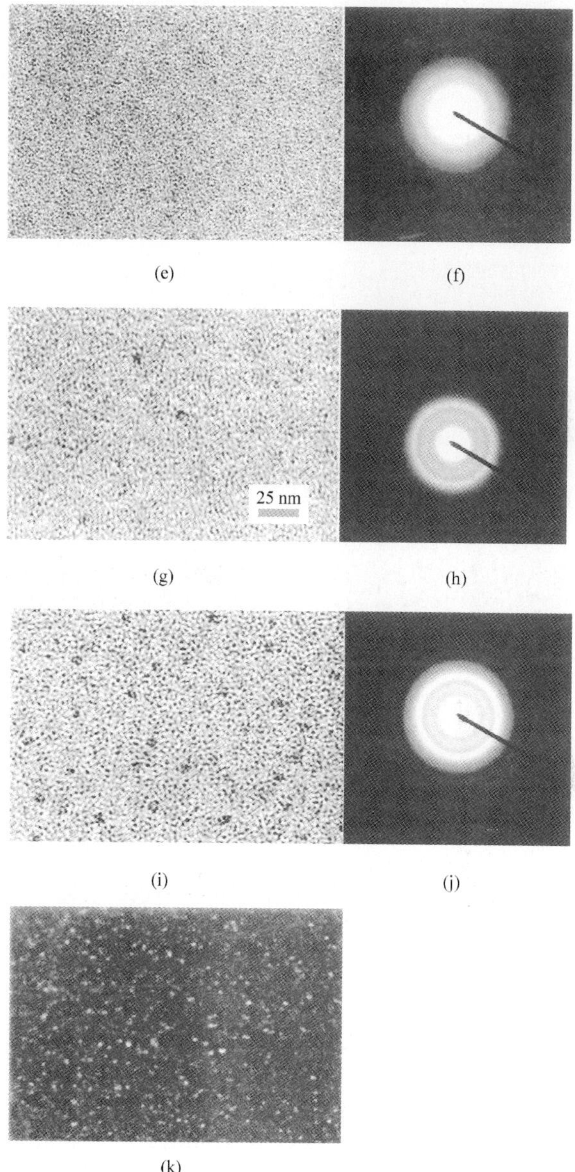

Fig.5.17 Continued

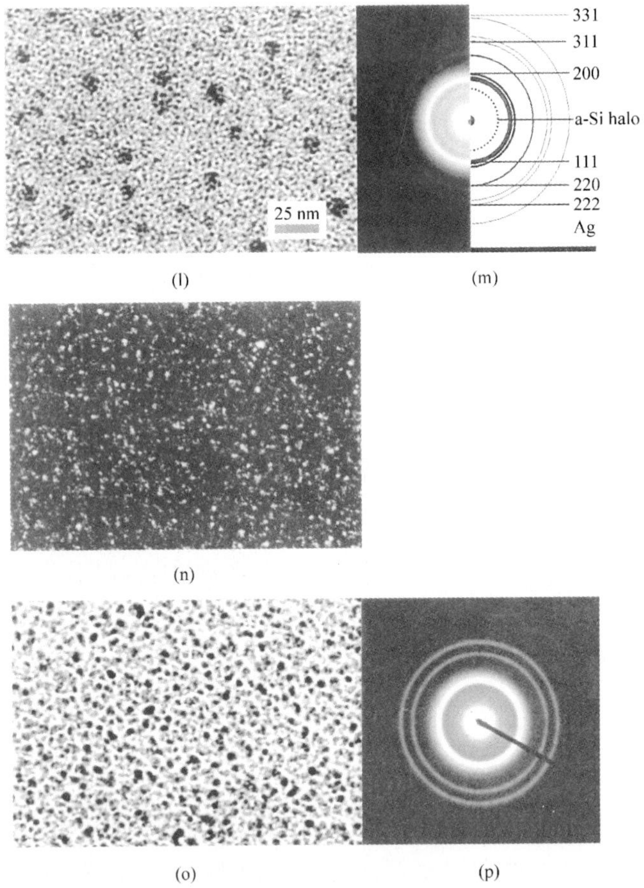

Fig.5.17 Continued

thin film, show typical uniform contrast and amorphous halos in amorphous materials. Figures 5.17c and d, taken from the Si-7 at % Ag thin film, show weak contrast and halo patterns, which suggests that the film was probably still amorphous. Figures 5.17e and f, taken from the Si-18 at %Ag thin film, show well-separated dark spots and an extra different ring superimposed on the amorphous halos. The spots grew in size and the extra diffraction ring increased in intensity when the Ag contents increased from 27 at %, 30 at % to 40 at % as shown in Figs.5.17g, h, i, j, l and m. The extra diffraction rings were indexed to belong to polycrystalline Ag(111), (220), etc. as shown in Fig.5.17h which proved that the film microstructure consisted of small Ag particles (dark spots) and an a-Si matrix (gray background). The uniform intensity of the Ag (111) rings was unchanged even when tilting the films in the TEM, indicating that there was no preferred orientation in the Ag particles. Figures 5.17k and n are

CDF micrographs corresponding to Figs.5.17i and l taken by using a portion of the Ag (111) and (220) rings, respectively. The bright spots further confirmed the presence of the small Ag particles.

Furthermore, some of the well-separated Ag particles as shown in Fig.5.17e began to form clusters, as shown in Fig.5.17g and the clusters became bigger when the Ag contents in the films increased. Figures.5.17o and p, taken from the Si-61 at % Ag thin film, showed larger Ag particles, somewhat interconnected.

Figures 5.18a and b show TEM cross-section (CSTEM) micrographs from the Si-18 and 40 at % Ag thin films, respectively, the parallel line at the film/Si substrate interface is a Fresnel fringe. For both films, the Ag particles were found to be randomly dispersed in the films without special arrangements such as alignment in arrays or columns.

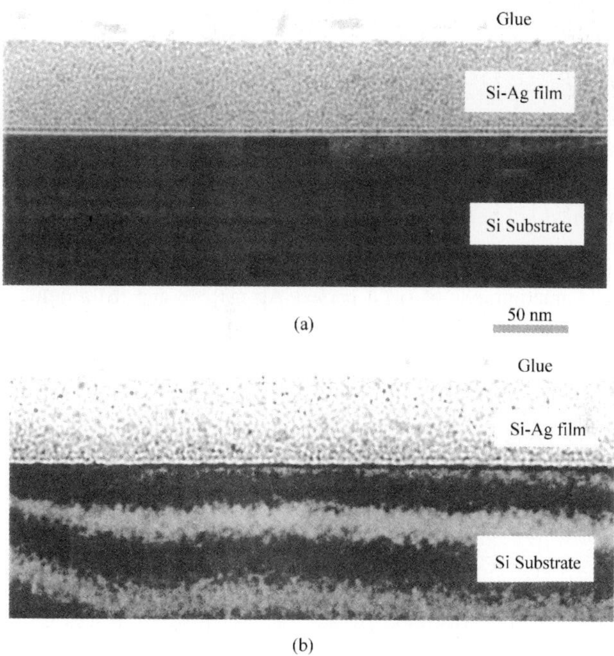

Fig.5.18　TEM cross-sectional micrographs from (**a**) the Si-18 at % Ag thin film; (**b**) the Si-40 at % Ag thin film

5.7.3.3　HREM Results

In addition, due to the poor resolution limit of the CTEM, the high resolution TEM technique was also used to examine the internal structure of the individual spots shown above. Figure 5.19a shows a HREM micrograph of a typical perfect Ag particle from the Si-18 at % Ag thin film. Most of the Ag particles had this spherical shape with good internal crystallinity as shown by the Ag(111) lattice fringes. Figure 5.19b shows a Ag particle from the same film with defects that are probably multiple twins or stacking faults. Figures 5.20a and b show HREM micrographs of the Ag particle distributions from the Si-18 at % Ag and 40 at % Ag thin films, respectively. The mean Ag particle sizes were emeasured accurately from these micrographs by counting many particles and then calculalely the average. Furthermore the two types of structural morphology, i.e., Ag particle clusters and separated Ag particles are clearly seen in Fig.5.20b.

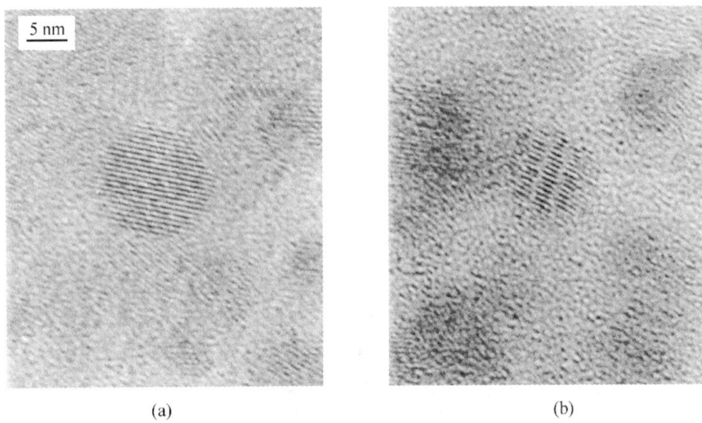

(a)   (b)

Fig.5.19 HREM micrographs of (**a**) a perfect Ag particle and (**b**) a defective Ag particle from the Si-18 at % Ag thin film

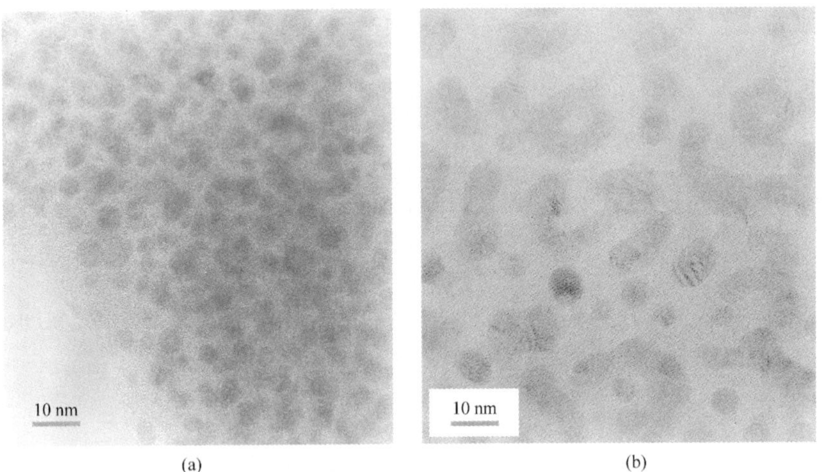

(a)   (b)

Fig.5.20 HREM micrographs from (**a**) the Si-18 at % Ag thin film; (**b**) the Si-40 at % Ag thin film

5.7.4 The Effect of Substrate Temperature

In order to investigate the influence of the substrate temperature on and consequently control, the film microstructure and the optical and electrical properties of the films, the thin film deposition was carried out on the substrate holder preheated to a certain temperature. Figure 5.21 shows BF and CDF micrographs and corresponding SADPs from the films with 40 at % Ag and 600 nm in thickness deposited at 20°C and 200°C, respectively, on the carbon-coated copper grid substrates. The film deposited at 20°C as shown in Figs.5.21a, c and e contained finer Ag particles with an average size around 4.5 nm and many large Ag particle clusters (the larger dots in Fig.5.21a). The film deposited at 200°C as shown in Figs.5.21b, d and f consisted of larger Ag particles, somewhat interconnected with a average size around 8 nm and some aggregated Ag particles on the film surface. The discontinuous diffraction spots on the SADP, as shown in Fig.5.21f confirmed the same trend towards increasing the Ag particle size when the substrate temperature was raised. The matrices, however, at both substrate temperatures remained a-Si, as indicated by the halos superimposed on the Ag rings on the SADPs, as seen in Figs.5.21e and f. It is therefore proved that the Ag particle sizes in the Si-Ag nanocomposite films can be controlled by adjusting the substrate temperature.

5.7.5 Optical Properties in the Near UV, Visible and Near IR

5.7.5.1 Optical Absorption in the Near UV, Visible and Near IR

Figure 5.22 shows optical absorption spectra from Si-Ag composite thin films with different Ag compositions over the wavelength region from 200 nm to 1500 nm. The spectra from films up to 40 at % Ag showed features similar to that of pure a-Si except that the addition of Ag reduced the absorption in the wavelength region shorter than 440 nm and enhanced the absorption in the wavelength region longer than 440 nm. The spectrum from the Si-61 at % Ag film, however, showed two additional absorption maxima at around 350 nm (P1) and 700 nm (P2).

5.7.5.2 The Effects of Substrate Temperature

Figure 5.23 shows a measured pair of absorption spectra from the Si-40 at % Ag thin films deposited on room temperature and 200°C substrates, respectively. The later spectrum showed a broad extra absorption peak around 700 nm by comparison with the spectrum from the former film.

5.7.5.3 Determination of Optical Constants of the Thick Film at 632.8 nm

A study of the effect of the film composition on optical constant, i.e., refractive index (n) and absorption index (κ) of the thick films deposited on Si wafer substrates was made by using spectroscopic ellipsometry. Figure 5.24 shows a plot

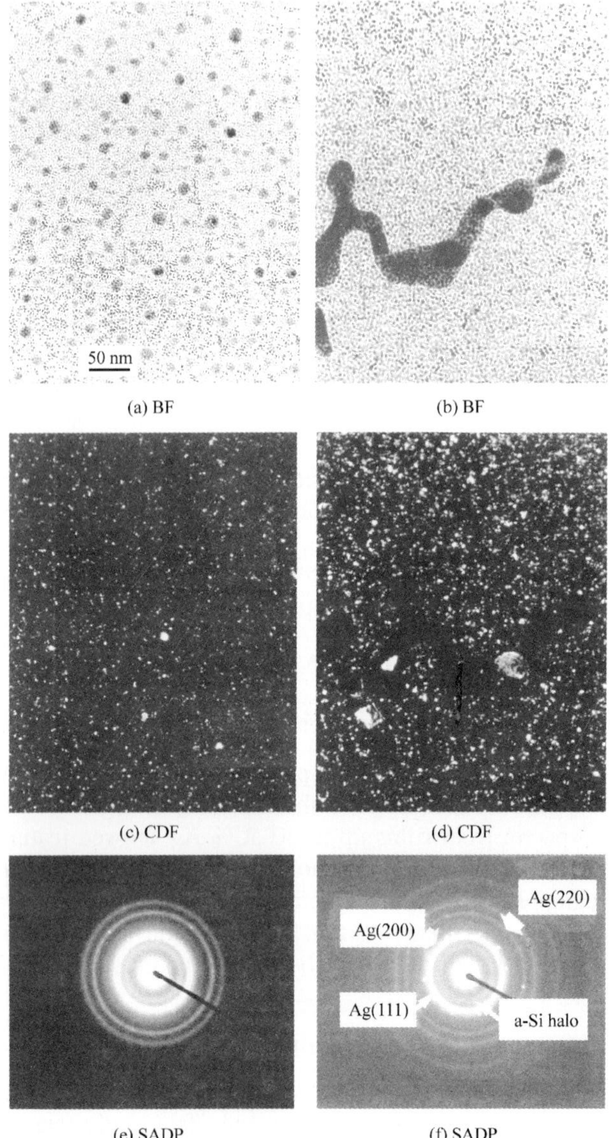

Fig.5.21 BF and CDF electron micrographs and corresponding SADPs from the Si-40 at %
Ag thin film deposited at different substrate temperatures (**a**), (**c**) and (**e**) 20 ℃; (**b**),
(**d**) and (**f**) 200 ℃

of n and κ against Ag concentration. Three n and three κ values in each
composition were presented that correspond to the measured maxima, minima
and mean values of n and κ from many measurements. The fact that κ is smaller

than n suggests that the non-metallic optical properties are dominant in the composition range used. But the trend that n decreased and κ increased and their difference got closer as the Ag content increased suggests a transition of optical properties from non-metallic to metallic at higher Ag contents.

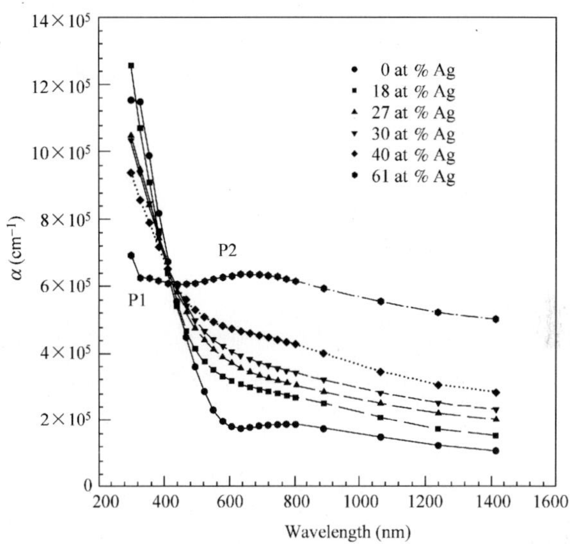

Fig.5.22 Measured absorption coefficients from Si-Ag nanocomposite thin films containing different Ag contents

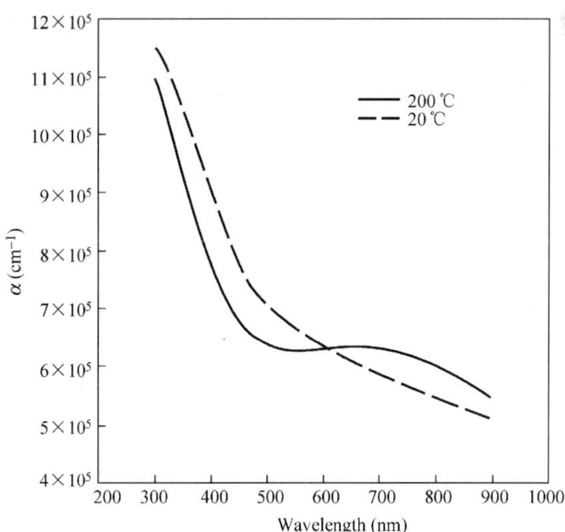

Fig.5.23 Measured absorption coefficients from a Si-Ag nanocomposite thin film containing 37 at % Ag deposited at different substrate temperatures

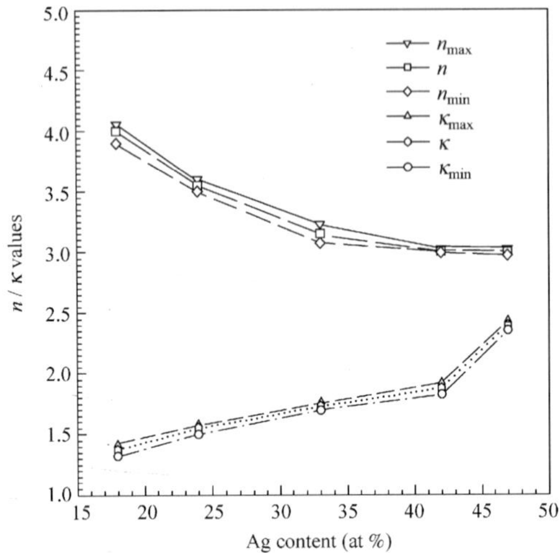

Fig.5.24 Measured refractive index (*n*) and absorption index (*κ*) of the Si-Ag nanocomposite thin films as a function of Ag content at wavelength of 632.8 nm

5.7.6 Electronic Transport Properties

5.7.6.1 Sheet Resistance of the Thin Films as a Function of Ag Content

Sheet resistance of each of the as-deposited Si-Ag thin films containing different Ag contents deposited on both the microscope glass slide and quartz substrates was measured by the four-point probe method and then converted to the resistivity. A question arose concerning the relative role of contact resistance and bulk resistance in the two-contact method. It is believed that the bulk resistance dominates the *I-V* measurements for the following reasons: (1) The *I-V* curves shown on an *I-V* curve tracer were fairly linear, which suggests an ohmic contact; (2) The measured results by the two contact methods as shown in the table were close to the ones on the same films by the four-probe contact method where the contact resistance is minimized, thus it is concluded that the contact resistance by the two-contact method was negligible. Figure 5.25 shows the measured log resistivity of the Si-Ag thin films with variation of Ag content at room temperature on both glass and quartz substrates. In general, the resistivity of the films decreased by almost 5 orders of magnitude with increasing Ag content from 0 at % to 61 at %. The resistivity of the a-Si film was beyond the sensitivity of the facility used, but estimated to be greater than 3 orders of magnitude, which is comparable with published data 3 of to 9 orders of magnitude in similar sputtering conditions (INSPEC, 1985).

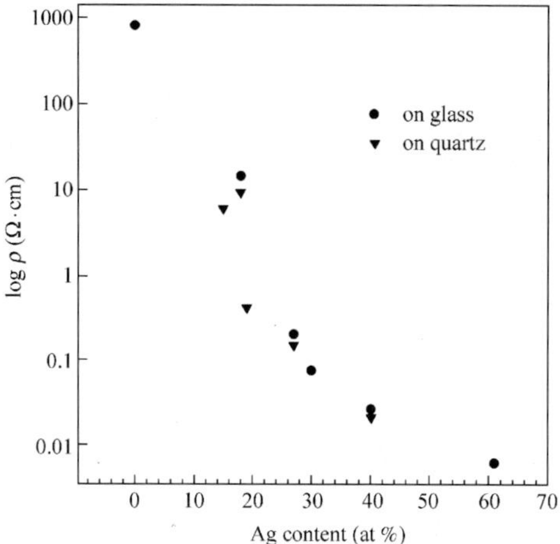

Fig.5.25 Resistivity of the Si-Ag nanocomposite thin films on different substrates as a function of the Ag content

5.7.6.2 Sheet Resistance as a Function of Temperature

In order to explore possible electric transport mechanisms that dominate in the Si-Ag nanocomposite films, the dependence of resistivity on temperature ranging from 373 K to 100 K was measured. The least squares nonlinear curve fitting method, based on the Levenberg-Marquardt method (Press, 1988), was used to fit the experimental results, assuming that the film resistivity follows the formula (5.43), i.e.,

$$\rho(T) = \rho_0 \exp\left(\frac{A}{T^a}\right) \tag{5.45}$$

This method uses a least-squares procedure to minimize the sum of the squares of the differences between the equation values and the measured data and to find the parameters that cause the equation to best fit the data.

Figure 5.26 shows all the fitting curves plotted by taking different α values from 0.25, 0.5 to 1 in comparison with the measured curve. Figures 5.26a, b and c correspond to the Si-27, 30 and 40 at% Ag thin films, respectively. The best fit curve is always the one by taking α=0.25. Figure 5.27 shows log resistivity-temperature ($T^{-1/4}$) curves for the thin films containing 27, 30, 40 and 61 at % Ag . All curves showed a "kink" around 100 K indicated by arrows. Above this temperature, all the curves were near linear with negative temperature coefficients of resistivity (θ) corresponding to the slopes of the straight lines, indicating a non-metallic electric transport property. The θ gradually decreased when the Ag content increased, which showed a tendency towards a metallic behavior.

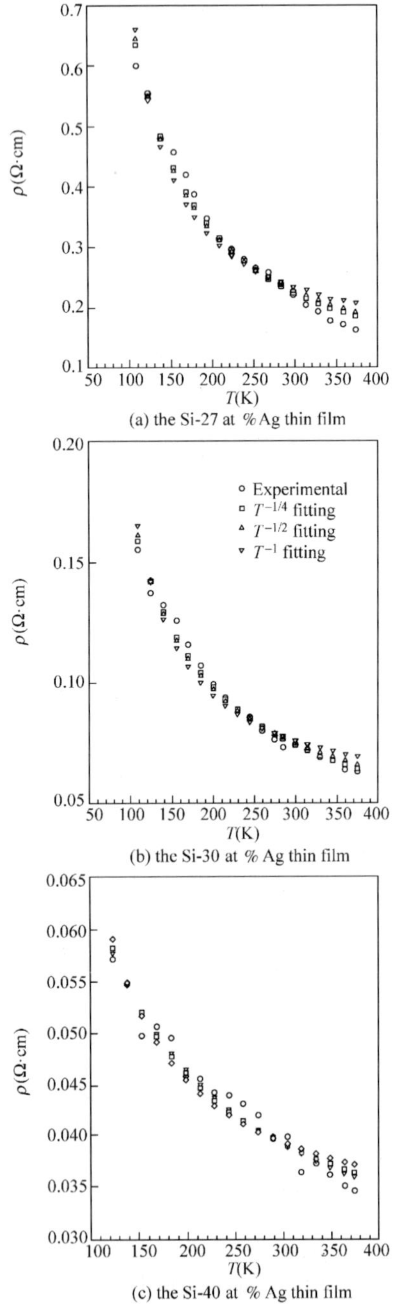

Fig.5.26 The best-fitting results of the temperature dependence relation of the resistivity for the Si-Ag nanocomposite thin films containing different Ag contents

The Si-61 at % Ag film shows a θ very close to zero, consistent with the percolation threshold, as shown in Fig.5.10. The fitting results in Fig.5.26 actually correspond to the temperature range 373-100 K. The "kinks" below 100K probably came from condensation of residual water or other impurities in the measurement structures.

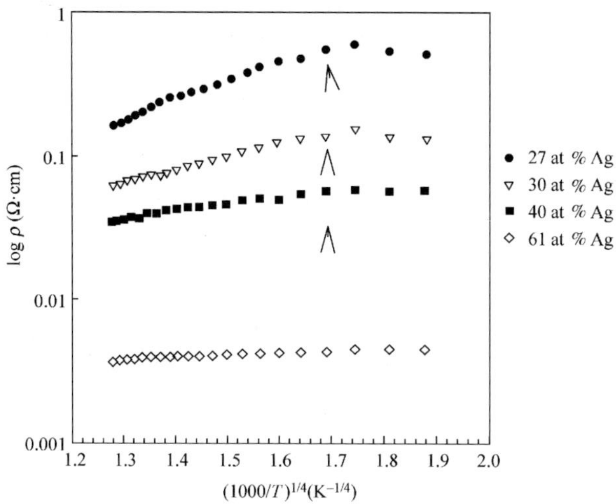

Fig.5.27 Log resistivity of the Si-Ag nanocomposite thin films with different Ag contents as a function of temperature (the arrows indicates kinks)

5.7.7 Discussion

5.7.7.1 Simulations of Optical Absorption by the Different EMTs

It is realized that the Si-Ag nanaocomposite thin films provide good test materials for the applicability of the various EMTs in general metal-semiconductor composite systems. The commonly used effective-medium theories like the Maxwell-Garnett theory (MGT) and Bruggeman theory (EMA) are based on different microstructure morphologies of composite materials, i.e., the MGT is suitable for a composite structure consisting of isolated metal particles surrounded by a dielectric matrix and the EMA is suitable for a composite structure consisting of a symmetrical interconnecting mixture of a metal phase and a dielectric phase as in a two-phase polycrystalline materials. However, it is found that the interfacial PRA wavelengths in the Si-Ag nanocomposite films can be approximately predicted by assuming that the matrix is pure a-Si and the embedded Ag particles are well-separated sphere-like particles. According to the surface PRA condition (5.35), or

$$\varepsilon_1^{Ag}(\omega) = -2\varepsilon_1^{a\text{-}Si}(\omega) \tag{5.46}$$

210

where ε_1^{Ag} and $\varepsilon_1^{a\text{-Si}}$ are the real parts of the complex dielectric functions of the Ag and the a-Si, two crossovers at around 310 nm and 810 nm are obtained by plotting $\varepsilon_1^{Ag}(\omega)$ and superimposing $-2\varepsilon_1^{a\text{-Si}}(\omega)$, which predict two interfacial plasma resonance absorption peaks, as shown in Fig.5.29. The strength of the resonance depends on the magnitudes of ε_2 of the Ag particles and the a-Si matrix. A large ε_2 gives a broad damped resonance. The red shift and sharpening of the peak (P1) could come from the segregation effect of the Ag particles onto the film surface. This effect should shift the resonance peak toward longer wavelengths and make the peak sharper because the damping is smaller. Nevertheless, the effective medium theories (EMTs) have to be considered when accurate predictions of the interfacial PRA frequencies and intensity are needed.

Simulation of the optical absorption spectra of the Si-Ag nanocomposite films has been done by all three EMTs and degrees of agreement of the three theories with the measured spectra have also been evaluated. It is found that the Ping Sheng effective medium theory (SPT), sometimes also called the symmetrical Maxwell-Garnett theory, is more suitable to describe the observed microstructures and gives the best fitting results. The SPT is based on two types of elementary microstructural units, i.e., B-phase coated A-phase spherical particles describing isolated metal particles (A) surrounded by dielectric matrix (B) and A-phase coated B-phase spherical particles describing metal particles clustered around the matrix. The effective dielectric function ε_{ps} can be obtained by solving (5.27).

In the present simulation by using the SPT, more details are given about how all the parameters are determined and how the equations are derived. First, it is assumed that multiple scattering from the Ag nanoparticles can be ignored because the surrounding matrix material a-Si is always a strong light absorber in the near UV and visible region, so the SPT can still be applied up to 61 at % Ag content.

It is well known that the values of bulk Ag may be decomposed into contributions from electronic interband transitions ε^b and free electrons ε^f (Ehrenreich and Philipp, 1962b):

$$\varepsilon = \varepsilon_1 + i\varepsilon_2 = \varepsilon^b + \varepsilon^f \tag{5.47}$$

The optical properties of a free-electron metal can be described by the Drude model, see (5.32) and a good approximation by taking $\omega^2 \gg \gamma_b^2$ at visible and ultraviolet frequencies (Bohren and Huffman, 1983) is

$$\varepsilon^f = 1 - \frac{\omega_{pf}^2}{\omega^2 + i\omega\gamma_b} = 1 - \frac{\omega_{pf}^2}{\omega^2} + i\frac{\omega_{pf}^2\gamma_b}{\omega^3} \tag{5.48}$$

When the metal particles are very small, the Drude model is no longer expected to be valid. Since there are very few published data about the optical constants for nanosized Ag particles so far (Ashrit, 1993; Kreibig, 1974; Kreibig, 1985), the dielectric function values of bulk Ag were modified by considering the mean free path limitation that takes into account the scattering of the conduction electrons at the metal particle boundaries (Kreibig, 1974; Kreibig, 1985). This can be done by re-adjusting the damping constant, see (5.36), or

$$\gamma_e = \gamma_b + 2\pi C \frac{\upsilon_F}{R} \tag{5.49}$$

where υ_F is the electron velocity on the Fermi surface, R is the particle radius and C is a coefficient of the order of unity. The physical meaning of C which is associated with the embedding matrix and spreads from 0.5 to 3.7 (Kreibig, 1974), is not really clear so far.

Following Kreibig's approximation approach (Kreibig, 1974 and 1985), both the real part ε_{p1} and imaginary part ε_{p2} of the dielectric function of small Ag particles can be derived and are given by

$$\varepsilon_{p1} = \varepsilon_1 \tag{5.50}$$

$$\varepsilon_{p2} = \varepsilon_2 + \left(2\pi C \frac{\upsilon_F}{R} \times \frac{\omega_{pf}^2}{\omega^3} \right) \tag{5.51}$$

In the present simulations, ω_{pf} =9.2 eV (Ehrenreich and Philipp, 1962a); υ_F=1.39$\times$ 10^6 m$\cdot$s^{-1} (Kittel, 1986), R from the measured mean particle sizes for the different Ag concentrations as listed in Table 5.2, and ε_1 and ε_2 for bulk Ag and a-Si from Palik (1985) and C=1 are taken. Figure 5.28 gives the real part (a) and imaginary part (b) of the effective dielectric function and the absorption coefficient α (c) calculated by the SPT. There is a good general agreement between the calculations and the experimental data as shown in Fig.5.28, i.e., at low Ag contents ($\leqslant$18 at %) the calculated absorption spectra, as seen in Fig.5.28c show features similar to that of pure a-Si and the addition of Ag reduces the absorption in the wavelength region shorter than 440 nm and enhances the absorption in the wavelength region longer than 440 nm. At a Ag content of 61 at % an additional absorption maximum appears at around 350 nm (P1) and a broad absorption peak appears between 700–800 nm (P2), the wavelength of which depends on Ag content. The features P1 and P2 are predicted reasonably well in terms of their positions, but the experimental data does not show a marked P2 feature between 18 to 40 at % Ag.

There are many probable reasons for this discrepancy, e.g., the doping effect of Ag in the a-Si matrix, surface electronic states at the Ag/a-Si interface,

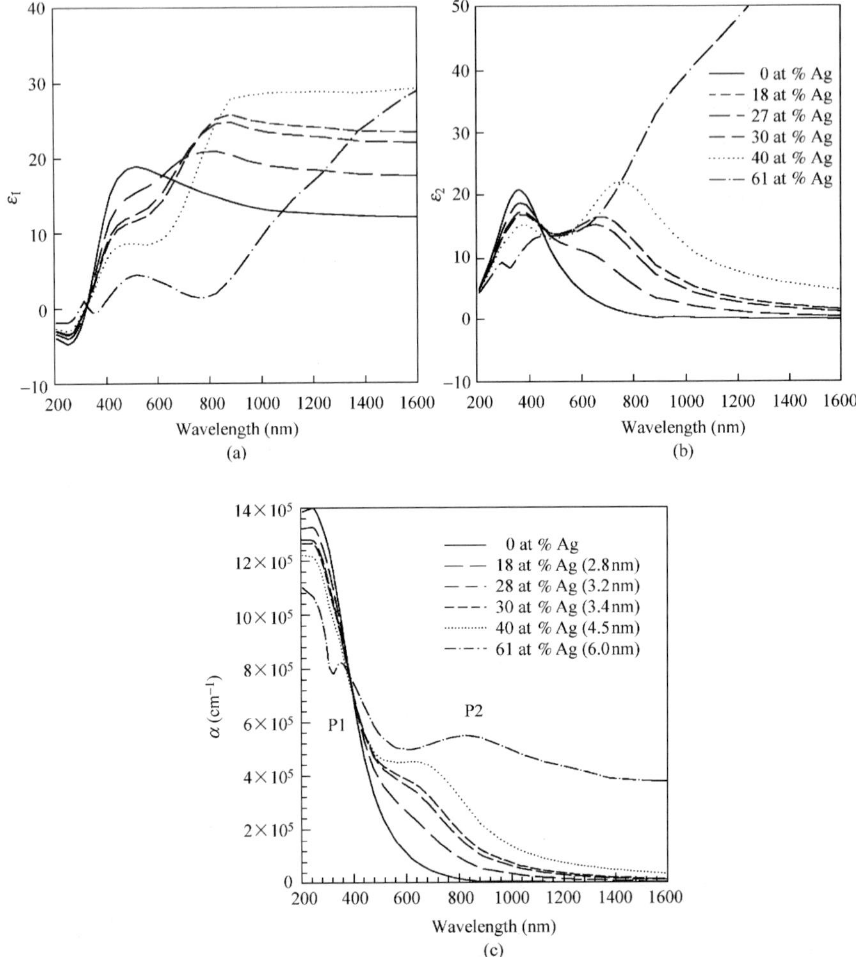

Fig.5.28 Calculated real (**a**) and imaginary (**b**) parts of the effective dielectric function and absorption coefficients (**c**) of the Si-Ag nanocomposite thin films containing different Ag contents by the SPT. Calculated absorption coefficients of the Si-Ag nanocomposite thin films containing different Ag contents by the SPT (the values in the brackets are the measured mean Ag particle sizes)

the electric and magnetic multipole effect and significant interactions between neighboring Ag particles at higher Ag content, etc. In view of all of these possible difficulties, we regard the agreement between the experimental data and the SPT as quite reasonable and it is much better over this Ag composition range than either the MGT or EMA.

According to the predictions from the interfacial PRA condition for spherical particles as shown in Fig.5.29 and the detailed calculations by the SPT, the observed absorption maxima P1 and P2 may be interpreted to originate from the interfacial PRA of the conduction electrons on Ag nanoparticles, as indicated by two peaks in ε_2 in Fig.5.28b and accompanied S-shaped feature in ε_1 as shown in Fig.5.28a which is small at low Ag content and crosses zero at higher Ag contents. P2 is present in the wavelength region dominated by the contribution from Ag free electrons, while P1 is present in the wavelength region dominated by the contribution from the 4d→Fermi surface interband electron transition (Ehrenreich and Philipp, 1962a). At low Ag contents ($\leqslant 40$ at %), the strong edge absorption determined by the energy bandgap from the a-Si matrix (< 400 nm) masks the PRA from the bound electrons (P1).

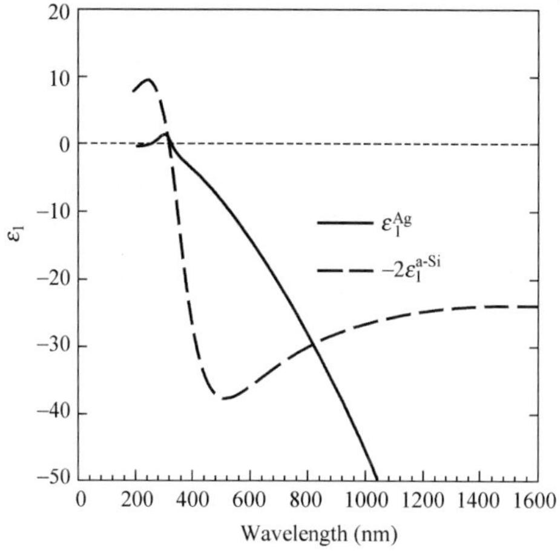

Fig.5.29 Predictions of the interfacial PRA wavelengths in the Si-Ag nanocomposite by the condition: $\varepsilon_1^{Ag} = -2\varepsilon_1^{a\text{-}Si}$, where ε_1^{Ag} and $\varepsilon_1^{a\text{-}Si}$ are the real parts of the dielectric functions of Ag and a-Si

However, there could still be some argument about the origin of P2. Bates (1984) observed the similar double plasma resonance absorption structure with a second absorption peak around 800 nm in the Ag-Cs$_2$O photocathode, where Ag nanoparticles and their aggregates were in vacuum and he attributed it to a PRA by the aggregated Ag particles. A PRA peak around 1 μm was reported by Chen and Wang (1988) in the same Si-Ag composite with 50 at % in Ag content and 100 nm average Ag particle size. It is suspected that the peak might have the same origin as P2 because the present calculations prove that P2 shift towards the near IR region and become sharper when the Ag volume fractions and mean particle sizes increase.

5.7.7.2 Electronic Transport Properties

Electronic transport properties of the Si-Ag nanocomposite thin films showed an evolution from non-metallic behavior to metallic behavior as the Ag content increased from 0 at % to 69 at % because the resistance temperature coefficient (θ) increased gradually from negative to zero. This result is consistent with the prediction from optical refractive index (n) and absorption index (κ) measurement by ellipsometry in which n-κ changes sign from negative to positive at certain high Ag content as shown in Fig.5.24, which should correspond to a percolation threshold composition in the Si-Ag nanocomposite films.

The $T^{-1/4}$ temperature dependence of the resistivity of the Si-Ag nanocomposite films suggests either a variable-range hopping electronic transport mechanism, which has been proved to be a dominant mechanism in amorphous semiconductors at very low temperature (Davis and Mott, 1979), or electron hopping between isolated metal particles as a result of combined thermal activation and tunneling, as has been seen in nanocermet films at low temperature (Sheng, 1994).

Schottky barrier *I-V* characteristics, which show a nonlinear *I-V* curve for forward bias (Sze, 1981), were not observed. On the contrary, the Si-Ag nanocomposite films always exhibited ohmic *I-V* characteristics. In bulk Ag/crystalline Si contacts, the measured Schottky barrier height is 0.54 eV (Sze, 1981). If this value still holds approximately between the Ag particles and a-Si matrix interfaces in the Si-Ag nanocomposite films, because of the existence of large numbers of interfaces, an absorption peak corresponding to an electronic transition from the Fermi level of metal particles to the conduction band of an a-Si by photo-excitation over the Schottky barrier should be expected in the near IR or middle IR (1–25 μm). However, no such absorption peak was observed. This implies that the Schottky barrier is possibly either greatly reduced or does not exist at all and thus has no effect on the electronic conduction process of the films. The former suggestion is consistent with the observed shrinkage effect of the optical energy gap when Ag was diffused into the a-Si matrices and an $Si_{1-x}Ag_x$ amorphous alloy was formed. In other words, the amorphous matrix is not pure silicon and should be considered as a heavily Ag- or Al- doped semiconductor (p-type). So a large amount of impurity states (impurity acceptors) should exist in the energy gap. It is thus suggested that for the Ag/a-Si nanosized SB junctions, the fact that the diffusion of Ag into the amorphous Si matrix reduces the optical energy gap of the a-Si subsequently reduces the SB height. Intermixing of Ag and Si at the local interfacial area because of the concentration gradient might cause further reduction of the SB height either because of shrinkage of the optical energy gap of the a-Si or because of the heavy p-type doping effect caused by the Ag(Al) in a-Si (Sze, 1981). All these possible factors make the Ag/a-Si interface ohmic or make the SB height too low to be detected even in the mid-IR. According to the above results, a simplified energy gap diagram between the metal particles and amorphous matrix was constructed as shown in Fig.5.30. The energy gap of the a-Si matrix E_g is given by the measured optical energy gap E_g=1.30 eV. The position of the

Fermi level (E_F) is still unknown. It could be pinned at the middle of the energy gap due to the high density of defect states or shift towards the valence band due to the p-doping effect of Ag or Al in the a-Si matrix. In this diagram the doping effect is assumed. It is also assumed that the impurity levels or defect levels are located very close to the Fermi level. E_c is the bottom of the conduction band and E_v is the top of the valence band.

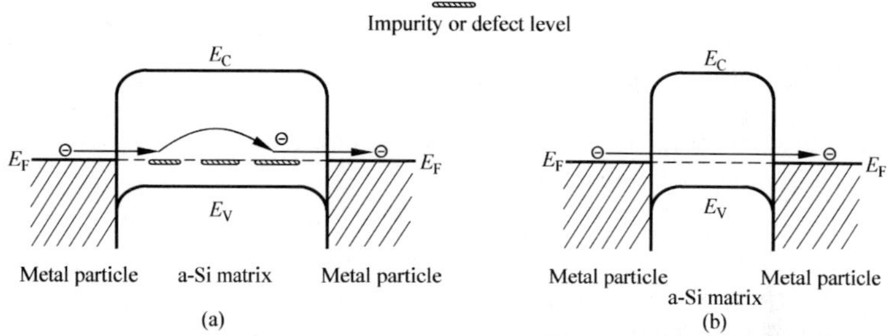

Fig.5.30 Schematic of energy-band diagram between metal particle particles and the amorphous Si matrix showing possible electron transport mechanism (**a**) low metal particle volume fraction; (**b**) high metal particle volume fraction

Two possible conduction mechanisms are suggested as shown in Figs.5.30a and b, respectively. At low metal content, the mean separation between the metal particle particles is large, say 50 nm, thus electrons are localized in the metal particles, and conduction between the metal particles is determined by an indirect hopping via the impurity or defect levels. Because of the low SB height and heavy doping in the amorphous silicon, the electrons in the metal particles may penetrate the metal/semiconductor interface easily by quantum-mechanical tunneling and then be captured by an impurity or defect state. The electron is then possibly transported through the impurity or defect levels (electron trapping centers) by variable-range hopping, which gives a $T^{-1/4}$ temperature dependence of conductivity σ_1 as shown by the term,

$$\sigma_1 = c_1 \exp\left(-\frac{A_1}{T^{1/4}}\right) \tag{5.52}$$

where A_1 is a hopping parameter associated with the indirect hopping mechanism as shown in Fig.5.30a.

When the metal content increases, the mean separation of the metal particles is decreased, the electrons in the metal particles could then transport by direct hopping between the particles as shown in the conductivity term σ_2,

$$\sigma_2 = c_2 \exp\left(-\frac{A_2}{T^{1/4}}\right) \tag{5.53}$$

where A_2 is a hopping parameter associated with the electron direct hopping mechanism as shown in Fig.5.30b.

So the total conductivity $\sigma = \sigma_1 + \sigma_2$. At low metal content σ_2 should provide some contribution but the conductivity is dominated by σ_1, while, at high metal content σ_1 should provide some contribution but conduction is dominated by σ_2. There should be a crossover at a certain temperature depending on the Ag content.

References

Abeles, F. Optical Properties and Electronic Structure of Metals and Alloys. (North-Holland Publishing Company Amsterdam, 1966)

Abeles, B. Granular Metal Films. Applied Solid State Science **6**, 1-109 (1976)

Alexander, N. V., C.W. Bates, Jr. Solid Sate Commun. **51(5)**, 331 (1984)

Alivisatios, A. P. Science **271**, 933 (1996)

Ashrit, P. V. J.Appl.Phys. **74**, 602 (1993)

Aspnes, D. E. Am.J.Phys. **50(8)**, 704 (1982)

Ausloos, M., P. Clippe, A.A.Lucas, Phys.Rev.B **18(12)**, 7176 (1978)

Bates, J., C.W. Appl.Phys.Lett. **45(10)**, 1058 (1984)

Bates, J., C.W. Mater.Lett. **18**, 128 (1993)

Bates, J., C.W., Q.Y. Chen. Mater.Lett. **23**, 7 (1995)

Bates, J., C.W. Mater.Lett. **23**, 1 (1995)

Bates, J., C.W., Q.Y.Chen. J.Appl. Phys. **81(3)**, 1457 (1997)

Bergman, D. J. Phys.Rev.Letts. **44(19)**, 1285 (1980)

Bohren, C. F.,D.R. Huffman. Absorption and Scattering of Light in Small Particles. (John Wiley & Sons, New York, 1993)

Brown, J., W.F. J. Chem. Phys. **23(8)**, 1514 (1955)

Bruggeman, D. A. G. Ann. Phys. (Leipzig) **5(24)**, 636 (1935)

Chen, Q. Y., L.Wang. Superlattices and Microstructures. **4(3)**, 265 (1988)

Chen, Q. Y., C.W.Bates, Jr. Phys.Rev. **B3(16)**, 9148 (1988)

Choudhury, M. G. M. J.Phys. **C20**, 2035 (1987)

Chylek, P. and Srivastava, V. Phys.Rev.B **27(8)**, 5098 (1983)

Clippe, P., Evrard, R. and Lucas, A.A. Phys.Rev. **B14(4)**, 1715 (1976)

Cooper, B. R., H.Ehrenreich. Phy.Rev. **138(2A)**, A494 (1965)

Cullity, B. D. Elements of X-ray Diffraction. (Addison-Wesley Publishing Company, Inc. Reading, 1956)

Davis, E. A., N.F.Mott. Electronic Process in Non-crystalline Materials. Clarendon Press (Oxford and London, 1979)

Doremus, R. H. J.Chem.Phys. **40**, 2389 (1964)

Doremus, R. H.J.Chem.Phys. **42**, 414 (1965)

Dupree, R., M.Smithard.J.Phys. **C5**, 408 (1972)

Dykhne, A. M., A.N.Lagarkov and A.K.Sarychev.In,Proceedings of the fourth international conference on electrical transport and optical properties of inhomogeneous media. Physica A **241(1-2)** (1997)

Ehrenreich, H. Phys.Rev.Letts. **8(2)**, 59 (1962a)

Ehrenreich, H. and H.R.Philipp. Phys. Rev. **128**, 1622 (1962b)

Ehrenreich, H. Phys.Rev. B **131**, 2469 (1963a)

El-Sahhar, S. A. Phys.Stat.Sol. (a)**104**, 755 (1987)

Frohlich, H. Physica(Utrecht) **4**, 406 (1937)

Garland, J. C., D.B.Tanner. Electrical Transport and Optical Properties of Inhomogeneous Media. (American Institute of Physics, New York, 1978)

Garnett, J. C. M. Phil.Trans.Roy.Soc.London **203**, 385 (1904)

Garnett, J. C. M. Phil.Trans.Roy.Soc.London **205**, 237 (1906)

Genzel, L. Physik B **21**, 339 (1975)

Gibson, U. J., R.A.Buhrman.Phys.Rev. **27**(8): B5046 (1983)

Gittleman, J. I., B.Abeles. Phys.Rev. **B.15(6)**, 3273 (1977)

Granqvist, C. G. Phys.Rev. B **18(4)**, 1555 (1978)

Hagemanm, H. J. J.Opt.Soc.Am. **65**, 742 (1975)

Halperin, W. P. Rev.Mod.Phys **58(3)**, 533 (1986)

Heilmann, A., J.Werner, O.Stenzel and F.Homilius. Thin Solid Films **246**, 77 (1993)

Heinrich, A., G.Gladun. Int. J. Electronics. **73(5)**, 883 (1992)

Henglein, A. Ber. Bunsenges. Phys. Chem. **99(7)**, 903 (1995)

Hoffman, D. W., R.Roy. and S.Komarneni. J.Am.Ceram.Soc. **67**, 468 (1984)

Huang, W. C., J.T.Lue. Phys.Rev. B **49(24)**, 17279 (1994)

Hutter, B. J.Phys, Condens. Mater. **7**, 907 (1995)

INSPEC.Properties of amorphous Silicon. London and New York, The institution of electrical engineers (1985)

INSPEC. Properties of Silicon. London and New York, The Institution of Electrical Engineers (1988)

JCPDS Powder Diffraction Files. Pennsylvania

Jezequel, G. J.Phys.F 7, 2613 (1977)

Johnson, P. B., R.W.Christy. Phys. Rev. **B 6(12)**, 4370 (1972)

Kawabata, A., R.Kubo. J. Phys. Soc. Japan **21(21)**, 1765 (1966)

Kawabata, A. J. Phys. Soc. Japan **29(4)**, 902 (1970)

Kerker, M. The Scattering of Light and Other Electromagnetic Radiation, (Academic Press, New York) (1969)

Kittel, C. Introduction to solid state physics. (Wiley, New York, 1986)

Komarneni, S. J.Mater.Chem. **2(12)**, 1219 (1992)

Koyama, R. Y. Phys. Rev. **B 8(6)**, 2426 (1973)

Kreibig, U. J.Phys. F, Metal Phys. **4**, 999 (1974)

Kreibig, U., L.Genzel. Surface Science **156**, 678 (1985)

Kreibig, U. Physica A **157**, 244 (1989)

Kubo, R. J.Phys.Soc. Japan **17(6)**, 975 (1962)

Kubo, R. Ann.Rev.Mater.Sci. **14**, 49 (1984)

Lafait, J., D.B.Tanner. In, Proceedings of the Second International Conference on Electrical Transport and Optical Properties of Inhomogeneous Media. Physica A **157** (1989)

Lamb, J. L., C.L.Nagendra. J.Appl.Phys. **76(11)**, 7195 (1994)

Landauer, R. Electrical Conductivity in Inhomogeneous Media. Electrical Transport and Optical Properties of Inhomogeneous Media, American Institute of Physics Ohio, (1978)

Leedy, K. D., J.M.Rigsbee. J.Vac.Sci.Technol. A 14(4), 2202 (1996)

Li, N., Q.D.Wu. Appl.Phys. 53, 172 (1991)

Liljenvall, H. G. Phil.Mag. 50(3), 243 (1970)

Maissel, L. I., R.Glang. Handbook of Thin Film Technology. New York, McGraw-Hill Book Company (1970)

Massalski, T. B. Binary Alloy Phase Diagrams, (William W.Scott, Jr., 1990)

Mclachlan, D. S. J.Phys.C 21, 1521 (1988)

Mclachlan, D. J.Am.Ceram.Soc. 73(8), 2187 (1990)

Mie, G. Ann.Physik 25, 377 (1908)

Milton, G. W. Appl.Phys.Lett. 37(3), 300 (1980)

Mochan, W.L., R.G.Barrera. In, Proceedings of the Third International Conference on Electrical Transport and Optical Properties of Inhomogeneous Media, Mexico, Physica A.207(1-3)(1993)

Mott, N. F. J.Non-Cryst. Solids 1, (Oxford University Press, Oxford) 1 (1968)

Nagendra, C. L., J.L.Lamb. Applied Optics 34(19), 3702 (1995)

Ndlela, Z., C.W.Bates, Jr. Mater. Lett. 9, 10, 465 (1991)

Niklasson, G. A., H.G.Graighead. Applied Optics 22(8), 1237 (1983)

Ohring, M. The Materials Science of Thin Films. (Academic Press, Boston, 1992)

Palik, E. D. Handbook of Optical Constants of Solids. (San Diego, Academic Press 1985, 1993)

Papavassiliou. Prog.Solid St.Chem. 12, 185 (1980)

Perenboom, J. A. A. J., P.Wyder. Phys.Rep. 78(2), 173 (1981)

Press, W. H. Numerical Recipes. Cambridge, Press Syndicate of the University of Cambridge (1988)

Quinten, M. Surface Science 156, 741 (1985)

Schonauer, D., U.Kreibig. Surface Science. 156, 100 (1985)

Sheng, P. Phys.Rev.Letts. 45(1), 60 (1980)

Sheng, P., J.Klafter. Phys.Rev.B 27(4), 2583 (1983)

Sheng, P. Electronic Transport in Granular Metal Films, (Kluwer Academic, New York, 1994)

Shoema, B. L. Metal Semiconductor Schottky Barrier Junctions and Their Applications New York, Plennm Press (1984)

Smith, G. B. J.Phys.D 10, L39 (1977)

Smith, G. B. Appl.Phys.Lett. 35(9), 668 (1979)

Smith, D. Thin Film Deposition, Principle & Practice. (McGraw-Hill, Inc, New York, 1995)

Smithard, M. A. Solid State Commun. 14, 407 (1974)

Stroud, D., F.P.Pan. Phys.Rev. B 17(4), 1602 (1978)

Sze, S. M. Physics of Semiconductor Devices. (John Wiley & Sons, New York, 1981)

Theye, M. L. Thin Solid Films 4, 205 (1969)

Wasa, K., S. Hayakawa. Handbook of Sputtering Deposition Technology. (Noyes Publications, Park Ridge, 1995)

Weller, H. Adv.Mater. 5(2), 89 (1993)

Wood, D. M., N.W.Ashcroft. Phys.Rev.B 25(10), 6255(1982)

6 Strength and Toughness of Functional Ceramic Matrix

Yongjian Sun

6.1 Introduction

Ceramic materials, in general, have a very attractive package of properties: high strength and stiffness at a very high temperature, chemical inertness, and low density. But, they are prone to catastrophic failure in the presence of flaws, and are extremely susceptible to thermal shock and other damages during fabrication and in service. Fiber-reinforced ceramic composites (FCMC), by incorporating fibers in ceramic matrices, however, not only exploit their attractive high-temperature strength but also enhance their toughness. A typical load-displacement curve for a ceramic composite subjected to tensile loading parallel to the fiber direction is shown as curve (b) in Fig.6.1 (Sun, et al., 1998). At the point A, the first matrix crack (FMC) initiates when the matrix reaches its elastic limit. If the fibers and matrix are weakly-bonded or frictionally coupled, the matrix crack propagates transversely across the fibers thus creating bridged fibers to carry additional load. When this happens, the matrix crack deflects at the fiber/coating/matrix interfaces because of the interfacial debonding and sliding. With the increase of the applied load, more matrix cracks are generated and interfacial debonding propagates with a larger sliding zone. This process leads to a nonlinear curve between A and B in Fig. 6.1. Point B indicates the saturation of the matrix cracking behavior. Beyond point B, the curve shows additional nonlinear behavior until the onset of fiber failures and fiber pull out at point C. This point C represents the ultimate strength of the composite, at which the breakage of a large number of fibers leads to decreased load-bearing capacity of the composite. Comparing to the monolithic ceramic material represented by curve (a) in Fig. 6.1, the mechanical properties of ceramic composite, especially the FMC stress, ultimate strength and work of fracture, are enhanced because of the reinforcing fibers (Hillig, 1987; Sun, et al., 1998).

Most research workers focused their interests on the study of first matrix-cracking stress, ultimate strength of composites, and the significant role of the interface on these two properties (Briggs and Davidge, 1989; Dutton et al., 1996; Marshall et al., 1985; Steif and Schwietert, 1990; Tredway and Prew, 1990;

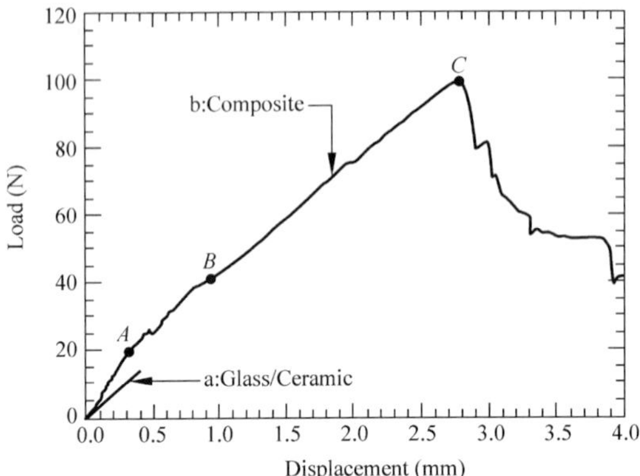

Fig.6.1 The typical load-displacement curves for a monolithic glass/ceramic and a composite

Wijeyewickrema and Keer, 1990; Xu, et al., 1994). The study of the initial nonlinear behavior of the composite after FMC, however, is not that extensive. This initial nonlinear behavior is usually associated with the phenomena of fiber bridging at the crack surfaces, debonding and sliding at the fiber/matrix interface, multiple matrix cracking, and sometimes fiber pullout upon fiber failure. The micromechanisms of these phenomena and interactions among them have not been thoroughly investigated yet because of the complexity of this initial nonlinear behavior. However, it is very important as it determines the fracture resistance of the materials at the early stage of the crack propagation. Therefore, the understanding of the micromechanisms of fiber bridging, matrix cracking, and interfacial debonding and sliding is necessary for the design and development of composites (Evans, 1990).

The primary contribution of the fiber-bridging effect to the fracture toughness of composites (Chou and Kelly, 1980; Wiederhorn, 1984), attracted more interest from scientists and engineers to make efforts on the analyses of the fiber-bridging toughening mechanism from theoretical modeling to experimental investigation. Based on the fracture mechanics fundamentals, the fiber-bridging toughness can be determined numerically from the distribution of the fiber-bridging stresses (traction) on the crack surfaces (Marshall et al., 1985). Several different techniques such as crack opening displacement (COD) measurement (Sakai, 1992) and Raman or fluorescence spectroscopy methods (Galiotis, 1991), have been developed to determine the distribution of fiber-bridging stresses. Comparatively, COD measurement is easier than Raman/fluorescence spectroscopy although the latter should provide more accurate results. The determination of fiber-bridging stress from COD has been proposed by several scientists. However, not many experimental results have been reported regarding the relationship in FCMCs. Therefore, it is very necessary to develop more new approaches and to obtain more experimental result.

The systematic studies included in this chapter are primarily focused on: 1) fabrication of a transparent fiber-reinforced ceramic composite, 2) determination of mechanical and interfacial properties of composites, 3) micromechanical study of matrix cracking, interfacial debonding, and fiber bridging, 4) development of the debond length measurement (DLM) technique and theoretical models to determine the distribution of fiber-bridging stress, 5) measurement of the fiber-bridging stress using the DLM and COD techniques, and 6) determination of the fracture resistance from the bridging-stress distribution (micro) and experimental SENB test (macro). Through these studies, the relationships among the processing, microstructure, properties, and performance of composites are thoroughly examined. These studies have made a significant contribution to the understanding and determination of the strengthening and toughening behavior in FCMCs.

6.2   Overview of Composite Technologies

Ceramic matrix composites (CMCs) are attracting increasing attention because of the broader diversity of, and especially improvement in, the properties they can provide. Ceramic matrix composites, currently 35% of the total market, will grow even faster at an annual rate of 14.5% to $400 million in year 2000 (Schwartz, 1997). These composite materials are offering broad applications in military, aircraft, automotive, and sporting goods. One of the examples of their applications is in aircraft and automotive engines. SiC_f/C composite secondary flaps and seals have been developed and used successfully, which have the potential of saving 5.9 kg of the weight per engine and even more air frame weight (Schwartz, 1997). This reduction in structural weight directly influences the reduction in fuel consumption and allows the downsizing of engines. In addition, the properties at high temperature of CMCs can meet the increasing requirement in temperature for engine components. As a major group of CMCs, fiber-reinforced ceramic matrix composites (FCMC) play an important role in industry. Retrospectively, the advantages of FCMCs were firstly reported by Crivelli-Visconti and Cooper (1969) in 1967. They showed that the fiber-reinforcement in ceramic matrix could substantially increase the work-of-fracture (WOF) relative to the unreinforced ceramic matrix. In their report, the WOF deduced from the area under the force-displacement curve for a carbon fiber-reinforced vitreous silica was 11 kJ/m^2 compared with 4 kJ/m^2 for the matrix only. In addition, there was no evident decrease in the strength of composite with an increase in temperature up to 800 ℃.

Since this major demonstration, fiber-reinforced ceramic matrix composites have aroused the interest of many scientists and researchers. Prewo and Brennan (Prewo, 1982; Prewo, 1988; Brennan and Prewo, 1982) and Phillips' group (Phillips, 1974; Phillips et al., 1972) have done many impressive studies on carbon, SiC, and oxide fibers-reinforced glass and glass-ceramic composites. At

222

the same time, SiC fiber-reinforced alumina (Tai and Chou, 1990), Si_3N_4 (Xu et al., 1994; Bhatt, 1988), and $ZrSiO_4$ (Singh, 1990; Llorca and Singh, 1991) composites, alumina fiber-reinforced Si_3N_4 (Shwab et al., 1995), carbon/carbon (Awasthi and Wood, 1988), and SiC/SiC composite systems (Gaeta et al., 1995) were successfully developed and studied as well. This increasing availability of ceramic composite systems was achieved because of the developments of high-temperature reinforcement fibers and composite processing technologies. Accompanying the ever-expanding FCMC systems and the development of new processing techniques, the mechanical and interfacial properties of the FCMCs have also been extensively characterized. The first matrix-cracking stress and ultimate strength of FCMCs, and the interfacial strength and debond energy at the interface have been examined in various environments at both room and elevated temperatures. The fatigue, creep, and thermal shock behaviors of FCMCs have been investigated. With the support from experimental studies, much progress has also been made in the fundamental and theoretical study on the mechanisms of the fiber reinforcement in FCMCs, which is accomplished by the strengthening and toughening effects of fibers.

The strengthening and toughening mechanisms of FCMCs have been studied since the early 1960s (Aveston and Kelly, 1973). Numerous concepts such as stress transfer, interfacial debonding, crack bridging, and fiber pullout, have been proposed for the improved mechanical properties of FCMCs. All these concepts can be generalized into the following three basic mechanisms (Gac, 1990): 1) increase the local driving force necessary to propagate a crack, 2) increase the mechanical energy consumed per unit area of crack propagation, and 3) decrease the local strain to reduce the crack-tip stress concentration. Other mechanisms, such as phase transformation, crack deflection, and microcracking, have also been proposed to toughen ceramic materials (Evans, 1990; Wiederhorn, 1984). Some of them have been applied to toughen the matrix in FCMCs as well. Of particular interest, fiber bridging became the focus of investigations because fiber bridging largely governed the fracture toughness of composites.

In this review, the mechanical properties of monolithic ceramics and FCMCs will be compared to show the advantages of FCMCs (Sect. 6.2.1). The influences of the fiber/matrix interface on the mechanical properties of FCMCs and the techniques for determining the interfacial properties will be discussed in Sect. 6.2.1. Several processing techniques of CMCs will also be briefly introduced (Sect. 6.2.2). Finally, and most importantly, the strengthening and toughening mechanisms will be elaborated (Sect. 6.2.3). Many relevant concepts such as the strength and toughness of composites, stress transfer, interfacial debonding, fiber bridging and fiber pullout, and several representative theories and models will thus be reviewed.

6.2.1 Mechanical and Interfacial Behavior of FCMCs

6.2.1.1 Mechanical Behavior of FCMCs

Ceramic matrix materials, with the exception of amorphous glasses, are crystalline. They mostly have strong ionic bonding and covalent bonding. For example, alumina (Al_2O_3) consists of ionic bonds and silicon carbide (SiC) consists of covalent bonds. This characteristic theoretically indicates the intrinsic high strength of ceramics. However, it turns out that the nature of bonding is not all that important in determining the strength of ceramics. The flaws inside and at the surface of the material play a more crucial role. The theoretical strength of normal solids computed by Griffith's law is in the order of $10 \, GN/m^2$, which corresponds to a fracture strain as large as 20% for an elastic body (Chawla, 1987). In practice, because of the presence of sharp flaws, ceramic materials fracture at applied stresses of only $100–300 \, MN/m^2$ with failure strains of less than 0.05% (Chawla, 1987). In addition, most ceramic materials possess complicated structures that determine that they, unlike metals, have very few slip systems to accommodate a large deformation, especially at room temperature. Therefore, they are brittle in nature with a particularly low failure strain and low value of WOF, hence a low toughness. Glasses are amorphous materials. They have a complex network structure. They have neither crosslinks nor crystal structures. Therefore, there is no slip system in glasses and they are also brittle. The curve (a) in Fig. 6.1 is the typical load-displacement curve for most glasses and monolithic ceramics (Sun and Singh, 1998). Because the low strength of ceramic materials can be overcome by reducing the flaw size and flaw density during processing and final machining, the low fracture toughness seems to be the only other detrimental characteristic for the application of these materials. Therefore, any approach and technique in processing and component design relating to the improvement of the fracture toughness of the ceramics will strongly influence the application of ceramics as engineering materials.

In fact, ceramic materials can almost reach their theoretical strength by reducing flaw size and flaw density. Griffith found that the tensile strength of glass fibers varied markedly with their diameters. The smaller the diameter, the lower the probability of large flaws, and the higher the strength. Following glass fibers, a variety of carbon, boron oxide, alumina, and silicon carbide fibers have been developed (Watt and Perov, 1985). Carbon fibers are commercially available with a variety of tensile moduli ranging from 207 GPa to 1035 GPa (Watt and Perov, 1985). The advantages of carbon fibers are their exceptional tensile strength-to-weight ratio as well as their low coefficient of thermal expansion (CTE) and high fatigue strengths. They are mostly used in the aerospace industry for weight saving. Al_2O_3 and SiC fibers are especially useful for their high temperature application because of their high melting point (2040 °C) and strength retention up to 1370 °C (Chawla, 1987). SiC fibers are

224

widely used also because of their diversity in microstructure and fiber coatings. The commercially used SiC fibers include SCS-2, SCS-6, and SCS-9 (Chawla, 1987). The differences between them are the fiber diameter, the types of coating (BN and carbon), and the thickness of coating. The SiC fiber used for this research is SCS-6 fiber with 142 μm in total thickness and 3 μm in carbon coating thickness (Bhatt and Hull, 1992a). Ceramic fibers have relatively high strength, high modulus, low density and high-temperature oxidation resistance, and so they serve as reinforcements in ceramic matrices by means of load transfer for strengthening and toughening by crack-bridging.

Monolithic ceramic materials and ceramic composites reinforced with either particles or short fibers fail in a typically brittle mode. In contrast, continuous fiber-reinforced ceramic composites fail in two different modes. Figure 6.1 shows a comparison of the typical load-displacement curves for both a monolithic ceramic and a fiber-reinforced ceramic composite (Sun and Singh, 1998a). Under the tensile load parallel to the fiber direction, when the fibers and matrix are well bonded, brittle failure occurs. For instance, a brittle fracture was observed in the Si_3N_4-SiC_f composites in which the SiC fibers and Si_3N_4 matrix were chemically bonded (Bhatt, 1988). The stress-strain behavior is linearly elastic up to the failure point at which a clean crack propagates suddenly through the sample. When the fibers and matrix are weakly bonded or loosely coupled by friction, the stress-strain behavior shown as curve (b) in Fig. 6.1 can be obtained. The first matrix crack appears once the linear elastic limit for the ceramic matrix is reached. The corresponding stress at point A is called the matrix cracking stress, σ_0. The matrix cracks develop at approximately regular spacing transverse to the fibers, but the fibers remain intact and continue to carry the tensile load. With the increase of the load, debonding happens at either the fiber-matrix, the fiber/fiber coating, or the fiber coating/matrix interfaces. The fibers begin to slip, and crack faces continue to separate. The overall increase of strain and stress in the fibers, especially the local high stress at the weak points of the fibers, leads to the onset of fiber failures and fiber pullout. The local stresses near the broken fibers redistribute, and the portion of load that is previously carried by the broken fibers is added to the unbroken fibers. Finally, more and more fibers are broken and pulled out. During this process, the overall stress in the composites and load capacity begins to drop after the onset of the coordinated failures of fibers. The stress at point C is called the ultimate strength of the composite, σ_{Cu}. The distinctive mechanical behavior of composites in comparison to the monolithic ceramics results from the enhanced fracture resistance because of the reinforcement of fibers. This enhanced fracture toughness indicated by the area under the load-displacement curve is the result of fiber bridging, interfacial debonding and sliding, and fiber, pullout which will be reviewed later.

As the three necessary components in a composite, the ceramic matrix, reinforcement fibers and the fiber/matrix interface should be well tailored for a structural design to meet the requirement of mechanical properties. The choice of ceramic material as the matrix in a composite is largely determined by its strength, thermal expansion, and ease of fabrication. The choice of fibers with the consideration of the interface, however, depends on their strength and diameters for design, and the compatibility between fiber and matrix.

6.2.1.2 Fundamentals of Fiber/Matrix Interface

Interface is defined as a discontinuous region through which material parameters, such as concentration of elements, crystal structure, atomic registry, elastic modulus, density and thermal expansion coefficient, change from one side to another (Kerans et al., 1989). Such discontinuity exists at the fiber-matrix interface. The fiber-matrix interface in ceramic composites may be one of following two general types of solid-solid interface (Kerans et al., 1989): 1) a grain boundary and 2) an interphase boundary. In most cases, an FCMC interface is an interphase boundary type. A coating as an interlayer, the segregation of impurities and reaction at interface, and the thermal expansion mismatch at the interface play important roles in the control of interfacial properties in composites.

A coating is taken to be a crystalline or noncrystalline phase deliberately introduced as an interlayer during processing. The coating can be a different phase from both the fiber and matrix, or the same phase but with a different microstructure. For instance, the SiC-enriched carbon coating and a BN coating on the SiC (SCS-6) fiber belongs to the former group. The importance of fiber coating to the mechanical response of composites was studied by Singh and Brun (1987a) on SiC-fiber-reinforced mullite composites. They measured the interfacial strength and found that BN-coated SCS-6 fiber formed a much weaker bond with mullite than did the uncoated SCS-6 fiber, and BN-coated SCS-6 fiber composite showed higher strain-to-failure.

The segregation of impurities and reaction at an interface can change the interfacial properties too. In a reactive fiber-matrix pair, a new interlayer with a different phase may be generated, which results in a different volume, surface energy, and stress state at the interface. The reaction that occurs at the interface between SiC fiber and Si_3N_4 matrix is one of the examples (Bhatt, 1988). The reaction at the interface results in a strong bond between the fiber and matrix. Such a strong bond in CMCs is not preferable because it hinders the interfacial debonding and fiber pullout and hence causes a catastrophic failure of composite (Bhatt, 1988; Kerans et al., 1989). Take the coefficient of thermal expansion (CTE) as another example (Chawla, 1987). If a fiber has axially a higher value and radially a lower value of CTE than the matrix, then in the axial direction, a tensile stress can arise at the fiber interface and a compressive stress can arise at the matrix interface. This generally happens when the composite cools from a high temperature to room temperature during fabrication and service. The thermally induced compression stress in the matrix will beneficially enhance the

226

first matrix cracking stress, σ_0. In the radial direction, a clamping stress results on the fiber, hence improving the stress transfer efficiency between fiber and matrix. Obviously, this situation is desired for an engineering composite design (Chawla, 1987).

Generally from the fracture toughening point of view, a desired fiber/matrix interface for FCMCs is a frictionally coupled or weakly-bonded interface so that when the matrix crack approaches the fibers, the crack can deflect or an interfacial debond can be generated at the fiber/matrix interface. On one hand, this crack deflection or debond at the interface affects the crack path reducing the singularity at the crack tip (Singh and Brun, 1987a). On the other hand, the energy dissipation associated with the debonding and interfacial frictional sliding increases the resistance to fracture (Evans, 1990). Hence, a composite becomes tougher, showing a higher fracture tolerance.

In order to develop and verify the micromechanics models or to empirically correlate composite mechanical behavior with the interfacial properties, the properties of the fiber/matrix interface, such as interfacial shear strength, frictional shear stress and debond energy, should be measured. Several different techniques have been developed for characterizing the interfacial properties. These techniques include fiber pushout (Singh and Sutcu, 1991; Bright and Shetty, 1989; Goettler and Faber, 1989; Jero, et al., 1991), microindentation (Fiber pushin) (Marshall, et al., 1987b), fiber pullout (Kerans and Parthasarathy, 1991a), matrix crack-spacing measurement (Aveston et al., 1971; Zok and Spearing, 1992), and Raman spectroscopy (Young and Yang, 1996; Ma and Clarke, 1993).

1) Microindentation (Fiber pushin)

Microindentation is a simple technique firstly developed by Marshall, et al. (1987). The individual fiber in a composite is the target of the experiment. A micro-indentor is forced on to the fiber and pushes the fiber into the matrix. In this test, the fiber will debond from the matrix at a certain value of load and locally slide. Accompanying the process of interfacial debond and sliding, a load-displacement curve of the indentor is recorded and then used to calculate the interfacial properties using the following relationship,

$$u = \frac{F^2}{4\pi^2 \tau_f r^3 E_f} - \frac{2\Gamma_d}{\tau_f} \tag{6.1}$$

where u is the displacement of the indentor, F is the force of the indentor applied on the fiber, Γ_d is the debond energy of the interface. By plotting the square of the force versus displacement, the frictional shear stress, τ_f, can be obtained from the slope, and the debond (fracture) energy, Γ_d, from the intercept. A similar expression was also derived by Singh and Brun (1987). One of the advantages of the microindentation technique is that it can be applied to composites with a larger thickness so that the sample preparation is much easier.

This method has been successfully used to evaluate a weak interface in Nicalon-LAS system (Grande et al., 1988) and other FCMCs.

2) Fiber pushout

Fiber pushout test is a modified technique based on microindentation. A thin slice sample is usually required for a test to obtain a load-displacement curve. The fiber pushout setup is schematically shown in Fig. 6.2. A schematic load-displacement curve for fiber pushout is shown in Fig. 6.3 (Kerans and Parthasarathy, 1991). From the load-displacement curve, the initiation point (point A in Fig. 6.3) of interfacial debonding, the peak load (point B) at which the interfacial debonding has propagated through the sample thickness, and the load drop (point C) corresponding to the beginning of a steady-state interfacial sliding can be measured. However, in many cases only points B and C are distinctively observable. The characterization of interfacial properties is obtained from the load values at points B and C. Based on the shear-lag model, the interfacial debond strength, τ_d, and frictional shear stress, τ_f, can be obtained from the following expressions (Bright and Shetty, 1989; Kerans and Parthasarathy, 1991a):

$$P_B = \frac{2\pi r}{\rho}\tau_d \tanh(\rho L) \approx 2\pi r L \tau_d \tag{6.2}$$

$$P_C = \frac{\pi r^2 \sigma_0}{k}\left[1 - \exp\left(-\frac{2\mu k L}{r}\right)\right] \approx 2\pi r L \tau_f \tag{6.3}$$

where P_B and P_C are characteristic forces in a load displacement curve, ρ in (6.2) is the shear-lag constant, k is a dimensionless parameter, and μ in (6.3) is the coefficient of friction. L is the thickness of a pushout sample.

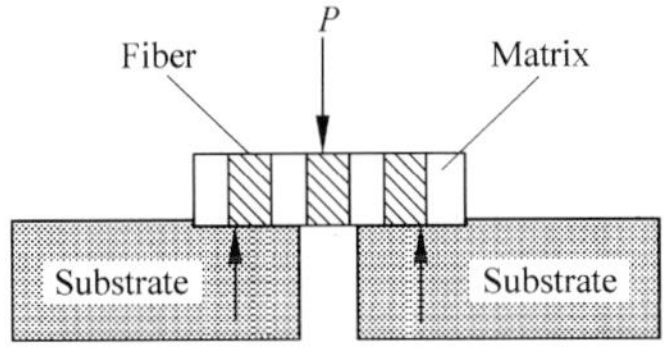

Fig.6.2 The schematic of fiber pushout set -up

This technique was extensively used by many researchers in several composite systems. Singh and Reddy (1996) used this technique on the $ZrSiO_4$-SiC_f composites to analyze to influence of coating on the interfacial properties. Bright and Shetty (1989) and Jero et al. (1991) applied this technique on the glass matrix composite to analyze the effect of the sample thickness on the experimental validity. The advantage of this method is that it can provide the most direct results from the load-displacement curve. The detailed description, including the photograph of a fiber pushout machine and the experimental

228

analysis will be illustrated and discussed in Sect. 6.5.2.

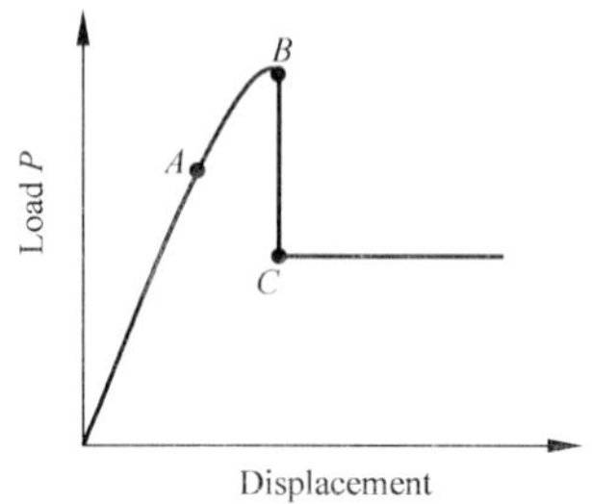

Fig.6.3　A representative load-displacement curve for the fiber pushout test

3) Fiber pullout

The fiber pullout test was developed to pull the fiber from a monofilament composite in a tensile test machine (Deshmukh and Coyle, 1988). The load-displacement curve is acquired for analysis. Similar to the fiber pushout technique, the characteristic load values in a load-displacement curve are used to obtain the interfacial properties. The fundamental analysis for the fiber pullout test is the same as the fiber pushout test except for the consideration of the fiber Poisson contraction during pullout. This utilizes the shear-lag model to correlate interfacial properties with the characteristic loads. Comparing with the above two techniques, the single fiber pullout method is not commonly used because of the difficulties in sample preparation, test alignment and handling.

It is, however, interesting to note that the above three methods, microindentation, fiber pushout, and fiber pullout, all rely on the analysis of the load-displacement curve. Therefore, the machine compliance should be subtracted from the experimental data to obtain the real results, which in return may introduce some errors to the analysis.

4) Matrix crack spacing measurement

This method uses a different approach from the previously described methods. It was introduced by Aveston, Cooper and Kelly (ACK) when they tried to explain the multiple matrix-cracking phenomenon in 1971 (Aveston et al., 1971). When a unidirectional continuous fiber-reinforced composite is strained in tension, the composite extends elastically until the matrix starts to crack. Beyond the first matrix cracking in a FCMC, the fiber at the crack surfaces takes over the additional stress thrown by the broken matrix. The stress in the bridging fiber again is transferred back to the matrix through the fiber/matrix interfacial shear stress. The stress in the matrix, however, re-established up from the free-crack surfaces due to the stress transfer. When the matrix stress reaches the matrix-cracking stress again, the new matrix cracks will be generated. The stress transfer is determined by the fiber/matrix interfacial shear strength. For a stronger interface, the stress in the matrix can be rebuilt very shortly from the crack surfaces (corresponding to a short stress transfer length). This characteristic length describing the efficiency of stress transfer was defined as (Aveston et al., 1971)

$$z' = \frac{V_m \sigma_{mu} r}{2 V_f \tau_f} \tag{6.4}$$

It was observed by ACK that the matrix crack spacing was between z' and $2 z'$. Later, some statistical analyses found that the average matrix crack spacing, L_{cs}, is equal to 1.43 z' (Kimber and Keer, 1982).

The concept of multiple matrix-cracking was adapted by ACK to analyze the dependence of the matrix-crack spacing, L_{cs}, on interfacial shear strength. This technique was later modified by Zok and Spearing (1992) and Curtin (1993) based on their simulation and statistical analysis. Although different formulas were available based on the studies of multiple matrix cracking in different composite systems to interpret the interfacial properties, the validity or accuracy of those expressions is still questioned because of the insufficient experimental data. But, in practice it is an interesting and easy technique.

By using the techniques described above, not only can the interfacial properties, such as interfacial shear strength, debond energy, and frictional shear stress, be obtained but other properties such as residual clamping stress at the interface, surface roughness and coefficient of friction can be analyzed as well (Singh and Reddy, 1996; Marshall et al., 1995).

6.2.2   Processing of Ceramic Matrix Composites

The processing choice of CMCs is based upon the character of the composite, its size and shape, and its cost. Design or tailoring of composite microstructures to achieve the improved or new properties presents processing challenges. A combination of the fineness of most dispersed phases and the high fiber volume fraction often desired commonly make achieving a high density and degree of homogeneity a challenge. These challenges of ceramic composite processing have resulted in a significant shift and development in emphasis of processing technologies in comparison to processes used with other ceramic materials. The processing methods for the fabrication of FCMCs include hot pressing, chemical vapor infiltration, liquid infiltration, gas metal reaction (Lanxide process), polymer pyrolysis, pressureless sintering, and injection molding (Schwartz, 1997; Strife et al., 1990). In this review only some of the techniques will be briefly addressed.

6.2.2.1   Diversity of Processing Methods

1) Hot Pressing (HP)

Hot pressing is one of the most commonly used processing technologies for making FCMCs in both industry and laboratory. The temperatures of HP range from 1200 °C to 2000 °C, which are based on the particular composite system (Schwartz, 1997; Chawla, 1987). The pressures used vary from 20 MPa to 50 MPa (Schwartz, 1997; Chawla, 1987). Hot pressing produces the highest density, lowest (near zero) porosity of any of the processing methods. It is applicable, in principle, to all ceramic materials as the predominant basis of conventional ceramic processing. Most FCMCs such as $ZrSiO_4$-SiC_f (Llorca and Singh, 1991), glass and glass ceramics-C_f (Phillips, et al., 1974a), and Si_3N_4-SiC_f (Bhatt, 1988) composites have been successfully made by this method. However, the HP method is limited by two factors. First, it is not a low-cost process and is mainly applicable to rather simple shapes. The other limitation stems from the high processing temperature required for substantial densification, which may lead to both reactions between fibers and matrix and degradation of the fibers.

2) Chemical Vapor Infiltration (CVI)

Chemical vapor infiltration is one of the most important reaction methods for generating ceramic matrices of FCMCs. In this technique, the fibers or whiskers are formed into the desired shape and orientation and then placed in a CVD/CVI reactor. With the application of heat and flow of the reactant gases, the matrix phase is formed in situ within the fiber preforms. The most recent advance on this technique is the special design of CVI reactor with the thermal gradient, which eliminates the formation of dense surface skins and enhances the rate of infiltration. By this method, both the carbon/carbon composites (Schwab et al., 1995; Lackey and Starr, 1990) and SiC/SiC composites (Roman and Stinton, 1995) have been made successfully, and carbon/carbon composites have been used for aircraft brake disks (Schwab et al., 1995). However, it is still difficult to obtain a very low porosity (<10%) with this method.

3) Polymer Pyrolysis

Polymer pyrolysis is another major chemical reaction process. Polymer pyrolysis consists of using an appropriate polymer containing the atoms of the desired ceramic product that can be pyrolyzed without excessive losses of polymer backbone. This technique has been applied to make $SiCw/SiO_2$ composites (Wu and Messing, 1990) and SiC/SiC composites (Russell-Floyd et al., 1993). Several factors strongly encourage the use of polymer pyrolysis, including low cost and low processing temperature, which are desired for making composite without fiber degradation. But this method is also limited by the low density, considerable shrinkage and long process time. Therefore, additional steps such as multiple impregnation are generally required.

4) Liquid (slurry) Infiltration

The slurry infiltration process is developed by preparing a stack of precursor tapes or cloth by dipping them in slurry or coated with slurry. The slurry is chosen to make the matrix for FCMCs. The technology is available for making carbon fiber-reinforced Si_3N_4 composites and SiC_f/SiC composites by using this method. Melt infiltration (Hillig, 1994; Filzer and Gadow, 1986), if placed in this group, however, is a high-temperature infiltration process, in which the solid reactant is melted and flowed into the fibrous or porous preform. The reaction occurs between the preform and the melt to form the desired composite. Such a method has been studied and applied to the fabrication of SiC/SiC composites (Hillig, 1994; Filzer and Gadow, 1986), where the silicon melt infiltrates into the SiC and graphite fibrous preform; the temperature for this processing is about 1450 ℃.

5) Injection Molding

Injection molding is the most promising, automated net shape manufacturing method for ceramic matrix composites with discontinuous reinforcements. This method has been investigated by several national laboratories. However, the control of the whisker orientation by selecting an appropriate aspect ratio, system rheology, shear rate, and mold geometry is still the most significant remaining problem for a real application in manufacturing.

Although many processing techniques for FCMCs have been developed, the processing parameters, such as processing temperature, pressure, time, environments, sintering additives and flux and processing kinetics of diffusion and reaction, are still under study. These parameters not only control the microstructure of FCMCs such as density, phases, and grain size but also influence their mechanical properties, such as the strength, fracture toughness, and characteristics of the interface.

6.2.2.2   Processing Methods for Glass Composites

Glass matrix composites show their most important attribute–the ease of fabrication. In contrast to the other CMCs discussed earlier, melt infiltration and injection molding have been successfully developed on glass and glass ceramic composites and commercially used in industries (Schwartz, 1997). In the laboratory, some small-scale processing techniques such as fiber sandwiching (Deshmukh and Coyle, 1988), sol-gel (Klug et al., 1993), and tape-casting plus HP (Gustafson et al., 1995; Gustafson and Dutton, 1998a) have also been developed. Using these technologies, carbon, Nicalon (SiC) and alumina (Al_2O_3) fiber-reinforced borosilicate, aluminosilicate (LAS), calcium-aluminosilicate (CAS) and other composites have been fabricated. The tape-casting plus hot pressing method was developed (Gustafson et al., 1995; Gustafson and Dutton, 1998) by making tapes from a slurry containing glass powders, laminating the fibers and tapes to form a composite green body, burning out the organic ingredients, and sintering or melting the glass matrix. The distinguishing character of this method is its ease of tailorability to control the distribution and orientation of fibers. A modified tape casting method was developed in this research to prepare SiC fiber-reinforced borosilicate glass composites (Sun and Singh, 1996b). The detailed procedure will be described in Sect. 6.4.1.

6.2.3 Strengthening and Toughening Mechanisms in FCMCs

6.2.3.1 Mechanisms of Strengthening in FCMCs

The strength of ceramics can be improved by several means, such as removing voids and etching or thermally compressing ceramic surfaces. Fiber reinforcement is another important and effective method. The fibers strengthen in the following two ways (Hsueh, 1989; Morley, 1987): a) increase the overall elastic modulus of the fiber-reinforced composite, and b) transfer load between matrix and fibers (load sharing) (Hsueh, 1989).

The elastic modulus of composites has been considered from both the isostrain and isostress situations (Chawla, 1987; Morley, 1987). Assuming that the fibers and matrix are perfectly adhered and that their Poisson ratios are identical, then for the isostrain situation, $\varepsilon_f=\varepsilon_m=\varepsilon_c$, the Young's modulus of composite in the longitudinal direction E_{cl} can be expressed as

$$E_{cl} = E_m V_m + E_f V_f \qquad (6.5)$$

where E is the Young's modulus, subscripts c, f, m denote composite, fiber, and matrix, respectively. V is the volume fraction, and $V_m+V_f=1$. For the isostress situation, $\sigma_{ct} = \sigma_f = \sigma_m$, the Young's modulus of composite in the transverse direction E_{ct} can be expressed as

$$\frac{1}{E_{ct}} = \frac{V_f}{E_f} + \frac{V_m}{E_m} \qquad (6.6)$$

Equations (6.5) and (6.6) are called the rule-of-mixtures for Young's modulus of composites in the longitudinal and transverse directions. The strengthening can be achieved by incorporating fibers of higher Young's modulus with a ceramic matrix in the load direction. The degree of strengthening depends on the fiber volume fraction, fiber orientation, and Young's modulus of the fiber. In this study, the dependency of the elastic modulus of composite on the volume fraction of fibers will be characterized experimentally to compare with the rule-of-mixtures.

The nature of strengthening in fiber-reinforced composites is dependent on the stress transfer between matrix and fibers. It is essential to describe the cases for both continuous fibers and short fibers. In composites with the continuous fiber reinforcement, when $\sigma_m < \sigma_{mu}$, no matrix cracking occurs and the stress transfer to fibers obeys the rule-of-mixtures for stress based on the isostrain situation (Chawla, 1987; Morley, 1987). That is,

$$\sigma_c = \sigma_f V_f + \sigma_m V_m \qquad (6.7)$$

and

$$\frac{\sigma_f V_f}{\sigma_m V_m} = \frac{E_f V_f}{E_m V_m} \tag{6.8}$$

When $\sigma_m = \sigma_{mu}$, the matrix fails. The load once carried by the matrix is then added onto the fibers. Thus, the total stress on the fibers for a steady-state crack is (Aveston and Kelly, 1973; Chawla, 1987; Morley, 1987)

$$\sigma_f = \frac{\sigma_{mu}}{E_m}\left(E_m V_m + E_f V_f\right) = \frac{\sigma_c}{V_f} \tag{6.9}$$

Equation (6.9) is a simple but very important relationship, which is often used for microanalysis and modeling in FCMCs. The stress transfer from fiber to matrix after matrix cracking has been well established (Aveston and Kelly, 1973; Chawla, 1987). The stress is transferred from the fiber to the matrix by interfacial shear stress τ_i:

$$\frac{dF}{dz} = \frac{2V_f \tau_i}{r} \tag{6.10}$$

where dF is the stress transferred from fiber to matrix over a distance dz along the fiber, and τ_i is the shear stress acting at the interface. Therefore, the efficiency of stress transfer is strongly dependent on the maximum value of τ_i, named interfacial shear strength τ_u, which depends on the nature of bonding at the interface. The general criterion for determining the interface type is established as (Sun and Singh, 1998; Budiansky et al., 1995)

$$\tau_u = \text{maximum of } \left(\tau_f, \tau_d\right) \tag{6.11}$$

where τ_f is the frictional interfacial stress and τ_d is the mode II debond shear strength. If τ_d is equal to zero, the interface is unbonded, and if $\tau_d \gg \tau_f$, the interface is strongly-bonded. Otherwise, the interface is considered as weakly-bonded. The stress redistribution in fiber and matrix after matrix cracking has also been analytically determined for both unbonded and partially bonded and debonded interfaces. For an unbonded interface, τ_i is equal to the constant frictional stress τ_f, and the stress distribution in fiber (τ_f) and matrix (τ_m) in the sliding area is determined by the following equations (Fig. 6.5) (Budiansky et al., 1995; Sutcu and Hillig, 1990):

$$\sigma_{\rm f}(z) = \frac{\sigma_{\rm a}}{V_{\rm f}} - \frac{2\tau_{\rm f}z}{r} \tag{6.12}$$

$$\sigma_{\rm m}(z) = 2\tau_{\rm f}\left(\frac{V_{\rm f}}{V_{\rm m}}\right)\left(\frac{z}{r}\right) \tag{6.13}$$

$$\tau_{\rm i}(z) = \tau_{\rm f} \tag{6.14}$$

Therefore, the characteristic length, z', in (6.4) can be derived from (6.13) by assuming that at this location (z'), the recovery of matrix stress from the crack surface reaches the strength of the matrix material, $\sigma_{\rm mu}$.

For a partially bonded and debonded interface, the stress distribution in the cracked area is dependent on the debond length $L_{\rm d}$, which is determined by the interfacial shear strength $\tau_{\rm u}$. Figure 6.4 shows a unit of composite with a debonded interface of length $L_{\rm d}$. The stresses in the debonded region ($0<z<L_{\rm d}$) follow (6.12), (6.13) and (6.14). According to the shear-lag solution given by BHE (Budiansky et al., 1986), the stresses beyond the debonded region ($z>L_{\rm d}$) are determined by the following expressions (Budiansky et al., 1995; Sutcu and Hillig, 1990):

$$\sigma_{\rm f}(z) = \sigma_{\rm f}^{\infty} + \left[\left(\frac{V_{\rm m}}{V_{\rm f}}\right)\sigma_{\rm m}^{\infty} - 2\tau_{\rm f}\frac{L_{\rm d}}{r}\right]\exp\left[-\frac{\rho(z-L_{\rm d})}{r}\right] \tag{6.15}$$

$$\sigma_{\rm m}(z) = \sigma_{\rm m}^{\infty} - \left[\sigma_{\rm m}^{\infty} - \left(\frac{V_{\rm f}}{V_{\rm m}}\right)2\tau_{\rm f}\frac{L_{\rm d}}{r}\right]\exp\left[-\frac{\rho(z-L_{\rm d})}{r}\right] \tag{6.16}$$

$$\tau_{\rm i}(z) = \left(\frac{\rho}{2}\right)\left[\left(\frac{V_{\rm m}}{V_{\rm f}}\right)\sigma_{\rm m}^{\infty} - 2\tau_{\rm f}\frac{L_{\rm d}}{r}\right]\exp\left[-\frac{\rho(z-L_{\rm d})}{r}\right] \tag{6.17}$$

where σ_f^{∞} and $\sigma_{\rm m}^{\infty}$ are stresses in the fiber and matrix far from the debonded region, respectively. ρ is the shear-lag parameter determined by the following formula, $\rho^2 = \dfrac{4E_{\rm c}G_{\rm m}}{V_{\rm m}E_{\rm m}E_{\rm f}\varphi}$, where $G_{\rm m}$ is the shear modulus of matrix and $\varphi = -\dfrac{\left(2\ln V_{\rm f} + V_{\rm m}\left(3-V_{\rm f}\right)\right)}{2V_{\rm m}^2}$. The above expressions are fundamentally identical for AK (Aveston and Kelly, 1973a), BEH (Budiansky, et al., 1995a), HJ (Hutchinson and Jensen, 1990), and Sutcu and Hillig (1990) (SH) models without considering the residual stresses in the composite. The stress distribution along the interface from the matrix crack surface to the far field beyond the debonded region is shown in Fig. 6.5. The interfacial shear stress is shown to have its maximum value at the debonded tip $(z=L_d)$ if τ_d is larger than τ_f.

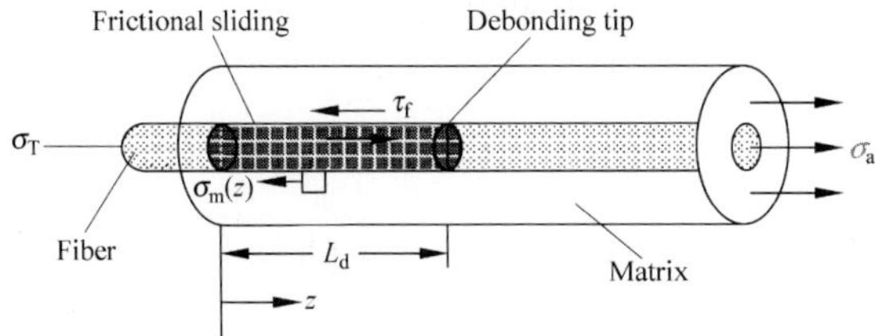

Fig. 6.4 Schematic of interfacial debonding and fiber bridging

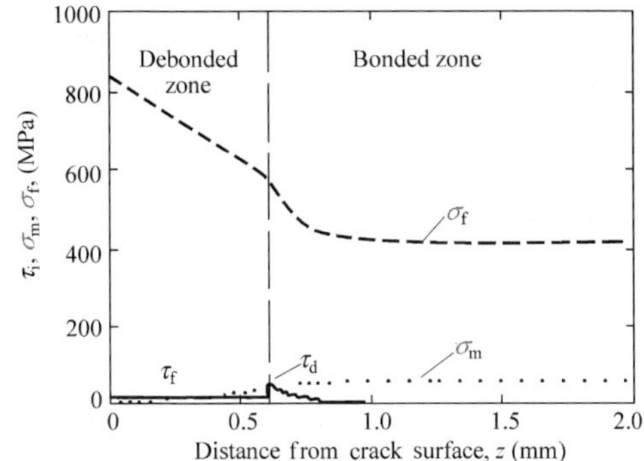

Fig. 6.5 Stress distributions in the fiber, matrix and at the interface

The above analysis is different for composites with the short fiber reinforcement, although the load transfer is still governed by the fiber/matrix interface. The tensile stress, σ_f, carried by the fiber, and the interface shear stress, τ_f, are dependent on the fiber length, l, by the following expressions (Chawla, 1987; Morley, 1987):

$$\sigma_f = E_f \varepsilon \left(1 - \frac{\cosh \beta (l/2 - x)}{\cosh (\beta l/2)} \right) \tag{6.18}$$

and

$$\tau_i = \frac{E_f r_f \varepsilon \beta}{2} \frac{\sinh \beta (l/2 - x)}{\cosh (\beta l/2)} \tag{6.19}$$

where ε is the maximum possible value of strain in the fiber, β is a constant, and x is the position along the fiber from one of the fiber ends. The tensile stress, σ_f, will increase from the two ends of the fiber to a maximum value, while the interfacial shear stress, τ_f, has a maximum value at the fiber ends and a minimum at the fiber central regions. The ratio of $\tau_{i\,max} / \sigma_{f\,max}$ is given by (Chawla, 1987)

$$\frac{\tau_{i\,max}}{\sigma_{f\,max}} = \left(\frac{G_m}{2E_f \ln\left(r_0 / r_f\right)} \right)^{1/2} \tag{6.20}$$

where r_0 is the distance between adjacent fiber centers. This ratio is dependent on the volume fraction of fibers, and the properties of the fibers and the matrix. A typical value of this ratio is of the order of 0.1 for various fiber-reinforced composites (Chawla, 1987). From this analysis, a large $\tau_{f\,max}$ is required in order to utilize the fiber strength effectively, but a very large $\tau_{f\,max}$ is not desired for toughening the composites. Thus, to optimize the value of interfacial shear strength is of great importance to the performance of composites (Chawla, 1987; Morley, 1987).

The tensile strength of FCMCs is a measure of the load-carrying capability of the composites. It includes: a) the first matrix cracking stress, σ_{FMC}, the onset stress of matrix cracking, which is one of the most important properties for designers of composite (Chawla, 1987; Morley, 1987). b) the ultimate strength, σ_{cu}, the stress corresponding to the saturation of matrix cracks and fiber bundle failures beyond which the load-carrying capacity of the composite is greatly diminished (Steif and Schwietert, 1990; He, et al., 1993).

The first matrix-cracking stress, σ_{FMC}, based on the rule-of-mixtures for stress, can be simply written as (Chawla, 1987; Aveston et al., 1971)

$$\sigma_0 = \sigma_f V_f + \sigma_{mu}(1 - V_f) \tag{6.21}$$

where σ_{mu} is the matrix stress at its breaking strain. Aveston et al. (1971) (ACK) derived an expression for the FMC stress, σ_{FMC}, for a long steady-state matrix crack using the fracture mechanics approach, and by assuming that the interfacial fracture energy is very small, and the fiber/matrix interfacial sliding behavior is characterized by a constant shear stress τ_f as

$$\sigma_{\text{FMC}} = \left\{ \frac{6E_{\text{f}} V_{\text{f}}^2 \tau_{\text{f}} E_{\text{c}}^2 \Gamma_{\text{m}}}{E_{\text{m}}^2 V_{\text{m}} r} \right\}^{1/3} \tag{6.22}$$

where Γ_{m} is the fracture energy of the matrix material. MCE (Marshall et al., 1985) and BHE (Budiansky et al., 1986) followed a similar approach, incorporating the effects of crack length, interfacial debonding energy, and residual stress on the FMC stress. They found that the first matrix crack in (6.22) is in fact the stress for initiating a steady-state crack (Marshall et al., 1985; Budiansky et al., 1986). A compressive stress in the matrix upon composite processing is expected to increase the first matrix-cracking stress.

The ultimate strength of composite, which corresponds to the initial failure the fiber bundle, can be written as (Aveston and Kelly, 1973)

$$\sigma_{\text{cu}} = \sigma_{\text{fu}} V_{\text{f}} + \sigma_{\text{m}} (1 - V_{\text{f}}) \tag{6.23}$$

where σ_{fu} is the fiber strength, and σ_{m} is the stress carried by the matrix at the ultimate failure strain of the fibers. A more complicated expression was derived by Curtin (Curtin, 1991) and (He et al., 1993) based on the statistical analysis by employing Weibull's statistics for the fiber strength,

$$\sigma_{\text{cu}} = S_{\text{u}} + V_{\text{m}} \sigma_0 \tag{6.24}$$

where S_{u} is the gauge-length-independent tensile strength, which is given by

$$S_{\text{u}} = V_{\text{f}} \sigma_{\text{fu}} \left[2/(m+2) \right]^{1/(m+1)} \left[(m+1)/(m+2) \right] \tag{6.25}$$

where m is the shape parameter (Weibull modulus). In practice, for most FCMCs, composite failure subject to global load sharing is proceeded by multiple matrix cracking. The matrix does not contribute directly to the composite failure. However, such an indirect influence is vital to the load/stress transfer along the fiber through the interfacial shear stress. The environmental attack on fiber and fiber/matrix interface after matrix cracking can result in the degradation of fiber strength hence a lower ultimate strength of the composite. It is also not difficult to see that both σ_{FMC} and σ_{cu} are insensitive to specimen size and component size. This is in distinct contrast to the monolithic ceramic materials, which show an extreme sensitivity to flaw sizes.

In this chapter, the measurements of FMC stress and the ultimate flexure strength in composites with various volume fractions of fibers will be reported, and these experimental values will be compared with the results obtained from various models, (6.22) and (6.24).

6.2.3.2 Mechanisms of Toughening in FCMCs

Besides the FMC stress and ultimate strength, the fracture toughness is another very important parameter, which is even more critical for the design of composite components. The toughening behavior of FCMCs is usually determined by four possible effects as shown in Fig. 6.6 (Evans, 1990; Wiederhorn, 1984). They are:

a) Crack-bridging fiber: When the fibers remain intact, they carry more load with a larger elastic strain, which results in the constraining stresses on the matrix crack surface; when the fibers fail, the elastic energy stored in the fibers is dissipated through acoustic waves. Both of them positively contribute to toughness.

b) Interfacial debonding: At the weak fiber/matrix interface, debonding generates a new surface and positively contributes to toughness.

c) Fiber pullout: Frictional sliding upon pullout results in local frictional heating, and positively contributes to toughness.

d) Matrix cracking: The relaxation of the residual strain energy because of the matrix crack extension and debonding increases. It lowers the toughness, but it is very small compared to other effects.

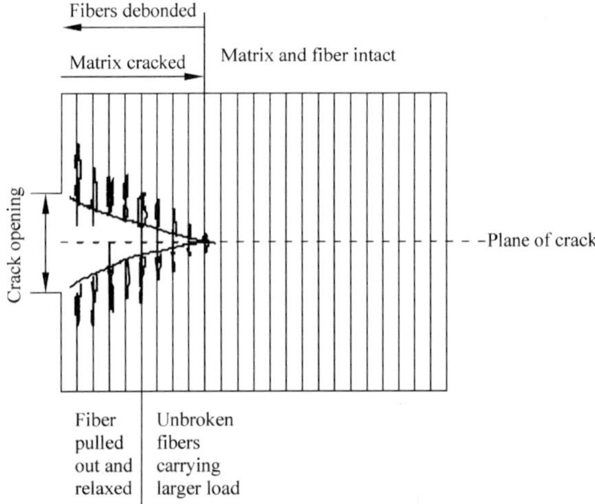

Fig. 6.6 Schematic of the toughening mechanisms in FCMCs

Aveston and Kelly (1973) analyzed the matrix crack-spacing phenomenon from a consideration of the maximum shear stress at the fiber/matrix interface. The stress-transfer mechanism and the critical conditions for the occurrence of debonding in a fiber-reinforced ceramic composite were first proposed. They attributed the enhanced matrix failure strain to the energetic dispersion of multiple cracking of the matrix and the debonding at the fiber/matrix interface. Marshall et al. (1985) presented an insight into the influence of the bridging fibers on the propagation of the matrix crack. The matrix cracking stress of composite with a purely frictional bonding between the fibers and matrix was evaluated for both long cracks in a steady state and small cracks in the nonsteady state. Cooper (1970) found that fiber fracture might occur near, but not at the matrix crack plane in the composite. He attributed this phenomenon to the statistical distribution of the weak points along the fiber length and then used it successfully to explain the fiber pullout in a continuous fiber-reinforced composite. He indicated that the energy absorbed in fiber pullout contributed directly to the work-of-fracture and hence to the toughness of the composite. Since their early work on the toughening analyses in FCMCs, more progress has been made both theoretically and experimentally.

1) Crack-Bridging Fiber

The most important advantage of fiber-reinforced ceramic matrix composites (FCMCs), over the monolithic ceramics, is their higher fracture toughness (higher flaw tolerance) (Hillig, 1987; Evans, 1990). The greatly enhanced toughness of FCMC is largely governed by the bridging (traction) behavior of fibers at the cracked matrix surfaces. As a matrix crack propagates transversely to the fiber, the intact fibers can take additional load because of the larger failure strain of the fiber. This results in the constraining stresses on the cracked matrix surfaces. More external mechanical work needs to be applied to open the matrix crack further. Fibers with a failure strain larger than the matrix material are usually desired in a continuous fiber-reinforced composites to create crack-bridging fibers that can prevent the catastrophic failure. The crack-bridging fiber phenomenon occurs right after the FMC occurs, when the strain in the matrix reaches its ultimate value and the stress in the composite reaches the FMC stress. The portion of load that was once shared by the matrix is thrown onto the fibers. The stresses in the fibers and matrix are locally redistributed in the cracked area. The stress is transferred from the fiber to the matrix by interfacial shear stress as described early in this chapter, where the stress distribution in fiber and matrix along the fiber direction is illustrated for composites with either unbonded or partially bonded and debonded interface.

However, it is not sufficient for understanding the toughening by fiber bridging if only the stress distribution along the fiber-direction (in Figs. 6.4 and 6.5 (Sun and Singh, 1998a)) is known. The fiber-bridging toughening is also determined by the fiber stress distribution in the direction of a matrix crack as shown in Fig. 6.7. The dependence of the fiber-bridging toughness K^L on the fiber stress distribution, $p(X)$, along the crack-propagation direction was defined by MCE (Marshall et al., 1985), as

240

$$K^L = 2(c/\pi)^{1/2} \int_0^1 \frac{\left[\sigma_\infty - p(X)\right]}{\sqrt{1-X^2}}$$ (6.26)

where $p(X)$ is the distribution of fiber-bridging stress (also called traction pressure in some other papers) on the fracture surfaces, c is the crack length, σ_∞ is the externally applied stress, and X is equal to x/c (where x is the real position along matrix crack shown in Fig. 6.7.) possessing a normalized value from 0 to 1. The determination of the in-situ distribution of the fiber bridging stress $p(X)$ or $p(x)$ along a matrix crack is the critical part for evaluation of the toughness. Therefore, any attempt to directly measure $p(x)$ will have an important impact on the fundamental understanding of factors contributing to toughness in composites.

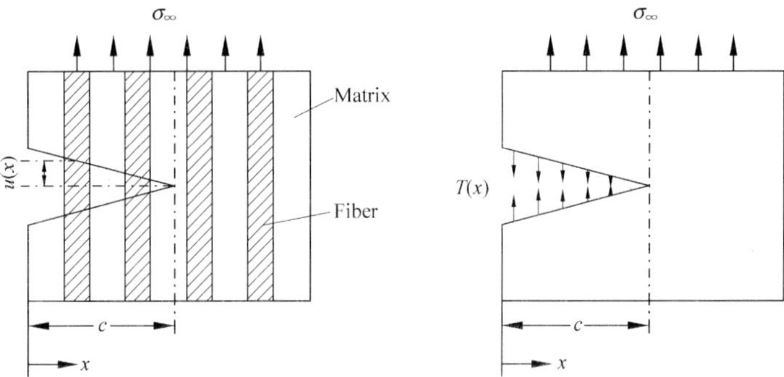

Fig. 6.7 Schematic diagrams showing fiber bridging and associated stress distributions in a FCMC

The conventional method for determining the distribution of fiber-bridging stress is the crack opening displacement (COD) measurement. The relationship between $p(x)$ and $u(x)$ can be generally expressed as (Sakai, 1992; Marshall and Cox, 1988e)

$$p = Au^n$$ (6.27)

where A is a constant related to material parameters such as elastic moduli of fiber and matrix, fiber volume fraction, and interfacial properties. Based on the shear-lag model in a cracked composite under tension with a frictionally-coupled interface, Marshall, Cox, and Evans (MCE) (Marshall et al., 1985) made the first derivation. Later, Marshall and Cox (MC) (1988), McCartney (1987), Xia et al. (1994) and Meda and Steif (1994) gave a modification to relate the fiber bridging stress $p(x)$ to COD:

$$p = 2\left[u\tau V_{\mathrm{f}}^2 E_{\mathrm{f}}(1+\eta)^2 / R\right]^{1/2} \tag{6.28}$$

where $\eta = E_{\mathrm{f}}V_{\mathrm{f}} / E_{\mathrm{m}}V_{\mathrm{m}}$, R is the fiber radius, and τ is the sliding frictional stress at the interface. They found that n was equal to 1/2. This relationship was used not only for determining the fiber-bridging stress and fiber-bridging toughness but also for determining the matrix cracking stress, σ_0, and critical crack size, c_0. The matrix cracking stress and critical crack size for a steady-state crack ($c>c_0$, c_0 is the transition crack length) are derived as (Marshall et al., 1985)

$$c_0 = w\sigma_\infty E_{\mathrm{c}} R / \tau V_{\mathrm{f}}^2 E_{\mathrm{f}}(1+\eta)(1-v^2) \tag{6.29}$$

and

$$\sigma_0 = \sigma_\infty = \delta'\left[(1-v^2)K_{\mathrm{c}}^{\mathrm{M}^2}\tau E_{\mathrm{f}}V_{\mathrm{f}}^2 V_{\mathrm{m}}(1+\eta)^2 / E_{\mathrm{m}}R\right]^{1/3} \tag{6.30}$$

where w is a dimensional constant, $\delta' = (wv^2)^{-1/3}$, and $K_{\mathrm{c}}^{\mathrm{M}}$ is the critical stress intensity factor for matrix. It is noted that the stress required to propagate a matrix crack is almost independent of the crack length for cracks larger than about $c_{\mathrm{m}}/3$. The extension stress will decrease with the crack size for a short crack.

On the other hand, Sakai (1992) employed the Dougdale model on the cracked region of composite by assuming a uniform traction stress. Thus, $n=0$ was defined in his simplification, which implied that COD may not strongly influence the fiber-bridging stress. The equation can be expressed by $p = A = \sigma_b$, the uniform stress over the entire bridging zone. In his analysis, the toughness increment ΔK_{b} by fiber bridging was defined from crack extension, $\Delta\alpha$, and uniform bridging stress,

$$\Delta K_{\mathrm{b}} = 2\sigma_{\mathrm{D}}\sqrt{\frac{a}{\pi}}\arccos\left(1-\frac{l_{\mathrm{b}}}{a}\right) \tag{6.31}$$

where a is the matrix crack length and l_{b} is the length of the crack-bridging zone. A rising R curve behavior of the fiber-reinforced composite during the fiber-bridging process was derived from (6.31) (Sakai, 1992). The toughness increases with the extension of matrix crack. Using this technique, the COD profile $u(x)$ is measured and related to the fiber-bridging stress profile. Although many studies have been conducted to extract the fiber-bridging toughness from the COD profile, no independent verifications have been achieved for these

242

models.

Another direct technique for measuring fiber-bridging stress using laser raman spectroscopy (LRS) has also been developed (Haerle et al., 1990; Sumarasiri, et al., 1994; Mendoza, et al., 1991). This technique has been successfully used to directly measure the fiber stress in polymer matrix composites, and some attempts have also been made in FCMCs (Mendoza and Cannon, 1991). Although the application of the LRS technique to FCMCs is limited by many factors, the LRS technique is still a promising approach.

2) Interfacial Debonding

In most FCMCs, the matrix cracking and fiber bridging are always associated with interfacial debonding, i.e., the failure of the cohesive bond between fiber/matrix, fiber/coating, and coating/matrix. The failures initiate at the fracture face of the matrix where the interfacial shear stresses have a maximum value (Aveston and Kelly, 1973). The debonded region will then propagate along the interface, provided sufficient work is done on the system to supply the surface energy of the newly created fracture.

The progress in understanding the debonding behavior was mostly done theoretically. Two fundamental approaches have been employed to establish the relationship between debond length and externally applied stress: the force balance approach (Budiansky, et al., 1995; Hutchinson and Jensen, 1990; Hsueh, 1996) and energy balance approach (Sutcu Hillig, 1990; Li et al., 1993). According to the force balance approach, the debond length L_d is related to the applied stress σ_a by the following equation (Budiansky et al., 1995; Hsueh, 1996b),

$$\frac{L_d}{r} = \left(\frac{V_m E_m}{V_f E_c} \right) \frac{\sigma_a - \sigma_b}{2\tau_f} \tag{6.32}$$

The external stress required to initiate the interfacial debonding, namely the debond initiation stress σ_d, can be determined by various expressions derived from different models as:

$$\sigma_d^{BEH} = 2V_f \sqrt{\frac{E_f E_c \Gamma_d}{V_m E_m r}} \quad \text{(BEH (Budiansky et al., 1995))} \tag{6.33}$$

$$\sigma_d^{ACK} = V_f \sqrt{\frac{4E_f \Gamma_d}{r}} \quad \text{(ACK (Aveston et al., 1971))} \tag{6.34}$$

$$\sigma_d^{HJ} = \frac{1}{c_1} \sqrt{\frac{E_m \Gamma_d}{r}} \quad \text{(HJ (Hutchinson and Jensen, 1990))} \tag{6.35}$$

where c_1 is a constant calculated from the material parameters. Despite the different expressions, these equations are consistent with the fact that the interfacial properties are controlled primarily by the interfacial debond energy Γ_d. The larger the interfacial debond energy, the larger is the debond stress. All the above expressions for σ_d have not considered the effect of interfacial frictional stress. In a composite with a bonded interface of nonzero coefficient of friction, σ_d can be defined from the shear-lag theory as (Budiansky et al., 1995)

$$\sigma_d = \left(\frac{2V_f E_c}{\rho V_m E_m} \right) \tau_d \tag{6.36}$$

where τ_d is available from (6.37) or (6.38). For a frictionally-coupled interface, τ_d is reduced to τ_f, and σ_d is reduced to σ_{slide}, a stress required to initiate an interfacial sliding (Budiansky et al., 1995). The mode II debond shear stress, τ_d, is equal to the interfacial shear strength τ_u for a bonded interface with zero coefficient of friction as estimated by BEH (Budiansky, et al., 1995a),

$$\tau_d^{BEH} = \sqrt{\frac{4G_m \Gamma_d}{r\varphi}} \quad \text{(BEH)} \tag{6.37}$$

where Γ_d is the interfacial debond energy. For a bonded interface with nonzero coefficient of friction, SH (Sutcu and Hillig, 1990) correlated the debonding energy Γ_d to the interfacial debond strength τ_d and frictional shear stress τ_f by the following expression:

$$\tau_d - \tau_f = \sqrt{\frac{4G_m \Gamma_d}{r\varphi}} \quad \text{(SH (Sutcu and Hillig, 1990))} \tag{6.38}$$

Equations (6.36) and (6.38) differ from (6.33), (6.34), (6.35), and (6.37) mainly because SH (Sutcu and Hillig, 1990) considered the contribution of shear strain to elastic strain energy at the appearance of frictional shear stress. The debond initiates when σ_a is equal to σ_d. With an increase of the external stress, the debond propagates until the interfacial shear stress at the debond tip is equal or less than the interfacial shear strength τ_u. Using an energy balance approach, Li et al. (1993) theoretically analyzed the influence of the dissipation energy of friction (G_s) and interfacial debonding energy (G_D) on the relationship between debond length and external stress, and obtained the following relationship,

$$\frac{L_{\mathrm{d}}}{r} = \frac{\sigma_{\mathrm{a}}}{2\tau_{\mathrm{f}}V_{\mathrm{f}}} \left\{ 1 - \left[\frac{V_{\mathrm{f}}E_{\mathrm{f}}}{E_{\mathrm{c}}} + \frac{V_{\mathrm{m}}E_{\mathrm{m}}}{E_{\mathrm{c}}} \left(\frac{\sigma_{\mathrm{d}}}{\sigma_{\mathrm{a}}} \right)^2 \right]^{1/2} \right\} \qquad \text{(Li et al. (1993))} \qquad (6.39)$$

This relationship differs from the results of the force balance approach in terms of the appearance of an extra offset term of $(\sigma_{\mathrm{a}}/\tau_{\mathrm{f}})^{1/2}$ (Li et al., 1993). However, Hsueh (Hsueh, 1996) found that the energy balance approach gave the same expression of the debond length-external stress relationship as the force balance approach. This implied the importance of theoretical assumption and modeling to the analysis in composites.

Although the measurements of interfacial properties such as τ_{f} and Γ_{d} are facilitated by the availability of new techniques (Ma and Clarke, 1993; Zhou et al., 1992; Jurewicz et al., 1989), the above theoretical relationships between debond length and external stress have not been closely examined by experiments because of the difficulty in measuring the debond length. Therefore, the attempt to experimentally verify those models will provide more insight into the understanding of fiber bridging and interfacial debonding. This has been done in this research.

3) Fiber Pullout

Fiber pullout is the process in which the work is done against frictional forces in extracting the broken fibers from matrix crack faces (Morley, 1987). It only happens if the fiber/matrix interface is initially unbonded or weakly bonded. The high shear stresses developed at the matrix crack surface can cause failure of the interfacial bond. Even as the debonded region propagates along the fiber/matrix interface, the interface can still be frictionally contacted. When τ is constant, the shear stress carried by the embedded fiber decreases linearly with increasing distance from the matrix crack face. The critical distance at which the ultimate tensile strength of the fiber is reached is $l_{\mathrm{c}}/2$ (Chawla, 1987; Morley, 1987). Assuming that the embedded fiber length is x and the total fiber length is l, then the two possibilities for fiber pullout will be (Morley, 1987): a) when $x<l_{\mathrm{c}}/2$, the embedded fiber will be extracted as the matrix crack faces continue to separate, and the force required to pull out the fiber decreases as the fiber is extracted. b) when $x>l_{\mathrm{c}}/2$, the embedded fiber will fracture before being extracted. In most cases, the fiber pullout process is a combination of these two situations, and the effective work of fracture continues to increase with the crack size.

However, in order to explain the substantial pullout phenomena in continuous fiber-reinforced composites, Cooper (1970) believed that, for a real continuous fiber, there are weak points along the fiber length. As a matrix crack propagates through the composite, a fiber will be broken either where it crosses the crack plane or at a flaw in the fiber that is near the plane. The tendency to failure at the weak points is determined by the frequency and degree of the weakness, and these two parameters have a direct influence on the mean pullout length of broken fibers. He introduced a critical space of flaws, d_{cr}, and the strength σ^* at flaws in a fiber. When $d<d_{cr}=(\sigma-\sigma^*)l_c/\sigma$, all fibers will fracture at the weak points; when $d>d_{cr}$, only a fraction d_{cr}/d of the fibers fracture at flaws, while the rest fracture at stress σ. Besides the weak points in fibers, other parameters such as Poisson's ratio, surface roughness of fiber and residual stresses produced by thermal contraction, can also influence the pullout behavior, and thus the toughness.

4) Matrix Cracking

The contribution of the matrix cracking to toughness is usually minor to the fiber-bridging toughness. Since Aveston and Kelly (1973) (AK) proposed the fundamental concepts and relationships among matrix crack spacing, interfacial debonding (or sliding) length, and interfacial shear stress of a continuous fiber-reinforced composite, Zok and Spearing (1992) developed a model for multiple matrix crack spacing showing both the periodic and random cracking patterns. Weitsman and Zhu (1993) employed an energy criterion to analyze the stress-strain relationship for the initial nonlinear curve (A–B). Curtin (1993), using the statistical approach, theoretically analyzed the evolution of multiple matrix cracking and related the crack spacing to the flaw distribution. Most of their analyses were done by assuming an interface loosely coupled by friction.

The influence of the multiple matrix cracking on the whole fracture toughness of the composite has been studied by Sih (1973) for a tensile test and Dharani and Tang (1989) for a pure bending. A function $F(a,c)$ was introduced to modify the stress intensity factor. The explicit expression for this factor will be given later in Sect. 6.7.

5) Toughness Determination

One of the objectives of engineering design is to determine the geometry and dimensions of the machine or structural elements and select materials in such a way that the elements perform their operating function in an efficient, safe, and economic manner (Gdoutos 1993; Sih and Faria 1984). The applications of ceramic materials in structures, electronics, and many other industries are limited by their brittle catastrophic failure characteristics because of the low fracture toughness. Consequently, attempts have been made to enhance the fracture resistance of ceramics so that the catastrophic failures of ceramic components can be prevented in service. Such an enhanced fracture resistance has been achieved by many techniques and mechanisms, such as fibers, whiskers, and particle reinforcements, phase transformation at the cracked zone, microcrack formation, and crack deflection and bowing (Evans, 1990; Wiederhorn, 1984). Fiber reinforcement is one of the promising methods not only because it can improve the toughness of ceramic materials considerably but also because both the microstructure and properties of the composites can be tailored for requirement of the engineering design.

In contrast to the stress-analysis-based engineering design of components, the fracture mechanics design is based on the realistic assumption that all materials contain initial defects that can affect the load-carrying capacity of engineering structures (Gdoutos 1993; Sih and Faria, 1984). In this approach the stress intensity factor, K, is the fundamental quantity that governs the stress field near the crack tip. The stress intensity factor depends on both the geometrical configuration and loading conditions of the body. The experimental determination of stress intensity factor can be performed using ASTM standard tests (ASTM E399-83) on notched (single-or double-edge-cracked) specimens subjected to a uniform tension, three-point flexure (Llorca and Singh, 1991) or wedge opening loaded (WOL) specimens (Sakai et al., 1991; Miyajima and Sakai, 1991). The stress intensity factor at the tip of a crack in a three point bending mode can be related to the external load P as (Llorca and Singh, 1991)

$$K_{\mathrm{p}}(a) = \frac{3PS}{2bw^2}\sqrt{\pi a}F(a/w) \qquad (6.40)$$

where S, b and w are span, thickness, and width, respectively, of the specimen, a is the crack length, and $F(a/w)$ is a dimensionless function (Tada, et al., 1985) but dependent on the crack length a. By measuring the applied load P and the corresponding crack length a, the in-situ stress intensity factor K_{p} can be calculated. The WOL fracture toughness test, however, utilizes a tensile load to open an initial crack. This technique was extensively used by Sakai group for their carbon/carbon composites (Sakai et al., 1991b; Miyajima and Sakai, 1991). As an alternative, the crack intensity factor can also be determined by a single-edge precracked beam in four-point bending method (SEPB-B and SENB-S in Europe) from (Damani et al., 1996)

$$K_{\mathrm{p}}(a) = \frac{P}{b\sqrt{w}}\frac{S_1 - S_2}{w}\frac{3\sqrt{\alpha}M(\alpha)}{2(1-\alpha)^{3/2}} \qquad (6.41)$$

where S_1 and S_2 are the outer and inner spans, respectively, of the four point bending fixture, α is equal to a_0/w, and $M(\alpha)$ is a dimensionless parameter given by (Damani et al., 1996)

$$M(\alpha) = 1.9887 - 1.326\alpha - \frac{(3.49 - 0.68\alpha + 1.35\alpha^2)\alpha(1-\alpha)}{(1+\alpha)^2} \qquad (6.42)$$

Equations (6.41) and (6.42) are applicable to long sharp cracks with α values between 0.4 and 0.6. These equations were found to be applicable to most ceramic systems such as partially stabilized zirconia (PSZ), Al_2O_3, sintered SiC, and hot pressed silicon nitride (Damani et al., 1996).

Despite the above experimental methods, several different theoretical analyses have been done to determine the toughness contribution of fiber-bridging stress to composite using the schematic diagram shown in Fig. 6.7. For a single crack, the equilibrium condition for a crack of length a is attained when the crack propagation driving force $K_p(a)$ is balanced by the crack growth resistance of the composite because of the fiber shielding effect. From a micro point of view, two cohesive forces are acting on the crack faces. One is the intrinsic atomic cohesive force at the crack tip defined by Brettnette (Barenblatt, 1962). It contributes to the fracture toughness of the matrix K^M. The other is the extrinsic shielding forcedue to the fiber-bridging effect in the wake of the crack. It contributes to the enhanced toughness K_b. Therefore, it has been well accepted that the crack growth resistance (fracture toughness K_R) of a composite is the sum of the intrinsic fracture toughness of the matrix K^M and the enhanced fiber-bridging toughness K_b, which can be expressed as (Llorca and Singh, 1991; Dharani and Tany, 1989)

$$K_P = K_R = K^M + K_b \tag{6.43}$$

When a local crack-bridging stress $\sigma(x)$ is acting on the crack surfaces, then the stress intensity factor at the crack tip dK can be given by (Llorca and Singh, 1991)

$$dK_\sigma(a,x) = \frac{2\sigma(x)}{\sqrt{\pi a}} H(a/w, x/a) dx \tag{6.44}$$

where x is the distance from the crack mouth to the point where the stress is applied. $H(a/w, x/a)$ is a geometrical dimensionless function. The fiber bridging toughness K_b can be given by (Llorca and Singh, 1991)

$$K_b(a) = \int_{a_0}^{a} \frac{2\sigma[u(a,x)]}{\sqrt{\pi a}} H(a/d, x/a) dx \tag{6.45}$$

where $u(a, x)$ is the crack-opening displacement function for a crack with length, a, at the position x. If the bridging stress in the crack wake, $\sigma(x)$, and the weight function, H, are known, then the enhanced fiber-bridging toughness K_b can be determined numerically.

Another approach for fracture toughness analysis, J-integral, has been applied to composites as well (Marshall and Cox, 1988; Hashida et al., 1994). Because of the singularity at the crack tip, the J-integral of the contour around a crack is equal to zero, which can be expressed as (Marshall and Cox, 1988; Hashida et al., 1994)

$$J_\infty + J_b + J_{tip} = 0 \tag{6.46}$$

where J_b is the integral resulting from the fiber bridging in the crack wake, and J_{tip} is the integral resulting from the crack-tip singularity. Both the intrinsic cohesive force at the crack tip and the fiber-bridging force in the crack wake pull the crack faces together and induce stress singularity at the crack end which cancels out the stress singularity introduced by the applied load. Consequently, the stresses are bounded and the faces of the crack join smoothly in cusp form at the ends. These two integrals are given by (Marshall and Cox, 1988; Hashida et al., 1994),

$$J_b = -2\int_0^u \sigma(u)\mathrm{d}u \tag{6.47}$$

$$J_{tip} = \frac{K_c^2(1-v_c^2)}{E_c} \tag{6.48}$$

It is worth noting that the relationship between the fiber-bridging stress and the crack opening displacement, the σ-u function, is critical to determining the fracture toughness of composites for both the J-integral and crack-intensity factor K approaches.

Fracture toughness of composites can also be characterized by WOF (work-of-fracture). WOF is evaluated from an energy approach. It is defined as the area under the load-displacement curve. It indicates how much energy is needed to make a crack per unit surface area. Numerically, WOF is equal to G, the elastic energy release rate. It includes the elastic strain energy contained in fibers and the energy consumed by interfacial debonding and fiber pullout. K_{Ic} is evaluated from a stress approach. It determines the extent of stress concentration in front of the matrix crack-tip at failure, which is largely reduced by the fiber-bridging stress. However, the well-known equation $K_c^2 = E_c G_c$ is not satisfied for fiber-reinforced ceramic composites because the assumption that the body is isotropic and homogenous for monolithic ceramics does not apply for fiber-reinforced ceramic composites (Hillig, 1987).

Recent advances in the fracture mechanics enhanced our understanding of the fibers toughening behaviors in ceramic composites. Fiber bridging applies a closure stress on the matrix crack faces and reduces the extent of stress concentration in front of the crack tip. Interfacial debonding along the fiber length produces a new crack surface vertical to the matrix crack and increases the work-of-fracture for matrix crack propagation. Fiber failure and pullout on one hand release the elastic energy once stored in the fiber through acoustic waves, and on the other hand dissipate the crack -propagation energy or work-of-fracture due to the frictional movement in the interface pullout region.

However, the theoretical analysis of toughening mechanisms in fiber-reinforced ceramic composites still needs more experimental support. In particular, the fiber-bridging mechanism is based on the assumption that the closure stresses at the matrix crack faces are related to the crack opening displacement by elastic fracture mechanics. This assumption is still open to debate and awaits verification by critical experiments.

6.3 Objectives and Approaches

The special interests of this study are focused on the initial nonlinear portion (*A-B*) of the load-displacement curve in Fig. 6.1. The curiosity and uncertainty of this behavior of composite pose many unanswered questions, such as: 1) When do the first matrix cracking and interfacial debonding occur? 2) How do the multiple matrix cracks interact with debond? 3) What are the relationships among the applied stress, debond length, amount of cracks, and interfacial properties? 4) How can the fiber-bridging stress be determined during matrix crack propagation? 6) What are the relationships among the fiber-bridging stress, debond length, and crack opening displacement? 7) How much does the fiber bridging contribute to the overall toughness of the composites? Answers to these questions will provide a better understanding of the strengthening and toughening mechanisms of fiber-reinforced ceramic matrix composites (FCMCs).

Therefore, the objectives of this study are to analyze the micromechanisms of matrix crack-interfacial debond interaction, develop a new technique and models for determining the fiber-bridging stress, and finally characterize the contribution of fiber bridging to the overall toughness of FCMCs. The following approaches/steps are taken to accomplish these goals:

1) Processing: preparing a transparent and toughened glass composite.

2) Characterization: measuring the mechanical and interfacial properties of composite.

3) Failure analysis: analyzing the micromechanisms of matrix cracking and interfacial debonding, investigating the relationship between applied stress and debond length, and introducing a novel debond length measurement (DLM) technique.

4) Analytical modeling: designing a simplified model for the DLM technique to establish the theoretical relationships (bridging law) among fiber-bridging stress, COD, and debond length.

5) Experimental validation: measuring in-situ the debond length, COD, and crack length, calculating the fiber-bridging stress distribution and fiber-bridging toughness, and comparing the experimental results with theoretical models.

Through these approaches, both the debond length measurement (DLM) technique and new fiber-bridging laws will be developed and applied for the determination of the fiber-bridging stress and toughness. It is worth mentioning that the debond length measurement (DLM) technique developed in this study is the technique initially used for determining the fiber-bridging stress, fracture toughness, and interfacial properties in FCMCs.

6.4 An Unconventional Process of Glass Composites

Most ceramic fiber-reinforced glass composites were made by the powder infiltration plus hot pressing method (Gustafson et al., 1995b; Gustafson and Dutton, 1998a). However, the sample dimensions were limited by the chamber size and die size of the hot pressing machine; and the fiber orientation was difficult to control under an applied load. The new fabrication route developed during this study is called "tape casting/binary sintering" ("TCBS") (Sun et al., 1996). This method was successfully applied to a SiC fiber-reinforced borosilicate glass composite system and produced a transparent and toughened composite.

6.4.1 Fabrication of Glass Composites

6.4.1.1 Selection of Materials

The composite system was chosen on the basis of the availability and properties of fibers and matrix materials. SCS-6 silicon carbide fibers (Textron Specialty Materials, Lowell, MA) were used as continuous reinforcement materials because of their high elastic modulus, high strength and ultimate strain, and thermochemical stability at elevated temperatures. This monofilament was made from a 37-μm-diameter carbon core by depositing the 50-μm-thick silicon carbide and about 3-μm-thick carbon and carbon-silicon coating layers resulting in an overall fiber diameter of 142 μm (Bhatt and Hull, 1992a). The density of SCS-6 fiber (ρ_f) is 3.11 g/cm^3. The double coatings play a very important role in controlling the fiber/matrix interfacial properties. The typical mechanical properties of SCS-6 fiber are listed in Table 6.1.

Table 6.1 Mechanical properties of SiC fiber and F glass

Materials	Elastic modulus (GPa)	Strength (GPa)	Failure strain (%)	α^* ($\times 10^{-6}/°C$)
SiC(SCS-6) fiber	400	3.4	0.8–1.0	4.3
F glass[**]	56	0.056	0.1	4.25

* The composition for F glass is 76% SiO_2 · 16% B_2O_3 · 8% K_2O;

** Coefficient of thermal expansion (25–500 °C)。

In order to obtain a transparent composite, borosilicate F glass (Corning Glass Works, NY) was chosen as the matrix material. The composition of F glass is 76% $SiO_2 \cdot 16\%$ $B_2O_3 \cdot 8\%$ K_2O. Its theoretical density (ρ_m) is 2.24 g/cm^3. F glass has its softening temperature at 710 ℃ and melting temperature at 1080℃. There is very little chance for it to crystallize. F glass was received in the form of powder with a particle size of about 10 μm. The glass properties are also listed in Table 6.1. This glass was selected especially because its thermal expansion coefficient is quite close to that of the SCS-6 fibers.

6.4.1.2 Fabrication Procedures

The fabrication procedure for SiC fiber-reinforced borosilicate glass composites is shown in Fig. 6.8.

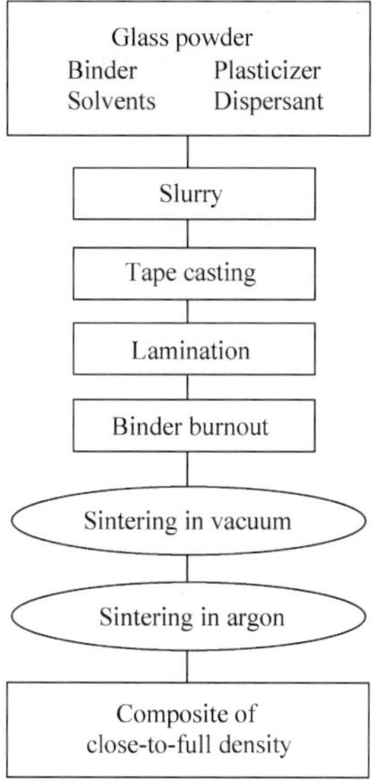

Fig. 6.8 Processing flowchart of TCBS (tape casting/binary sintering)

1) Binder, Dispersant, Plasticizer, and Solvents
Polyvinyl butyral resin (Butvar) powder with a particle size of 100 μm was used as a polymeric binder to provide a bridge between glass powders and increase the strength of green tape. Disodecyl glutarate (Plasthall DIDG) was added as dispersant to defloculate the slurry system. It is very important to prevent the agglomeration of the glass particles in slurry before tape casting. Dipentaerythritol penta-acrylate (Monomer DPE) was used as a plasticizer. Ethanol and toluene were used as solvents. They were chosen because the binder, dispersant and plasticizer can all be dissolved in toluene and ethanol.

2) Tape Casting from Slurry

The glass and binder powder, dispersant, plasticizer and solvents were weighed and added into a milling bottle. The amount of binder, dispersant, plasticizer and solvents was determined experimentally based on the principle of making a uniform slurry with minimum organic materials to reduce the potential of a large shrinkage, gas blowout and cracking during binder burnout. The mixture underwent milling for as short as one hour. This step provided a defloculated slurry. The slurry was deaired for 5–10 min. before tape casting. The doctor blade for tape casting was set at 0.32 mm above the glass platform of the tape caster. After drying, the thickness of the tape was about 0.22 mm. The thickness of an after-dry tape was controlled by adjusting the gap and velocity of the doctor blade. The shrinkage of green tape due to volatility of solvents was about 30%. A slow drying speed was chosen to achieve a strong green body with no cracks. The tape was cut into smaller pieces of 55 mm × 40 mm for lamination.

3) Fiber Alignment and Lamination

The reinforcement SCS-6 fibers with a length of 60 mm were aligned to form a fiber mat. The distance between fibers was changed to obtain a specific volume fraction of fibers. Typically, six fiber layers were prepared for a bulk composite sample. The green tapes were sprayed with a binder solution on both surfaces to enhance both fiber layer-green tape and green tape-green tape adhesion. The fiber layers and green tapes were stacked and pressed before lamination. The lamination was performed in a two ton hydraulic laminating press machine (Tetrahedron Inc., San Diego, CA). Both stages of the machine were heated to a temperature of 125 ℃. After reaching this temperature, the sample was put into the machine and were held under a pressure of 100 psi (1 psi=6894.8 Pa) for 10 min to obtain a uniform temperature across the thickness of the green body. The pressure was increased to 1000 psi and maintained at this level for about 15 min. Then, the laminate was taken out from the machine. This is one of the important steps in fabrication because the temperature, pressure, and hold time strongly influence the uniformity of fiber distribution, uniform deformation of green state matrix, and interlayer delamination after binder burnout. The laminate was cut with a diamond blade to dimensions of 52 mm × 33 mm.

4) Binder Burnout

Thermogravimetric analysis (TGA) and differential thermal analysis (DTA) were conducted on a green tape. It was found that the weight loss started at 420 ℃ and was complete at around 580 ℃. Therefore, the laminated composite was put into an air furnace, heated in air with a heating rate of 2 ℃ /min to 420 ℃ and kept at 420 ℃ for an hour to complete the partial decomposition. The sample was heated with the same heating rate to 580 ℃ and kept at this temperature for about 1.5 h. The color of the composite during binder burnout changed from white green (RT) through brown (420 ℃) to white (580 ℃). Experimental observation showed that the slow heating rate is very important for binder burnout because that process involved the decomposition and oxidation of organic materials. Finally, the laminate was cooled to room temperature with a cooling rate of 4 ℃/min. The organic ingredients such as binder, plasticizer, and dispersant were removed during this step.

5) Binary Sintering

The sample after binder burnout was carefully placed on a ZrO_2 rectangular crucible because the sample was fragile after the removal of organic materials such as binder and plasticizer. A sheet of graphite foil was placed between the sample and crucible to prevent sticking or reaction of the glass matrix to the crucible. The crucible was then inserted in a horizontal alumina tube of a high-temperature vacuum furnace. The temperature/time (T/t) diagram for sintering is shown in Fig. 6.9. The term of binary sintering stemmed from the two characteristic sintering phases. The sample initially was heated to 725 ℃ (just beyond the softening point of F glass) in vacuum at a rate of 2 ℃/min and kept at that temperature for about 2 h. The vacuum condition was 60 mTorr (7.99932 Pa). Then the furnace temperature was increased to 940 ℃ with the same heating rate, but at 900 ℃ argon gas was introduced into the furnace with a pressure of about one atmosphere. The sample was kept at 940 ℃ for about half an hour, and finally cooled to room temperature at a rate of 4 ℃/min.

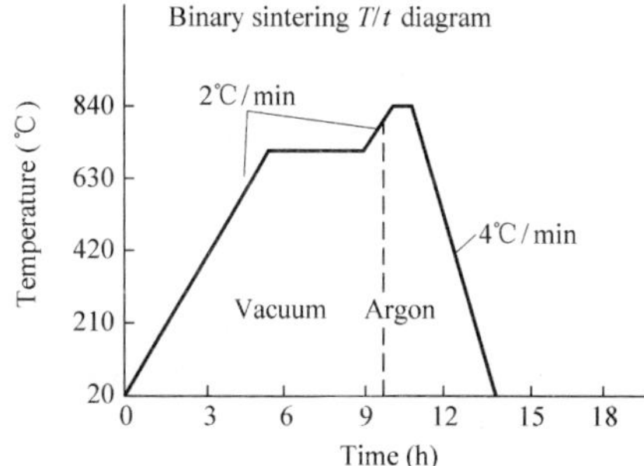

Fig. 6.9 The temperature / time (T/t) diagram for binary sintering of glass composites

6.4.2 Microstructure of Composites

6.4.2.1 Density, Uniformity and Transparency of Composites

The sample after sintering was mounted and ground with a diamond wheel on both surfaces. It was polished on the automatic vibratory polishing machine for the last step (0.03 μm). The rough edges of samples were cut with a diamond wafering saw with a total useful surface dimension of about 50 mm × 30 mm. The thickness of the typical sample varied from 0.9 mm to 1.4 mm. The further sample preparation was dependent on the requirement of different tests. Bar-shaped specimens were prepared for four-point flexure test and in situ microanalysis, and thin slice samples were made for the fiber pushout test.

The as-fabricated composites with different volume fraction of fibers had a fairly good transparency as shown in Fig. 6.10. Figure 6.11 shows the cross section of composite with uniformly distributed fibers. It was also found that a few small voids existed at the fiber/matrix interface because of the lower wettability of the carbon-enriched coating on the fiber with the glass.

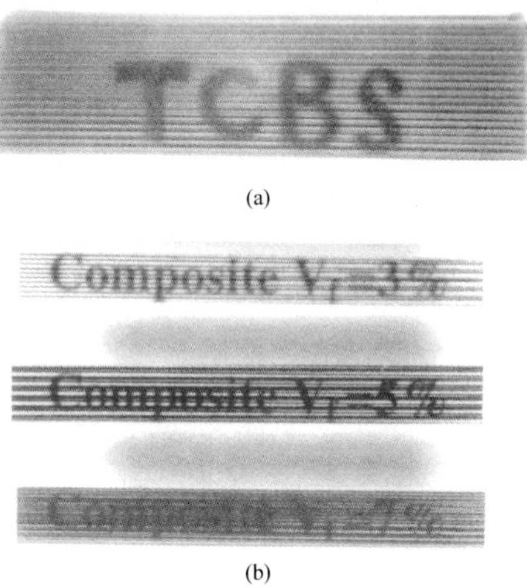

Fig. 6.10 Optical micrographs of (**a**) the as-fabricated and (**b**) the polished transparent glass composites made by TCBS method

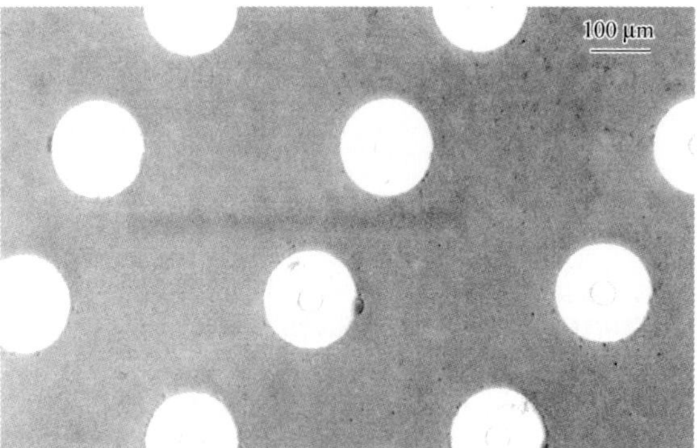

Fig. 6.11 Cross section of composite with uniformly distributed fibers

The densities of both the pure glass and glass matrix in the composite were calculated using the following formula:

$$\rho_{\mathrm{m}} = \frac{W_{\mathrm{c}} - N_{\mathrm{f}} L_{\mathrm{f}} A_{\mathrm{f}} \rho_{\mathrm{f}}}{L_{\mathrm{f}}(wh - N_{\mathrm{f}} A_{\mathrm{f}})} \tag{6.49}$$

where W_c is the weight of composite, N_f, L_f, A_f are the amount, length and area of cross section of fibers in composite, respectively. w and h are the width and height of the composite. The fiber length L_f is also the length of composite. It was assumed that the sample after binder burnout had the same dimensions as the as-laminated sample. The density variation of the composite matrix and pure glass during processing is listed in Table 6.2.

Table 6.2 Density of the glass and glass composite matrix made by TCBS

Sample condition	Density of matrix (g/cm³)	Percentage to full density* (%)
Lamination	1.50	67
Burnout	1.20	54
Sintering	2.19	97
Glass by TCBS	2.21	98–99

* The full density of F glass is 2.24 g/cm³.

A close-to-full density of composite was achieved using this TCBS method.

The transparency of composite was determined from the absorption coefficient of the pure glass sample processed under similar conditions to processing of composites with varying sintering temperatures. The absorption coefficient of glass samples was measured using an optical spectrometer. The higher the sintering temperature, the better the transparency of the glass samples, as shown in Fig. 6.12.

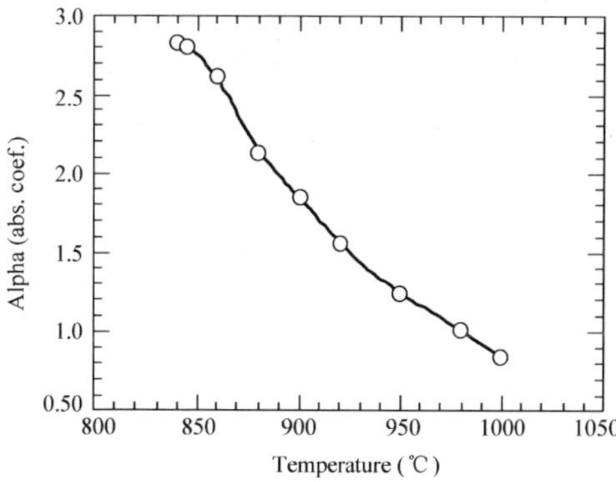

Fig. 6.12 Dependence of the absorption coefficient (α) on the sintering temperature for a glass fabricated by TCBS

6.4.2.2 Influences of Processing Parameters

The properties of composites such as transparency, density, interfacial properties, and ultimate strength are influenced by the processing parameters such as temperature, time, heating rate, pressure, and environment.

Temperature is one of the most important processing parameters. For binder burnout, the temperature was chosen as low as possible in order to reduce the possibility of oxidation of SCS-6 fiber coating (SiC/C coating). A long hold at 425 ℃ completed the partial decomposition of the organic materials and provided sufficient channels for the in-diffusion of oxygen and out-diffusion of CO_2. Therefore, the exposure time at 580 ℃ for oxidation was shortened. For sintering, it was found that the transparency of pure glass made by TCBS was increased with the increasing temperature as shown in Fig. 6.12. However, the fiber degradation would also occur at a high processing temperature. This tradeoff between transparency and mechanical properties was optimized by choosing a temperature of 940 ℃(the viscosity of F glass at this temperature is about 1.0×10^5 poise (1 poise=0.1 Pa·s) for a short period of time to obtain a good transparency without degrading the strong fibers. The two-hour dwell at 725 ℃(the viscosity of F glass at this temperature is about 4.0×10^7 poise) in vacuum during sintering provided a good condition for diffusion and the initiation of plastic flow for the glass matrix.

The heating rate for binder burnout was selected to control the decomposition/pyrolysis rate of the gas, reduce the thermal stress between fiber and matrix, and minimize the degree of oxidation of fiber coating. If the heating rate is too high, the explosion of the decomposed gas and the thermal mismatch stress between fiber and matrix could generate cracks and delamination in the sample. The heating rate also played an important role in sintering. A higher heating rate would trap the gas or close the channel for gas to escape because in that case the sample surface was always heated much quicker than the sample internal. Therefore, usually a low sample density was unexpectedly obtained. However, too low a heating rate is also not recommended because of the longer processing time and more energy consumption.

The influences of processing environment such as vacuum condition and argon purity on the properties of samples were found during processing. A higher vacuum provided a higher density of final sample, but it was limited by the pumping system. The larger void was found at the interface in samples with a lower vacuum condition 500 mTorr (66.661 Pa). The backfill of argon during processing was found to be very crucial to obtain a high density. The backfill of argon crushed the surface bubbles, enhanced the mobility of internal bubbles to surface, and thus increased the sample density. Another important factor is the purity of the argon. The differences in fiber/matrix interfacial properties were found for samples prepared under different argon gas purity. Argon gas with a higher purity for sintering resulted in a very weak interface whereas the argon gas with a lower purity showed interfacial bonding at the fiber/matrix interface. The details will be discussed in Sect. 6.5.2.

6.4.3 Testing Techniques

1) Mechanical Testing

Mechanical properties of the composites were characterized using the four-point flexure test. The schematic diagram of the four-point flexure test is shown in Fig. 6.13. The samples were cut from a well-polished bulk sample into small pieces with dimensions of 50 mm $\times$ 3.5 mm $\times$ 1.4 mm. The Instron 4201 machine was set up with an outer span of 40 mm and an inner span of 20 mm. It was calibrated before a real test using a very thick ceramic bulk. A crosshead speed of 0.2 mm/min was chosen for all the tests. The stress and strain on the tensile surface of a specimen are expressed as

$$\sigma = \frac{3PL}{4bh^2} \tag{6.50}$$

and

$$\varepsilon_{max} = \frac{6h}{L^2} y_L \tag{6.51}$$

where P is applied load, L is the outer span, b and h are the width and height of the specimen, respectively, and y_L is the displacement of the loading pins. For each volume fraction of fibers, five samples were tested.

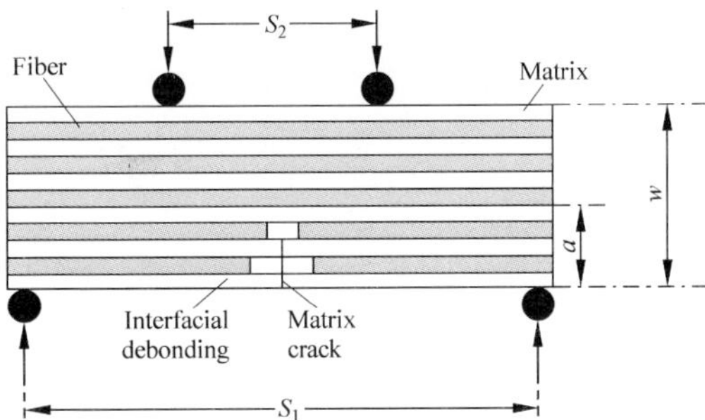

Fig. 6.13 Schematic diagram showing specimen configuration for four - point bending test, SENB test, debond length, and crack opening displacement measurements

The load-displacement behavior, FMC stress/strain, ultimate strength/strain, elastic modulus, and their dependencies on the fiber volume fraction were examined. After test, some samples were photographed and coated with gold by sputtering to analyze the fractography of composite using SEM.

2) Fiber Pushout

The thin slices from a composite were cut along the cross section. The thickness of the samples varied from 0.2 mm to 1.0 mm after polishing. The polishing was carefully done on diamond wheels (rough, medium, and fine polish) and vibratory polishing machine to achieve the 0.03 μm surface finish. The fiber pushout tests were performed on the Micro Measure Machine (Process Equipment Company, OH). Figure 6.14 is a photograph of this machine consisting of a fiber pushout machine and microscope/video recording system. A schematic diagram of the fiber pushout setup is shown in Fig. 6.2. The probe of fiber pushout is made of tungsten carbide with a 100 μm in diameter. The thin slice sample was glued to the substrate using a cement glue. The gap between the two substrate blocks was about 200 μm. A flat sample on the substrate was required in order to minimize bending stresses. The test was performed under load control at a rate of 10 N/s. The tests were performed on two groups of samples (PA and PB). Samples PA differed from PB in processing environment, PA was sintered in low-purity argon whereas PB was sintered in high-purity argon. These treatments produced different interfacial properties. Normally ten to twenty fibers were pushed under each condition. After fiber pushout, the samples were coated with a gold film by sputtering to observe the coating at the fiber/matrix interface.

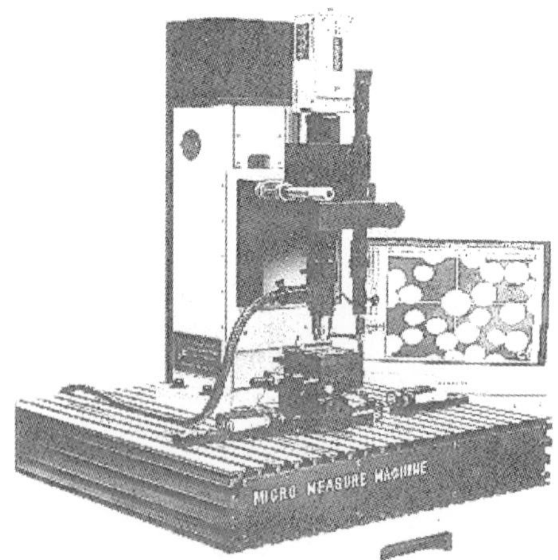

Fig. 6.14 A photograph of the fiber pushout and Micro Measure Machine

3) Matrix Crack-Spacing Measurement

The matrix crack spacing measurements were obtained from tests conducted in four point flexure using a mechanical testing system (Instron). The samples were removed from the machine after the saturation of the multiple matrix cracks but before the total failure of the composite (fiber pullout). The average crack spacing was measured for samples with different volume fractions of fibers. The average matrix crack spacings were used to calculate the interfacial shear stress from (6.4).

4) In-situ Measurements of Matrix Cracking and Interfacial Debonding

The purpose of these tests was to analyze the evolution of the multiple matrix cracking and interfacial debonding and dynamic interactions of the matrix crack with the process of debonding. These results were then used to determine the relationship between applied stress and debond length. These measurements utilized an Instron testing machine and the four-point flexure mode at the same crosshead speed. The experimental set up and sample preparation was identical to what was described in mechanical tests. However, the sample was loaded to several different levels of stress/load and then unloaded. The data on the number of cracks and crack spacings of the samples were used for the statistical distribution of crack density and crack spacing. The changes in the light transmission pattern (white band) on either sides of a matrix crack were observed. It was confirmed that the white band was caused by the light scattering and absorption at the debonded portion of the fiber/matrix interface. The change in the white band was attributed to the process of relative displacement between the fiber and matrix at the interface which led to debond propagation, as shown in Fig. 6.13. Each white band corresponds to a fiber/matrix interfacial debonding with a matrix crack running across in the middle of the band. The debond length was thus measured as half of the length of the white band. After each load-unload cycle, the number and position of matrix cracks, and the debond length around each crack surface were recorded and measured by photography and video recording methods. The externally applied stress was calculated as the stress on the outermost tension surface of a rectangular specimen assuming that the elastic beam theory for the flexure mode (MIL-STD-1942A) was valid.

5) Crack Opening Displacement (COD) Measurement

The COD measurements were made on composites using a Micro Measure Machine (Process Equipment Company, OH) equipped with CCD camera as shown in Fig. 6.13. The highest magnification of this microscope system was 2000. An image-freezing function was used to achieve the highest resolution of 0.4 µm. The COD images were recorded in a computer with respect to the position of each measurement. The sample was loaded using a fixture as shown in Fig. 6.15. The COD value at the crack tip was not accessible because of the limited resolution of the microscope. The crack length was also measured at a certain load/stress level using a lower magnification of about 10. The COD measurements were made under the same load/stress level as the DLM in order to obtain data under identical conditions.

6) Debond Length Measurement (DLM)

Based on the models experimentally examined and modified by Sun and Singh (1998a; 1998d), the fiber-bridging stress can be calculated from the measured value of debond length, L_d, using various equations based on the force balance and energy balance approaches. The debond length measurements were made on samples stressed in four-point bending mode using a fixture shown in Fig. 6.15.

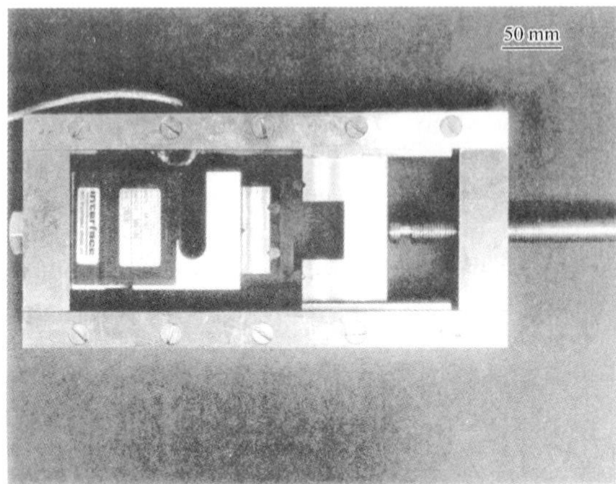

Fig. 6.15 A picture of the fixture used for DLM and COD measurements in flexure mode

A schematic of the specimen setup is also illustrated in Fig. 6.13. The outer and inner spans were 40 mm and 20 mm, respectively. The loading rate was controlled manually and all tests were done under displacement control. The load was measured using a load cell. Matrix cracking occurred on the tensile surface of a flexure bar when the applied load increased beyond the matrix-cracking stress. The debond length was in-situ measured using the photography and video recording methods without unloading the specimen. The magnitude of the debond length was taken as half of the length of the white band in Fig. 6.13. The debond length for each fiber layer along a matrix crack was obtained with respect to the fiber positions along the crack path. Therefore, both the debond length and COD profiles were obtained for each matrix crack under the same load/stress level. These experimental data were then used for determining the fiber bridging stress along the matrix crack.

     7) Fracture Toughness Test

The SENB test was conducted in the same fixture as described in Fig. 6.15. The applied load values were read from a load cell. The crack length was measured at a certain load/stress level using a Micro Measure Machine (Process Equipment Company, OH). A lower magnification of about 10 was used for this purpose. The measurements of the applied load and crack length were then used to calculate the fracture toughness K_P from (6.41) and (6.42). Such measurements of fracture toughness were then compared with the results obtained from COD measurements and DLM methods.

6.5 Mechanical and Interfacial Properties of Composites

6.5.1 Mechanical Behavior of Composites

Mechanical properties of the composites were characterized using the four-point flexure test. The load-displacement behavior, elastic modulus, FMC stress/strain, ultimate strength/strain, and their dependencies on the fiber-volume fraction were examined (Sun and Singh, 1996).

6.5.1.1 Load-Displacement Responeses

The load-displacement behaviors of the pure F glass and SiC fiber-reinforced glass composites with different volume fractions of fibers were measured. Five samples of each type (glass or composite samples with the varied fiber-volume fraction of 3%, 7%, 9%, 12%, and 16%) were tested. The typical stress-strain plots after subtracting the machine compliance are shown in Fig. 6.16. The pure glass samples failed at a strain of about 0.1% and a stress of 56 MPa. In contrast, composite samples showed a toughening and strengthening behaviors even when the matrix cracked at the same strain value as the monolithic glass. The ultimate strength of the composites was considerably increased in comparison to the strength of F glass. The WOF (work-of-fracture: the area under the stress-strain curve) was also increased due to the reinforcement by strong SiC fibers. The characteristic toughening phenomena such as multiple matrix cracking, interfacial debonding, and fiber pullout were observed throughout the experiments. Figure 6.17 shows a pullout SCS-6 fiber in a fractured composite with smooth coatings. Figure 6.18 shows the multiple matrix cracking on the tensile surface of a glass composite sample after a flexure test.

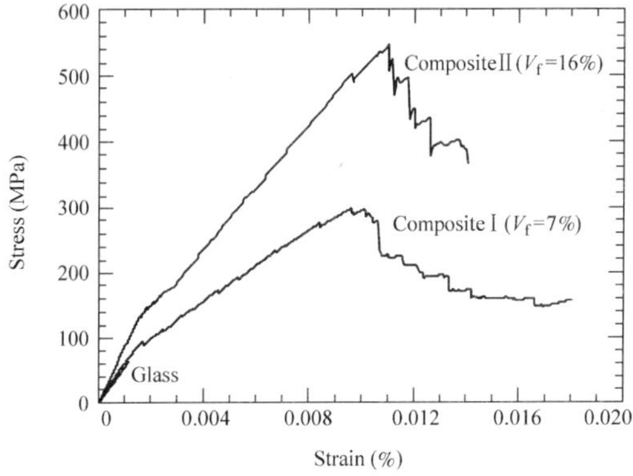

Fig. 6.16   Stress-strain curves for composites with different fiber volume fractions

6.5.1.2 Mechanical Properties of Glass and Composites

Mechanical properties such as elastic modulus, FMC stress and strain, and ultimate strength and strain were determined from stress-strain curves.

1) Elastic Modulus

The elastic moduli of the glass and composites were obtained both by dynamic resonance (Grindosonic) method and from the initial linear portion of the stress-strain plots. Figure. 6.19 shows the variation of the elastic moduli with fiber-

262

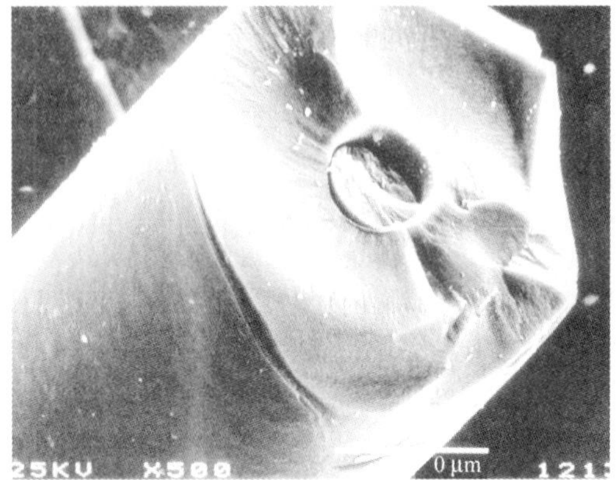

Fig. 6.17 A pullout SCS-6 fiber in a broken composite with smooth coatings

Fig. 6.18 The multiple matrix cracking phenomenon in glass composites

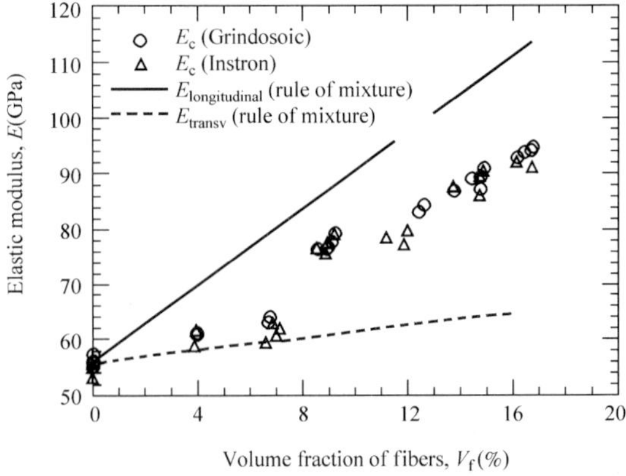

Fig. 6.19 Dependence of the elastic moduli on the fiber volume fractions

volume fraction. The experimental measured data from both methods were consistent in that the elastic modulus increases with the increase of the volume fraction of fibers. The data resided within the area bounded by two curves that correspond to the predicted values from the rule-of-mixtures in the longitudinal and transverse directions. These lower values of the elastic moduli obtained experimentally could result from the weak fiber/matrix interface and voids in real samples.

2) First Matrix Cracking (FMC) Stress and Strain
The first matrix cracking (FMC) occurred at a strain of $(0.1\pm0.01)\%$ for samples with a lower fiber-volume fraction of 3%, which is consistent with the failure strain of the pure glass matrix. With the increase of fiber-volume fraction, the FMC point was observed at a higher strain of about $(0.18\pm0.03)\%$. This behavior verified the relationship between the matrix failure strain and fiber volume fraction of the ACK model (Aveston et al., 1971). It was also found that usually more than one crack was generated in this glass composite system when the FMC occurred, as shown in Fig. 6.18. Figure 6.20 shows the dependence of the FMC stress on the fiber-volume fraction. An increase of 80% in FMC strain was shown because of the incorporation of strong fibers, as shown in Fig. 6.21.

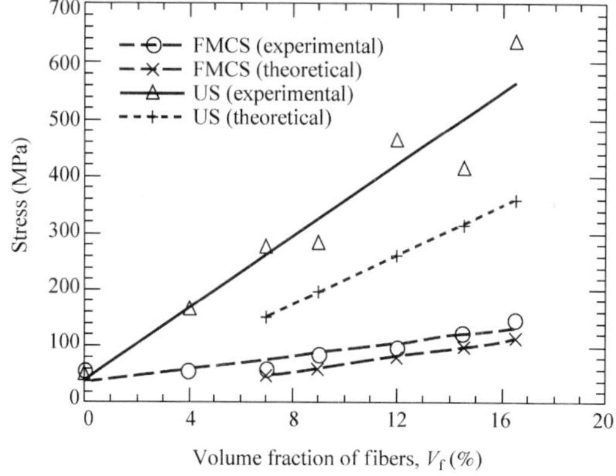

Fig. 6.20 Comparison of the experimentally measured FMCS and US with theoretical predictions

3) Ultimate Strength and Strain
The flexure strength of the composite was improved by incorporating SiC fibers into the glass matrix. Figure 6.20 also shows the overall increase in flexure strength of composite with the fiber-volume fraction. These strength values were comparable to the reported strength values from other studies (Gustafson et al., 1995; Gustafson et al., 1998). It was interesting to see that the corresponding strain did not vary significantly with the fiber-volume fraction, as shown in Fig. 6.21. Therefore, the ultimate strain of composite appears to be determined by the failure strain of the fibers. The comparison of the mechanical properties of the

glass and composite samples made by TCBS is listed in Table 6.3. Therefore, by using the TCBS fabrication method, strong and tough composites were achieved.

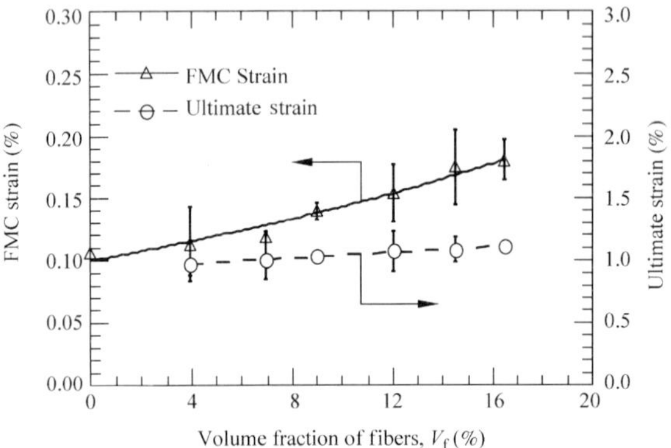

Fig. 6.21 Dependence of the FMC strain and ultimate strain on the fiber-volume fraction

Table 6.3 Mechanical properties of the glass and PA composites made by TCBS

| Materials | V_f (%) | E_c [■] | Material properties in flexure | | | | | Interfacial shear strength[*] |
			σ_{cr} (MPa)	ε_{cr} (%)	σ_u (MPa)	ε_u (%)	E_c [♣] (GPa)	τ (MPa)
Glass	0	56	56	0.10	56	0.1	56	0
Composite I	7	64	60	0.12	273	1.0	61	9.1
Composite II	16	94	146	0.18	632	1.1	89	10.2

■ Measured by grindosonic method;

♣ Calculated from the slope of the linear part of the stress-strain plots;

∗ Determined by matrix crack-spacing measurement.

The dependence of FMC stress on the fiber-volume fraction was calculated using (6.22) based on the ACK model. The material parameters used were obtained from Table 6.3. The interfacial frictional shear stress τ_f was determined as 30 MPa for PA samples (processed in low-purity argon) from the fiber pushout test described in Sect. 6.5.2. A comparison of the experimental data with the predictions indicates that the ACK model is an appropriate model for estimating the FMC stress of composite as shown in Fig. 6.21.

The ultimate strength of composites was calculated from Curtin's model as expressed by (6.23). The Weibull's modulus m was taken as 3. The experimental and calculated data are also compared in Fig. 6.20. The estimated stress values from Curtin's model are much lower than the experimental results. This can be explained by the fact that the ultimate strength measured in flexure test is generally 1.64 times higher than the results obtained from a tensile test (Xu et al., 1994).

From the measured mechanical properties of composites, it can be concluded that the greatly enhanced elastic modulus, FMC stress, ultimate strength, and WOF are observed. The FMC and US values predicted from the ACK and/or Curtin's models are found to be in good agreement with the measured values. In addition, the phenomena of multiple matrix cracking, interfacial debonding, crack-fiber bridging, and fiber pullout also show that the toughened composites are fabricated by the TCBS method.

6.5.2   Fiber/Matrix Interfacial Properties

The mechanical properties of FCMCs such as the first matrix-cracking stress and fracture toughness (WOF) are strongly dependent on the properties of the fiber, matrix, and fiber/matrix interface. Generally, a weakly bonded interface is desirable for a tough ceramic matrix composite because a weak interface can facilitate the toughening processes of fiber bridging, interface debonding, and fiber pullout. Therefore, the methods for characterizing interfacial properties are important for the evaluation of composite performance. A number of different techniques including microindentation (Marshall and Oliver, 1987; Grande et al., 1988), fiber pushout (Singh et al., 1991; Bright and Faber, 1989), single fiber pullout (Kerans et al., 1989; Goettler, et al., 1989), matrix crack-spacing measurement (Aveston and Kelly, 1973; Zok and Spearing., 1992) and laser Raman spectroscopy (Young and Yang, 1996; Ma Clarke, 1993) techniques have been developed for measuring the fiber/matrix interfacial properties. Among these techniques, fiber pushout is most easily performed and gives the most direct measurements. Therefore, it was used to determine the fiber/matrix interfacial properties. The matrix crack-spacing measurement was also used based on the crack-spacing calculation. The results from these measurements were compared.

6.5.2.1   Fiber Pushout Tests

These tests were performed on two groups of composites, PA and PB. Composite PA differed from PB in processing environment. PA was sintered in low-purity argon whereas PB was sintered in high-purity argon, which resulted in different interfacial properties. Normally five to ten fibers were pushed in each sample. Figure 6.22 shows six fibers after test. Figure 6.23 shows the top surface of the sample with a pushed-in fiber. The typical load-displacement curves for fiber pushout of PA and PB, after subtracting the machine compliance, are shown in Fig. 6.24 (PA) and Fig. 6.25 (PB). The machine was calibrated to obtain its compliance before test. There are three characteristic points in Figs 6.24 and 6.25: *A*, *B*, and *C*. The load-displacement curve showed a linear dependence up to point *A*, the debond initiation point. After point *A*, the data started to deviate from the linear behavior. This deviation from linear behavior of sample PA is clearly shown in Fig. 6.26, which has the same data as in

266

Fig. 6.24 (PA) but on an expanded scale. The debond initiation stress σ_d at point A was calculated using the following expression (Sun and Singh, 1998b):

$$\sigma_d = \frac{P_A V_f}{\pi r^2}, \tag{6.52}$$

where P_A is the load value at point A.

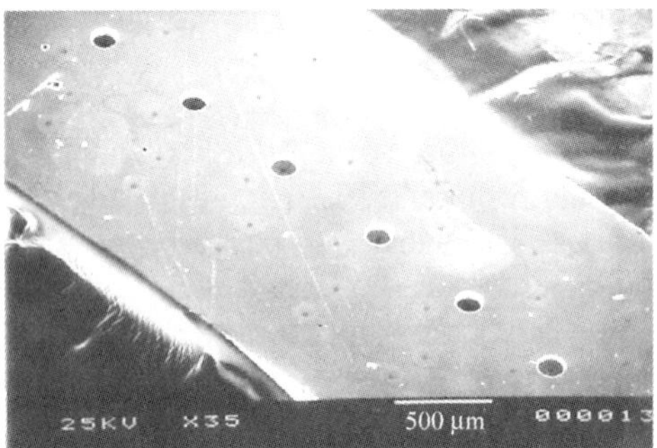

Fig. 6.22 SEM micrograph showing fibers after pushout tests

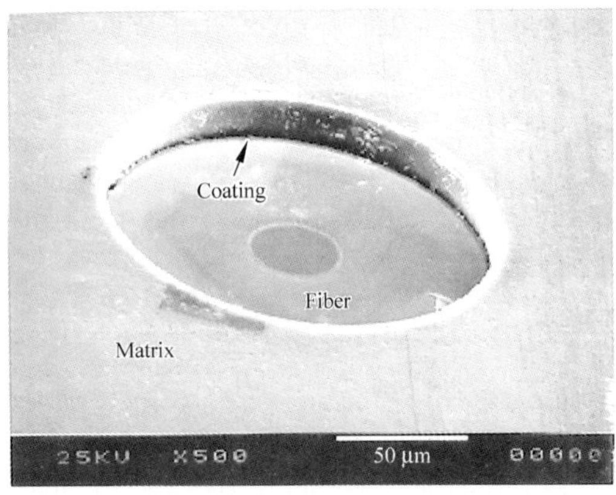

Fig. 6.23 Top surface of a sample with a fiber being pushed in

With the increase of pushin stress, the interfacial debond propagated. This was verified by the continuous decrease in the slope of the load-displacement curve after point A, as shown in Fig. 6.26. The peak load at point B corresponded to the complete debonding of the fiber/matrix interface, and was dependent on the debond energy, sample thickness, and end effect of the thin sample. A shear-lag analysis of a pushout test was used to determine the interfacial shear strength, τ_d, from the peak load using (6.2).

Because of the total failure of the interface bond, the energy that was

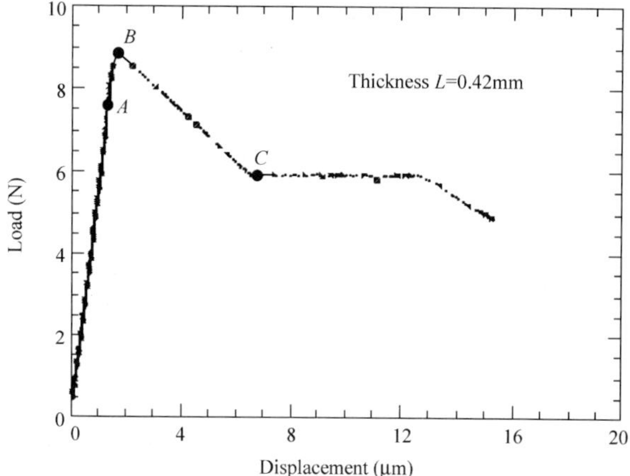

Fig. 6.24 Typical load-displacement curve for the fiber pushout test on PA composites

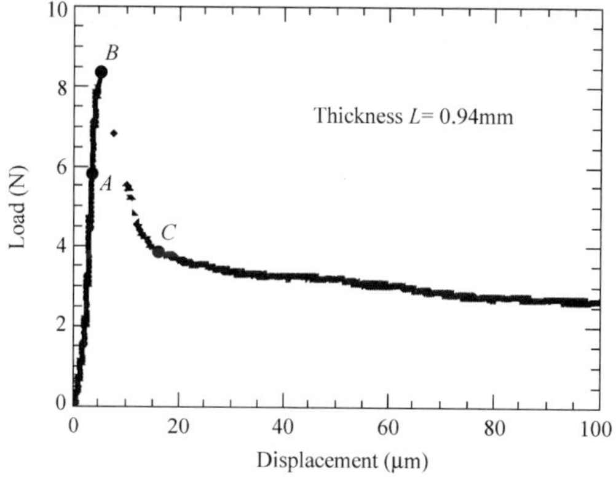

Fig. 6.25 Representative load-displacement curve during fiber pushout for PB composites

associated with the bond and stored in the elastic body was suddenly released, which led to a large increase in displacement and a sudden drop in load. This unstable state was finally stabilized by the interfacial frictional force and the dissipation of frictional energy, which created a steady state shown by point C. This steady state was established basically by the balance between the pushin force and frictional shear resistance force, which is expressed by (6.3). The approximation given by (6.3) is valid because the high exponential terms of a series expansion are negligible because L is relatively small in this study. Therefore, the interfacial frictional stress τ_f was calculated. The interfacial properties of composites PA and PB are given in Tables 6.4 and 6.5, which show that PA composites had a stronger interface than PB composites. The only difference between PA and PB was that PA was sintered in a lower purity argon gas whereas PB was sintered in a higher purity argon gas. The stronger bond in

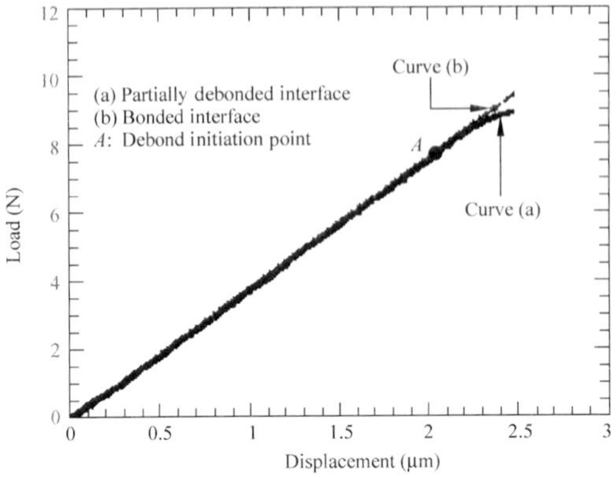

Fig. 6.26   Initiation of interfacial debonding during fiber pushout

PA composites might result from the partial oxidation of fiber coating and its reaction with the glass matrix. The different interfacial properties of composites PA and PB are consistent with their response during fiber pushout tests.

Table 6.4   Material parameters and properties for PA composites
(V_f =12%, sintered in 96% argon)

Fiber volume fraction, V_f	12%
Elastic modulus of composite, E_c(GPa)	97.3 [a]
Matrix porosity(%)	1–2
K_{Ic}(matrix) (MPa/ $\sqrt{m}$)	0.77
First matrix-cracking stress, σ_{FMC} (MPa)	90
Ultimate strength (composite), σ_{cu} (MPa)	440
Interfacial frictional stree, τ_f (MPa)	33±4 [b]
Interfacial debonding energy, Γ_d (J/m^2)	1.2±0.3 [b]
Debond initiation stress, σ_d (MPa)	57±5 [b]

a   Calculated by the rule-of-mixtures;

b   Measured by the fiber pushout technique.

Table 6.5   Material parameters and properties for PB composites
(V_f =7%, sintered in 99.9% argon)

Fiber-volume fraction, V_f	7%
Elastic modulus of composite, E_c(GPa)	80 [a]
First matrix-cracking stress, σ_{FMC} (MPa)	74
Ultimate strength, σ_{cu} (MPa)	273
Interfacial frictional stree, τ_f (MPa)	8.5±1.5 [b]
Debond initiation stress, σ_d(MPa)	16±2.1 [b]
Interfacial shear strength, τ_u (Mpa)	36±4 [b]
Debond energy, Γ_d (J/m^2)	0.85±0.25 [b]

a   Calculated by the rule-of-mixtures;

b   Measured by the fiber pushout technique.

A few radial matrix cracks were observed on the bottom surface of PA samples but no matrix crack was seen in PB samples. It was also found that some fibercoating debris remained on the glass matrix wall of the PA composite sample after the fiber pushout because of the strong interfacial bonding between the fiber coating and matrix, as shown in Fig. 6.27. Therefore, it was not surprising to observe radial matrix cracks because of the weak glass matrix and some type of bonding between the fiber and matrix. A close look at the pushout sample revealed that the interfacial debonding occurred and propagated between the outermost fiber coating (carbon) and glass matrix, as shown in Fig. 6.28.

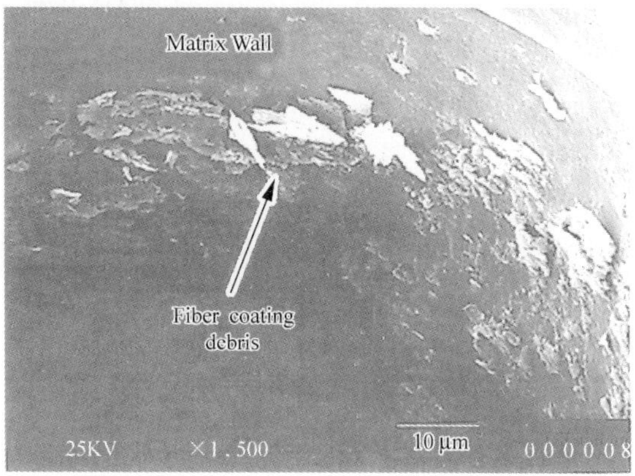

Fig. 6.27   Remaining fiber coating on the matrix wall after fiber pushout on a PA composite

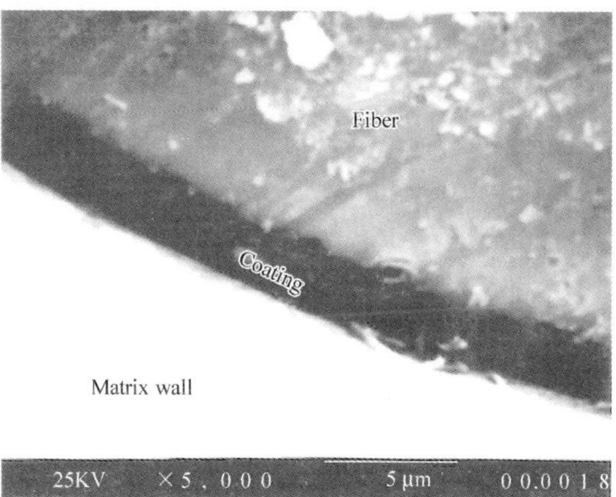

Fig. 6.28   Debond initiation between the fiber outermost carbon coating and matrix wall

The interfacial properties of PA and PB samples are listed in Tables 6.4 and 6.5. The typical interfacial frictional stress τ_f for glass matrix composites is reported between 4–14 MPa (Bright et al., 1989; Grande, et al., 1988), the interfacial shear strength τ_d is between 4–30 MPa (Goettler and Faber, 1989; Jurewicz et al., 1989). Compared to the reported values, PA composites showed a strong bond at the interface whereas the PB composites demonstrated a weak interface.

6.5.2.2 Measurement of Matrix Crack Spacing

SCS-6 fiber reinforced borosilicate glass composite showed desired mechanical properties and toughening behaviors such as multiple matrix cracking in Fig. 6.18 and fiber pullout in Fig. 6.17. The relatively uniform matrix crack spacing was observed when the fiber-volume fraction was larger than 5%. Figure 6.18 shows the uniformly distributed transverse matrix cracks with an average spacing of 1.1 mm in the PA composite with a fiber volume fraction of 12%. The interfacial shear strength was calculated by the ACK model (Aveston et al., 1971):

$$
X = \left(\frac{\sigma_c^f}{2\tau V_f} r_f \right) \left(\frac{E_m V_m}{E_m V_m + E_f V_f} \right)
\tag{6.53}
$$

where X is the average matrix crack spacing, σ_c^f is the composite stress at which the matrix cracks-appear. The interfacial shear strengths determined from the matrix crack-spacing measurement for PA composites are also listed in Table 6.3. For PA samples, the results obtained from the matrix crack-spacing, measurement are quite different from the data measured by fiber pushout. This difference resulted from the limitation of the model for matrix crack spacing, which is based on the assumption that the fiber and matrix are weakly coupled. This apparently may not have been the case for PA composites.

The interfacial properties of PA and PB samples determined from both the fiber pushout test and matrix crack-spacing measurement show that a very weak interface was achieved for PB samples (processed in argon with a very high purity 99.9%) and a relatively stronger interface was obtained for PA samples (processed in argon with a relatively lower purity 96%). Therefore, a controllable fiber/matrix interface is achieved by changing the processing environment of TCBS.

The comparison of results from fiber pushout and matrix crack-spacing measurements reveals that the matrix crack spacing method may not be a universally applicable method. It can provide only an approximation of the interfacial properties for the frictionally coupled interfaces. The fiber pushout test, however, not only provided the direct values of the debond initiation stress σ_d, interfacial shear strength τ_d, and interfacial frictional stress τ_f, but also gave the opportunity to directly observe the fiber/matrix interface.

6.6 Micromechanisms of Multiple Matrix Cracking and Interfacial Debonding

6.6.1 In-situ Analysis of Matrix Cracking and Interfacial Debonding

As mentioned earlier, the mechanical behavior of composites associated with the nonlinear portion (*A–B*) of Fig. 6.1 (Sun and Singh, 1998) is complex because it is related to the process of fiber bridging, multiple matrix cracking and interfacial debonding/sliding. One of the limitations of the micromechanical studies related to region *A–B* has been in directly observing and measuring the debond or sliding length in most ceramic composites. Because of the availability of transparent glass composites made by TCBS method developed during this study, the number of matrix cracks and the length of interfacial debonding were directly measured. Therefore, the objective of this part of the research is a study of the evolution of matrix cracks and their relationship to interfacial debonding. By measuring the debond length and matrix crack spacing directly, the interrelationship among the multiple matrix cracking, interfacial debonding, and external stress is analyzed.

The sample was stressed using an Instron testing machine in the four-point flexure mode. The experimental set-up and sample preparation was described in Sect. 6.4.2. The loading curves during a typical test are shown in Fig. 6.29. The changes in the light transmission pattern (white band) on either sides of a matrix crack were observed. It was believed that the white band was caused by the light

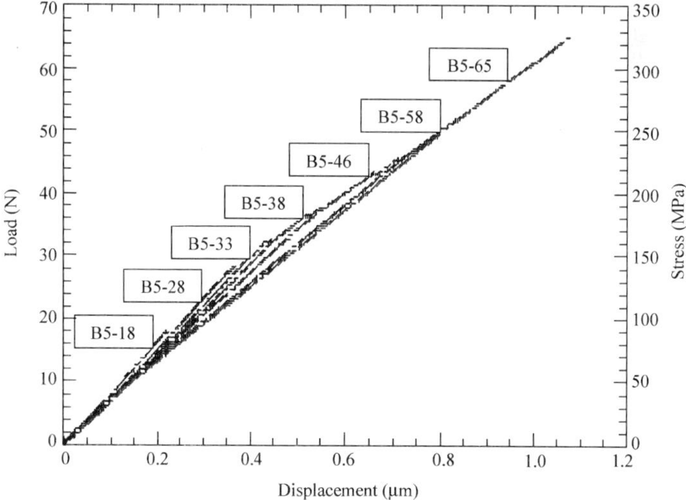

Fig. 6.29 Load-displacement curves for a composite loaded to different stress levels   (For example, B5-18 denotes the sample B5 loaded at 18 N)

scattering and absorption at the debonded portion of the fiber/matrix interface. The change in the white band was attributed to the process of relative displacement between the fiber and matrix at the interface that led to debond propagation as shown in Fig. 6.30 a–e. Each white band corresponds to a fiber/matrix interfacial debonding with a matrix crack running across in the middle of the band. The debond length was thus measured as half of the length of the white band. After each load–unload cycle, the number and position of matrix cracks, and the debond length around each crack surface were recorded and measured by photography and video recording methods. The magnification of the pictures in Fig. 6.30 was about 16x. Experiments were conducted on a total of ten PA samples of the same fiber volume-fraction. Debond length measurement (DLM) techniques were applied to only five of them. All of them showed a similar behavior. These data on the number of cracks and crack spacings from the ten

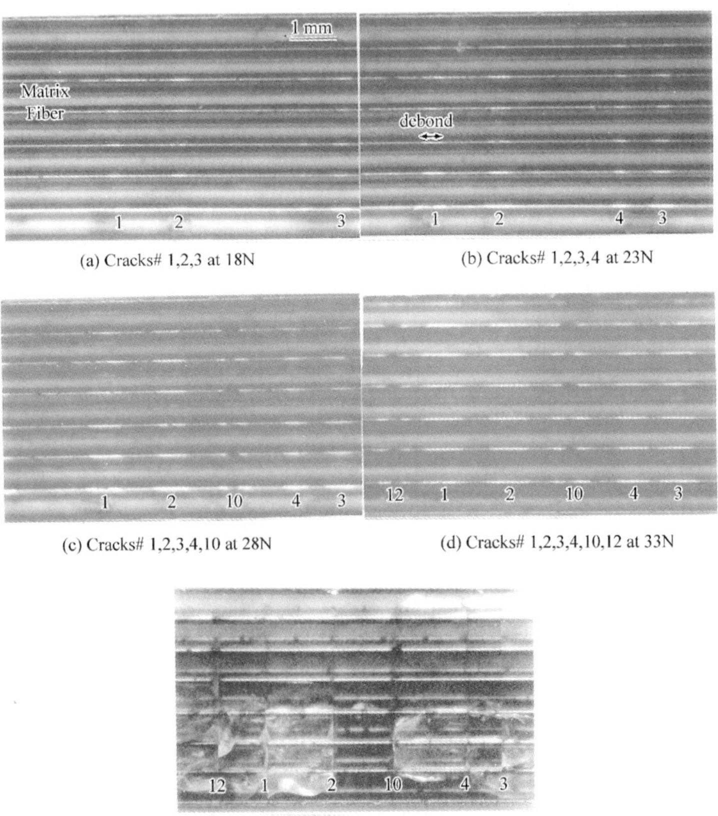

(a) Cracks# 1,2,3 at 18N

(b) Cracks# 1,2,3,4 at 23N

(c) Cracks# 1,2,3,4,10 at 28N

(d) Cracks# 1,2,3,4,10,12 at 33N

(e) Matrix crack saturation and associated debond length at 58N

Fig.6.30 Optical micrographs showing the progression of debonding (patterns) with increasing applied load

samples were used for determining the statistical distribution of crack density and crack spacing.

6.6.1.1 Multiple Matrix Cracking

Before the appearance of the first matrix crack (FMC), the stresses in fiber and matrix obeyed the rule-of-mixtures. The stress distribution along the fiber direction (sample longitudinal direction) was uniform. However, after FMC, the stresses in fiber and matrix were redistributed. These redistributed stresses can be expressed by (6.12) – (6.17). The stress distribution is shown in Fig. 6.5. The characteristic length z was defined as the value when σ was equal to σ_{FMC}. The first matrix crack for PA sample under a four-point flexure test occurred at a stress of 85 MPa. A close look at the matrix under an optical microscope, as shown in Fig. 6.30a, revealed that more than one crack was generated, which also included an incompletely propagated crack. This behavior was also observed by Dutton et al. (1996). Therefore, an evidence of a first load drop (also called "the proportional drop") in the load–displacement curve may not represent the real FMC point. After FMC, the stress distribution along the fiber length, but in the matrix phase, varied from zero at the matrix cracks surfaces to a new pre-crack value as shown in Fig. 6.5. Subsequent formation of new cracks always occurred at progressively higher stress levels because the flaw size is smaller than that for the previous matrix cracks. This preference of cracks for a larger flaw demonstrated that the matrix crack stress could vary over a range of values, as discussed by Zok and Spearing (1992) and Curtin (1993). The stiffness of the composite indicated by the slope of the linear portion of each curve in Fig. 6.29 decreased with the generation of additional new matrix cracks.

Figure 6.31 shows the sequence and position of the matrix cracks generated because of the increasing load. The numbers on the dashed lines from 1 to 16 in Fig. 6.31 indicate the sequence of matrix crack generation. The magnitude of the load/stress, written horizontally along the various crack numbers, indicates the load/stress at which these cracks were created. For example, the first three cracks 1, 2, 3 were detected at a load of 18 N (85 MPa, corresponding to the first matrix cracking). The next cracks 4, 5, 6 were found at a load of 23 N, and cracks 15, 16 were generated at a load of 46 N (217 MPa, corresponding to the saturation of matrix cracking). No more matrix cracks were generated between the inside two pins after the saturation load of 46 N, the saturated state of multiple matrix cracking of composites. Figure 6.32 shows the influence of the applied stress on the crack density. The crack density was obtained as the ratio of the number of cracks at a certain stress level to the total number of cracks at saturation. The saturation stress was about twice as much as the FMC stress for this composite. This range of stress was much larger than the prediction of Zok and Spearing's (1992) model, in which the saturation stress was about 30% higher than the FMC stress. This difference can be explained by the fact that Zok and Spearing's model was valid under the assumption of a frictionally-coupled fiber-matrix interface. But, the current PA composite showed a higher interfacial shear strength (a large debond initiation stress σ_d) as shown in Table 6.4. According to Eqs. (6.32) and (6.39), the saturation stress for a bonded interface will be higher than the frictionally-coupled interface ($\sigma_d > \sigma_{\text{slide}}$) provided other properties remain the same. Another explanation from Li et al. (1993) and Weitsman and Zhu (1993) is that the amount of mechanical energy absorbed during the multiple fracture is larger because of the appearance of an additional interfacial debond energy Γ_d term for a bonded interface. Therefore, the saturation stress will be higher because of the contribution from interfacial debonding.

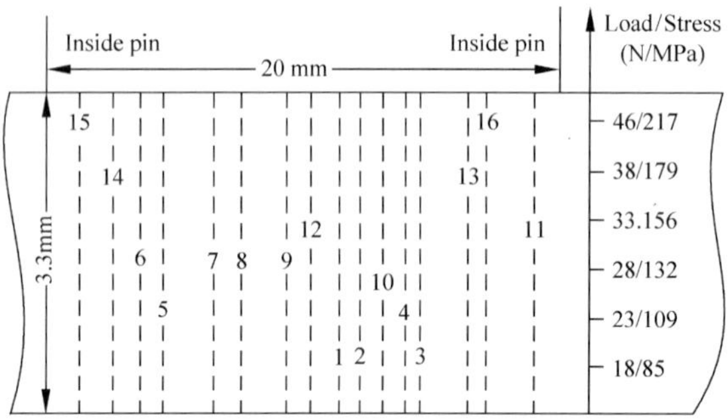

Fig. 6.31 Evolution and distribution of matrix cracks on the tensile face of the specimen

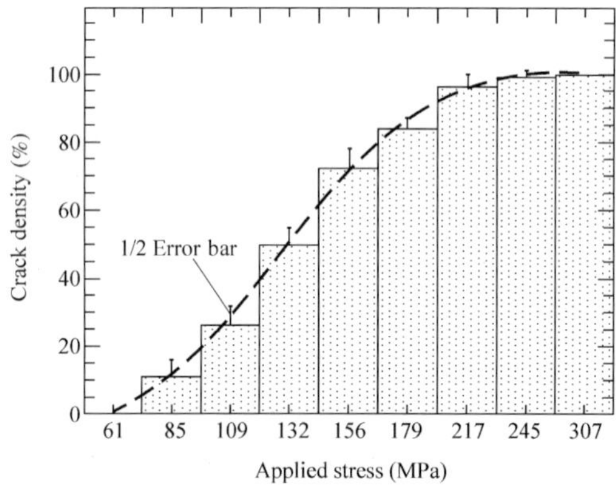

Fig. 6.32 Dependence of matrix crack density on the applied stress

Another interesting phenomenon was the nonuniformity of matrix crack distribution as shown in Fig. 6.33. The crack spacing did not follow (6.4) strictly. The experimental average value from five specimens was about 1.1 mm. The ratio of a particular crack spacing to the average crack spacing was taken as the normalized crack spacing. The statistical distribution of matrix crack spacing was then obtained by sorting the crack spacing, counting the number of occurrences, and calculating their probability of occurrence. Figure 6.33 shows the statistical distribution of the normalized crack spacings. The predicted matrix crack spacing calculated from (6.4) with statistical modification is about 0.65 mm by taking the interface as a frictional case with a value of τ_f equal to 30 MPa (Sun and Singh, 1998b). The experimental average crack-spacing value was about 1.1 mm, which was much higher than the calculated crack-spacing value. This was because firstly the real interface of the composite shows debonding rather than a purely frictional coupling, and secondly the statistical modification parameter α may not be equal to 1.34 in the case of a partially-bonded and debonded interface. In the present study, α is experimentally found to be 2, and the overlap of debonded zones was not commonly observed in these composites. The value of α was determined by the measurements of matrix crack spacing and debond length as described later. In addition, the nonuniformly distributed cracks implied that the matrix cracks and crack spacings were dependent on the distribution of flaw and flaw size in the matrix.

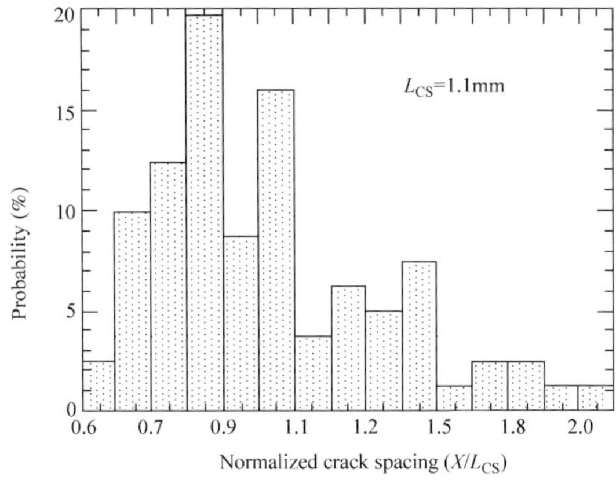

Fig. 6.33 Histogram of matrix crack spacing

6.6.1.2 Interfacial Debonding

The initiation and propagation of interfacial debonding were the most prominent phenomena accompanying the multiple matrix cracking that were visually observed in this investigation. Figure 6.30 a–e shows the debonded regions (lengths) on either side of the matrix cracks at five different load levels. The measured FMC stress for this composite with a fiber volume fraction of 12% was 85 MPa, and the initial debond stress obtained from a fiber pushout test (Sun and Singh, 1998c) was about 55 MPa, as shown in Table 6.4. Therefore, upon FMC, the additional load which was thrown onto the bridging fibers locally around the matrix crack was sufficiently large to initiate the debond as well. This was why debonding was observed at the interface when the FMC occurred. This observation was even true for an incompletely propagated crack, #4 in Fig. 6.30a. Such a weakly-bonded interface was desired for toughening brittle matrix materials. As expected, the debond propagated further with an increase of applied load and finally reached their neighbors, as shown in Fig. 6.30e.

The experimental data of measured debond length as a function of the external stress for different matrix cracks are illustrated in Fig. 6.34. A close-to-linear relationship between the debond length and applied stress was observed before the matrix crack saturation, which is quite similar to the trend predicted by ACK (Aveston et al., 1971b), BHE (Budiansky et al., 1986b), HJ (Hutchinson and Jenser, 1990) and Li et al. (1993) models. But, a suppression of debond length was also observed before the saturation of the matrix cracking. Therefore, the application of these models is restricted to some extent in real cases. The cracks #1, 2, 3 and 10 showed similar debond lengths because they were in the same region of the sample and showed similar crack spacing. However, cracks #9 and 12 differed greatly from other cracks in the measured values of the debond length. This unusual behavior resulted from a nonuniform crack spacing and stress distribution along the length of the bend specimen after a large displacement.

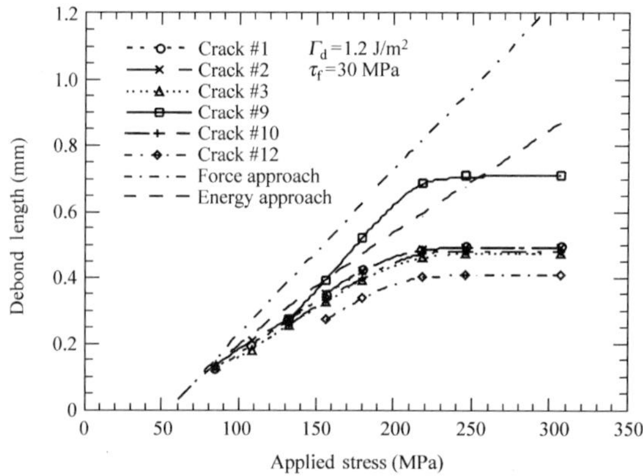

Fig. 6.34 Experimental and theoretical dependencies of debond length on the applied stress

Figure 6.34 also showed a comparison of the calculated (using (6.32) and (6.39)) and measured dependencies of debond length on the applied stress. The calculated values from both force balance (BEH et al., 1995; Hsueh, 1996) and energy balance approaches (Li et al., 1993) were close to each other for the short debond length, but deviated markedly for the long debond length corresponding to the evolution of multiple matrix cracks. It was found that the model from the energy balance approach seemed to fit the data better than the force balance approach for the debond length in the cases of multiple matrix cracking. This difference can be explained by the fact that the two models are based on different assumptions in their derivations (see more details in Sect. 6.6.2). It is known that two conditions are necessary for interfacial debonding to form and propagate. Firstly, the fiber/matrix interfacial shear stress must be greater than the shear strength of the interface. Secondly, the work done by the externally applied stress must supply sufficient energy for the debonding process. In Li's model, the elastic strain energy change in bonded region of a composite was neglected. Therefore, by Li's model a larger applied stress was required for a debond to propagate to be same length. This difference in applied stress increased with the debond length. As explained by Li's model, for a small debond length, the frictional dissipation energy plays little role compared with the interfacial debond energy. But, with the increase of debond length, the frictional dissipation energy increases while the debond energy term still remains the same. The frictional dissipation energy cannot be ignored after a certain extension of the debond length, which then contributes partly to the suppression of the debond length for long debond length, as observed in Fig. 6.34. However, it is worth noting that Li's model is not a perfect model. The fitting of experimental data may result from the overestimation of applied stress by the elastic beam theory. Thus, further investigations on the debond length-applied stress relationship are necessary.

6.6.1.3　Interaction between the Matrix Cracks and Interfacial Debond

The saturation of matrix cracks for a composite with a weakly-bonded interface is determined by the interaction of two dynamic processes without considering the delamination effect: a) Size-reduction of the uncracked block of matrix because of the generation of new cracks, and b) Longitudinal growth of the interfacial debond because of the increased stresses in fibers.

The interactions between these two processes were clearly observed in all experiments of this study. Figure 6.35 illustrates these two dynamic processes upon increase of the external stress. The generation of new matrix cracks was dependent on both the interfacial properties, such as τ_f and τ_d, and the distribution of flaws in the matrix. For matrix cracking to occur, the size of the uncracked matrix block must be sufficiently large so that the stress in the matrix due to load transfer from fiber to matrix can occur up to the matrix-cracking stress. Saturation in the matrix cracking occurs when the matrix stress cannot rise sufficiently because of the complete propagation of the debonded zones from either side of the uncracked matrix blocks.

Both the force and energy balance approaches have not considered the saturation phenomena in the multiple matrix cracking situation which is commonly seen in fiber-reinforced ceramic composites. Therefore, based on the knowledge of the interrelationship between the interaction of matrix cracks and debond length, the saturation stress σ_s for a composite displaying multiple matrix cracking can provide an upper bound for the application of these models. Because the model from the energy balance approach seemed to provide a better

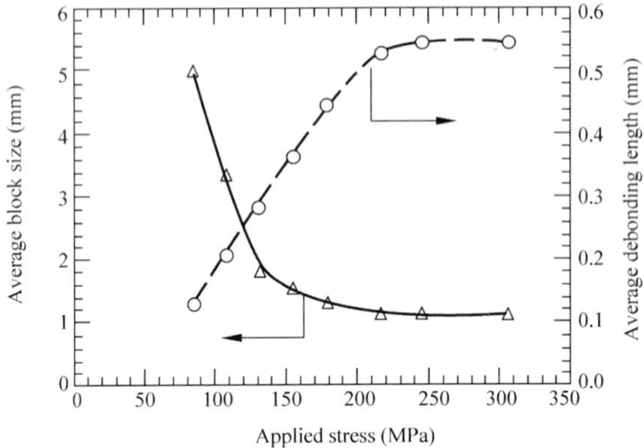

Fig.6.35 Interrelationships among average block size, interfacial debonding, and the applied stress

fit to the experimental data, especially for a long debond length as shown in Fig. 6.34, (6.39) is used for determining the saturation stress σ_s. This saturation stress σ_s is the critical applied stress corresponding to the maximum average debond length, which is half of the matrix crack spacing L_{cs}. Thus, in (6.39), σ_a is substituted by σ_s, and L_d by $\dfrac{L_{cs}}{2}$, where L_{cs} can be statistically approximated as $\alpha\, z'$, and z' is obtained from (6.4). It needs to be mentioned that the parameter α is not necessary equal to 1.34 (Zok and Spearing, 1992; Curtin, 1993), especially for a composite showing debonding. Rearranging (6.39) then gives the saturation stress σ_s as

$$\sigma_s = \frac{\alpha\, E_c}{2\, E_m}\sigma_{mu} + \frac{2}{2E_m}\left(\alpha^2 \sigma_{mu}^2 E_c E_f V_f + 4E_m^2 \sigma_d^2\right)^{1/2} \tag{6.54}$$

Taking the pertinent material parameters and properties from Table 6.4, the saturation stress σ_s calculated from (6.54) is about 190 MPa assuming α is equal to 2 which provides a closer value of matrix crack spacing for these composites showing debonding. In the case of the frictionally-coupled interface, σ_d is reduced to σ_{slide}, which is normally negligible. Equation (6.54) then can be simplified as

$$\sigma_s = \frac{\alpha\, E_c}{2\, E_m}\sigma_{mu}\left(1+\left(\frac{E_f V_f}{E_c}\right)^{1/2}\right) \tag{6.55}$$

where α is 1.34, a statistically-derived parameter for matrix crack spacing in composites with a pure frictional interface. Equation 6.55 with these assumptions then gives the same prediction as Zok and Spearing (1992) that the saturation stress is about 30% higher than FMC stress σ_{FMC}, which can be approximated as $\dfrac{E_c}{E_m}\sigma_{mu}$.

The investigation of the role of the interfacial debonding on the multiple matrix cracking has shown that the debond initiated when the first matrix cracking occurs in a composite with a weakly-bonded interface that is desirable for toughening. The debond length increases linearly with the applied stress and then saturates as the debonded zones from either side of the uncracked matrix block approach each other. The matrix crack density also increases with the applied stress, and saturatess, at some critical stress level. The matrix crack spacings are not uniformly distributed, indicating their dependence on the distribution of flaws and flaw size.

The relationship between debond length and applied stress shows a close-to-linear relationship. For a very short debond length, both approaches give an estimate close to the experimental data. But, for a long debond length, the energy balance seems to provide a better fit. It is also found that the saturation stress for matrix cracking can be determined by the interaction between multiple matrix cracking and interfacial debonding.

6.6.2   A New Theoretical Model for Interfacial Debonding

Several theoretical models are available for the interfacial debonding analysis. BEH (Budiansky et al., 1995), HJ (Hutchinson and Jensen, et al., 1990) and Hsueh (Hsueh, 1996) (hereafter denoted as BEH, HJ, Hsueh) employed lame solutions to simplify the debonding analysis by assuming a very weak fiber/matrix interface. Li et al. (1993) (thereafter denoted as Li, Shah & Mura) used an energy balance approach for the treatment. However, they ignored the strain energy change in the bonded region during debonding process. Very few models, except SH (Sutcu and Hillig, 1990), had considered the influence of the change of shear strain (energy) in the debonding process. In a real composite, the fiber/matrix interface could be tailored to have desired properties. A very weak interface may not always be the case. In another words, in the case of strong interface (a large debond energy and or frictional shear stress), the shear deformation could not be ignored. Therefore, it is quite desired to provide a new model suitable for more comprehensive situations.

A single fiber cylindrical unit ("the discrete fiber model" (McCartney, 1987)) in Fig. 6.4 is used for modeling. The fiber/matrix interface is characterized by the debond energy, Γ_d, and frictional shear stress, τ_f. To differentiate the modeling equations from the others, the equations appearing in this part of the chapter are denoted by prefix M.

6.6.2.1 Stress Distribution in a Composite Unit

For the debonded/slipped region ($0<z<L_d$) in Fig. 6.4, interfacial shear stress τ_i is equal to the constant frictional stress τ_f, and the stress distribution in fiber (σ_f) and matrix (σ_m) in the sliding area is determined by the following equations:

$$\sigma_f(z) = \sigma_T - \frac{2\tau_f z}{a} \tag{M1}$$

$$\sigma_m(z) = 2\tau_f \left(\frac{V_f}{V_m}\right)\left(\frac{z}{a}\right) \tag{M2}$$

$$\tau_i(z) = \tau_f \tag{M3}$$

$$\tau_m^{rz}(r,z) = \frac{V_f(b^2 - r^2)}{V_m ar}\tau_f \tag{M4}$$

where z is the distance measured from the crack surface parallel to the fiber direction, σ_T is the bridging stress in the fiber at the crack surfaces, $\sigma_f(z)$ and $\sigma_m(z)$ are stresses in the fiber and matrix, respectively, at the interface, $\tau_i(z)$ is the interfacial shear stress.

According to the shear-lag solution given by BHE, the stresses beyond the debonded region ($z>L_d$) are determined by the following expressions:

$$\sigma_f(z) = \frac{V_f E_f \sigma_T}{E_c} + \left(\frac{2\tau_s}{\rho}\right)\exp\left[-\rho(z - L_d)/a\right] \tag{M5}$$

$$\sigma_m(z) = \frac{V_f E_m \sigma_T}{E_c} - \left(\frac{2\tau_s}{\rho}\right)\left(\frac{V_f}{V_m}\right)\exp\left[-\rho(z - L_d)/a\right] \tag{M6}$$

$$\tau_i(z) = \tau_s \exp\left[-\frac{\rho(z - L_d)}{a}\right] \tag{M7}$$

$$\tau_m^{rz}(r,z) = \frac{V_f(b^2 - r^2)}{V_m ar}\tau_s \exp\left[-\frac{\rho(z - L_d)}{a}\right] \tag{M8}$$

where a is the fiber radius, b is the radius of composite unit, and τ_s is interfacial shear strength. Because of the continuity of stress in fiber and matrix, τ_s can be expressed as

$$\tau_s = \frac{V_m E_m \rho \sigma_T}{2E_c} - \frac{L_d \rho \tau_f}{a} \tag{M9}$$

ρ is the shear-lag parameter determined by the following formula,

$$\rho^2 = \frac{4E_c G_m}{V_m E_m E_f \varphi} \tag{M10}$$

where G_m is the shear modulus of matrix and $\varphi = -\dfrac{[2\ln V_f + V_m(3 - V_f)]}{2V_m^2}$.

6.6.2.2 Displacement Calculations

The displacements in fiber and matrix are obtained by integrating the strains with respect to z as

$$w_f(z) = \int_z^{L_d} \frac{\sigma_f(z)}{E_f}\,dz + \int_{L_d}^{L} \frac{\sigma_f(z)}{E_f}\,dz$$

$$= \frac{\sigma_T(L_d - z)}{E_f} - \frac{\tau_f(L_d^2 - z^2)}{aE_f} + \frac{V_f \sigma_T}{E_c}(L - L_d) + \frac{V_m E_m a \tau_s \varphi}{2E_c G_m} \times$$

$$\{1 - \exp[-\rho(L - L_d)]\} \tag{M11}$$

$$w_m(z) = \int_z^{L_d} \frac{\sigma_m(z)}{E_m}\,dz + \int_{L_d}^{L} \frac{\sigma_m(z)}{E_m}\,dz$$

$$= \frac{\tau_m V_f(L_d^2 - z^2)}{E_m V_m a} + \frac{V_f \sigma_T}{E_c}(L - L_d) - \frac{V_f E_f a \tau_s \varphi}{2E_c G_m}\{1 - \exp[-\rho(L - L_d)]\} \tag{M12}$$

The crack opening displacement U_{COD} (half the real crack opening displacement) can be obtained from the difference between the fiber displacement $w_f(z)$, and matrix displacement $w_m(z)$.

$$U_{COD}(z) = w_f(z) - w_m(z)$$

$$= \frac{\sigma_T}{E_f} \cdot (L_d - z) - \frac{\tau_f E_c}{E_m V_m a} \cdot (L_d^2 - z^2) + \frac{\tau_s a \varphi}{2G_m}\{1 - \exp[-\rho(L - L_d)/a]\}$$

$$\tag{M13}$$

$$U_{COD}(0) = \frac{\sigma_T}{E_f} \cdot L_d - \frac{\tau_f E_c}{E_m V_m a} \cdot L_d^2 + \frac{\tau_s a \varphi}{2G_m}\{1 - \exp[1 - \rho(L - L_d)/a]\}$$

$$\tag{M14}$$

The additional displacement of composite due to debonding, U_{debond}, is defined as the difference between the fiber displacement, $w_f(0)$, and composite displacement w_c in the absence of debonding.

$$U_{\text{debond}}(0)$$

$$= w_f(0) - w_c$$

$$= \frac{V_m E_m \sigma_T L_d}{E_f E_c} - \frac{\tau_f}{a E_f} L_d^2 + \frac{V_m E_m a \tau_s \varphi}{2 E_c G_m}\{1 - \exp[-\rho(L - L_d)/a]\} \qquad \text{(M15)}$$

6.6.2.3 Energy Balance Approach for Bridging Stress Calculation

From the fracture mechanics point of view, the interfacial debonding and frictional sliding along the fiber/matrix interface can be treated as the propagation of a debond crack. For a semi-infinite unit, the stresses beyond the debonding front remain unchanged. Therefore, the energy of this unit changes as the debonded surface at the interface enlarges by an incremental amount $2\pi r \cdot d(L_d)$. There are four energy terms associated with the debonding process: (1) the stored elastic strain energy in the fiber and matrix, dW_E, including the shear strain energy in the matrix, (2) the work done by the external force (the bridging force in the fiber) at the crack surfaces, dW_B, (3) the frictional work, dW_F, along the debonded zone, and (4) the surface energy, dW_s, for the newly-generated debonded zone, which is associated with the debond energy Γ_d (approximately $2\gamma^*$, where γ^* is the surface energy). From the energy consideration, the surface energy of a debonded crack must be balanced by the increase of the released strain energy, work done by the external load, and the amount of frictional dissipation energy (Li et al., 1993; Hsueh, 1996).

$$dW_s = dW_B - dW_E - dW_F \qquad \text{(M16)}$$

These different terms are evaluated as followings:

(1) dW_E: The elastic strain energy W_E consists of two portions: U_d, debonded region ($0 < z < L_d$), and U_b, bonded region ($z > L_d$).

$$W_E = U_b + U_d \qquad \text{(M17)}$$

$$U_b = \int_{L_d}^{L}\int_0^a \frac{\sigma_f^2(z)}{E_f}\pi r dr dz + \int_{L_d}^{L}\int_a^b \frac{\sigma_m^2(z)}{E_m}\pi r dr dz + \int_{L_d}^{L}\int_a^b \frac{\tau_m^2(r,z)}{G_m}\pi r dr dz$$

$$\text{(M18)}$$

$$U_d = \int_0^{L_d}\int_0^a \frac{\sigma_f^2(z)}{E_f}\pi r dr dz + \int_0^{L_d}\int_a^b \frac{\sigma_m^2(z)}{E_m}\pi r dr dz + \int_0^{L_d}\int_a^b \frac{\tau_m^2(r,z)}{G_m}\pi r dr dz$$

$$\text{(M19)}$$

and

$$W_E = \frac{\pi a^2 V_f \sigma_T^2 L}{2 E_c} + \frac{\pi a^2 E_m V_m \sigma_T^2 L_d}{2 E_c E_f} - \frac{\pi a \tau_f \sigma_T L_d^2}{E_f} + \frac{2\pi E_c \tau_f^2 L_d^3}{3 V_m E_m E_f} + \frac{\pi a^2 \varphi \tau_f^2 L_d}{2 G_m} +$$

$$\frac{\pi a^3 \varphi \tau_\text{s}^2}{4\rho G_\text{m}}\{1 - \exp[-2\rho(L - L_\text{d})/a]\} +$$

$$\frac{\pi a^3 \tau_\text{s} \sigma_\text{T} E_\text{m} E_\text{f} (V_\text{f} V_\text{m} - V_\text{f}^2)\varphi}{2 G_\text{m} E_\text{c}^2}\{1 - \exp[-\rho(L - L_\text{d})/a]\} +$$

$$\frac{\pi a^3 (E_\text{m} V_\text{m}^2 + E_\text{f} V_\text{f}^2)\varphi \tau_\text{s}^2}{4 E_\text{c} V_\text{m} G_\text{m} \rho}\{1 - \exp[-2\rho(L - L_\text{d})/a]\} \qquad \text{(M20)}$$

It is important to see that when $L\text{-}L_\text{d}$ is much larger than the fiber diameter a, then all the exponential terms in Eq. (M20) can be disregarded. Therefore,

$$W_\text{E} = \frac{\pi a^2 V_\text{f} \sigma_\text{T}^2 L}{2 E_\text{c}} + \frac{\pi a^2 E_\text{m} V_\text{m} \sigma_\text{T}^2 L_\text{d}}{2 E_\text{c} E_\text{f}} - \frac{\pi a \tau_\text{f} \sigma_\text{T} L_\text{d}^2}{E_\text{f}} + \frac{2\pi E_\text{c} \tau_\text{f}^2 L_\text{d}^3}{3 V_\text{m} E_\text{m} E_\text{f}} + \frac{\pi a^2 \varphi \tau_\text{f}^2 L_\text{d}}{2 G_\text{m}} +$$

$$\frac{\pi a^3 \varphi \tau_\text{s}^2}{4\rho G_\text{m}} + \frac{\pi a^3 \tau_\text{s} \sigma_\text{T} E_\text{m} E_\text{f} (V_\text{f} V_\text{m} - V_\text{f}^2)\varphi}{2 G_\text{m} E_\text{c}^2} + \frac{\pi a^3 (E_\text{m} V_\text{m}^2 + E_\text{f} V_\text{f}^2)\varphi \tau_\text{s}^2}{4 E_\text{c} V_\text{m} G_\text{m} \rho} \qquad \text{(M21)}$$

where τ_s should be substituted by the expression in (M9). Thus,

$$\frac{dW_\text{E}}{dL_\text{d}} = \frac{\pi a^2 E_\text{m} V_\text{m} \sigma_\text{T}^2}{2 E_\text{c} E_\text{f}} - \frac{2\pi a \tau_\text{f} \sigma_\text{T} L_\text{d}}{E_\text{f}} + \frac{2\pi E_\text{c} \tau_\text{f}^2 L_\text{d}^2}{V_\text{m} E_\text{m} E_\text{f}} + \frac{\pi a^2 \varphi \tau_\text{f}^2}{2 G_\text{m}} -$$

$$\frac{\pi a^2 \sigma_\text{T} E_\text{m} V_\text{m} \varphi \rho \tau_\text{f}}{2 G_\text{m} E_\text{c}} + \frac{\pi a \varphi \rho \tau_\text{f}^2 L_\text{d}}{G_\text{m}} \qquad \text{(M22)}$$

(2) dW_B: The work done by the applied force can be determined by multiplying the fiber-bridging force half the value of the debond displacement, U_debond.

$$W_\text{B} = \pi a^2 \sigma_\text{T} \cdot \left(\frac{V_\text{m} E_\text{m} \sigma_\text{T} L_\text{d}}{E_\text{f} E_\text{c}} - \frac{\tau_\text{f}}{a E_\text{f}} L_\text{d}^2 + \frac{V_\text{m} E_\text{m} \tau_\text{s} a \varphi}{2 E_\text{c} G_\text{m}}\{1 - \exp[-\rho(L - L_\text{d})/a]\} \right)$$

$$\text{(M23)}$$

After disregarding the exponential term and substituting τ_s by expression in (M9),

$$\frac{dW_\text{B}}{dL_\text{d}} = \frac{\pi a^2 V_\text{m} E_\text{m} \sigma_\text{T}^2}{E_\text{f} E_\text{c}} - \frac{2\pi a \tau_\text{f} \sigma_\text{T}}{E_\text{f}} L_\text{d} - \frac{\pi a^2 \sigma_\text{T} V_\text{m} E_\text{m} \tau_\text{f} \varphi \rho}{2 E_\text{c} G_\text{m}} \qquad \text{(M24)}$$

(3) dW_F: The work done against the frictional sliding is determined by multiplying the frictional sliding force with the fiber/matrix interfacial sliding

284

distance, δ_{slide}. The contribution of the matrix shear deformation to the apparent crack opening displacement can be expressed by (Sutcu and Hillig, 1990)

$$U_{\text{shear}} = \int_0^{r\varphi} \frac{\tau_{\text{f}}}{2G_{\text{m}}}\, \mathrm{d}r = \frac{\tau_{\text{f}} r \varphi}{2G_{\text{m}}} \tag{M25}$$

where φ is a nondimensional parameter determined only by the fiber-volume fraction, G_{m} is the matrix shear modulus, and $\varphi \cdot r$ is the effective radius of the cylindrical unit of composite with a V_{f} fiber-volume fraction (Budiansky et al., 1995a; Sutcu and Hillig, 1990). Therefore, it is necessary to subtract this elastic shear displacement from the real sliding displacement.

$$\delta_{\text{slide}}(z) = U_{\text{COD}}(z) - \frac{a\tau_{\text{f}}\varphi}{2G_{\text{m}}} \tag{M26}$$

$$W_{\text{F}} = 2\pi a \tau_{\text{f}} \cdot \int_0^{L_{\text{d}}} \left[U_{\text{COD}}(z) - \frac{a\tau_{\text{f}}\varphi}{2G_{\text{m}}} \right] \mathrm{d}z$$

$$= \frac{\pi a \sigma_{\text{T}} \tau_{\text{f}}}{E_{\text{f}}} \cdot L_{\text{d}}^2 - \frac{4\pi E_{\text{c}} \tau_{\text{f}}^2}{3 E_{\text{m}} V_{\text{m}} E_{\text{f}}} \cdot L_{\text{d}}^3 - \frac{\pi a^2 \tau_{\text{f}}^2 \varphi}{G_{\text{m}}} \cdot L_{\text{d}} +$$

$$\frac{\pi a^2 \tau_{\text{s}} \varphi \tau_{\text{f}} L_{\text{d}}}{G_{\text{m}}} \{ 1 - \exp[-\rho(L - L_{\text{d}})/a]\} \tag{M27}$$

and

$$\frac{\mathrm{d}W_{\text{F}}}{\mathrm{d}L_{\text{d}}} = \frac{2\pi a \sigma_{\text{T}} \tau_{\text{f}}}{E_{\text{f}}} \cdot L_{\text{d}} - \frac{4\pi E_{\text{c}} \tau_{\text{f}}^2}{E_{\text{m}} V_{\text{m}} E_{\text{f}}} \cdot L_{\text{d}}^2 +$$

$$\frac{\pi a^2 \varphi \tau_{\text{f}}^2}{G_{\text{m}}} - \frac{2\pi a\ \varphi \tau_{\text{f}}^2 \rho L_{\text{d}}}{G_{\text{m}}} + \frac{\pi a^2 \tau_{\text{f}} \varphi V_{\text{m}} E_{\text{m}} \rho \sigma_{\text{T}}}{2 G_{\text{m}} E_{\text{c}}} \tag{M28}$$

(4) $\mathrm{d}W_{\text{s}}$: The surface energy associated with the interfacial debonding energy Γ_{d} is expressed as

$$\frac{\mathrm{d}W_{\text{s}}}{\mathrm{d}L_{\text{d}}} = \Gamma_{\text{d}} \cdot 2\pi r \tag{M29}$$

Based on the energy balance criterion, (M16) can be rearranged by substituting the above energy terms obtained in (M22), (M24), (M28), and (M29); and the fiber bridging stress after considering the shear strain (energy) in composite can be solved as

$$\sigma_{\mathrm{T}} = \frac{2E_c\tau_f L_d}{aV_m E_m} + \frac{2E_c\tau_f}{E_m V_m \rho} + \sqrt{\frac{4E_f E_c \Gamma_d}{aE_m V_m}} \tag{M30}$$

Assuming that the fiber-bridging stress is related to the distributed bridging stress $T(x)$ (traction) on the crack surface by the following relationship,

$$T(x) = \sigma_{\mathrm{T}} V_f \tag{M31}$$

By applying (M30), the distributed fiber-bridging stress $T(x)$ can be obtained as

$$T(x) = \frac{2V_f E_c \tau_f}{aV_m E_m} \cdot L_d(x) + \frac{2V_f E_c \tau_f}{E_m V_m \rho} + V_f \cdot \sqrt{\frac{4E_f E_c \Gamma_d}{aE_m V_m}} \tag{M32}$$

It is very interesting to find that $T(x)$ in (M32) consists of three terms: a frictional term $\dfrac{2V_f E_c \tau_f L_d}{aV_m E_m}$, which is identical to the expression of BHE, HJ Hsueh, a debond energy term $V_f\sqrt{\dfrac{4E_f E_c \Gamma_d}{aE_m V_m}}$, which corresponds to the initiation stress σ_d in BHE, HJ Hsueh, and the shear energy term $\dfrac{2V_f E_c \tau_f}{E_m V_m \rho}$. In comparison with BHE, HJ Hsueh, the consideration of shear strain energy in this model brings an additional term σ_s into the expression of fiber-bridging stress.

Another important result obtained from (M30) and (M9) is that the shear strength of the fiber-matrix interface can be redefined as

$$\tau_s = \tau_f + \sqrt{\frac{4\Gamma_d G_m}{a\varphi}} \tag{M33}$$

The result turns out to be identical to the derivation of Sutcu and Hillig (1990). It also implies that the shear strain (energy) could play an important role in interfacial debonding. In the absence of interfacial debonding, the interfacial shear strength τ_s is equal to the interfacial frictional stress τ_f. In the case of a zero friction interface, the interfacial shear strength is only determined by the debond energy (approximately two twice the surface energy). Both conditions are also illustrated in detail by BEH (Budiansky, et al., 1995a).

The crack opening displacement U_{COD} based on the energy balance approach can be obtained by substituting the fiber stress σ_{T} and interfacial shear strength τ_s in (M9) with the expressions (M30) and (M33), respectively.

$$U_{COD} = \frac{E_c \tau_f}{E_m V_m E_f a} \cdot L_d^2 + \left(\frac{2E_c \tau_f}{E_m V_m E_f \rho} + \sqrt{\frac{4E_c \Gamma_d}{a V_m E_m E_f}} \right) \cdot L_d + \frac{a \varphi \tau_f}{2G_m} + \sqrt{\frac{a \varphi \Gamma_d}{G_m}}$$

$$(M34)$$

Because both the crack opening displacement U_{COD} and fiber-bridging stress $T(x)$ can be expressed in terms of the debond length as given in (M32) and (M34), the relationship between the $T(x)$ and U_{COD} can be derived. However, the discrete fiber model differs from the continuum composite model (Marshall and Evans, 1988; McCartney, 1987; Marshall and Cox, 1988e; Danchaivijit and Shetty, 1993) in the expression of $u(x)$. For a real composite case, the energy dissipation takes place when a crack of finite length grows as shown in Fig. 6.7. It was found that the crack opening displacement predicted by continuum models is less than that for the discrete model by a factor of $\dfrac{V_m E_m}{E_c}$, which can also be expressed as $\dfrac{1}{1+\eta}$ (Marshall and Shetty, 1988; McCartney, 1987; Marshall and Cox, 1988; Danchaivijit, et al., 1993). The details were discussed by McCartney (McCartney, 1987), and Marshall and Evans (1988). This factor had been assigned to make the two models consistent by the following expression.

$$u(x) = \frac{1}{1+\eta} \cdot U_{COD} \tag{M35}$$

Therefore, the distributed fiber-bridging stress $T(x)$ can be related to crack opening displacement $u(x)$ by

$$u(x) = \frac{1}{1+\eta} \left(\frac{a V_m E_m T(x)^2}{4 E_f E_c \tau_f V_f^2} - \frac{\Gamma_d}{\tau_f} + \frac{a \varphi \tau_f}{4 G_m} \right) \tag{M36}$$

In the case of no debonding and low friction at the interface, the debond energy term equals zero, and the shear energy term $\dfrac{\varphi a \tau_f}{4 G_m}$, can be dropped. Therefore, (M36) can be simplified to

$$u(x) = \frac{r}{4 E_f \tau_f V_f^2 (1+\eta)^2} \cdot T^2(x) \tag{M37}$$

This expression is identical to the MC model for the COD analysis. It is not difficult to see that the newly derived model is more comprehensive than the MC model of COD measurement and BHE, HJ, Hsueh by including the influences of both interfacial debonding and shear deformation. Therefore, the newly derived model can be applied to both a strong and a weak interface.

6.7 A Novel Technique for Studying Fiber Reinforcement

The most important advantage of fiber-reinforced ceramic matrix composites (FCMC) over monolithic ceramics is their higher fracture toughness (higher flaw tolerance). The greatly enhanced toughness of FCMC is largely governed by the bridging (traction) behavior of fibers at the cracked matrix surfaces. This fiber-bridging behavior is also associated with the processes of fiber/matrix interfacial debonding /sliding and fiber pullout after the failure of fiber bundles. The enhanced fracture toughness K^{L} is related to the fiber-bridging stress by (6.26). Therefore, the determination of in-situ distribution of the fiber-bridging stress $p(x)$ along a matrix crack becomes the critical part for evaluation of toughness.

6.7.1 Determination of Fiber-Bridging Stress by Debond Length Measurement

One of the conventional methods for determining the distribution of fiber-bridging stress is the crack opening displacement (COD) measurement using the relationship between $T(x)$ and $u(x)$ as expressed by Eq. (6.28), which is referred to hereafter as MC COD (Marshall and Cox, 1988e). Using this technique, the COD profile $u(x)$ is measured and related to the fiber bridging stress profile. Although many studies have been conducted to extract the fiber-bridging toughness from COD profile, no independent verifications have been achieved.

Besides COD, the debond length is another parameter that is more directly associated with both the process of fiber bridging and interfacial debonding/sliding. Considerable progress has made in the area of the theoretical modeling related to interfacial debonding/sliding (Budiansky et al., 1995; Sutcu, and Hillig, 1990; Li, et al., 1993; Hutchinson and Jensen, 1990; Hsueh, 1996). The relationships among the externally applied stress, interfacial properties, and debond length were derived from both the force balance and energy balance approaches. However, these relationships had not been experimentally examined in FCMCs until the recent work done by Sun and Singh (1998a; 1998d). The difficulty in measuring debond length at the fiber/matrix interface in FCMCs was overcome by carefully choosing a transparent composite system (Sun and Singh, 1996). The observation and measurement of debond length provided useful information about the processes of fiber bridging and interfacial debonding/sliding. The accessibility of the debond length measurement along the fiber/matrix interface in a transparent composite system prompted the development of a new approach, namely the debond length measurement (DLM), for directly measuring the fiber bridging stress associated with a bridged matrix crack.

Therefore, the objective of this portion of the study is to develop a new technique for directly measuring the fiber-bridging stress by debond length measurement (DLM) in FCMCs. Firstly, a theoretical model for DLM is developed to establish the relationships among the fiber-bridging stress, debond length, and COD. Secondly, the DLM is performed in-situ to obtain the debond length profile along a matrix crack. Based on the DLM models, the distribution of fiber-bridging stresses is quantified. Finally, the COD measurement is made to independently calculate the fiber-bridging stress and compare these stresses with the results obtained from the DLM.

The composites PB used for this study had 7% fibers by volume. After grinding, cutting and polishing, specimens with dimensions of 1.95 mm $\times$ 5.0 mm $\times$ 50 mm were prepared for four-point flexure tests. The DLM and COD measurements were performed in-situ on the same specimen at the same time. The material parameters and mechanical properties of composites are listed in Table 6.5.

6.7.1.1 Techniques for Determining Fiber-Bridging Stress

The crack opening displacement (COD) measurement and a new method using the debond length measurement (DLM) are used to measure fiber-bridging stress.

1) COD Measurement

The relationship between the fiber-bridging stress, T, and COD value $u(x)$ in Fig. 6.7 can be expressed by (6.28) (Marshall and Cox, 1988; Danchaivijit and Shetty, 1993; Meda and Steif, 1994). From the in-situ measurement of $u(x)$ along a matrix crack, the fiber-bridging stress distribution $T(x)$ can be determined providing that the interfacial properties are available. This equation is used for relating the COD measurement (MC COD) to the bridging stress, $T(x)$, in this study. However, whether this formula is suitable or not for the real composites with a partially-bonded/debonded interface needs to be examined. The experimental approaches and test set-up were described in Sect. 6.4.2.

2) Debond Length Measurement (DLM)

As newly modeled in Sect. 6.6.2, the relationship among the fiber bridging stress $T(x)$ and debond length L_d can be expressed from both force balance and energy balance approaches. Two existing models and one new derivation are introduced and compared.

a) Budiansky, Evans and Hutchinson (BHE) (1995), Hutchinson and Jensen (HJ) (1990), and Hsueh (1996)'s models (hereafter denoted as BHE, HJ, Hsueh):

$$T(x) = \frac{2V_f E_c \tau_f}{aV_m E_m} \cdot L_d(x) + V_f \cdot \sqrt{\frac{4E_f E_c \Gamma_d}{\cdot aE_m V_m}} \qquad (6.56)$$

where Γ_d is the debond energy for the fiber/matrix interface in a composite. Primarily based on a force balance approach, they employed the lame solutions in the analyses. Therefore, no shear effect was considered for simplicity. Hsueh (1996) also proved that the expression derived from force balance was the same as that obtained from energy balance.

b) Li, Shah and Mura's model (hereafter denoted as Li, Shah & Mura) (1993)

$$T(x) = \frac{2V_f E_c \tau_f}{aV_m E_m} \cdot L_d(x) + V_f \cdot \left[\frac{4E_c E_f \tau_f^2 V_f}{E_m^2 V_m^2 a^2} \cdot L_d^2 + \frac{4E_f \Gamma_d}{a} \right]^{1/2} \qquad (6.57)$$

Equation (6.57) is a derived result from (33) in Li et al. (1993). It differs from (6.56) because they are based on different assumptions during theoretical analyses. Li, Shah & Mura was proposed from the energy balance approach. The change of elastic strain energy in the bonded portion was ignored, also no shear effect was considered.

c) Sun and Singh's model (1998) (hereafter denoted as Sun & Singh) (see details in Sect. 6.6.2)

$$T(x) = \frac{2V_f E_c \tau_f}{aV_m E_m} \cdot L_d(x) + \frac{2V_f E_c \tau_f}{E_m V_m \rho} + V_f \cdot \sqrt{\frac{4E_f E_c \Gamma_d}{aE_m V_m}} \qquad (6.58)$$

where ρ is the shear lag parameter, and φ is a constant determined by composite volume fractions. In this model, an energy balance approach with the consideration of shear strain energy change was used. Comparing with BEH, HJ, Hsueh, an additional term appears in the debond length-applied stress relationship due to the contribution of shear strain energy to the debonding process.

Therefore, the fiber-bridging stress can be calculated using either (6.56) (BEH, HJ, Hsueh) or (6.57) (Li, Shah & Mura) or (6.58) (Sun & Singh) if the debond length L_d can be measured. Comparisons among (6.56), (6.57) and (6.58) show that the fiber bridging stresses determined from different models may vary significantly.

The debond length measurement were made by the approach described in Sect. 6.4.2. The matrix cracking occurred on the tensile surface of a flexure bar when the applied load increased beyond the matrix cracking stress. Accompanying matrix cracking, interfacial debonding was also created because of the weakly-bonded interface (Sun and Singh, 1998a). This debond phenomenon was observed throughout the experiments. The debond length was in-situ measured using the photography and video recording methods without unloading the specimen. The magnitude of the debond length was taken as half of the length of the white band in Fig. 6.13. The debond length for each fiber layer along a matrix crack was obtained with respect to the fiber positions along the crack path. Therefore, both the debond length and COD profiles were obtained for each matrix crack under the same load/stress level. These experimental data were then used for determining the fiber-bridging stress along the matrix crack.

6.7.1.2 Debond Length and COD Profiles

The flexure tests were performed under the displacement control. The interfacial debonding was observed when FMC was generated because of a weak fiber/matrix interface. This was consistent with the previous observations (Sun and Singh, 1998). A triangular debond profile along the propagation direction of each matrix crack was observed for the first time in this study. Figure 6.36 is an optical micrograph from a sample loaded to 134 N or 85 MPa stress. The debond length profiles of the four different matrix cracks shown in Fig. 6.36 are measured and plotted in Fig. 6.37. The variation of the debond length values among different cracks is small because of the uniformity of the stress distribution created upon loading in the four-point flexure mode. This observation on the debond length measurement is consistent with the earlier observation of the interfacial debonding/sliding (Sun and Singh, 1998a).

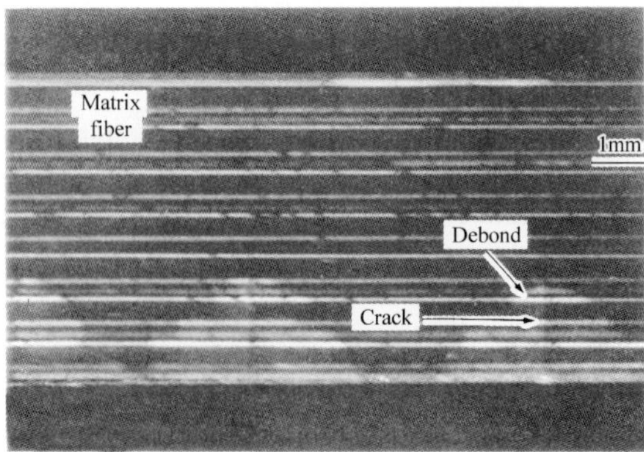

Fig. 6.36 Micrograph showing the debond profiles along matrix cracks subjected to a force of 134 N

The COD profiles of these four matrix cracks are also measured using a high-magnification microscope (2000x). Because of the amorphous structure of the glass matrix, relatively straight cracks, are observed instead of zigzag or

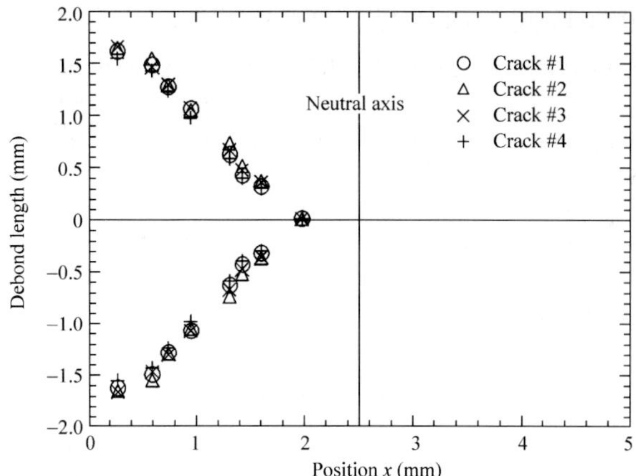

Fig. 6.37   Debond length profiles of the four matrix cracks subjected to a force of 134 N

torturous cracks, which are commonly observed in polycrystalline ceramics (Gilbert et al., 1998). Therefore, the COD measurement in this composite system produced less scatter and error. The COD profiles measured for these four matrix cracks are shown in Fig. 6.38. It is interesting to observe that the shapes and dimensions of these cracks are very similar, which is in contrast to the results of Li, Shah and Mura's on a steel-cement composite system (Li et al., 1993). This might be explained partly by their different loading method. Li, Shah and Mura's tests were performed in a tensile mode, whereas this study utilized the four-point flexure. The crack lengths are also measured and found to be very close to each other, as shown in Fig. 6.38. At a stress of 76 MPa the matrix cracks did not reach the neutral axis of the flexure specimen, which was about 2.5 mm form both ends of the sample.

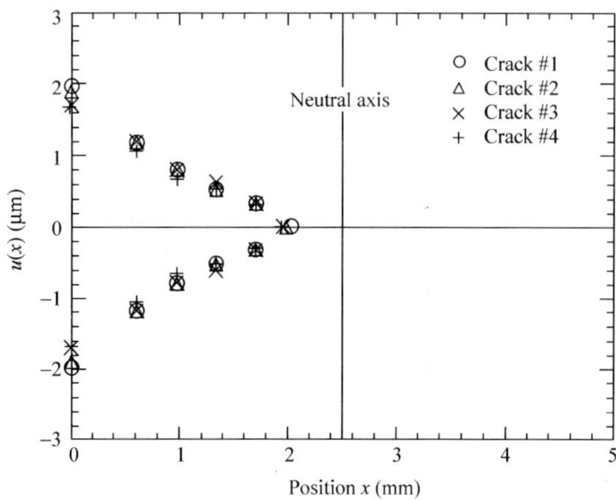

Fig. 6.38   COD profiles of the four matrix cracks subjected to a force of 134 N

6.7.1.3   Predicted and Measured COD Profiles

As the relationships between fiber-bridging stress and debond length are derived from both the force balance and energy balance models, the relationships between COD and debond length are also analyzed. Therefore, the COD values can be predicted from the experimentally measured debond length based on the following formulas:

a)   Relation based on the derivation from BEH, HJ, Hsueh:

$$U_{COD} = \frac{E_c \tau_f}{E_m V_m E_f a} \cdot L_d^2 + \sqrt{\frac{4E_c \Gamma_d}{a V_m E_m E_f}} \cdot L_d \qquad (6.59)$$

b)   Relation based on the derivation from Li, Shah and Mura:

$$U_{COD} = \frac{E_c \tau_f}{E_m V_m E_f a} \cdot L_d^2 + \left[\frac{4E_c V_f \tau_f^2}{V_m^2 E_m^2 E_f a^2} \cdot L_d^2 + \frac{4\Gamma_d}{aE_f} \right]^{1/2} \cdot L_d \qquad (6.60)$$

c) Sun & Singh expression (see details in Sect. 6.6.2)

$$U_{COD} = \frac{E_c \tau_f}{E_m V_m E_f a} \cdot L_d^2 + \left(\frac{2E_c \tau_f}{E_m V_m E_f \rho} + \sqrt{\frac{4E_c \Gamma_d}{a V_m E_m E_f}} \right) \cdot L_d + \frac{a\varphi\tau_f}{2G_m} + \sqrt{\frac{a\varphi\Gamma_d}{G_m}}$$

$$(6.61)$$

In all those expressions ((6.59), (6.60) and (6.61)), U is half of the COD value for a discrete fiber mode (the crack surfaces are parallel to each other). In the case of a continuum fiber mode, a factor of $\dfrac{1}{(1+\eta)}$ is applied to modify (6.59), (6.60) and (6.61) (as given by (M35) in Sect. 6.6.2). This factor was introduced by MC (Marshall and Cox, 1988) to modify the model of MCE (Marshall et al., 1985) for a non-through crack. The matrix crack in the four-point flexure mode is also considered as a nonthrough crack. Therefore, the modified COD profile $u(x)$ obtained from DLM was used to predict the COD profile for a certain matrix crack and then to compare with the experimental COD profile. This approach provided a substantial support and validity to the model and new approach based on the DLM technique to analyze the data.

The DLM models were tested by predicting the COD profile from the experimentally measured debond length. Equations (6.59), (6.60), and (6.61) were used for the calculations. The matrix crack #2 was chosen as a model crack for this calculation. The real COD profile $u(x)$ for crack #2 is plotted together with the predicted profiles in Fig. 6.39. It should be mentioned that $u(c)$ (COD at the crack tip) must be zero. This statement is consistent with (6.59) and (6.60) when L_d is equal to zero, but conflicts with (6.61). It can be realized that the term $\dfrac{\varphi \tau_f r}{2G_m}$ (related to interfacial sliding) for an uncracked matrix at the crack tip should be dropped because no interfacial friction occurs at the crack tip. However, it is also important to know that zero debond length may not be necessarily related to zero COD. For a very weak interface, $\dfrac{\varphi \tau_f r}{2G_m}$ can be ignored; but, this term cannot be ignored for a relatively strong interface. The $u(x)$ values calculated from Li, Shah & Mura is found to be larger than the results from BEH, HJ, Hsueh, and Sun & Singh. This can be explained from the assumption of Li, Shah & Mura. Because of the ignorance of strain energy release in the bonded region during debond, a larger force will be expected to drive the propagation of debond at interface. The COD values determined from Sun & Singh are always larger than those predictions from BEH, HJ, Hsueh because the shear displacements are included in Sun & Singh analysis. It is, however, interesting to find that such a difference is not that significant in the case of a weak fiber/matrix interface. In that case, both BEH, HJ, Hsueh and Sun & Singh gave a reasonable prediction of COD profile of this study (small σ_d and/or τ_f for a weak interface). Findings confirmed the validity of the expressions of BEH, HJ, Hsueh and Sun & Singh. However, a large deviation is expected in the case of a stronger fiber/matrix interface (larger σ_d and/or τ_f), and Sun & Singh will provide a better estimate.

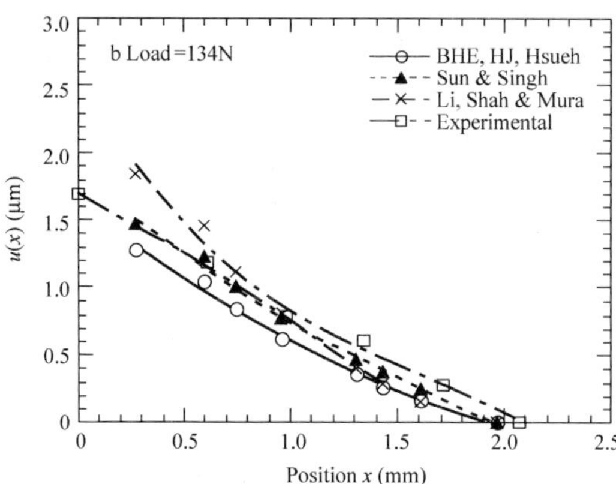

Fig. 6.39 Comparison of the predicted COD profiles obtained from the DLM with the experimentally obtained COD for Crack #2 subjected to a force of 134 N

294

6.7.1.4 Fiber-Bridging Stress Distributions

The distribution of fiber-bridging stress $T(x)$ along the matrix crack #2 was calculated from the experimentally measured debond length profile and using DLM models ((6.56), (6.57) and (6.58)). Figure 6.41 shows a comparison among the calculated values for crack #2 from BEH, HJ, Hsueh, Li, Shah & Mura, and Sun & Singh. It is noted from (6.56), (6.57), and (6.58) that the fiber-bridging stress $T(c)$, the fiber-bridging stress at the crack tip, for a zero debond length is not zero, but is equal to σ_d. However, in Fig. 6.40 the fiber-bridging stress at the crack tip (corresponding to a zero COD and debond length) was made zero in order to be compatible with the COD measurement. This adjustment is reasonable because there is no fiber bridging effect when $u(c)$ is zero, although stress in the fiber/matrix may not really be zero. This nonzero fiber stress at the crack tip with a zero crack opening displacement was also emphasized by Media and Steif (1994).

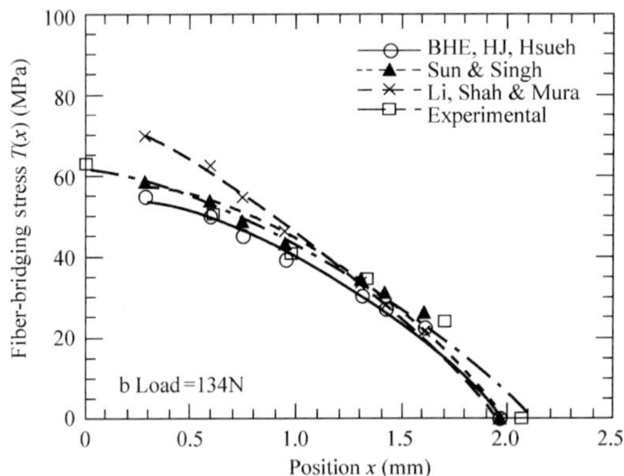

Fig. 6.40   Comparison of the fiber-bridging stress profile obtained from the DLM with the estimate from the COD for crack #2 subjected to a force of 134 N

The fiber-bridging stresses determined by Li, Shah & Mura are mostly larger than the values obtained by the other two models. This discrepancy results from the ignorance of release of elastic strain energy in the bonded region by assumption, as explained early. Such an assumption brings an imperfection to the model. The fiber-bridging stresses obtained from Sun & Singh are always larger that those values obtained from BEH, HJ, Hsueh. It is important to note that the composite once without any shear deformation has changed into a composite with two components (fiber and matrix) partially combined through interfacial shear. Therefore, from the viewpoint of fracture mechanics, the stress to initiate and propagate a debond crack at the interface should be larger after considering the additional energy requirement for shear deformation. For a composite with a low friction interface, the shear strain energy plays a

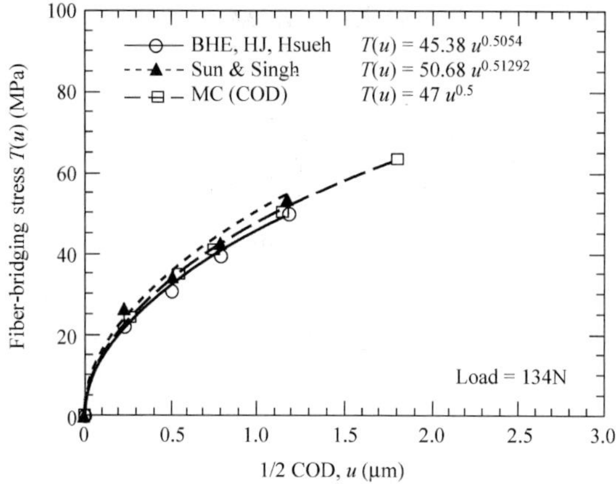

Fig. 6.41 Comparison of the $T(x)$-$u(x)$ relationship calculated from the DLM and COD measurements for crack #2 subjected to force of 134 N

comparatively small role, and in this situation the extra term $\dfrac{2V_f E_c \tau_f}{E_m V_m \rho}$ can be ignored. BEH, HJ, Hsueh is a good model for the analysis. However, in a composite with a high friction interface, the shear strain energy associated with the deformation of matrix plays an important role as well. Sun & Singh will show its advantage over the other two models in the debonding and fiber bridging analysis.

The fiber-bridging stresses were also calculated from the measured COD for crack # 2 using (6.28) (MC COD Model). These results were plotted against the position x in Fig. 6.40 which shows a nonlinearly decreasing fiber-bridging stress starting from the outermost tensile surface towards the neutral axis of a bend bar. This observation is also interesting from the viewpoint of an elastically loaded beam which is supposed to show a linearly varying stress from the outer surface towards the neutral axis. The fiber-bridging stress distributions measured from the DLM and COD measurements displayed the same trend along the matrix crack, although the differences in the fiber-bridging stress values between DLM and COD measurements cannot be ignored. The fiber-bridging stress determined from the COD is close to the stress values obtained from BEH, HJ, Hsueh and Sun & Singh of the DLM but inconsistent with the trend predicted by Li, Shah & Mura. This is because the COD measurement relies on the same derivation as BEH, HJ and Hsueh except that MC COD ignores the extra debond energy term.

In an attempt to relate the distribution of fiber-bridging stresses to the COD, and to identify the power law exponent n in (6.27), the fiber-bridging stresses $T(x)$ calculated from the DLM (both the force balance and energy balance approaches) were plotted against the experimentally measured COD values $u(x)$

in Fig. 6.41. The curve based on (6.27) of MC Model (COD measurement) is also shown. In addition, curve fitting of the data of $T(x)$ vs $u(x)$ using a simple power law as described by (6.27) was made that gave an n value of 0.513 for Sun & Singh and 0.505 for BEH, HJ, Hsueh with very close constant A values. Both n values are close to the value of 0.5 as suggested by MC (Marshall and Cox, 1988e) but not zero as indicated by Sakai's model (Sakai, 1992).

The experimental examination of the relationship between $T(x)$ and $u(x)$ in Fig. 6.41 indicated that the $T(x)$-$u(x)$ relationship should be modified for real composite systems in which various interfacial conditions could be tailored. Based on the energy balance approach concerning the interfacial debond energy and shear strain energy, a new bridging law for describing the fiber-bridging stress can be expressed as (see details in Sect. 6.6.2) (Sun and Singh, 1998d)

$$u(x) = \frac{1}{1+\eta}\left(\frac{aV_\mathrm{m}E_\mathrm{m}T(x)^2}{4E_\mathrm{f}E_\mathrm{c}\tau_\mathrm{f}V_\mathrm{f}^2} - \frac{\Gamma_\mathrm{d}}{\tau_\mathrm{f}} + \frac{a\varphi\tau_\mathrm{f}}{4G_\mathrm{m}} \right) \tag{6.62}$$

where η is defined early in the paper. For a low-friction interface in the absence of interfacial debonding (i.e., $\Gamma_\mathrm{d} = 0$) (6.62) becomes identical to the MC model for COD measurement (6.28). Comparing to BHE, HJ, Hsueh (6.62) (Sun & Singh) has an additional term, $\dfrac{a\varphi\tau_\mathrm{f}}{4G_\mathrm{m}}$, which corresponds to the shear deformation in composites.

Conclusively, a new approach for determining the distribution of fiber bridging stress based on the debond length measurement (DLM) was developed. This approach utilized the in-situ study of fiber-matrix crack interaction, measurements of the interfacial debonding, and modeling of the fiber-bridging and fiber-matrix interfacial debonding/sliding. The debond length profiles along a fiber-bridged matrix crack are observed and measured in-situ in this study. The fiber-bridging stresses are calculated from the measured debond lengths. The fiber-bridging stresses are also determined from the independently measured COD using the MC model. Both the DLM and COD measurements provided similar estimates of the distribution of the fiber-bridging stresses in this study. However, discrepancies in the measured values of the-fiber bridging stresses between the DLM and COD methods are also found.

The power law relationship between fiber bridging stress and crack opening displacement is examined experimentally. It is found that the MC model of COD measurement (power $n=0.5$) is acceptable in the case of a very weak interface. But, for a stronger interface, it is expected that the newly derived fiber-bridging laws should provide a better estimate of the crack-bridging stress than the MC COD measurement. This bridging law is applicable to both a weak and a strong fiber/matrix interface.

6.7.2 Contribution of Fiber-Bridging to Fracture Toughness

In an attempt to understand the fiber-bridging effect to the composite toughness and verify the validity of the COD and DLM approaches for determining fiber-bridging stress and fracture toughness, three independent methods are utilized for determining the fracture toughness of a composite. They are: 1) SENB test ((6.41) and (6.42)), 2) COD measurements including MC COD ((6.28), (6.45), and (6.43)) and New COD ((6.62), (6.45), and (6.43)), and 3) DLM ((6.58) and (6.45)). The objective of this study is to compare the fracture toughness of the composites obtained from these three methods in which the in-situ measurements of the crack length, applied load, COD, and debond length are made. Thus, the contribution of the crack-bridging fibers to the toughness of a composite can be identified. Especially, method (1) gives the macroscopic determination of toughness, whereas methods (2) and (3) provide the microscopic material properties such as fiber-bridging stress distribution. As a result, the bridging-stress function obtained in the previous study can be further verified by comparing the fracture toughness values determined from the macro and micro approaches (Sun and Singh, 1998e).

6.7.2.1 SENB Test, DLM and COD Measurements

The experiments were conducted on an unnotched sample with dimensions of $1.95 \text{ mm} \times 5.0 \text{ mm} \times 50 \text{ mm}$ in the four-point bending mode. A schematic of the specimen setup is illustrated in Fig.6.2. The specimen was loaded in a mini-flexure fixture as shown in Fig. 6.15. The outer and inner spans were 40 mm and 20 mm, respectively. The loading rate was controlled manually and all tests were done under displacement control. The sample was not removed from the fixture during the tests.

The SENB test (method (1)) was conducted in the same fixture. The applied load values were read from a load cell. The crack length was measured at a certain load/stress level using a Micro Measure Machine (Process Equipment Company, OH) equipped with an optical microscope and video imaging facility. A lower magnification of about 10x was used for this purpose but magnification of up to 2000x was possible with this system. The measurements of the applied load and crack length were then used to calculate the fracture toughness K_P from (6.41) and (6.42).

The COD measurements (method (2)) were made under the same microscope but with the highest magnification of 2000x. An image-freezing function was used to achieve the highest resolution of 0.4 μm. The COD images were recorded in a computer with respect to the crack position of each measurement. The COD value at the crack tip was not accessible because of the limited resolution of the equipment. The COD measurement determined the $u(x)$ profile along a matrix crack under different load or stress levels. These COD values were firstly used to calculate the fiber-bridging stress $\sigma(x)$ from (6.28)

298

and (6.62), and then the crack intensity factor K_b from (6.43) and (6.45).

The debond length measurements (DLM) (method (3)) were performed in-situ along with the measurement of COD. Accompanying matrix cracking, interfacial debond was also created because the interface is weakly bonded. The debond length was measured using the photography and video recording methods without the removal of specimen. The value of debond length was determined as half the length of the white band. The debond length for each fiber layer along a matrix crack was obtained with respect to the fiber positions. The fiber-bridging stress $\sigma(x)$ was determined from the debond length profile $L_d(x)$ using (6.58) (which has been confirmed in an earlier study (Sun and Singh, 1998)). The fracture toughness K_T was then calculated from both (6.45) and experimental COD data.

6.7.2.2 Debond Length and COD Profiles

The flexure tests were performed under four different load/stress levels. The first matrix cracking (in fact a total of four cracks were observed at the same time) occurred at a load of 120 N, which corresponded to a stress of 76 MPa at the outermost tensile face. Interfacial debondi ng was observed upon the FMC was generated. Triangular debond length profiles along the direction of matrix crack propagation were observed. The debond length increases with the crack growth. Figure 6.42 a–d shows the optical micrographs of a sample from which the dependency of the debond length profiles on the applied loads can be observed. The triangular profiles are distorted in Fig. 6.42 d because of the deflection of matrix cracks. The measured debond length profiles, from four different matrix

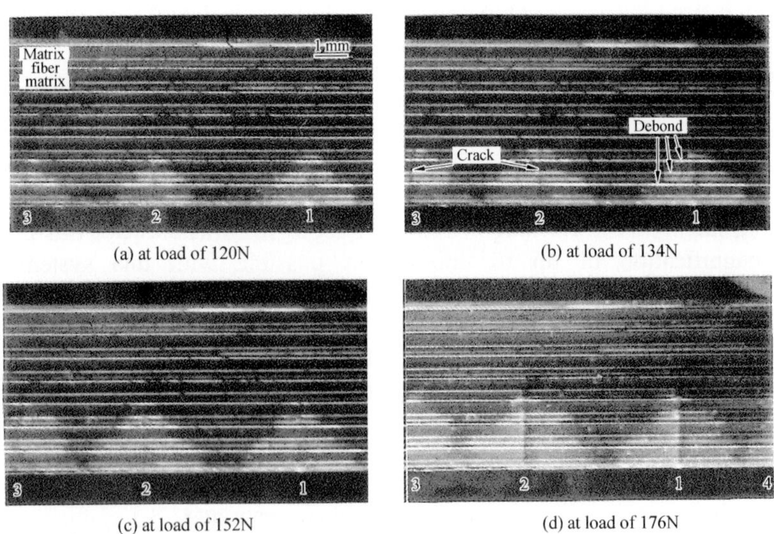

Fig. 6.42 Triangular debond profiles associated with advancing cracks

cracks subjected to four different loads, are plotted in Fig. 6.43 a–d. Figure 6.43 also shows that the debond and crack lengths for all of these matrix cracks are quite similar at a given load level indicating the relative uniformity in stress distribution in spite of loading in the four-point flexure mode. This phenomenon was also observed in a previous study of the interfacial debonding/sliding (Sun and Singh, 1998a, 1998d).

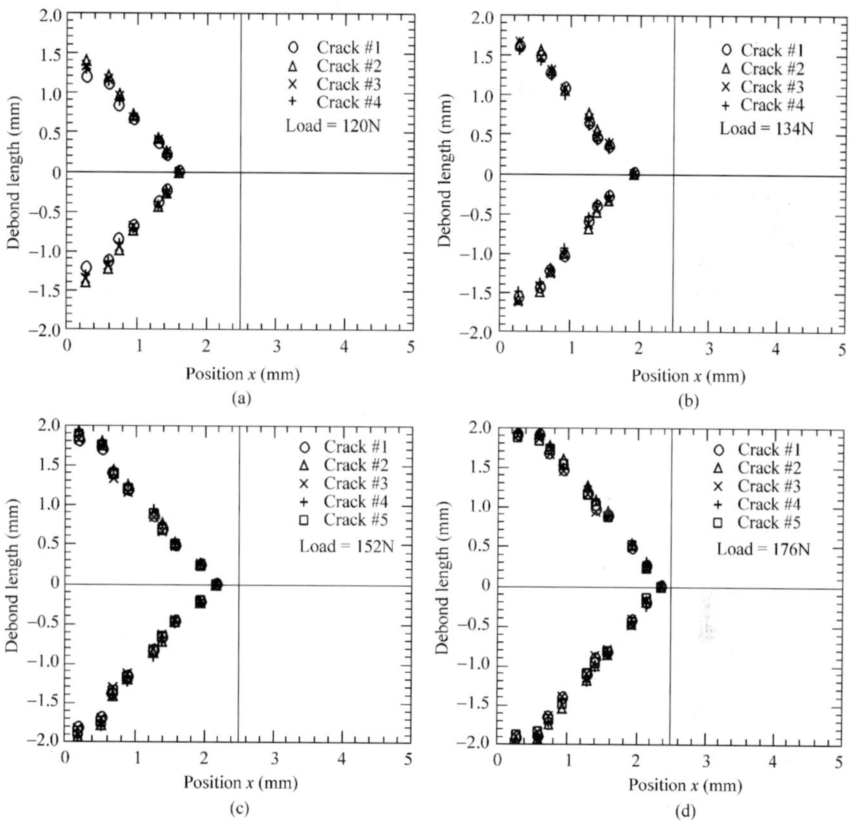

Fig. 6.43   Influence of the applied load on the debond profiles

The COD profiles from different matrix cracks were also measured a using a high-magnification microscope (2000x). Relatively straight cracks are observed in this composite instead of zigzag and torturous cracks, which were commonly observed in polycrystalline ceramics (Gilbert et al., 1998). This situation was beneficial for the COD measurement because it gives a more accurate measurement. The COD profiles for these matrix cracks subjected to four different loads are displayed in Fig. 6.44 a–d. The COD increased with the load and crack length. It is very interesting to see that even different cracks are very similar in terms of shapes and dimensions at the same load.

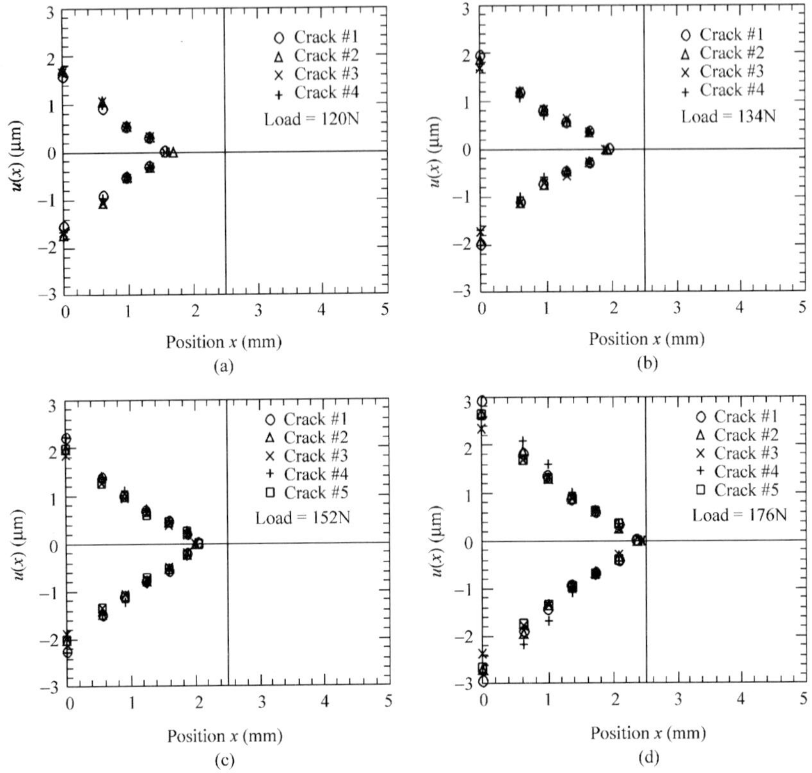

Fig. 6.44 Influence of the applied load on COD profiles

The crack length, debond length measured at the outermost fiber, and COD measured at the outermost surface for crack # 2 are also plotted against applied load/stress in Fig. 6.45. The saturation phenomenon of debond length is observed at a load of 210 N at which the crack almost reached the neutral axis.

6.7.2.3 Determination of Fiber-Bridging Stress

To obtain the fiber-bridging toughness K_b, the distribution of fiber-bridging stress along the matrix crack must be determined first. Two methods, COD and DLM, were applied to determine the fiber-bridging stress in this study. Since different matrix cracks showed similar COD and debond length profiles and crack length, the crack # 2 was used as a model crack for calculating the fiber-bridging stress, which was then used to calculate the fiber-bridging toughness.

The distribution of fiber-bridging stress, $\sigma(x)$, along the matrix crack #2 at four different load levels was calculated from the experimentally measured debond length profiles using (6.58) (DLM). Figure 6.46 a–d shows the dependency of the fiber-bridging stress on the applied load for crack # 2. It is noted from (6.58) that the fiber-bridging stress $\sigma(0)$ for a zero debond length and

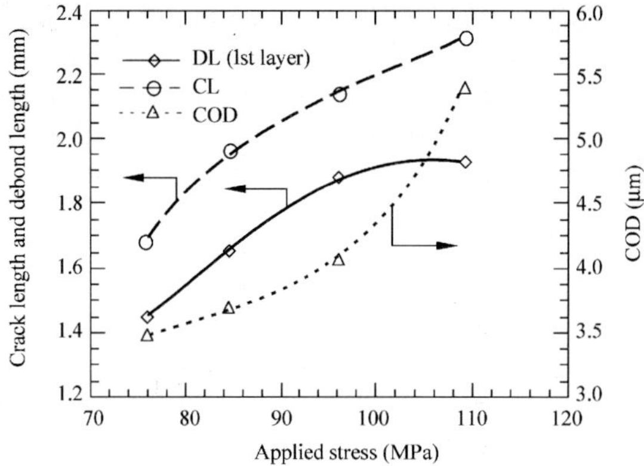

Fig. 6.45 Dependencies of crack length (CL), debond length (DL),   and COD on the applied stress

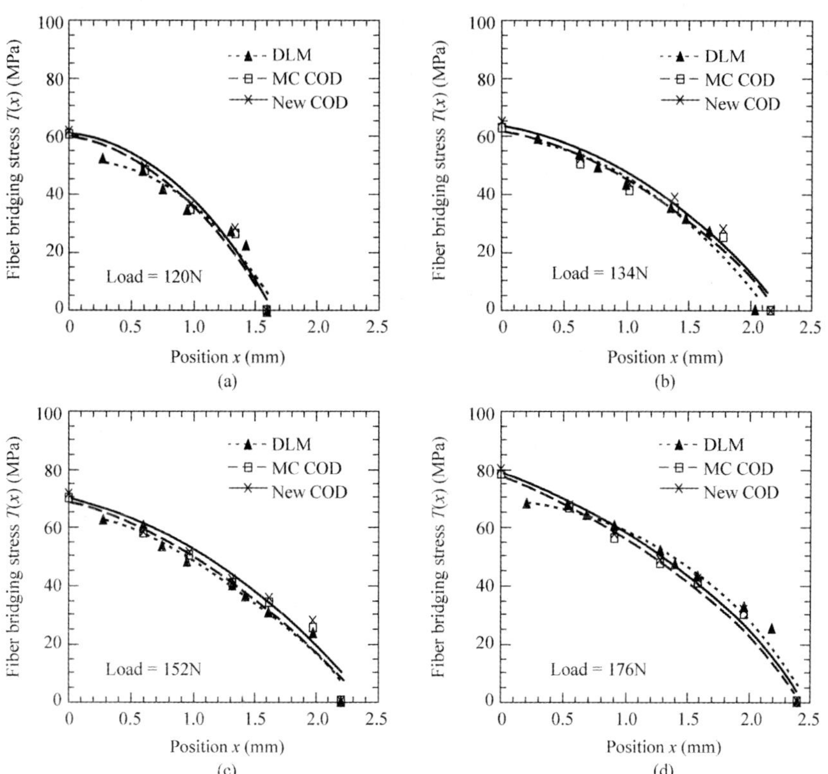

Fig. 6.46 Influence of the applied load on the fiber bridging stress distributions along the direction of matrix crack propagation

COD is not zero. But, in Fig. 6.46 the fiber-bridging stress at the crack tip (corresponding to a zero COD and debond length) was made zero in order to be compatible to COD measurement. These results show that the fiber-bridging stress increased with applied load. But, such a fiber-bridging effect is observed to saturate at the outermost fibers when the debond from either side of an uncracked matrix block reached each other as shown in Fig. 6.42d. This phenomenon was also observed in previous research (Sun and Singh, 1998a).

Based on the COD profiles for crack # 2, the fiber-bridging stresses at different loads were also calculated using (6.28) (MC COD) and (6.62) (New COD). The distribution of fiber-bridging stress obtained from COD measurement is also shown in Fig. 6.46 a–d for comparison with the results obtained from the DLM. The fiber-bridging stress distributions obtained from DLM and COD measurement show the same trend along the matrix cracks. The differences in the fiber-bridging stress values obtained from DLM, MC COD, and new COD measurements are also observed. Generally, the bridging stress calculated from new COD are always higher than those calculated from MC COD because the new bridging law in (6.62) has included the influences of both interfacial debond energy and shear strain energy. The stress values obtained DLM, however, scattered around the main trend. This is mainly attributed to the scatter of measurements of debond length.

The DLM and COD approaches to determine fiber-bridging stress are different when using different models for analysis. For instance, the MC COD method is based on a force balance approach with an assumption of a very weak fiber/matrix interface in composites, whereas the DLM and New COD are based on the energy balance approach. This difference is expected to be greater for a stronger interface (larger Γ_d and/or τ_f). Therefore, it is suggested that the application of the MC COD measurement should be limited to composites with a very weak interface.

The fiber-bridging stress $\sigma(x)$ calculated from both the DLM and COD measurement are plotted against the experimentally measured COD values $u(x)$ for four different load levels in Fig. 6.47. After curve fitting the data of $T(x)$ vs, $u(x)$ using a power law, the exponent n was found to range from 0.499 to 0.510 for DLM, 0.502 to 0.504 for new COD and 0.5 for MC COD (Marshall and Cox, 1988e). Figure 6.48 displays all the experimental data for four different load levels in one figure. After curve fitting these data using (6.27), the exponent n and constant A for DLM are 0.510 and 47.75, respectively. The exponent n and constant A for new COD are 0.505 and 49.20, respectively. As expected, the fiber-bridging stresses calculated from MC COD, DLM, and new COD are close to one another. It is interesting to see in Fig. 6.48 that the fiber-bridging stress determined from the debond length (6.58) were close to those calculated from the new bridging law (6.62). They were both larger than the values calculated from COD measurement. The power law expressions, as shown in Fig. 6.48, were used for the fracture-toughness calculations.

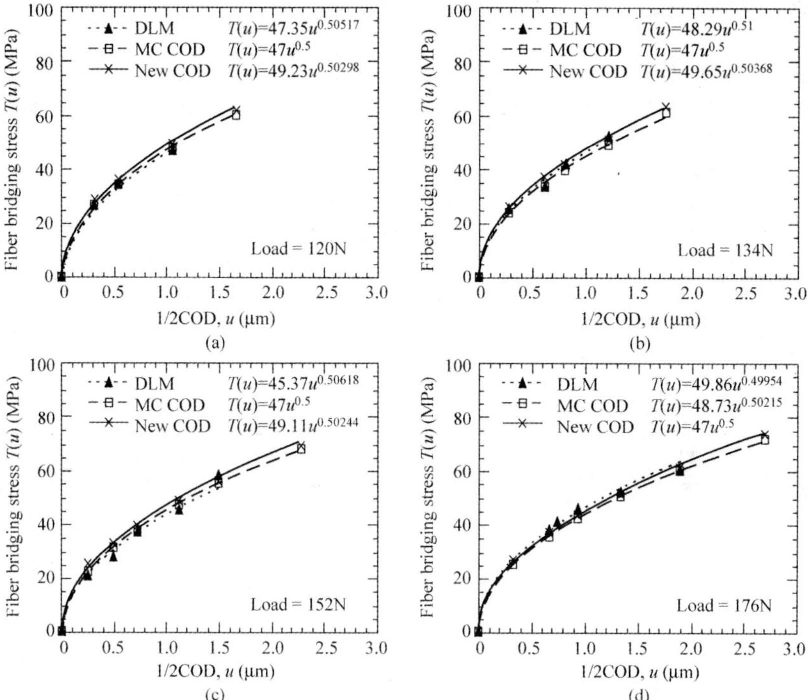

Fig. 6.47 Dependencies of fiber-bridging stress on COD subjected to different loads

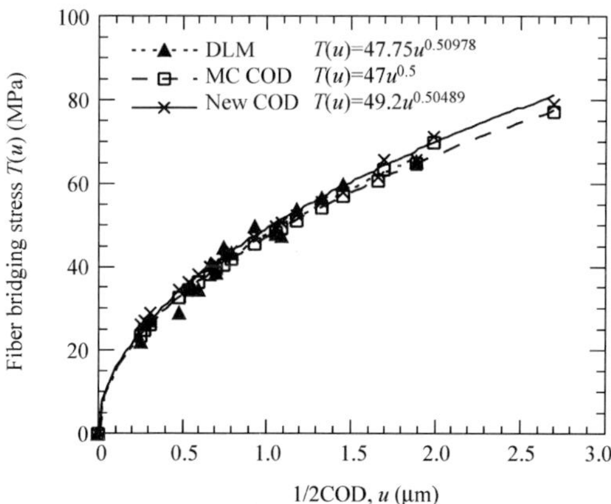

Fig. 6.48 Comparison of the fiber-bridging stress profiles determined from DLM, the new bridging law (6.62), and the MC COD measurements

6.7.2.4 Determination of Fracture Toughness

1) Stress Intensity Factor K_p and K_p^m Determined from SENB Test

The stress intensity factor K_p induced by the applied load for this composite in a four-point flexure mode was determined from (6.41) and (6.43). Because of the generation and propagation of multiple matrix cracks in this composite, another function $F(a, c)$ was used to modify K_p for the case of multiple matrix cracking as (Dharani and Tang, 1989; Tada et al., 1985)

$$K_p^m = K_p F(a,c) \tag{6.63}$$

An explicit expression for the factor $F(a, c)$ was given by Dharani and Tang (1989). For the case of the multiple parallel edge cracks subjected to remote uniform tension, which is assumed to apply also for a beam with a multiple cracks subjected to bending load as

$$F(a,c) = \frac{1}{1.12}\sqrt{\frac{c}{a+c}}\,k(a,c) \tag{6.64}$$

where c is the average crack spacing of the specimen, and $k(a, c)$ is given in the literature (Dharani and Tang, 1989). Both the values of K_p (without consideration of multiple matrix cracking) and K_p^m (with consideration of the multiple matrix cracking) are calculated and listed in Tables 6.6 and 6.7, respectively. The influence of the multiple matrix cracking on the fracture toughness of the composite is explicitly demonstrated. With the increase of the matrix cracks, the compliance of the composite decreased, and more matrix strain energy was released. Both factors lower the measured fracture toughness of the whole specimen compared to the case of the single-crack system. However, in either case, the fracture toughness of the composite increases with the propagation of matrix cracks because of the toughening by the fiber bridging, as shown in Fig. 6.49.

Table 6.6 The fracture toughness determined from the four-point bending tests without consideration of multiple matrix cracking

Load	$K_p /$ $(MPa \cdot m^{1/2})$	$K_T / (MPa \cdot m^{1/2})$			$K_R / (MPa \cdot m^{1/2})$		
		DLM	MC COD	New COD	DLM	MC COD	New COD
120	6.23	3.26	3.15	3.53	4.03	3.92	4.30
134	8.35	5.89	5.57	6.42	6.66	6.34	7.19
152	10.38	8.10	7.60	8.81	8.87	8.37	9.58
176	13.70	11.34	10.60	12.37	12.11	11.37	13.14

Table 6.7 Fracture toughness determined from the four-point bending tests with consideration of multiple matrix cracking

Load	No. Of cracks	Crack length	$K_{\mathrm{P}}^{\mathrm{m}}/$ (MPa·m$^{1/2}$)	$K_{\mathrm{R}}^{\mathrm{m}}$/(MPa·m$^{1/2}$)		
				DLM	New COD	MC COD
120	4	1.63	4.04	2.63	2.81	2.56
134	4	1.98	4.89	3.88	4.19	3.70
152	5	2.15	5.17	4.35	4.71	4.11
176	5	2.38	6.45	5.68	6.16	5.33

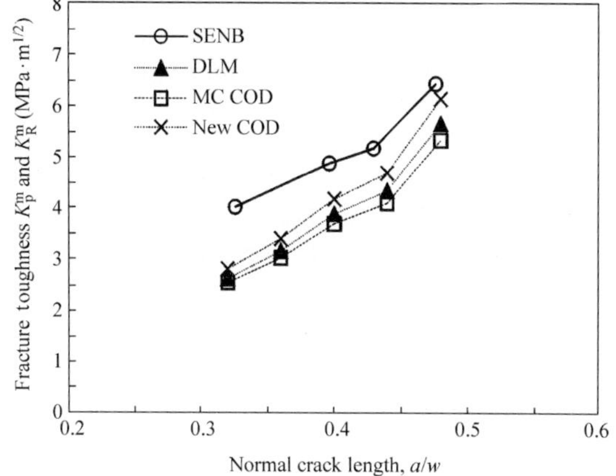

Fig. 6.49 Comparison of the fracture resistance curves obtained from SENB, DLM, the new bridging law, and COD measurements

2) Stress Intensity Factors K_{T}, K_{R} and $K_{\mathrm{R}}^{\mathrm{m}}$

Based on the calculation of the fiber-bridging stress from DLM and COD methods, the stress intensity factor at the crack tip due to the local shielding stress, $\sigma(x)$, acting on the crack surfaces was calculated from (6.44) and (6.45). A numerical calculation was performed using (6.44) and (6.45) by substituting $\sigma(x)$ or $\sigma(u)$ with the fiber-bridging stress expressions given in Figs. 6.47 and 6.48. The calculated values of the fiber-bridging toughness K_{T} without considering the multiple matrix cracking are listed in Table 6.6. It is seen that the fiber bridging toughness K_{T} obtained from DLM is larger than the COD measurement by 0.11 to 0.74 MPa·m$^{1/2}$ with the increasing crack length. This higher fracture toughness was caused by higher fiber-bridging stress obtained for the DLM or New COD, as discussed before. The fracture toughness K_{R} (without considering the multiple matrix cracking) was calculated from (6.43) and is also listed in Table 6.6. It is shown that the fiber-bridging toughness K_{T} contributes about 90% to the overall fracture toughness of the composite, K_{R}. Therefore, the toughening by fiber bridging governs the mechanical properties of the composites.

To be compatible, the fracture toughness K_R^m (for multiple matrix cracking) was also calculated by multiplying K_R with the factor $F(a,c)$. Both the fracture toughness K_P^m and K_R^m listed in Table 6.7 increase with the increasing crack length. The rate of increase in fracture toughness is dependent on the geometry of the test specimen, interfacial bonding strength, and statistical strength of fibers. The composite specimen of this study was relatively small, and the fiber/matrix interface was very weak. No fiber failure was observed up to the load of 176 N. Consequently, the rate of increase in toughness increased. Such a K_R curve was also observed in $ZrSiO_4$-SiC_f composites (Wang et al., 1998). It should be noted that the K_R curve obtained in this study is different from those obtained by many other researchers (Llorca and Singh, 1991; Sakai et al., 1991; Fett and Munz, 1993). In their cases, the fracture toughness increased with a decreasing rates due to the continuous failure of the reinforcing fibers. Both the increasing and decreasing rates in fracture toughness were supported by the computation from Cox and Marshall (1991) and Cox et al. (1989). In their modeling, the fracture toughness for the composite systems with small elastic modulus of the matrix but very strong fibers showed the same trend as our experimental K_R curve results, provided that no fiber fracture occurred.

It is interesting to observe that the fracture toughness values K_R^m obtained from the DLM and new COD are closer to K_P^m than the value obtained from MC COD in Fig. 6.49. This discrepancy is expected because of the difference in these two models (MC model and Sun & Singh model). The DLM and New COD (a new fiber-bridging law) are based on the energy balance approach, which takes into account the interfacial debond energy and the shear strain energy in matrix (Sutcu and Hillig, 1990; Sun and Singh, 1998). Therefore, the DLM and new COD techniques should be better approaches for determining both the fiber-bridging stress and fracture toughness.

Through this part of the study, the contribution of fiber bridging to the toughness of FCMCs was investigated. Three techniques including SENB, COD and DLM were used experimentally to obtain the fracture toughness and KR curve of the composite. Among these techniques, the DLM is a new approach developed for the first time in this study.

The fiber-bridging stresses and fracture toughnesses were calculated from the experimentally obtained debond length and crack opening displacement profiles using the DLM and COD (micro) approaches. A comparison of these results showed that the fiber-bridging stress, $T(u)$, and fracture toughness, K_R or K_R^m, obtained from DLM and new fiber-bridging laws was higher than the values calculated from COD (MC model) which was consistent with the theoretical expectation. The fiber-bridging toughness K_T was compared with the total toughness of the composite K_R and toughness of the matrix K_M. For both the DLM and COD methods, it was confirmed that the fiber bridging played a dominant role in toughening the ceramic matrix. The fracture resistance curves (K_R curves) illustrated that an increasing fiber-bridging toughening effect with the propagation of matrix crack was achieved in this composite system.

The fracture toughness K_p was also experimentally measured using the SENB (macro) approach. This macro approach provided a trend for the stress intensity factor-which was similar to results calculated from the DLM and COD (micro) approaches. This indicated that both the DLM and COD techniques can be applied for determining the fiber-bridging stress and fracture toughness of composites. It was also noted that the fracture toughnesses determined from DLM were closer to the values obtained from the SENB approach than those calculated from the MC COD method. This suggests that the DLM and new fiber-bridging laws should be better approaches for determining fiber-bridging stress and fracture toughness in FCMCs.

6.8 Summary

A systematic study comprising processing, mechanical testing, in-situ failure analysis, and fracture toughness characterization was conducted on a transparent SiC fiber-reinforced borosilicate glass matrix composite. This study concentrated on the analysis of mechanical behavior, interfacial debonding, multiple matrix cracking, and fiber-bridging, which provided much significant information about the micromechanics of fiber-strengthening, fracture, debonding, and fiber-toughening mechanisms in fiber reinforced ceramic matrix composites (FCMCs).

A unique tape casting/binary sintering processing technique was developed for fabricating a transparent and toughened SiC fiber-reinforced glass composites. Improved mechanical properties such as higher elastic modulus, first matrix cracking stress, and ultimate strength were achieved and a desired weak fiber/matrix interface was created.

The micromechanisms of multiple matrix cracking, interfacial debonding, and fiber bridging were theoretically and experimentally studied. The interfacial debond was observed and the debond length was measured. The relationship between the debond length and applied stress was experimentally obtained for the first time in FCMCs. The dynamic interaction between the multiple matrix cracking and interfacial debonding was observed and utilized for the analysis of the initial nonlinear load-displacement behavior of composites.

A debond length measurement (DLM) technique was developed for measuring the interfacial properties and fiber-bridging stress in FCMCs. This technique, based on an energy balance model illustrated a fiber-bridging stress profile for the first time from a debond profile of a matrix crack. Comparatively, this DLM technique turned out to be a novel approach for measuring fiber-bridging stress because it can be applied to composite with either a strong or weak fiber/matrix interface. A new fiber-bridging law was theoretically derived and experimentally verified from both fiber-bridging stress and fiber-bridging toughness standpoints. This new fiber-bridging law showed its advantages in its applicability to both strong and weak interfaces.

The fracture-resistance curves (K_R) were determined from SENB, COD, and DLM approaches, showing an increasing toughness with the propagation of matrix cracks. The significant contribution of fiber bridging to the toughness of composites was confirmed through this study. The consistency of the results obtained from SENB, COD, and DLM approaches verified that the newly developed debond length measurement technique and bridging law could be an effective tool and model for the analysis of fiber reinforcement.

Acknowledgement

The author is very grateful to Dr R.N. Singh, at University of Cincinnati for his guidance through the research, to Dr Chun-Hway Hsueh at Oak Ridge National Laboratory, Dr William Hillig at Rensselaer Polytechnic Institute, and Dr Susmit Kumar at University of Cincinnati for discussions on the theoretical model. The author also thanks Dr James Webb at Corning for helping with the interfacial shear strength measurement using the Micro Measure Machine., and to Dr Yu-lin Wang for conducting the numerical calculation. The research described in this chapter was supported by National Science Foundation of China.

References

Aveston, J. and A. Kelly. J. Mater. Sci. **8**, 352 (1973)

Aveston, J., G. A. Cooper and A. Kelly. National Physical Laboratory Conference, Nov. 4th, UK (1971 b)

Awasthi, S. and J. L. Wood. Ceram. Eng. Sci. Proc. **9** (**7 and 8**), 553 (1988)

Barenblatt, G. I. Advances in Applied Mechanices. Academic Press **7**, 55 (1962)

Bhatt, R. T. and D. R. Hull. NASA Technical Memorandum. 107169. (1992)

Bhatt, R. T. Whisker-and Fiber-Toughened Ceramics: Proceedings of an International Conference, (ASM International, 1988)

Brennan, J. J. and K. Prewo. J. Mater. Sci. **17**, 2371 (1982)

Briggs, A. and R. W. Davidge. Materials Science and Engineering. A **109**, 363 (1989)

Bright, J. D. and D. K. Shetty. J. Am. Ceram. Soc. **72** (**10**), 1891 (1989)

Budiansky, B., A. G. Evans and J. W. Hutchinson. Int. J. Solids Structures **32** (**3/4**), 315 (1995)

Budiansky, B., J. W. Hutchinson and A. G. Evans. J. Mech. Phys. Solids **34** (**2**), 167 (1986)

Chawla, K. K. Composite Materials, (Springer-Verlag, New York Inc., 1987)

Chou, T. W. and A. Kelly. Ann. Rev. Mater. Sci. **10**, 229 (1980)

Cooper, G. A. J. Mater. Sci. **5**, 645 (1970)

Cox, B. N. and D. B. Marshall. Acta Metall. Mater. **39** (**4**), 579 (1991)

Cox, B. N., D. B. Marshall and M. D. Thouless. Acta Metall. Mater. **37** (**7**), 1933 (1989)

Crivelli-Visconti, I. and G. A. Cooper. Nature. **221**, 754 (1969)

Curtin, W. A. Acta Metall. Mater. **41** (**5**), 1369 (1993)

Curtin, W. A. J. Am. Ceram. Soc. **74 (11)**, 2837 (1991)

Damani, R., R. Gstrein and R. Danzer. J. Europ. Ceram. Soc. **16(7)**, 695 (1996)

Danchaivijit, S. and D. K. Shetty. J. Am. Ceram. Soc. **76 (10)**, 2497 (1993)

Deshmukh, U. V. and T. W. Coyle. Ceram. Eng. Sci. Proc. **9 (7 and 8)**, 627 (1988)

Dharani, L. R. and H. Tang. J. Composite Materials. **23**, 308 (1989)

Dutton, R. E., N. J. Pagano and R. Y. Kim. J. Am. Ceram. Soc. **79 (4)**, 865 (1996)

Evans, A. G. J. Am. Ceram. Soc. **73 (2)**, 187 (1990)

Fett, T., and D. Munz. J. Mater. Sci. **28**, 742 (1993)

Filzer, E. and R. Gadow. Am. Ceram. Soc. Bull. **65**, 326 (1986)

Gac, F. D. Ceram. Eng. Sci. Proc. **11 (7 and 8)**, 551 (1990)

Gaeta, P. J., R. D. Sisson, M. Singh and J. I. Eldridge. Ceramic Matrix Composites-Advanced High Temperature Structural Materials, Materials Research Society. **365**, 365 (1995)

Galiotis, C. Composite Science and Technology **42**, 125 (1991)

Gdoutos, E. E. Fracture Mechanics. (Kluwer Academic Publishers, Netherlands 1993)

Gilbert, C. J., J. W. Ager III and R. O. Ritchie. submitted to Acta Materialia. **46**, 609 (1998)

Goettler, R. W. and K. T. Faber. Composites Science and Technology. **37**, 120 (1989)

Grande, D. H., J. F. Mandell and K. C. C. Hong, J. Mater. Sci. **23**, 311 (1988)

Gustafson, C. M. and R. E. Dutton. Private discussion (1998)

Gustafson, C. M., R. E. Dutton, and R. J. Kerans. J. Am. Ceram. Soc. **78 (5)**, 1423 (1995)

Haerle, A. G., W. R. Cannon and M. Denda. J. Am. Ceram. Soc. **74 (11)**, 2897 (1990)

Hashida T., V. C. Li and H. Takahashi. J. Am. Ceram. Soc. **77 (6)**, 1553 (1994)

He, M. Y., A. G. Evans and W. A. Curtin. Acta Metall Mater. **41 (3)**, 871 (1993)

Hillig, W. B. American Ceramic Society Bulletin. **73 (4)**, 56 (1994)

Hillig, W. B. Ann. Rev. Mater. Sci. **17**, 341 (1987)

Hsueh, C. H. J. Mater. Sci. **24**, 4475 (1989)

Hsueh, C. H. J. Acta Mater. **44 (6)**, 2211 (1996)

Hutchinson, J. W. and H. M. Jensen. Mechanics of Materials. **9**, 139 (1990)

Jero, P. D., R. J. Kerans, and T. A. Parthasarathy. J. Am. Ceram. Soc. **74 (11)**, 2793 (1991)

Jurewicz, A., R. Kerans and J. Wright. Ceram. Eng. Sci. Proc. **10 (7 and 8)**, 925 (1989)

Kerans, R. J. and T. A. Parthasarathy. J. Am. Ceram. Soc. **74 (7)**, 1585 (1991a)

Kerans, R., R. Hay, N. Pagano and T. Parthasarathy. Ceramic Bulletin. **68 (2),** 429 (1989)

Kim, Y. W. and J. G. Lee. J. Mater. Sci. **26 (5)**, 1316 (1991)

Kimber, A. C. and J. G. Keer. J. Mater. Sci. Lett. **1**, 353 (1982)

Klug T., J. Reichert and R. Bruckner. J. Mater. Sci. **28**, 6303 (1993)

Li, S., S. P. Shah, Z. Li and T. Mura. Int. J. Solids Structures. **30 (11)**, 1429 (1993)

Llorca, J. and R. N. Singh. J. Am. Ceram. Soc. **74 (11)**, 2882 (1991)

Ma, Q. and D. R. Clarke. Acta Metall. **41(6)**, 1817 (1993)

Marshall, D. B. and A. G. Evans. Materials Forum. **11**, 304 (1988)

Marshall, D. B. and B. N. Cox. Mechanics of Materials. **7**, 127 (1988)

Marshall, D. B. and W. C. Oliver. J. Am. Ceram. Soc. **70 (8)**, 542 (1987)

Marshall, D. B., B. N. Cox and A. G. Evens. Acta metall.mater. **33 (11)**, 2013 (1985)

Marshall, D. B., M. C. Shaw, and W. L. Morris. Acta Metall. Mater. **43 (5)**, 2041 (1995)

McCartney, L. N. Proc. R. Soc. Lond. A **409**, 329 (1987)

Meda, G. and P. S. Steif. J. Mech. Phys. Solids **42 (8)**, 1323 (1994)

Mendoza, E. A. and W. R. Cannon. Ceram. Eng. Sci. Proc. **12 (7 and 8)**, 1448 (1991)

Miyajima, T. and M. Sakai. J. Mater. Res. **6 (3)**, 539 (1991)

Morley, J. G. High-Performance Fibre Composites, (Academic Press Limited, 1987)

Phillips, D. C. J. Mater. Sci. **9**, 1847 (1974)

Phillips, D. C., R. A. Sambell and D. Bowen. J. Mater. Sci. **7**, 676(1972)

Prewo, K. M. and J. J. Brennan. J. Mater. Sci. **17**, 1201 (1982)

Prewo, K. M. J. Mater. Sci. **23**, 2745 (1988)

Roman, Y. G. and D. P. Stinton, Ceramic Matrix Composites-Advanced High Temperature
Structural Materials, Mater. Res. Sol. **365**, 343 (1995)

Russell-Floyd, R. S., B. Harris and r. G. Cooke. J. Am. Ceram. Soc. **76 (10)**, 2635 (1993)

Sakai, M. ISIJ International **32 (8)**, 937 (1992)

Sakai, M., T. Miyajima and M. Inagaki. Composites Science and Technology. **40**, 231 (1991)

Schwab, S. T., R. A. Page, D. L. Davidson, and R. C. Graef. Ceram. Eng. Sci. Proc. **16 (5)**,
743 (1995)

Schwartz, M. M. Composite Materials Vol. II: Processing, Fabrication, and Applications,
(Prentice-Hall, Inc., NJ 1997)

Sih, G. C. and L. Faria. Fracture Mechanics Methodology, (Martinus Nijhoff Publishers, The
Hague, Netherlands 1984)

Sih, G. C. Handbook of Stress-Intensity Factors, Institute of Fracture and Solid Mechanics,
(Lehigh University, USA, 1973)

Singh, R. N. and M. K. Brun. Ceram. Eng. Sci. Proc., **8 (7 and 8)**, 636 (1987)

Singh, R. N. and M. Sutcu. J. Mater. Sci. **26**, 2547 (1991)

Singh, R. N. and S. K. Reddy. J. Am. Ceram. Soc. **79 (1)**, 137 (1996)

Singh, R. N. J. Am. Ceram. Soc. **73 (8)**, 2399 (1990)

Steif, P. S. and H. R. Schwietert. Ceram. Eng. Sci. Proc. **11 (9 and 10)**, 1567 (1990)

Strife, J. R., J. J. Brennan and K. M. Prewo. Ceram. Eng. Sci. Proc. **11 (7and 8)**, 871 (1990)

Sumarasiri K. E. D. and O. Van der Biest. Scripta Metall. **30**, 79 (1994)

Sun, Y. and R. N. Singh. Acta Materialia **46 (5)**, 1657 (1998)

Sun, Y. and R. N. Singh. Advances in Ceramic Matrix composites III, Ceramic Transactions.
74, 141-151, (1996)

Sun, Y. and R.N. Singh. J. Mater. Sci., **35**, 5681 (2000c)

Sutcu, M. and W. B. Hillig. Acta Metall.Mater. **38 (12)**, 2653 (1990)

Tada, H., P. Paris and G. Irwin. The Stress Analysis of Crack Handbook, (Del Research
Corporation, Hellertown, PA, 1985)

Tai, N. H. and T. W. Chou. J. Mater. Res. **5 (10)**, 2255(1990)

Tredway, W. K. and K. M. Prewo, Materials Research Society Symposium Proceedings 170,
Interfaces in Composites (Materials Research Society, USA, 1990)

Wang, Y. L., U. Anandakumar and R.N. Singh. (1998) to be published

Watt, W. and B. V. Perov. Strong Fibers, Handbook of Composites, Vol. 1, (Elsevier Science
Publishers, B.V. Netherlands, 1985)

Weitsman, Y. and H. Zhu. J. Mech. Phys. Solids. **41 (2)**, 351 (1993)

Wiederhorn, S. M. Ann. Rev. Mater. Sci. **14**, 373 (1984)

Wijeyewickrema, A. C. and L. M. Keer. Int. J. Solids Structures. **30 (1)**, 91 (1990)

Wu, M. and G. L. Messing. J. Am. Ceram. Soc. **73 (11)**, 3497 (1990)

Xia, Z. C., J. W. Hutchinson, A. G. Evans and B. Budiansky. J. Mech. Phys. Solids **42 (7)**, 1139 (1994).

Xu, H. K., C. P. Ostertag and L M. Braun. J. Am. Ceram. Soc. **77 (7)**, 1889 (1994)

Young, R. J. and X. Yang. Composite Part A **27A (9)**, 737 (1996)

Zhou, L., J. Kim and Y. Mai. J. Mater. Sci. **27**, 3155 (1992)

Zok, F. W. and S. M. Spearing. Acta Metall. Mater. **40 (8)**, 2033 (1992)

7 Applications of HRTEM in Functional Materials

Xiaojing Wu

7.1 Theory of High-Resolution Transmission Electron Microscopy (HRTEM)

Transmission electron microscopy (TEM) was invented by E. Ruska and M. Mnoll in 1931. In 1939, Siemens Halske Co. produced the first commercial TEM. Sicne that time, TEM has become a powerful research tool in biology, chemistry, physics, metallurgy, mineralogy and materials science, etc. fields. In 1986, Ruska, after more than a half century delay, was awarded the Nobel Physics Prize for his contribution to developing the electron microscope, indicating the importance of TEM for the modern scientific research community.

The first image contrast theory of high-resolution transmission electron microscopy (HRTEM), weak-phase object approximation (WPOA) theory, was established by Scherzer (1949), which explained the correspondence between the projection structure of the specimen and the image contrast formed by the phase difference produced by the electron wave passing through a very thin specimen. Experimentally, Menter (1956) succeeded in taking a lattice fringe image. In the 1950s, however, the line resolving power of TEM was only 1.2nm, and the point resolving power was much lower than that. The poor resolving power of TEM restricted the application of HRTEM until the 1970s when the manufacturing technique of TEM made significant progress. Cowley and Iijima (1972) and Ueda et al. (1972) succeeded in taking HRTEM images with a point resolving power better than 0.4 nm from thin specimens, indicating the appearance of a new era for HRTEM.

7.1.1 Phase Contrast

The imaging mechanism of HRTEM is phase contrast. Assuming the specimen is so thin that after passing through it the only change for the incident electron beam is its phase, then the specimen is called a phase-object.

The wavelength of the electron beam in vacuum without relative modification is:

$$\lambda = h / (2m_0 eE)^{1/2} \qquad (7.1)$$

where h is Planck's constant, m_0 the electron static mass and E the accelerating voltage. When the beam is affected by the specimen, a potential energy term $e\varphi_0(\mathbf{r})$ ($\varphi_0(\mathbf{r})$: electric-potential distribution function of the specimen) is added, and the wavelength becomes:

$$\lambda' = h/\{2m_0 e[E + \varphi_0(\mathbf{r})]\}^{1/2} \tag{7.2}$$

The phase difference between the beam passing through the specimen and the beam transmitting in vacuum is:

$$\alpha = 2\pi\int_0^t (1/\lambda' - 1/\lambda)\mathrm{d}z \tag{7.3}$$

where t is the thickness of the specimen. For most TEM, $E \gg \varphi_0(\mathbf{r})$, and (7.3) can be simplified as:

$$\alpha \approx (\pi/\lambda E)\int_0^t \varphi_0(\mathbf{r})\mathrm{d}z = \sigma\varphi_{\mathrm{p}}(\mathbf{r}) \tag{7.4}$$

Here, $\varphi_{\mathrm{p}}(\mathbf{r}) = \int_0^t \varphi_0(\mathbf{r})\mathrm{d}z$ is the projection potential along the electron beam incidence direction and $\sigma = \pi/\lambda E$ a constant under a given accelerating voltage.

If the incident beam is planar, then the transmission function can be expressed as:

$$\Psi(\mathbf{r}) = \exp(-\mathrm{i}\sigma\varphi_{\mathrm{p}}(\mathbf{r})) \tag{7.5}$$

and the diffraction function on the back focal plane can be written as:

$$\Phi(\mathbf{H}) = \mathscr{F}[\Psi(\mathbf{r})] \tag{7.6}$$

Here, $\mathbf{H}$ is a vector on the diffraction plane (reciprocal space), and $\mathscr{F}$ the Fourier transform.

If TEM is perfect, namely, there is no information distortion during the imaging process, according to the Abell theory of imaging, under correct focus the image function on the image plane becomes:

$$\Psi(\mathbf{r}) = \mathscr{F}^{-1}[\Phi(\mathbf{H})] = \mathscr{F}^{-1}\{\mathscr{F}[\Psi(\mathbf{r})]\} = \Psi(m\mathbf{r}) \tag{7.7}$$

where $\mathscr{F}^{-1}$ is the inverse Fourier transform. This image wave is only an enlarged object wave, and the intensity on the image plane is

$$I(r) = |\Psi(r)|^2 = |\exp(-i\sigma\varphi_{\mathrm{p}}(mr))|^2 = 1 \tag{7.8}$$

In other words, there is no image contrast at all on the image plane for an ideal objective lens under correct focus.

However, no perfect objective lens exists, and every TEM will modify the transmission wave by its transfer function $A(H)$, where

$$A(H) = P(H)\exp[i\chi_1(H)]\exp[-\chi_2(H)]2J_1[\chi_3(H)]/\chi_3(H) \tag{7.9}$$

Here, $P(H)$ is the aperture function and is defined as

$$P(H) = \begin{cases} 1; & H \leqslant H_{\mathrm{p}} \\ 0; & H > H_{\mathrm{p}} \end{cases}$$

$J_1(\chi_3)$ the first-order Bessel function, and

$$\chi_1(H) = \pi\Delta f\lambda H^2 + \pi C_s\lambda^3 H^4/2 \tag{7.10}$$

$$\chi_2(H) = \pi D\lambda^2 H^4/2 \tag{7.11}$$

$$\chi_3(H) = |2\pi\alpha[\Delta f H + \lambda(\lambda C_s - i\pi D^2)H^3]| \tag{7.12}$$

Here, Δf is the defocus amount, C_s the spherical aberration constant, D the standard difference of the Gauss distribution function caused by the chromatic aberration and α the illumination angle.

The second term of the transfer function, $\exp[i\chi_1(H)]$, is a vibration function, while the third and fourth ones, $\exp[-\chi_2(H)]$ and $J_1[\chi_3(H)]/\chi_3(H)$, attenuate at high angle (large H), and form an envelope curve, which acts as a "virtual aperture", and thus restricts the resolving power of TEM.

Therefore, the image function on the image plane should be

$$\Psi'(r) = \mathscr{F}^{-1}\{\mathscr{F}[\Psi(r)]A(H)\} = \Psi(r) * \mathscr{F}^{-1}[A(H)] \tag{7.13}$$

Here, $*$ expresses convolution. The image function includes the information of the specimen modified by the transfer function of TEM. The electron path through an objective lens is shown in Fig. 7.1. In this procedure, all diffraction beams on the back focal plane may contribute to the final image. Therefore, the phase contrast theory is a multibeam imaging theory.

When the specimen is a crystal with periodicity r_0, we have

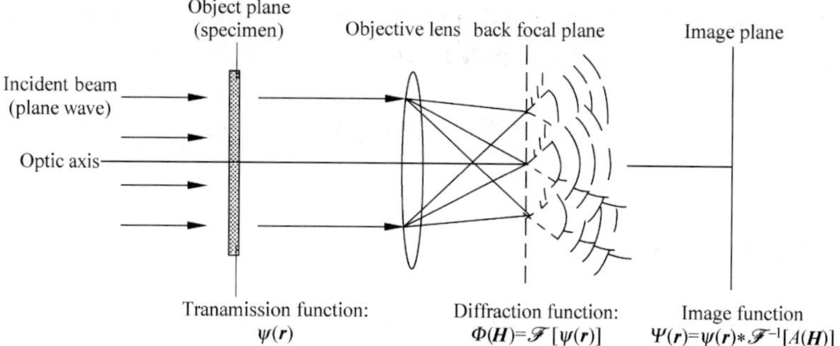

Fig. 7.1 Schematic to show the electron path under an objective lens

$$\varphi_{p}(r) = \varphi_{p}(r + r_{0}) \tag{7.14}$$

Substituting this relationship into (7.5), we obtain

$$\Psi(r) = \exp(-i\sigma\varphi_{p}(r)) = \exp(-i\sigma\varphi_{p}(r + r_{0})) = \Psi(r + r_{0}) \tag{7.15}$$

The image function then can be expressed as

$$\Psi'(r) = \Psi(r) * \mathscr{F}^{-1}[A(H)] = \Psi(r + r_{0}) * \mathscr{F}^{-1}[A(H)] = \Psi'(r + r_{0}) \tag{7.16}$$

and finally we have

$$I(r) = |\Psi'(r)|^{2} = |\Psi'(r + r_{0})|^{2} = I(r + r_{0}) \tag{7.17}$$

Therefore, the intensity of the phase contrast image should show a periodic change with the lattice periodicity. The HRTEM image showing the periodicity of the crystal lattice is called the lattice image.

7.1.2 Weak-Phase Object Approximation

The correspondence between the specimen structure and the image contrast can not be simply found from (7.13).   However, it may become much simpler under the following assumptions:

1) the specimen is so thin that $\sigma\varphi_{p}(r) \ll 1$ (weak-phase object approximation, WPOA), and then the expression of (7.5) can be written as:

$$\Psi(r) = \exp(-i\sigma\varphi_{p}(r)) \approx 1 - i\sigma\varphi_{p}(r) \tag{7.18}$$

2) the attenuation terms of $\exp[-\chi_{2}(H)]$ and $J_{1}[\chi_{3}(H)]/\chi_{3}(H)$ in the transfer

function are absorbed by the aperture function $P(H)$, which is equal to unity as $H<H_p$, and zero as $H>H_p$, and then the transfer function becomes:

$$A(H) = \exp[i\chi_1(H)] = \exp[i\pi\Delta f\,\lambda H^2 + i\pi C_s\lambda^3 H^4/2] \qquad (7.19)$$

when $H<H_p$.

Under these two assumptions, (7.13) becomes

$$\Psi'(r) = [1 - i\sigma\varphi_p(r)] * \mathscr{F}^{-1}[\exp(i\chi_1)] \qquad (7.20)$$

The image intensity on the image plane is calculated as

$$I(r) = |\Psi'(r)|^2 \approx 1 - 2\sigma\varphi_p(r) * \mathscr{F}^{-1}[\sin\chi_1(H)] \qquad (7.21)$$

For each TEM, C_s and λ are fixed. Noticing that $\chi_1(H)$ is a function of Δf, we can select a proper Δf and make

$$\sin\chi_1 \approx 1 \qquad (7.22)$$

in the range of $H<H_p$, and then the image intensity will be

$$I(r) = |\Psi'(r)|^2 \approx 1 - 2\sigma\varphi_p(r) \qquad (7.23)$$

Now, we can see the relationship between the HRTEM image and the specimen structure: the image contrast is directly proportion to the projection potential. If the electron beam is incident along the proper crystal zone, the heavy atom column with a large projection potential will show dark contrast, while the light atom column will show light contrast. Therefore, in principle we can identify the positions of the heavy and light atomic columns directly from the HRTEM image.

When the HRTEM image reveals the crystal structure directly, the image is called the structure image. Obviously, the structure image is a special lattice image. The most important condition for taking a structure image is to realize the condition of (7.22). It was found that under Scherzer focus,

$$\Delta f_s = -1.2 C_s^{1/2}\lambda^{1/2} \qquad (7.24)$$

(7.22) will be closest to 1 (minus means under-focus). Namely, Scherzer focus is the optimum condition for obtaining the structure image.

7.1.3 Multislice Method

Formula (7.23) shows an ideal correspondence between the structure projection and the HRTEM image. However, the necessary condition for (7.23) holding is too harsh. For a TEM with an accelerating voltage of 200 kV, WPOA holds under the condition of $\varphi_p(r) \ll 1.2 \times 10^3$ VÅ. For most inorganic crystals, the mean inner potential, $<\varphi_0(r)>$, is about 20V. From

$$\varphi_p(r) = t <\varphi_0(r)> \ll 1.2 \times 10^3 \text{ VÅ}$$

we obtain

$$t \ll 60 \text{ Å} \tag{7.25}$$

This condition means that we can use (7.23) to explain the image contrast only when the specimen is thinner than, say, 15Å along the beam direction. Generally, the signal/noise (S/N) ratio from such a thin area is very low. In additional, such a thin region is easier to become contaminated in TEM. Therefore, experimentally WPOA theory does not work well for most inorganic materials.

Comparing with X-ray, the interaction between electron with substance is about four orders of magnitude stronger. Therefore, it becomes possible to receive structure information from a tiny particle by TEM. This is a major advantage of TEM. On the other hand, however, such strong interaction suggests that the kinematical theory (WPOA) does not hold in most cases, and the dynamical effect must be taken into consideration in a general HRTEM imaging theory. The multislice method is one such theory.

A specimen with thickness t is divided into $N+1$ slices. The thickness of each slice is Δt_n. The potential is projected onto the bottom of each slice along the incident direction, and the transmission function of the n-th slice is q_n. Let the wave function of incident beam be $\Psi_0(r)=1$, then the wave function leaving the first slice is

$$\Psi_1(r) = q_1(r)\Psi_0(r) = q_1(r) \tag{7.26}$$

and leaving the second slice is

$$\begin{aligned}\Psi_2(r) &= q_2(r)[\Psi_1(r) * p_1(r)] \\ &= q_2(r)[q_1(r) * p_1(r)]\end{aligned} \tag{7.27}$$

Here, $p_n(r)$ (n=1, 2, 3, $\cdots$) is the Fresnel propagation function

$$p_n(r) = (i/\lambda\Delta t_n)\exp(-i\pi r^2/\lambda\Delta t_n) \tag{7.28}$$

The wave function leaving the third slice is

$$\begin{aligned}
\Psi_3(r) &= q_3(r)[\Psi_2(r) * p_2(r)] \\
&= q_3(r)\{q_2(r)[q_1(r) * p_2(r)]\} * p_1(r)
\end{aligned} \tag{7.29}$$

Generally, we have

$$\begin{aligned}
\Psi_{n+1}(r) &= q_{n+1}(r)[\Psi_n(r) * p(r)] \\
&= q_{n+1}(r)\{q_n(r)[[\Psi_{n-1}(r) * p_{n-1}(r)] * p_n(r)]\}
\end{aligned} \tag{7.30}$$

Therefore, the wave function leaving the whole specimen should be

$$\begin{aligned}
&\Psi_{N+1}(r) + q_{N+1}(r)[\Psi_N(r) * p_n(r)] \\
&= q_{N+1}(r)[q_N(r)\cdots[q_2(r)[q_1(r) * p_1(r)] * p_2(r)] * \cdots * p_N(r)]
\end{aligned} \tag{7.31}$$

At the back focal plane of the objective lens, the diffraction function is the Fourier transform of (7.31),

$$\begin{aligned}
\Phi(H) &= \mathscr{F}[\Psi_{N+1}(r)] = Q_{N+1}(H) * [\Phi_N(H)P_N(H)] \\
&= Q_{N+1}(H) * [Q_N(H) * \cdots * [Q_2(H) * [Q_1(H)]P_1(H)]P_2(H)] \cdots P_N(H)]
\end{aligned} \tag{7.32}$$

Here, $Q_n(H) = \mathscr{F}[q_n(r)]$, $\Phi_n(H) = \mathscr{F}[\Psi_n(r)]$ and $P_n(H) = \mathscr{F}[p_n(r)]$. Considering the effect of the transfer function $A(H)$, the wave function on the image plane is

$$\begin{aligned}
\Psi(r) &= \mathscr{F}^{-1}[\Phi_{N+1}(H)A(H)] \\
&= \mathscr{F}^{-1}[\Phi_{N+1}(H)] * \mathscr{F}^{-1}[A(H)]
\end{aligned} \tag{7.33}$$

and the image intensity can be written as:

$$\begin{aligned}
I(r) &= |\Psi(r)|^2 \\
&= |\mathscr{F}^{-1}\{Q_{N+1}(H) * [Q_N(H) * \cdots * [Q_2(H) * \\
&\quad [Q_1(H)P_1(H)]P_2(H)] \cdots P_N(H)]\} * \mathscr{F}^{-1}[A(H)]|^2
\end{aligned} \tag{7.34}$$

The multislice method is a good model to describe the actual physical process of electron beam propagation in TEM. Though the relationship between the specimen structure and the image contrast is not so clear in (7.34), it does not cause any difficult for us to perform HRTEM image calculation using a computer. Nowadays, most computer programs for the HRTEM image simulation were based on this method.

7.1.4 Pseudo-Weak-Phase Object Approximation

The multi-slice method gives an exact intensity function on the image plane, if the slice is thin enough. Due to the presence of dynamical diffraction, we can no longer find a simple relationship between the structure projection and the image contrast.

Taking dynamical effect partially into consideration, the pseudo-weak-phase object approximation (PWPOA) theory can explain the correspondence between the projection structure and the HRTEM image contrast well for a specimen with thickness less than a critical value, t_c.

The specimen is divided into $N+1$ slices equally with the thickness of Δt. If each slice has the same projection potential $\varphi_p(r)$, and can be treated as a weak-phase object, neglecting the double scattering term in the wave function for each slice, then under the optimum defocus condition (Scherzer focus condition) the image intensity can be expressed as:

$$I(r) = 1 - 2\sigma\varphi_{N+1}(r) = 1 - 2\sigma\varphi_p(r) - 2\sigma\Delta\varphi_{N+1}(r) \tag{7.35}$$

Here,

$$\Delta\varphi_p(r) = \varphi_p(r) * \sum_{j=1}^{N} S_j(r) - \sigma\varphi_p(r)\left[\varphi_p(r) * \sum_{j=1}^{N} S_j(r)\right]$$

$$-\sigma\left[\varphi_p(r) * \sum_{j=1}^{N} S_j(r)\right]^2 \bigg/ 2 - \sigma\left[\varphi_p(r) * \sum_{j=1}^{N} C_j(r)\right]^2 \bigg/ 2$$

$$S_j(r) = [\sin(\pi r^2 / j\lambda\Delta z)] / j\lambda\Delta z$$

$$C_j(r) = [\cos(\pi r^2 / j\lambda\Delta z)] / j\lambda\Delta z$$

Comparing with (7.23) it is found that an extra term, $-2\sigma\Delta\varphi_{N+1}(r)$, is added in (7.35). As a function of $\varphi_p(r)$ and N, $\Delta\varphi_{N+1}(r)$ includes four terms, the first one is positive, while the others are negative. Calculation indicates that depending on N the behavior of $\Delta\varphi_{N+1}(r)$ is quite different for the relative heavy and light atomic columns in the crystal. When N is small enough, the linear relationship between the contrast and the projection potential still roughly holds. On increase of N to N_c, $\varphi_{N+1}(r)$ will become near to zero for the heavier atom columns, but will still be positive for the lighter atom columns. When $N > N_\mathrm{c}$, $\varphi_{N+1}(r)$ will become negative for the heavier atom columns. This means that the heavier atom columns will be shown as bright dots, rather than dark ones as suggested by WPOA, in the image. In other words, the contrast reverse due to the thickness effect occurs.

As the double diffraction terms are ignored in PWPOA, there are some differences in the simulated image by the multislice method and the PWPOA method. However, (7.35) shows a clear correspondence between the image intensity and the projection potential, which can help us to understand qualitatively what will occur in the HRTEM image when the dynamical effect is taken into account.

WPOA is available only for extremely thin specimen. This request seriously restricts the application range of the WPOA theory. In the PWPOA theory, on the other hand, we know that as the specimen is thinner than t_c ($t_\mathrm{c} = (N_\mathrm{c}+1)\Delta t$), all atom columns should appear as dark dots at optimum focus, though the contrast of the dots is no longer proportion to the projection potential. Therefore, for many materials we can practically identify the positions of the atomic column from the image taken from a region with the thickness being several times greater than that in the case of WPOA.

Generally, for taking a structure image, there are two necessary conditions: one is the amount of defocus, and the other is the specimen thickness. The structure image can be obtained only when it is taken under the optimum focus from a region where the thickness is thinner than t_c.

7.1.5   Resolving Power of Phase Contrast

Resolving power is one of the most important parameters for TEM. There are three kinds of resolving power for phase contrast: (a) the point-to-point resolving power (or point resolving power); (b) the information resolving power; and (c) the line resolving power.

7.1.5.1   Point-to-Point Resolving Power

Point resolution is defined as:

$$d_\mathrm{p} = 1/H_\mathrm{p} \tag{7.36}$$

Here, H_p is the smallest value in the solutions for equation of

$$\sin[\chi_1(H)]\exp[-\chi_2(H)]J_1[\chi_3(H)]/\chi_3(H) = 0 \tag{7.37}$$

Under the Scherzer focus condition, (7.11) becomes

$$\chi_1 = \pi(-1.2C_s^{1.2}\lambda^{1/2})\lambda H_s^2 + \pi C_s \lambda^3 H_s^4 s/2 \tag{7.38}$$

When the value of $\sin \chi_1$ becomes the first-order zero, $\chi_1=0$, in (7.38), we have

$$H_s = (2.4)^{1/2} C_s^{-1/4} \lambda^{-3/4} \tag{7.39}$$

The resolution limit of phase contrast at Scherzer focus is defined as

$$d_s = 1/H_s = 0.65 C_s^{1/4} \lambda^{3/4} \tag{7.40}$$

Similarly, the resolution limits due to the beam convergence and the chromatic aberration, d_j and d_D, are defined as

$$d_j = 1/H_j, \qquad \text{where} \quad J_1[\chi_3(H_j)]/\chi_3(H_j) \approx 0$$
$$\text{and} \qquad d_D = 1/H_D, \qquad \text{where} \quad \exp[-\chi_2(H_D)] \approx 0 \tag{7.41}$$

Both $\exp[-\chi_2(H_D)]$ and $J_1[\chi_3(H_j)]/\chi_3(H_j)$ terms are envelop function, and induce a virtual aperture. In most cases, we have $d_j < d_D < d_s$. Therefore, the point-to-point resolving power of a TEM is usually determined by the Scherzer resolution limit, and thus we have

$$d_p = d_s = 0.65 C_s^{1/4} \lambda^{3/4} \tag{7.42}$$

Due to $\lambda \propto E^{-1/2}$ (see (7.1)), the point-to-point resolving power becomes

$$d_p \propto C_s^{1/4} E^{-3/8} \tag{7.43}$$

7.1.5.2 Information Limit

The information limit of a TEM is defined by $d_i=1/H_i$, where H_i satisfies

$$\exp[-\chi_2(H_i)]J_1[\chi_3(H_i)]/\chi_3(H_i) = 0 \tag{7.44}$$

7.1.5.3 Line Resolving Power

The minimum spacing of lattice fringes observed by TEM experimentally defines the line resolving power. The mechanical vibrations coming from evacuation pump, sound wave, etc., have a strong effect on the line resolving power. The resolving power is also restricted by the information limit.

7.2   What Can be Achieved by HRTEM?

Comparing with other methods, the major advantage of HRTEM is that we can "see" the projection structure in the atomic scale directly under the optimum conditions. Nowadays, for materials science, HRTEM is mainly used in three aspects: (a) crystal structure determination; (b) defects observation and (c) thin film and multi-layer structure observation.

7.2.1   Crystal Structure Determination

As an equipment that can "see" atoms directly, HRTEM is considered to be a useful tool for crystal structure determination. As mentioned above, the contrast of the HRTEM image was influenced by several parameters, such as specimen thickness and defocus amount etc. Therefore, we must be very careful to choose the structure image as we do not know the real imaging conditions at the beginning. In order to know these conditions for each taken image, image simulations must be carried out. Here, we meet a paradox: for determining the crystal structure, we need to know the atom positions from the structure image; on the other hand, for selecting the structure image, we must compare the observed images with the calculated images, while to do image simulation requires us to know the atom positions in advance. Hence, there is no straightforward way to determine the crystal structure.

Generally, the structure determination should include four steps.

1) Diffraction Data Collection

First, we need to collect diffraction data from the electron diffraction (ED) or X-ray diffraction (XRD) method. Based on the diffraction data, the lattice parameters and, sometimes, the space group can be determined. It is worth pointing out here that the information from XRD is very important for the structure determination. In many cases, we can find some similar diffraction patterns from powder diffraction file (PDF) cards. Such information from the PDF database will be very helpful for constructing the structure model.

2) Taking HRTEM Images

The contrast of the HRTEM image does not correspond to an individual atom position, but the projection of the atom columns. To isolate a particular atom column, we must select a proper projection direction. If the direction of the incident electron beam is not along the proper crystallographic axis, the different atoms may overlap in the image. The crystal structure is 3-dimensional, while the image is only 2-dimensional. For constructing a 3-dimensional structure model from the 2-dimensional images, usually we need to take the HRTEM images along at least two different crystallographic axes.

The contrast of the HRTEM image strongly depends on the specimen thickness as well as the focus condition. For taking the structure image, we must select an area with a proper thickness. If the specimen is too thin, the S/N ratio will be too low to identify the atomic columns and the specimen will be easily contaminated under the electron beam illumination. If the specimen is too thick, contrast reversal will occur in the image, bringing some difficulties in the structure analysis. Generally, we can take good structure images at a thickness of $30 - 60$Å for most substances using a conventional 200 kV TEM. As the image contrast may have a dramatic variation with the change of focus, it is rather difficult to identify the structure image under TEM, unless the structure is already well known. Therefore, in most cases, we need to take a series of images under different focus conditions.

The quality of the series image for structure determination is very important. If the astigmatism of the TEM is not corrected properly, or the incident beam is not exactly along the zone axis, image distortion will occur.

3) Constructing the Structure Model

Combining the information from the diffraction data and the HRTEM images, we can now construct a possible structure model for the specimen. For doing this, the crystal chemistry knowledge is necessary. The difficulty here is that we do not actually know which HRTEM image is the so-called structure image. If we select a wrong image, the atomic coordinates determined from the image will be incorrect. For constructing the structure model, the information about any similar structure can help us to avoid some unnecessary mistakes. If the specimen has a completely new structure and we can not refer to any known structure, the structure model construction will become a very difficult task in many cases.

4) Image Simulation

To confirm whether the model is correct, image simulation should be carried out. Most programs for the HRTEM image simulation are made based on the multislice method. During the simulation, several variable parameters, such as the specimen thickness t and the defocus amount Δf, must be adjusted. The model is considered to be correct only when we can obtain a good match between the observed and simulated images under different focus or thickness. It can explain nothing if only one of the simulations has a good agreement with the observation.

When the agreement for every series of the simulations and the observations is acceptable, we can conclude that the proposed structure model is correct. Otherwise, we have to check: (a) if the images were taken properly; (b) if the simulations were performed for all possible conditions. If there is no problem in the experimental and the image simulation, we have to return to step (3) to modify the proposed structure model. Steps (3) and (4) should be done repeatedly until we find the good correspondence between the series of the simulated and observed images.

Therefore, the crystal structure determination by HRTEM is carried out by the trial-and-error method. In this procedure, personal experience is important.

The simulation of the HRTEM images is not so sensitive for some

324

crystallographic parameters, such as the atom positions, the occupation factors and the thermal vibration factors. Comparing with the data obtained by XRD, the precision of the structure parameters determined by HRTEM is, unfortunately, rather low. Therefore, the use of HRTEM in the crystal structure determination mainly occurs in three cases: (a) to offer an approximate model for further structure refinement; (b) to determine the structure of a minor phase that mixes with other phases; (c) to determine the modulation structure.

Actually, when we determine a structure by the diffraction method, we need to start from an initial structure model, too. The correctness of the initial model is a key point for any further structure refinement. If a wrong model is selected, the following refinement may converge to a wrong direction, and we may never reach the real solution. The structure information obtained by the diffraction method is in reciprocal space. In some cases, it is rather difficult to construct a correct structure model based only on the diffraction data. HRTEM, on the other hand, reveals the atom positions, normally the heavier atom positions, directly. Such information is very helpful for constructing a correct initial structure model.

In newly synthesized compounds, it is quite common that several phases coexist in one specimen. Such a phase mixture causes many difficulties for the structure analysis by XRD. Comparing with X-rays, electrons have much stronger interaction with substances. It is not difficult to pick up the structure and composition information by TEM from a grain with a size of several hundreds of nanometers. Under some special conditions, we can even determine the structure of a new phase with a scale of about several nanometers. A famous example was the discovery of the high-T_c superconducting phase of $YBa_2Cu_4O_9$.

In 1987, Wu et al. reported their historic discovery of $YBa_2Cu_3O_{6+y}$ (i.e., Y-123 phase, T_c>90 K), which was the first known compound to show superconductivity above 77 K. The stacking order of Y-123 was confirmed to be ... CuO_y/BaO/CuO_2/Y/CuO_2/BaO/CuO_y ... along the c-axis. In 1988, Zandbergen et al. (1988) found a stacking fault in the Y-123 matrix by HRTEM. The stacking fault occurs only in a region of several nanometers, being far from perfect. Based on the observation, they proposed a structure model with the stacking order of ... CuO/CuO/BaO/CuO_2/Y/CuO_2/BaO/CuO/CuO ..., which contains a double CuO chain, and carried out the image simulation. The agreement between the simulations and the observations was rather good along the [100] and [010] directions, indicating their model is correct. Therefore, they concluded the presence of a new phase, $YBa_2Cu_4O_9$ (i.e., Y-124 phase). Soon afterwards, the Y-124 single-phase sample was synthesized, and its critical temperature, T_c, was confirmed to be higher than 80 K. The ratio of Y-124 in the Y-123 matrix is so low that it is very difficult to identify the presence of this phase by XRD in their original specimen. In this case, without the help of HRTEM, it was almost impossible to find Y-124, and to determine its structure. The discovery of Y-124 showed the importance of HRTEM for studying new phases.

The modulation structure exists in many materials. The modulation could be formed by the atomic position displacement or the element substitution (substitution: including vacancy), etc. When the ratio of the lattice parameters of the superstructure to those of the basic structure is integer, the modulation is said to be commensurate; if the ratio is irrational, the modulation is incommensurate. Generally, the reflection intensity from the superlattice is weaker than that from the basic structure. Sometimes, the intensity differences between the basic structure and the superstructure could be very large, leading to some difficulties for the structure determination by XRD where the kinemical diffraction is prominent. When an electron beam passes through the specimen, on the other hand, the dynamical effect can not be ignored. This means that the double diffraction effect becomes an important factor, which leads to a relative intensity variations between the main reflections from the basic structure and the satellites from the superstructure. Generally, due to the dynamical effect, the intensities will increase for the satellites and decrease for the main reflections. Therefore, comparing with XRD, it is much easier to observe the superstructure by TEM.

7.2.2 Microstructure and Defect Observation

In materials science, it is well known that the material properties strongly depend on microstructure. To study the microstructure and the structural defect is one of the most important subjects in materials science. TEM is one of the most powerful tools to study such structural phenomena. Since the HRTEM image can reveal the structure detail on the atomic scale, it becomes much simpler to classify the defects, such as dislocation, stacking fault, microtwin, antiphase domain, grain boundary and so on, in the specimen. This information can help us to understand the relationship between the microstructure and the physical properties, and to improve material synthesis techniques.

7.2.3 Thin Film Structure Observation

Thin film growth is a very important technique in the electronics industry. The physical properties of the thin film strongly depend on its microstructure, such as structural defect, grain boundary, perfection, homogeneity etc. The thickness of the thin film is usually submicrometer, being much thinner than that of the substrate. In order to grow the thin film epitaxially, in most cases we select a substrate that is isomorphic to the film. Therefore, many reflections from the film will overlap with those from the substrate in the XRD pattern. The reflection intensity from the film is normally much weaker than that from the substrate. Moreover, the information obtained by the XRD method is only from an average structure, while in many cases we need to know the local structure on an atomic scale. Obviously, XRD can do little for such scale studies.

Another interesting subject is multilayer structures which are grown by stacking different composition layers alternately to form a superstructure. The thickness of each layer may be as small as several nanometers only. For studying such a complex structure, the information gained from HRTEM can not be replaced by any other method.

7.3 Specimen Preparation

Specimen preparation is one of the most important techniques in HRTEM, specially for thin film specimens. For taking good HRTEM image, there must be some very thin areas ($t<100$Å) in the specimen. Meanwhile, we must avoid structural phase transform and specimen contamination induced by specimen preparation. These are three essential requirements for preparing a good HRTEM specimen.

Depending on the material and the purpose of the study, we can select different methods, including (a) crushing; (b) ion milling; (c) electro-polishing; (d) chemical polishing; (e) cleavage and microtome and (f) evaporation or sputtering in vacuum, etc., to prepare the TEM specimen. Nowadays, the first three methods are most commonly used in materials science.

7.3.1 Crushing Method

The crushing method is widely used for brittle substances like ceramics, minerals, semiconductors, etc. The substances could be in bulk, powder, or single crystal forms.

The original sample is put in an agate mortar, and crushed into small fragments, with size about a few micrometers, by a pestle. In the edge part of the fragments, the thickness is expected to be less than 100Å. Fragments are dispersed in a solvent, like ethanol or acetone, etc., which does not react with the sample, then a droplet of the solution is dropped onto a microgrid, which is made by Cu (or Mo, Ni, Al etc.) mesh covered by a thin polymer supporting film containing many holes in size of 0.1–1 μm, and finally the specimen is made available for the TEM observation.

Specimen preparation by the crushing method can be done in a few minutes. Commonly, the specimen prepared by this method is less contaminated. When the substance does not have any strongly preferred cleavage plane, we can find all the concerned orientations from different grains, which is considered to be very important for the structure determination.

Crushing is the simplest way for preparing TEM specimens, but it has two major unavoidable weaknesses: (a) it destroys the original microstructure in the bulk or single crystal form substance and (b) it is not suitable for the highly extensive substance such as metal and alloy materials.

7.3.2 Ion-Milling Method

The original bulk or single crystal form substance is polished mechanically into several tens of micrometers, and the thin foil is stuck onto a Cu (or Mo etc.) of diameter~3 mm. The specimen is then polished by Ar ions in an ion-milling machine until some very thin regions (thinner than 10 nm) are formed. Depending on the substance, the foil thickness and the ion-milling conditions , such as acceleration voltage, beam current, ion beam incident angle, etc., ion milling may need several to a hundred hours.   If the sample is a single crystal, we must select a proper polishing orientation at the beginning.

When the original sample is thin film (including multilayer thin film) grown on the substrate, we can prepare the planar view or cross-sectional specimen. The cross-sectional specimen is prepared by the following procedure (see Fig.7.2): (a) cutting the sample into slices with thickness and width about 1mm$\times$3mm (possibly wider than 3mm); (b) sticking two slices face-to-face (i.e., film-to-film); (c) polishing the stuck slices to 10–50 μm in thickness mechanically (the thinner, the better if the sample is not broken during polishing); (d) sticking the thin foil on a Cu (or Mo, etc.) ring and confirming that the film locates in the center of the ring; (e) polishing the specimen by ion milling until a hole occurs in the center of the foil.

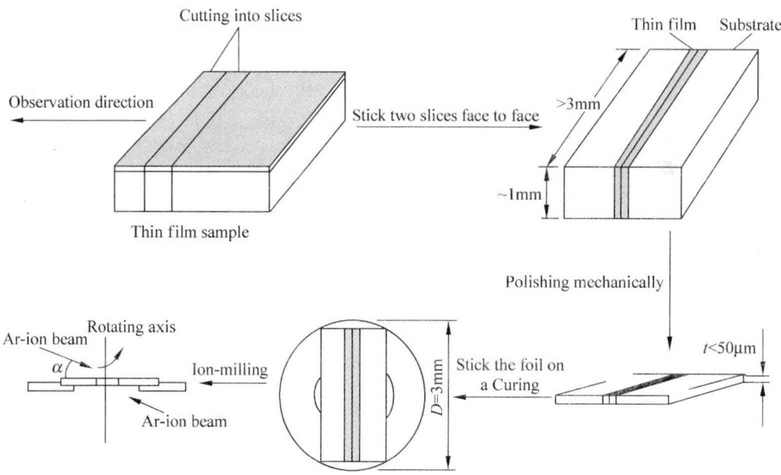

Fig. 7.2   Procedure to prepare a cross-sectional thin film specimen for TEM observation

In the ion-milling polished specimen, the original microstructure still remains, so that we can observe structural defects, grain boundary, interface, etc., i.e. many kinds of structure phenomena.   In principle, ion milling can polish any material. This is a major advantage of the ion-milling method.

The ion-milling method, however, has several weaknesses. First, ion milling may damage the original structure, or cause a structural phase transformation. This problem could be partially avoided by selecting the proper milling conditions, like lower accelerating voltage and weaker beam current, or using a cooling well to protect the specimen from the damage by Ar ion beam. Secondly, contamination is almost unavoidable, especially, in the thinner area. Thirdly, this method requests quite a long time to prepare one TEM specimen. Finally, the thin region, which is suitable for the HRTEM observation, usually is not large.

7.3.3 Electropolishing Method

If the substance is a conductor, the specimen can be prepared by electro-polishing. This method has been widely used since the 1950s.

In the electropolishing method, the specimen foil is made in a disk shape with thickness of 0.1– 0.5 mm and diameter of 3 mm. The platinum or stainless steel plates are placed as the cathode, while the sample is treated as the anode. An electrolyte beam with diameter of 0.1–1.0 mm is jetted on the center of the disk. When the voltage is supplied between the cathode and the anode, the target area becomes thinner. A light source is placed on one side of the specimen which is observed with a telescope on the opposite side. When the light can be seen by the telescope, indicating the presence of a hole on the specimen, the current should be cut immediately. The specimen for the TEM observation is obtained after washing the disk carefully in water.

Under different conditions, including electrolyte, current density, voltage and temperature, the polishing speed is different. Generally, a specimen can be prepared within 10 – 60 min, being much quicker than that prepared by the ion-milling method.

There are several weaknesses for this method, too. Firstly, it can only be used only for conductors. This restricts its application in many materials like ceramics, semiconductors, minerals, etc. Secondly, the thin area is easy broken during polishing for substances with a higher brittleness. In such specimens, it is rather difficult to find a thin region to do the HRTEM observation. Thirdly, when the specimen contains the particles with different phases, the polishing rate may be different from phase to phase. The phase with the higher polishing rate may disappear more quickly, showing an artificial microstructure. Finally, we must note that a chemical reaction may occur between the specimen and the electrolyte to form some new phases. For solving the final two problems, the electrolyte must be selected carefully for each particular specimen.

7.4 Applications of HRTEM for High T_c Superconducting Cuprates

Since the 1970s, HRTEM has been widely used in materials research, and people succeeded in using it to solve problems in crystal structure determination, defect observation and thin film microstructure studies. Some applications of HRTEM in dealing with the latter two issues are shown in other chapters of this book. In this section, we would like to show three particular examples in the crystal structure determination of cuprates by HRTEM (Wu et al., 1995a, b; 1996; 1998).

In 1986, Bednorz and Müller (1986) reported their discovery of a possible superconducting compound in the La-Ba-Cu-O system with T_c above 30 K, which was much higher than the highest T_c record, about 23 K in Nb_3Ge, at that time. Their result was soon confirmed by several groups, and the superconducting phase was isolated to be $(La, Ba)_2CuO_4$ in their sample. In the next few years, the T_c record was renewed at an astonishing speed: $YBa_2Cu_3O_{6+y}$ with T_c above 90 K was found in 1987 (Wu et al., 1987), $Bi_2Sr_2Ca_2Cu_3O_y$ with T_c above 110 K was found in 1988 (Maeda et al., 1988), $Tl_2Ba_2Ca_2Cu_3O_y$ with T_c above 120 K was found in 1988 (Sheng and Hermann, 1988a; 1988b), and $HgBa_2Ca_2Cu_3O_y$ with T_c above 130 K was found in 1993 (Putilin et al,. 1993; Schilling et al., 1993), in only seven years, the T_c record was enhanced by more than 110 K. From the discovery of superconductivity by Onnes in 1911 to 1985, during more than seventy years the T_c record was enhanced by only about 19 K. Comparing the first seventy years with the later seven years, we can easily understand what great progress occurred since 1986. So far, more than a hundred superconducting cuprates phases have been reported, and more than several hundred compounds with related structures were synthesized by many methods.

Depending on the carrier type, the high-T_c cuprates are classified into two groups, i.e., p-type superconductors, in which the carriers are holes, and n-type superconductors, in which the carriers are electrons. So far, the known highest T_c in the n-type superconductors is only about 40 K, being much lower than that in the p-type ones.

Essentially, the crystal structures of all p-type high-T_c cuprates can be divided into two blocks along the c-axis: one is the so-called charge reservoir block, and the other is the so-called current transmission block (see Fig.7.3). The charge reservoir block consists of at least two more atomic sheets. By changing oxygen content or substituting elements in this block, we can change the hole density in the crystal, resulting in a variation of the superconductivity. The structure of the current transmission block consists of the n CuO_2 sheets and the $(n–1)$ M sheets (n=1, 2, 3, $\cdots$). Until very recently, we knew that M must be Y, alkali earth (Ca, Sr, Ba) or rare earth elements (it may also be occupied by some other elements as we will see in Sect. 7.4.1). Alternate stacking of the CuO_2 and M sheets forms a so-called infinite-layer (IL) structure, which actually is an oxygen-deficient perovskite-type structure. The CuO_2 sheet is considered to be a basic unit to carry the superconducting current, and any structural damage in this sheet, such as the occurrence of oxygen vacancies, the substitution of Cu by the other elements or the presence of a large structure distortion, may seriously degrade or completely destroy the superconductivity.

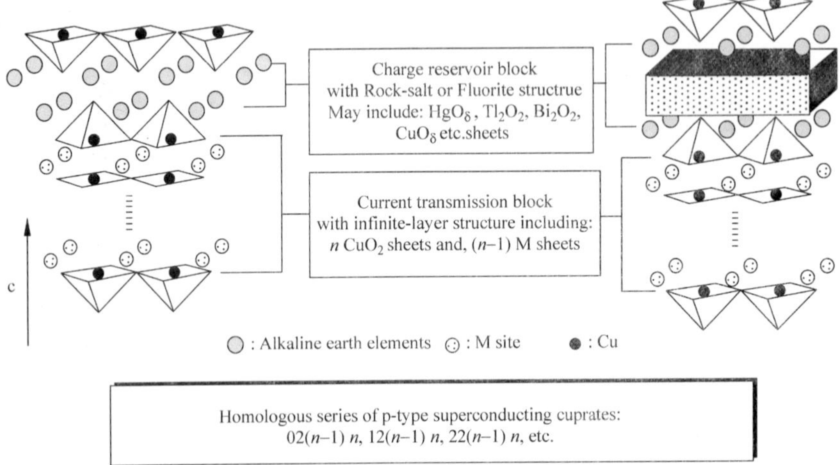

Fig. 7.3 Schematic diagram showing the crystal structure of p-type superconducting cuprates

7.4.1 Structure Determination of $(Hg,Tl)_2Ba_2(Hg,Tl)Cu_2O_y$ in a Multiphase Sample

In 1993, a homologous series of mercury-based compounds $HgBa_2Ca_{n-1}Cu_nO_{2(n+1)+\delta}$, i.e., Hg-12$(n-1)n$ ($n=1$, 2, 3), was discovered (Putilin et al., 1993; Schilling et al., 1993). The third member of this series, Hg-1223, shows superconductivity at 135 K, being the highest so far discovered for all superconductors. This discovery broke the T_c record held by $Tl_2Ba_2Ca_2Cu_3O_y$ (Tl-2223) for five years (Sheng and Hermann, 1988a; 1988b), and accordingly many studies soon focused on this system. In 1994, a new high-pressure-formed cuprate containing mercury-oxide double layers, $Hg_2Ba_2Y_{1-x}Ca_xCu_2O_{8-\delta}$ (Hg-2212, $x = 0$–0.5) was reported (Radaelli et al., 1994). The sample is an insulator when $x = 0$, and becomes a superconductor by partially replacing Y by Ca. The substituted compounds with $x=0.4$ exhibited bulk superconductivity at 45 K. Later, it was found that Tl doping on the Hg site may stabilize the structure in the (Hg,Tl)-22$(n-1)n$ system (Tokiwa-Yamamoto et al., 1996a). With Tl doping, several numbers, $n=1$, 2, 3, 4 and 5, in this homologous series were synthesized by employing a high-pressure technique. T_c of (Hg,Tl)-2223 is the highest in this homologous series and may exceed 120 K (Tokiwa-Yamamoto et al., 1996a; 1996b; Tatsuki et al., 1996a; 1996b; 1997). In these phases, the charge reservoir block consists of the BaO/(Hg,Tl)O/(Hg,Tl)O/BaO sheets, having the so-called rock-salt structure, while the current transmission block consists of $(n–1)$ $CuO_2/Ca(Y)/CuO_2$ infinite-layer unit cells.

In studying the $(Hg,Tl)–22(n–1)n$ family, the sample with nominal composition of Tl ꞉ Hg ꞉ Ba ꞉ Cu ꞉ O=2 ꞉ 1 ꞉ 2 ꞉ 2 ꞉ 10, being Ca and Y-free, was found to be multiphase. One major phase was identified to be (Hg, Tl)-2201, the first member in the $(Hg,Tl)–22(n–1)n$ family, while the other phases could not be identified by XRD except for a few weak peaks from CuO. Such a sample was then investigated by TEM.

The sample was synthesized using a cubic-anvil-type apparatus under a pressure of 6 GPa at 1000 ℃ for 1 h. Starting materials were commercial powders of HgO, Tl_2O_3, BaO_2, and CuO. The nominal composition was Tl ꞉ Hg ꞉ Ba ꞉ Cu ꞉ O=2 ꞉ 1 ꞉ 2 ꞉ 2 ꞉ 10. The sample was examined by a JEM-2010 F analytic transmission electron microscope (ATEM) equipped with an energy dispersive X-ray (EDX) detector operating at an accelerating voltage of 200 kV. HRTEM images were taken by a JEM-4000 EX operating at an accelerating voltage of 400 kV. The specimen for TEM observation was prepared by the crushing method, and a thin golden foil was vaporized on the microgrid for calibrating the electron diffraction camera length.

The sample contains several unknown phases, while our interest here is the perovskite-relating structure phase. Figure 7.4 shows the electron diffraction patterns (EDPs) taken from one grain with the perovskite-related structure. The electron beam is parallel to (a) the [100], (b) the [120] and (c) the [110] directions, respectively. This phase has a body-centered tetragonal unit cell, and its possible space groups are: I422(97), I4mm(107), I4m2(119), I42m(121) and I4/mmm(139). Among these space groups, only I4/mmm has a mirror plane

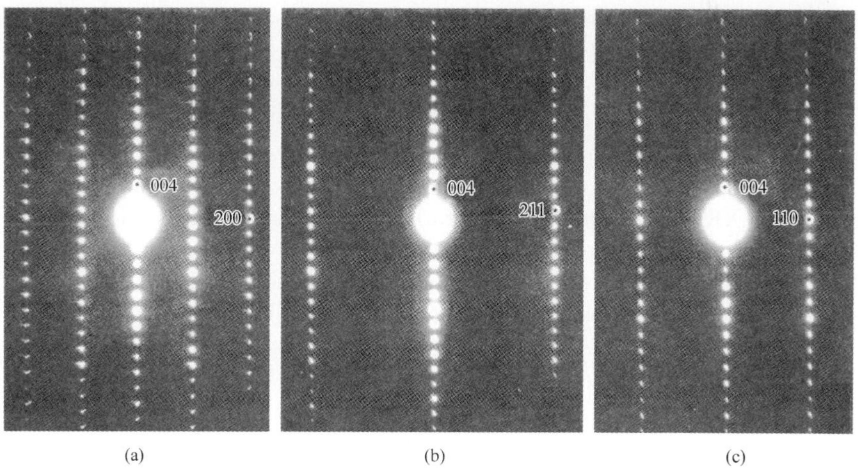

(a)   (b)   (c)

Fig. 7.4 Electron diffraction patterns taken from one grain along (a) the [100] , (b) the [120] and (c) the [110] directions

vertical to the c-axis. Using the diffraction rings from a gold foil as an internal standard, the lattice parameters are determined to be a=3.88Å and c=29.1Å.

The compositional analysis for the phase with c=29.1Å was carried out by EDX. Several grains were measured, and the results are listed in Table 7.1. The average atomic ratio of Tl : Hg : Ba : Cu was found to be 25.14 : 13.74 : 31.40 : 29.72, being not far from the nominal composition.

Table 7.1   Composition analysis for $Tl_2HgBa_2Cu_2O_{10}$ by EDX

Grain	Ba (at%)	Cu (at%)	Hg (at%)	Tl (at%)
1	30.33	30.85	14.54	24.27
2	30.69	29.65	14.45	25.22
3	28.78	30.94	14.9	25.39
4	30.57	30.68	14.48	24.27
5	35.02	28.73	14.62	21.63
6	30.70	27.37	13.79	28.14
7	32.10	29.07	12.13	26.70
8	30.51	30.32	13.16	26.01
9	31.55	28.24	14.01	26.21
10	33.72	31.37	11.36	23.55
Average	31.40	29.72	13.74	25.14

Figure 7.5 shows an HRTEM image taken with the electron beam parallel to [100]. A mirror plane vertical to the c-axis can be clearly seen in this figure, indicating that the space group of the phase is I4/mmm(139). Using the a-axis length as a standard, the length of infinite-layer (IL) and rock-salt (RS) blocks along the c-direction can be determined as 3.3Å and 7.0Å, respectively. These values are in good agreement with those reported for $(Hg_{0.7}, Tl_{0.3})_2Ba_2(Y_{0.8}, Ca_{0.2})Cu_2O_y$ (Wu et al., 1997). In the right part of this figure, a schematic 2212 structure projected onto the (100) plane is presented. The correspondence between the image and the structure is quite good, indicating that the phase has a so-called "2212" structure.

Figure 7.6a shows an XRD pattern from the sample, and Fig. 7.6b is a reference XRD pattern from a nearly single phase $(Hg_{0.7}, Tl_{0.3})_2Ba_2(Y_{0.8}, Ca_{0.2})Cu_2O_y$ sample. More than ten peaks indicated by upside-down triangles in Fig. 7.6a can be indexed by 2212, and the fraction of the 2212 phase is substantial in the sample. Using these peaks, the lattice parameters can be calculated as a=3.889(5)Å and c=29.04(4)Å, which agree with those obtained by the ED analysis.

Now, an important question arises here: which elements occupy the M-site (see Fig. 7.5) in the present 2212 phase?

This phase includes four kinds of cations: Cu, Ba, Tl and Hg. The M site can not be occupied by Cu, because it is impossible to form an IL block like $CuO_2/Cu/CuO_2$. A Tl-1212 ($TlBa_2Ba_{1-x}Cu_2O_y$) phase has been reported

(Hasegawa et al., 1996), in which the M site was completely occupied by Ba.

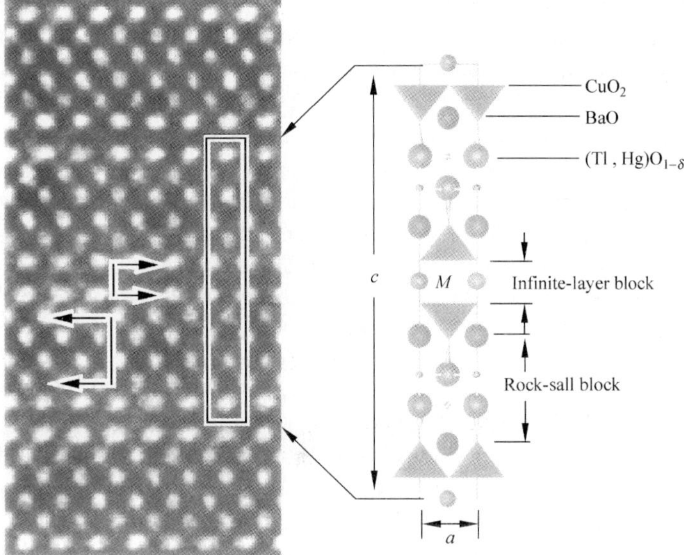

Fig. 7.5 HRTEM image taken with the electron beam parallel to the [100] direction. The unit cell is shown by the rectangle. The right-forward and left-forward arrow pairs indicate the infinite-layer (IL) block and the rock-salt (RS) block, respectively. The projection of so-called 2212 structure is drawn on the right, which shows good correspondence with the image

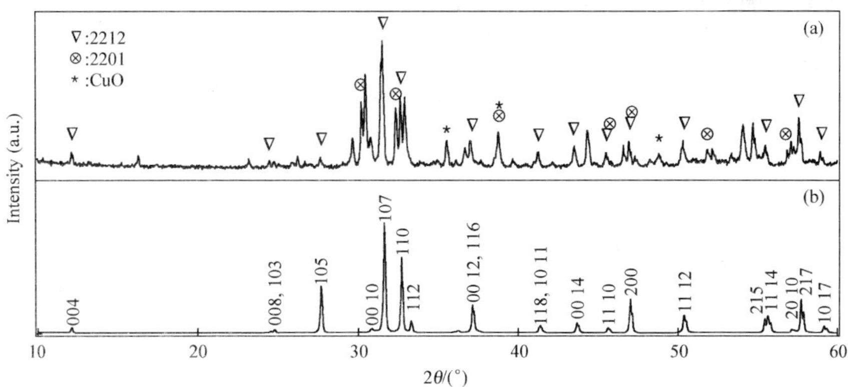

Fig. 7.6 X-ray diffraction patterns (**a**) from the multiphase Tl-Hg-Ba-O sample and (**b**) from a nearly single-phase 2212 sample (Hg$_{0.7}$, Tl$_{0.3}$)$_2$Ba$_2$(Ca$_{0.2}$, Y$_{0.8}$)Cu$_2$O$_{8-2\delta}$. The reflections from 2212, 2201 and CuO are labeled in (**a**), and the unlabeled reflections are from some unknown phases

For this phase it was found that an ordering between a vacancy and Ba in the M site occurs, resulting in a lattice dimension increase of $a = a_p \sqrt{2}$ and $c = 2c_0$. (a_p: the a-axis of perovskite structure; c_0: the c-axis of normal 1212 structure). Meanwhile, because Ba^{2+} has a much larger ionic size than that of Ca^{2+} or Y^{3+} ($r_{Ba}^{2+(VIII)} = 1.42$ Å, $r_{Ca}^{a2+(VIII)} = 1.12$ Å and $r_{Y}^{3+(VIII)} = 1.019$ Å) (Shannon 1976), the length of the IL block, $CuO_2/Ba/CuO_2$, is about 3.8 Å, being much longer than that of about 3.3 Å for $CuO_2/Ca/CuO_2$ or $CuO_2/Y/CuO_2$. For the present 2212 phase, no superstructure was observed by the TEM studies, and the length of the IL block is only about 3.3 Å. Moreover, EDX analysis showed that the atomic ratio of Ba is almost the same as that of Cu. There fore, combining the structural information by HRTEM with the result of composition analysis by EDC, we conclude that in the present 2212 phase the M site is not occupied by Ba. The remaining choices for the M site are Tl and /or Hg.

The ionic sizes of Tl^{3+} and Hg^{2+} are $r_{Tl}^{3+(VIII)} = 0.98$ Å and $r_{Hg}^{2+(VIII)} = 1.14$ Å, respectively (Shannon, 1976). Comparing these values with those of Ca and Y, it is found that the differences among them are not so large. This means that if Tl and/or Hg occupy the M site, the change in the length of the IL block should be very small. It has been reported that the length of the IL block is 3.305 Å in $(Tl_{0.75}, Pb_{0.25})Sr_2(Ca_{0.86}, Tl_{0.14})Cu_2O_{6.9}$ (Otto et al., 1989) and 3.316 Å in $Tl_{0.97}Sr_2(Ca_{0.82}, Tl_{0.18})Cu_2O_{6.85}$ (Shimakawa et al., 1995). As expected, the substitution of Tl for Ca less than 20% in the M site does not cause any evident change in the length of the IL block. By the EDX measurement it was found that the sum of Tl and Hg is about 40 at%, being more than the amount of Ba or Cu. Considering these results, we propose that in this 2212 phase the M site is occupied by Tl and Hg, and the stacking order should be ... $\{(Tl,Hg)O_{1-\delta}$-BaO-CuO_2-(Tl,Hg)-CuO_2-BaO-$(Tl,Hg)O_{1-\delta}\}\cdots$.

Based on this consideration, a simulation for the HRTEM images taken along [100] was carried out. The atomic positions for this 2212 phase are assumed to be the same as those determined for $(Hg_{0.7}, Tl_{0.3})_2Ba_2(Y_{0.8}, Ca_{0.2})Cu_2O_y$ (Wu et al., 1997), and are listed in Table 7.2. Figure 7.7 shows two observed images together with their simulations under the conditions of $t = 40$ Å (t: specimen thickness) and (a) $\Delta f = -600$ Å; (b) $\Delta f = -800$ Å (Δf: defocus),

Table 7.2 Atomic positions of $(Tl_{0.65}, Hg_{0.35})_2Ba_2(Tl_{0.65}, Hg_{0.35})Cu_2O_8$

	Site	Occupancies	x	y	z
Tl/Hg-1	2b	0.65/0.35	1/2	1/2	0
Cu	4e	1	0	0	0.0560
Ba	4e	1	1/2	1/2	0.1258
Tl/Hg-2	4e	0.65/0.35	0	0	0.2126
O-1	8g	1	1/2	0	0.052
O-2	4e	1	0	0	0.140
O-3	4e	1	1/2	1/2	0.217

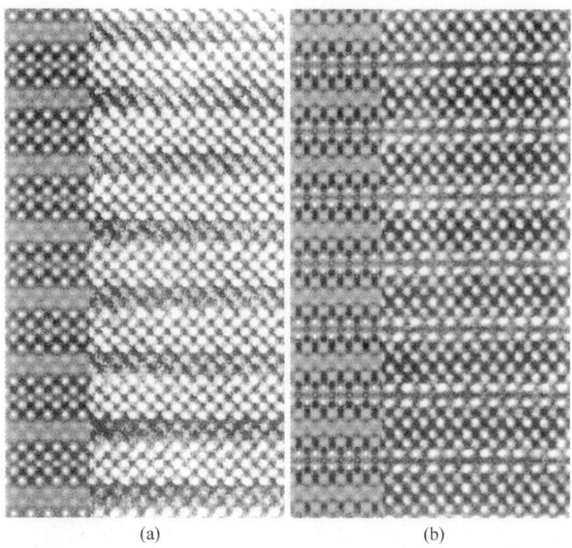

Fig. 7.7 Observed HRTEM images taken with the electron beam parallel to the [100] direction together with their corresponding simulations under conditions of $t = 40$ Å and (**a**) $\Delta f = -600$ Å; (**b**) $\Delta f = -800$ Å

respectively. The simulations show very good agreement with the observations, indicating that the proposed structural model is basically correct.

Because the atomic weights of Tl and Hg are very close, it is impossible to distinguish Tl from Hg by the HRTEM image simulation. Therefore, in the present structural model, it was assumed that Tl and Hg in the 2b and 4e sites have the same atomic ratio as determined by EDX. However, because of the different chemical properties of these two elements and the different co-ordination numbers in these two sites (2b : CN=VIII and 4e : CN=VI), the distribution of Tl and Hg in the two sites may not be uniform.

Though the composition of the 2212 phase does not deviate from the nominal one very much, the sample contains many impurity phases with a rather high fraction. Actually, many efforts to synthesize a pure phase sample have been made, but the present one is the best so far. It seems that this 2212 phase can be formed only under some very critical conditions.

7.4.2 Superstructure Formed by Oxygen Vacancies in (Bi, Pb)-2223

In 7, 7, 8, 8-tetracyanoquinodimethane (TCNQ)-treated $(Bi_{1.85}, Pb_{0.35})$ $Sr_2Ca_2Cu_{3.1}O_{10+\delta}$ ((Bi, Pb)-2223, $T_c \sim 120$ K) superconductors, a superstructure with the superlattice of $2a_0 \times 2b_0 \times c_0$ was observed by TEM. Superconductivity is virtually destroyed in the TCNQ-treated (Bi,Pb)-2223 cuprate, but it may be recovered, with T_c upto 110 K, by annealing the sample in flowing oxygen gas.

Then the $2a_0 \times 2b_0 \times c_0$ superstructure is found to disappear. This suggests that the superstructure is caused by an ordering of oxygen vacancies. HRTEM observations reveal that the oxygen vacancies reside not only in the CuO_2 sheets, but also probably in the SrO sheets. Based on the HRTEM observations, a model for the superlattice in both CuO_{2-x} and SrO_{1-y} sheets is proposed.

The sample was prepared by conventional solid-state reaction with nominal composition of $(Bi_{1.85}, Pb_{0.35})Sr_2Ca_2Cu_{3.1}O_{10+\delta}$. XRD indicated that the sample was a nearly single phase of (Bi,Pb)-2223. A pellet sample and $TCNQ(C_{12}H_4N_4)$ were sealed in an evacuated pyrex ampoule and the sample was kept at 300°C for 14 day (sample A). Sample B was obtained by annealing the sample A in a flowing oxygen gas at 300°C for 100 h. T_c was measured by a conventional four-probe technique. TEM observations were made at an accelerating voltage of 400 kV. The specimens for TEM observation were prepared by the crushing method.

Figure 7.8a shows the XRD patterns from the samples A and B. In the XRD pattern from the sample A some extra peaks can be seen. The peak indicated by an arrow for the sample A corresponds to a "d-spacing" of 3.28Å. This peak vanishes for the sample B. When this structural change occurred, the sample A became virtually nonsuperconductive, though a resistivity drop was observed at opproximately 12 K. As shown in Fig. 7.8b, the 110 K superconductivity (with zero resistivity at about 92K) recovered for the sample B.

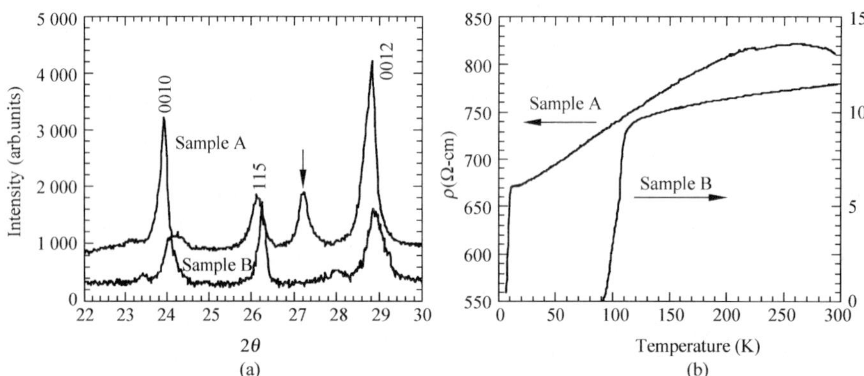

Fig. 7.8 (a) X-ray diffraction patterns from samples A and B. Structural change is identified by the peak indicated by an arrow, which can be indexed as 209 based on the superlattice. (b) resistivity (ρ)-vs. -T curves for samples A and B

Figures 7.9 a–c show EDPs taken, respectively, along [001], [010] and [110] for the sample A. In Fig. 7.9a additional spots located at (1/2 0 0) and (0 1/2 0), etc., are seen, indicating that $2a_0 \times 2b_0$ superstructure has formed in the a-b plane, with $a_s \approx b_s = 2 \times a_0 = 2a_p\sqrt{2} \approx 10.80$ Å (where a_s, b_s: a, b axes of the superstructure, a_0: a axis of the pristine phase, a_p: distance between two adjacent Cu atoms in the CuO_2 sheet). Meanwhile, the wavelength of the Bi-type modulation increased from $4.6a_0$ to~$5.6a_0$. Several tens of grains were carefully examined by ED, and all of them showed the same superstructure. In Fig. 7.9b no 10l diffraction spots appear, and in Fig. 7.9c some diffraction streaks around *hhl* spots are visible. From these observations the diffraction conditions for the superstructure may be written as follows: $h=2n$ for *h0l*; $h=2n$ for *h00*; and no conditions for *hkl* ($h, k, l \neq 0$). There is no obvious change for the lattice parameter of c. Hence, based on both XRD and ED analysis, it is derived that the TCNQ-treated bulk sample has shown a $2a_0 \times 2b_0 \times c_0$ superstructure. Figure 7.9d shows a EDP taken along [001] for the sample B. In this case, no diffraction spots that are related to the $2a_0 \times 2b_0 \times c_0$ superstructure, are observed any longer. For nearly half of the grains, the wavelength of the Bi-type modulation was still close to $5.5a_0$, but the rest had a wavelength of ~$4.6a_0$.

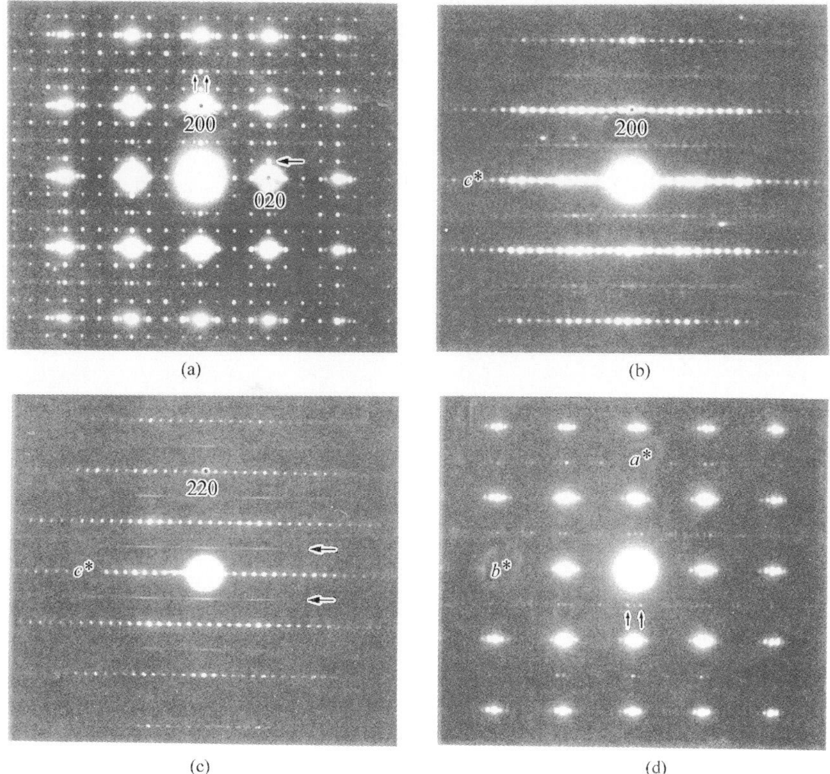

Fig. 7.9 Electron diffraction patterns taken along (**a**) [001], (**b**) [010] and (**c**) [110] for sample A. (**d**) is taken along [001] for sample B

338

Combining Fig.7.8 and Fig.7.9, we know that TCNQ-treatment makes the formation of a $2a_0 \times 2b_0 \times c_0$ superstructure in a bulk form (Bi,Pb)-2223 sample. Moreover, this superstructure can be removed by reoxidation, indicating that it closely relates to poor oxygen content in the sample. Therefore, we can conclude that this $2a_0 \times 2b_0 \times c_0$ superstructure is originated from the ordering of oxygen vacancies in the sample. It is well known that in (Bi,Pb)-2223 oxygen may reside in the (Bi,Pb)-O, Sr-O and Cu-O sheets. Consequently, we can judge in which sheet the oxygen vacancy locates by the modulation structure.

Figure 7.10a is a micrograph taken along [110] from sample A. In its left part, which corresponds to a thin area, the (Bi, Pb)-O sheet, the Sr-O sheet and the Cu-O/Ca infinite layer block can be clearly seen. In the right part, on the other hand, some fringes are seen along the c-axis. Figure 7.10b is an enlarged micrograph from the right part of Fig. 7.10a. Though the resolution is poor in Fig. 7.10b, referring to the left part of Fig. 7.10a, we still can identify that the fringes are located at the Cu-O/Ca IL blocks, and also most likely at the Sr-O sheet, but not at the (Bi, Pb)O bilayers. The separation distance between adjacent fringes is ~7.7 Å$=2a_p$. The fringes can be observed only in thick area, implying that the contrast is related to light atoms, like O, rather than heavy atoms. Clearly, the superstructure of $2a_0 \times 2b_0 \times c_0$ is formed by the ordering of oxygen vacancies, while we know that there is no oxygen in the Ca sheet. Therefore, though the Cu-O sheet can not be distinguished from the Ca sheet in Fig. 7.10b ((Bi,Pb)-O can be recognized without doubt), we can conclude that the oxygen vacancy resides in the Cu-O, probably also the Sr-O sheet.

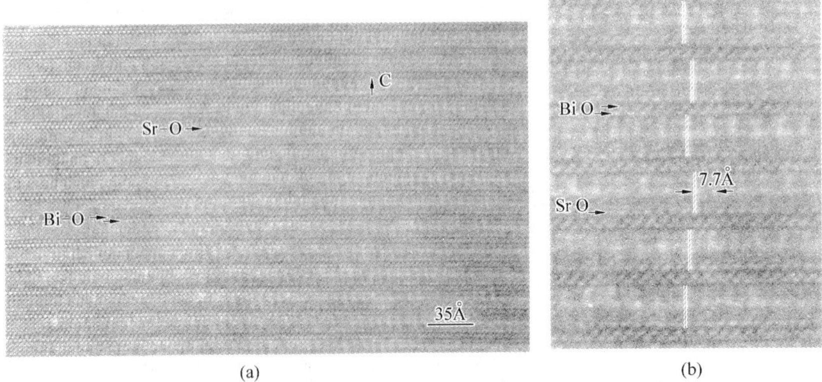

(a)   (b)

Fig. 7.10 (a) HRTEM image taken along [110] for sample A. In thinner left part, the (Bi, Pb)-O sheet, which is marked by two Bi-O arrows instead of (Bi, Pb)-O, the Sr-O sheet, which is marked by a Sr-O arrow, and the Cu-O/Ca infinite layer block can be identified clearly. In thicker right part, which is enlarged in (**b**), a group of fringes along the c direction appears. These fringes locate at the Cu-O/Ca infinite layer block, also probably extend to the Sr-O sheet

From this image we can also find that the contrast of fringes is inhomogenous along the c-axis, and the phases of the fringes are also not fixed for different Cu-O/Ca stacks in the IL block, which is sandwiched by the SrO/(Bi, Pb)O/(Bi, Pb)O/SrO blocks. The unfixed phases for different IL blocks seem to be why streaks but not spots have appeared in the EDP shown in Fig.7.9c.

The superstructures exposed by $2a_p\sqrt{2}\times 2a_p\sqrt{2}$, have been reported for several cuprates including Y-Ba-Cu-O (Krekels et al., 1990) and La-Sr-Cu-O (Fu et al., 1992; Errakho et al., 1998). In the case of $La_2Sr_6Cu_8O_{16}$, the fundamental structure is quite similar to a primitive perovskite cell, and a superstructure can form by taking away one quarter of the oxygen atoms from the CuO_2 sheets and a half of the oxygen atoms from the (La, Sr)O layers (Fu et al., 1992). In $YBa_2Cu_3O_{6+\delta}$, a superstructure forms by removing a half of the oxygen atoms from the CuO chains and one quarter of the oxygen atoms from the BaO sheets (Krekels et al., 1990). However, in (Bi, Pb)-2223 the situation is slightly different from these cases: between two adjacent blocks consisting of SrO/(Bi, Pb)O bilayer/SrO along the c-axis there are three CuO_2 sheets and two Ca sheets. For the three CuO_2 sheets, there are two types of circumstance: both the top and the bottom CuO_2 sheets are neighboring a SrO sheet on one side and a Ca sheet on the other side, and the middle CuO_2 sheet is neighboring Ca sheets on both sides. Supposing that oxygen vacancies are ordered not only in the CuO_2 sheets but also in the SrO sheets, the models proposed for the superstructure are shown in Fig. 7.11a and b for SrO_{1-y}/CuO_{2-x} and for CuO_{2-x}, respectively. In these models, one quarter of the oxygen atoms are lost in both CuO_2 and SrO sheets, namely, $y=0.25$ and $x=0.5$. Then, the space group for the

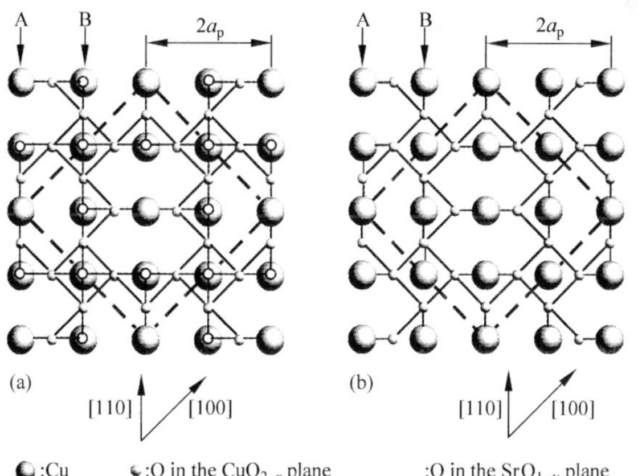

Fig. 7.11 Structural models containing oxygen vacancies: (**a**) $SrO_{0.75}/CuO_{1.5}$ sheets (Sr atoms are not represented), and (**b**) $CuO_{1.5}$ sheet

$SrO_{0.75}/CuO_{1.5}/Ca/CuO_{1.5}/Ca/CuO_{1.5}/SrO_{0.75}$ subcrystal should be $P\bar{4}b2$. If such a structure is observed along [110], since rows A and B in Figs.7.10 a and b have different oxygen contents, it becomes possible to observe different contrasts for the A and B rows. Note that, in these models, we simply took oxygen atoms away but did not take any lattice relaxation into account. It is plausible that when some oxygen atoms are gone, the remained atoms should suffer from lattice distortion, as suggested by Krekels et al. (1990) for $YBa_2Cu_3O_{6+\delta}$.

7.4.3 Modulation Structure Determination for "Pb"-1212 and "Pb"-1223

Two high-pressure-synthesized superconducting "Pb"-1212 (T_c–70 K) and "Pb"-1223 (T_c–115 K) phases were studied by means of ED, XRD and HRTEM. The lattice parameters of the basic structures are a_{20}=3.81, b_{20}=3.83 and c_{20}=12.10 Å for "Pb"-1212, and $a_{30}\approx b_{30}$=3.82, c_{30}=15.3 Å for "Pb"-1223, respectively. EDX analysis indicated that the crystals are Sr-, Cu- rich, and Pb-deficient, Ca-deficient. Superstructures were observed with superlattice parameters of a_{2s}=4a_{20}, b_{2s}=b_{20} and c_{2s}=2c_{20} for "Pb"-1212, being commensurate modulation, and a_{3s}=3.1a_{30}, b_{3s}=b_{30} and c_{3s}=2c_{30} for "Pb"-1223, being incommensurate modulation. Based on the HRTEM observations as well as simulations, the origins of these superstructures were determined.

The samples were synthesized under 5GPa at 900 ℃ for 30min with nominal composition, Pb：Sr：Ca：Cu：O=0.5：2：1：2.5：6.8 for "Pb"-1212, and Pb：Sr：Ca：Cu：O=0.5：2：2：3.5：8.9 for "Pb"-1223. The specimens for the TEM observation were prepared by conventional cursh method. Gold was vapor-deposited on a microgrid for correcting the camera length.

Figure 7.12 a–c show EDPs taken along the a, b and c axes of "Pb"-1212, respectively. From these diffraction patterns, we know that the fundamental structure belongs to the orthorhombic system with a primitive unit cell, and has a space group of Pmmm. The lattice parameters are a_{20}=3.81, b_{20}=3.83 and c_{20}=12.1 Å. The satellites from a superstructure can be found in Fig. 7.12b and c. No satellites are observed along b^* indicating that the superstructure is B-face centered. For the superstructure, the lattice parameters can be determined as a_{2s}=4a_{20}, b_{2s}=b_{20} and c_{2s}=2c_{20}. Here, the first subscript 2 indicates n=2 phase in the "Pb"-12(n–1)n family, while the second one, 0 indicates a fundamental lattice, and s a superlattice. For "Pb"-1212, only this superstructure has been detected.

Figure 7.13a, b and c show EDPs taken along the a-, b- and c-axes for "Pb"-1223 (n=3), respectively. The fundamental structure of this phase belongs to a pseudo-tetragonal system with a primitive unit cell and a space group of P4/mmm, and lattice parameters $a_{30}\approx b_{30}$=3.82 Å and c_{30}=15.3 Å. The satellites

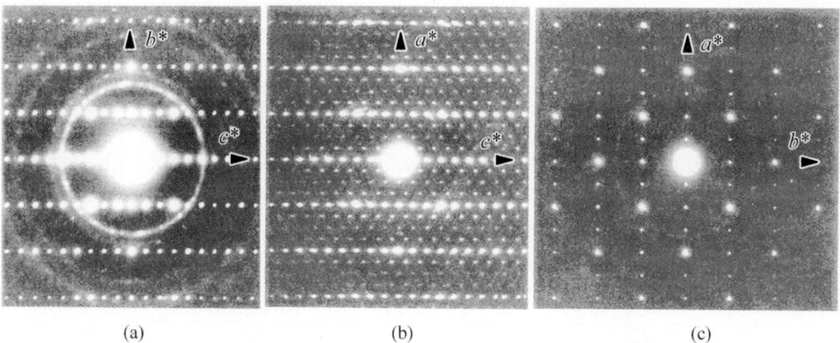

(a)   (b)   (c)

Fig. 7.12 ED patterns taken along (**a**) [100] (**b**) [010] and (**c**) [001] for "Pb"-1212

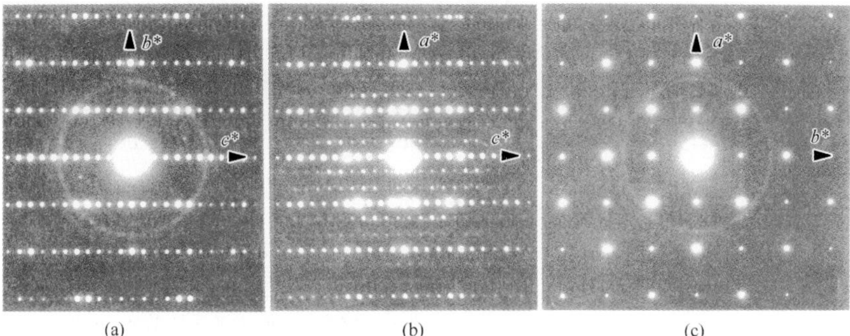

(a)   (b)   (c)

Fig. 7.13 ED patterns taken along (**a**) [100], (**b**) [010] and (**c**) [001] for "Pb"-1223

relating to an incommensurate modulation were found along a^*, and the lattice parameters of the superstructure are $a_{3s}=3.1a_{30}$, $b_{3s}=b_{30}$ and $c_{3s}=2c_{30}$.

For both phases the superstructures were rather stable against electron beam irradiation. In fact, no marked change was observed in EDPs. When even the grains were exposed under a strong electron beam for over ten minutes.

Several tens of EDX spectra were obtained from different grains. The average of atomic percentages of "Pb"-1212 and "Pb"-1223 phases are listed in the first lines of Tables 7.3 and 7.4, respectively (I). If the phases follow the chemical formulas $PbSr_2CaCu_2O_y$ and $PbSr_2Ca_2Cu_3O_z$, respectively, the atomic percentages can be calculated as shown in the second lines of Tables 7.3 and 7.4 (II). Comparing the measurements with the calculations, it is recognized that both phases are Sr- and Cu-rich, and Ca- and Pb-deficient. We propose that the excess Cu should be in the Pb site, while the excess Sr should be in the Ca as well as the Pb sites. Therefore, the chemical formula for "Pb"-1212 and "Pb"-1223 phases can be written as $(Pb_{0.5},Cu_{0.3},Sr_{0.2})Sr_2(Ca_{0.6}Sr_{0.4})Cu_2O_y$ and $(Pb_{0.5},Cu_{0.2},Sr_{0.3})Sr_2(Ca_{0.6}Sr_{0.4})_2Cu_3O_z$, respectively. Strictly speaking, these phases should be written as (Pb,Cu,Sr)-1212 and (Pb,Cu,Sr)-1223, respectively. For simplicity, (Pb,Cu,Sr) is written as "Pb".

Table 7.3 Average atomic percentage for $Pb_{0.5}Sr_{2.0}CaCu_{2.5}O_{6.8}$: measured by EDX and calculated from $PbSr_2CaCu_2O_y$

Elements	Cu (at%)	Pb (at%)	Sr(at%)	Ca(at%)
Measured	38.17	8.34	42.03	11.44
Calculated	33.33	16.67	33.33	16.67

Table 7.4 Average atomic percentage for $Pb_{0.5}Sr_{2.0}Ca_{2.0}Cu_{3.5}O_{8.9}$: measured by EDX and calculated from $PbSr_2Ca_2Cu_3O_y$

Elements	Cu (at%)	Pb (at%)	Sr(at%)	Ca(at%)
Measured	40.15	6.23	37.58	16.05
Calculated	37.50	12.50	25.00	25.00

Figure 7.14a and b show the HRTEM images taken along the a-axis for "Pb"-1212 and "Pb"-1223, respectively. Clearly, these images show that the fundamental structures of the phases present are quite similar to those from A-1212 and A-1223 (A=Tl, Hg, Cu, etc.). Therefore, we can conclude that the two phases present belong to n=2 and 3 members in the "Pb"-12(n–1)n homologous series. Figure 7.15a and b show the structural projection of "Pb"-1212 and "Pb"-1223 along [100], respectively.

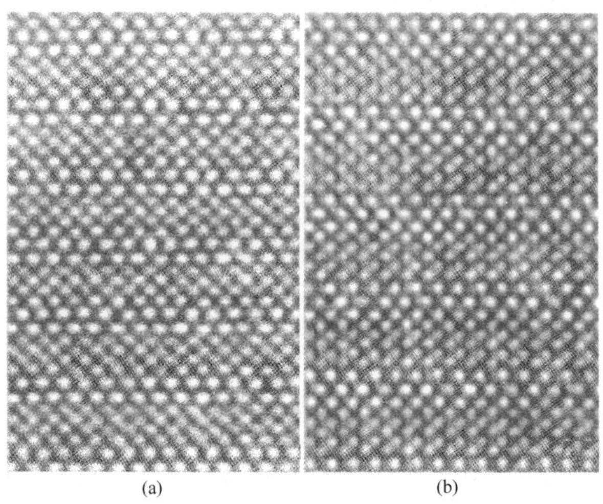

(a)　　　　　　　(b)

Fig. 7.14 HRTEM images taken along the [100] direction for the "Pb"-1212 phase (**a**) and the "Pb"-1223 phase (**b**)

A HRTEM image taken along [010] for "Pb"-1212 is shown in Fig. 7.16. In the thinner left part, almost no contrast relating to the $4a_{20}\times 2c_{20}$ superstructure can be observed. When the specimen becomes thicker from left to right, however, the contrast from the $4a_{20}\times 2c_{20}$ superstructure gradually

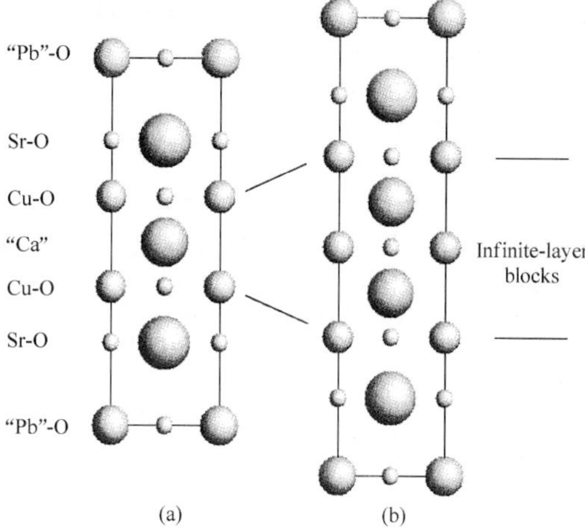

Fig. 7.15 Structural projections along [100] for (**a**) "Pb"-1212 and (**b**) "Pb"-1223

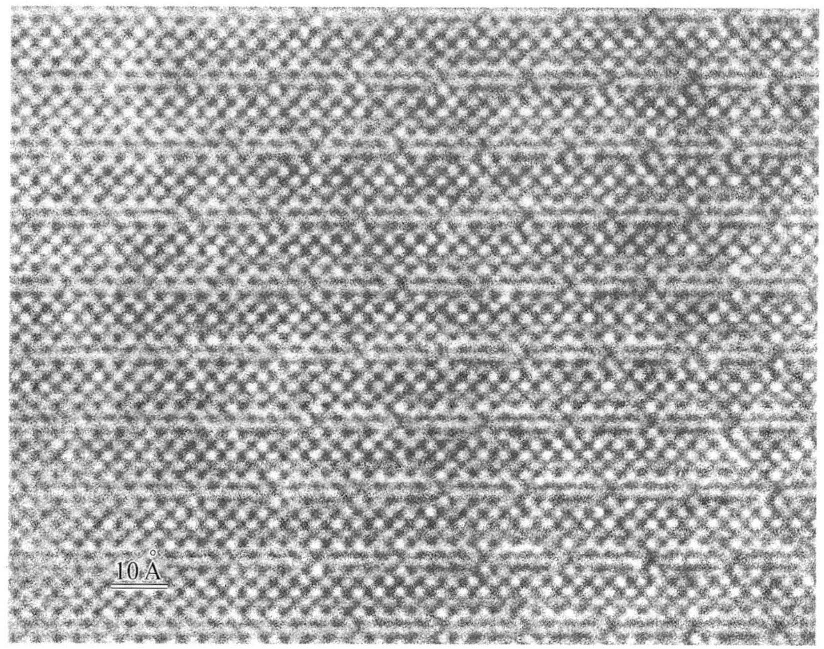

Fig. 7.16 HRTEM image taken along [010] from "Pb"-1212. The contrast of superstructure becomes clearer as the thickness increases

becomes visible in the "Pb"-O sheet. Clearly, the thicker the area, the stronger the contrast form the $4a_{20} \times 2c_{20}$ superstructure.

In fact, such a superstructure is not new at all in the "Pb"-1212 structure.

For (Pb, Cu)-1212, Ono and Horiuchi (1994) proposed that the superstructure was formed by the ordering of Pb and Cu in the (Pb, Cu)-O sheet. For (Pb, Sr)-1212, on the other hand, Rouillon et al. (1989; 1990) suggested that the superstructure is originated from the ordering of oxygen vacancies in the Sr-O sheet, which somehow relates to the presence of the $6s^2$ lone pair of Pb(II). Usually, a superstructure formed by the oxygen vacancy is unstable under electron beam irradiation. Besides, our HRTEM observation shows that the contrast of the superstructure probably locates in the "Pb"-O rather than the Sr-O sheet. Therefore, for the present sample it is proposed that the superstructure originates mainly from the ordering of cation/cation or vacancy/cation in the "Pb"-O sheet rather than the ordering of oxygen vacancies in the Sr-O sheet.

Three possible models could be considered for the present modulation structure in the "Pb"-O sheet. (A) the ordering of Sr/(Pb,Cu), where Pb and Cu reside in the same site randomly, and Sr resides in a special site; (B) the ordering of Cu/(Pb, Sr), where Pb and Sr reside in the same site randomly, and Cu resides in a special site; (C) the ordering of vacancy/(Pb,Sr,Cu), where Pb, Cu and Sr reside in the same site randomly, and the vacancy resides in a special site. All these models can be presented as Fig.7.17.

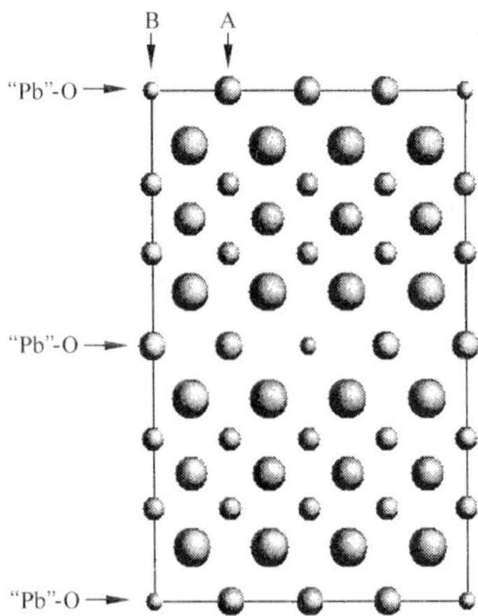

Fig. 7.17 Scheme to show the formation of superstructure in "Pb"-1212 by three possible models in the "Pb"-O sheet (O is not presented): (A) Sr/(Pb, Cu) ordering, smaller dots B present Sr, larger dots A (Pb, Cu); (B) Cu/(Pb,Sr) ordering, smaller dots B present Cu, larger dots A(Pb,Sr); (C) vacancy/(Pb,Cu,Sr) ordering, smaller dots B present vacancy, larger dots A(Pb, Cu, Sr)

Based on these three models, the HRTEM image simulations were carried out. The calculations were performed at three defocuses, − 600, −700 and −800 Å, and under three thicknesses, 20, 40 and 60 Å, for each model. The results are shown in Fig. 7.18a, b and c, which correspond to model (A), (B) and (C), respectively. In Fig. 7.18a the contrast of superstructure is invisible at a thickness of 20 Å, and becomes visible at 40 Å, and finally becomes the strongest at 60 Å. On the other hand, both Fig.7.18b and c show a similar character that the strongest contrast is visible at a thickness of 20 Å, and then becomes weaker as the thickness increases. This tendency is even cleares for model (C). Comparing the simulation with the observation, we found that only model (A), namely Sr/(Pb,Cu) ordering, showed a good agreement with the observation.

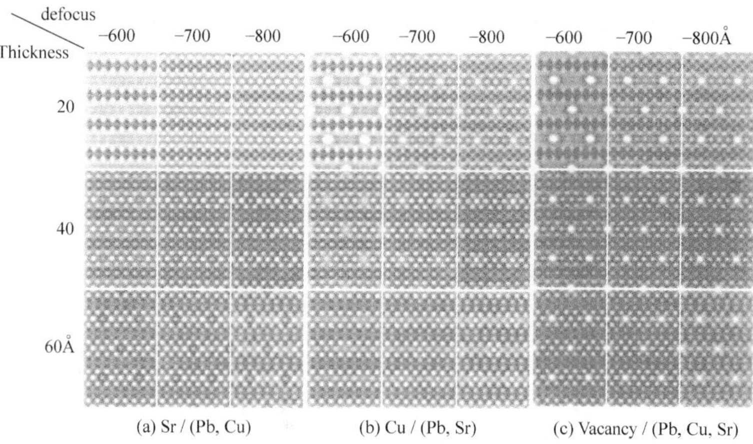

Fig. 7.18 HRTEM image simulations by (**a**) Sr/(Pb,Cu) ordering model, (**b**) Cu/(Pb,Sr) ordering model and (**c**) vacancy/(Pb,Cu,Sr) ordering model. Simulation was performed at defocuses of −600, −700 and −800 Å under thicknesses of 20, 40 and 60Å

According to Shannon (1976), the six coordinate ionic radii of Pb^{+2}, Pb^{+4}, Cu^{+2} and Sr^{+2} are 1.18, 0.78, 0.73 and 1.16 Å, respectively. Since the size of the Pb^{+2} ion is much larger than that of the Cu^{+2} ion, it is difficult to imagine that these two ions occupy the same site. Hence, the most probable situation is that Pb exists in this phase as Pb^{+4} rather than Pb^{+2}. The formation of superstructure probably relates to the ion size. Such ordering of Sr/(Pb, Cu) may also cause the modulation of oxygen distribution, and a little distortion of the fundamental structure. The present discussion, however, does not exclude the possibility of the presence of Pb^{+2} in the phase. In fact, Pb may have different valence depending on the synthesis conditions, such as oxygen partial pressure, heating temperature, and so on. If Pb^{+2} appears, it should reside with Sr, and then the modulation should be formed by the $(Pb^{+2},Sr)/(Pb^{+4},Cu)$ ordering. In this case, even the ratio of Pb, Sr and Cu is fixed in the "Pb"-O sheet, the superperiod of the modulation may change from sample to sample. Practically, such a relationship has been reported by Zandbergen et al. (1990) in $Pb_2SrLaCu_2O_{6+\delta}$ and $Pb_2Ba_2YCu_3O_{8+\delta}$. Ono and Horiuchi (1994) also reported the change of modulation wavelength versus the synthesizing conditions in $(Pb, Cu)Sr_2(Y, Ca)Cu_2O_z$.

In the "Pb"-1223 sample, an HRTEM image similar to Fig. 7.16 is observed along the [010] direction, as shown in Fig. 7.19. In the thinner region near the edge, no contrast relating to the superstructure can be seen. In the thicker region far from the edge, however, some strong white dots located in the "Pb"-O sheets can be clearly observed. No corresponding contrast modulation in the (Ca, Sr) or/and Sr-O sheets can be found anywhere. Here, as the modulation is incommensurate, the image simulations like Fig.7.16 can not be performed easily for this phase. However, taking the structural and the compositional similarity between "Pb"-1212 and "Pb"-1223 into account, it is suggested that the origin of the superstructures in "Pb"-1212 and "Pb" -1223 is the same, i.e., the contrast change in the "Pb"-O sheets is caused by an

Fig. 7.19 HRTEM image taken along [101]. In the thinner region (lower right part of this image), no contrast relating to the superstructure can be seen. When the specimen becomes thicker (upper left part of this image), some ordered strong white dots, which are considered to be Sr column positions, appear in the "Pb"-O sheets

ordering of Sr/(Pb, Cu), and the strong white dots in Fig.7.19 must correspond to the projection of Sr columns.

Figure 7.20 shows an HRTEM image taken along [010] in a thicker area. In most places, the intervals between two neighboring strong white dots along the a-axis are $3a_{30}$. However, in some places indicated by a pair of small rightward arrows, $4a_{30}$ intervals can be also found. Counting the intervals between two neighboring strong white dots in this figure carefully, the average interval can be calculated as:

$$<a_{3s}> =[(138\times3a_{30})+(16\times4a_{30})]/(138+16) = 3.104\ a_{30}.$$

Obviously, this average value is the a-axis length of the superstructure determined by the ED analysis. In other words, the ED analysis actually only revealed an average interval between two neighboring Sr columns. Consequently,

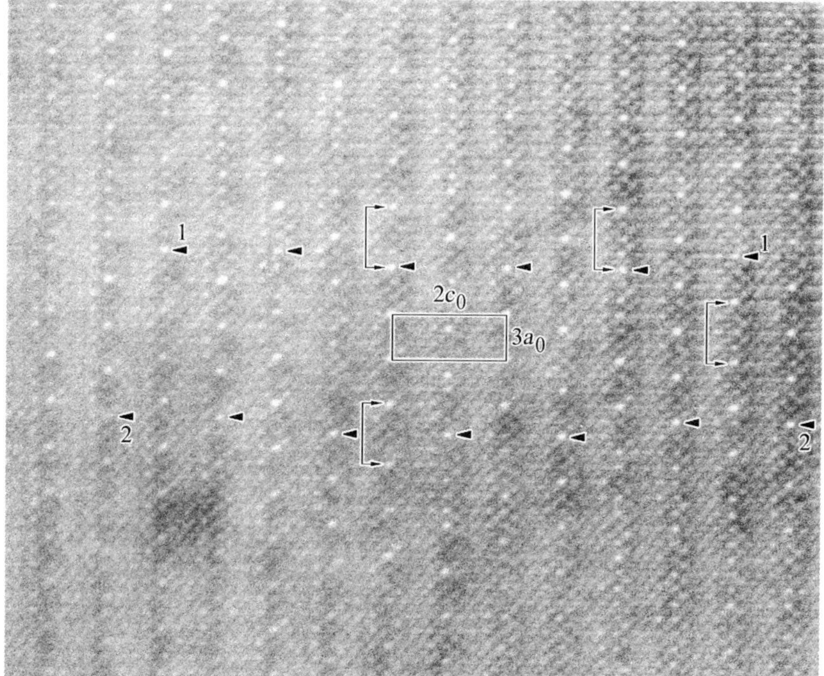

Fig. 7.20 HRTEM images showing the distribution of strong white dots in the thicker region. The intervals between two neighboring strong white dots are $3a$ usually, but sometimes are $4a$ as indicated by the right-forward small arrow pairs. Besides, along some rows, for example, the rows labeled with 1 and 2 by the left-forward large arrows, it is found that in some places the position of strong white dots shift by a along the a-axis

348

we can not identify an incommensurate "average superlattice" in Fig.7.20; instead, a commensurate superlattice of $3a_{30} \times 2c_{30}$ can be plotted.

For simplification, we will hereafter use the commensurate superlattice $3a_{30} \times 2c_{30}$ instead of the incommensurate one. from the ED analysis it is known that the superlattice is a B-face centered one. This means that if the Sr site in the "Pb"-O sheet is selected as the origin $(0,0,0)_{3s}$, then the symmetric operation will create another Sr site, $(1/2,0,1/2)_{3s}$, automatically. Since $a_{3s}/2=3a_{30}/2=1.5a_{30}$ is not an integral times of a_{30}, physically a Sr atom can not reside in such a position. Practically, in the superlattice drawn in Fig.7.20, we can find a Sr atom, i.e. the strong white dot, residing in $(1/3,0,1/2)_{3s}$ rather than $(1/2,0,1/2)_{3s}$. Taking a careful look along the rows of white dots indicated by large left-ward arrows in Fig.7.20, we can find that along rows 1 and 2 there are some places where strong white dots shift a_{30} along the a-axis. This means that the position of Sr atom in the B-face may locate in $(1/3,0, 1/2)_{3s}$ or $(2/3,0, 1/2)_{3s}$ with an equal possibility, and then the average position is $(1/2,0, 1/2)_{3s}$. Again, the ED analysis showed only an average character.

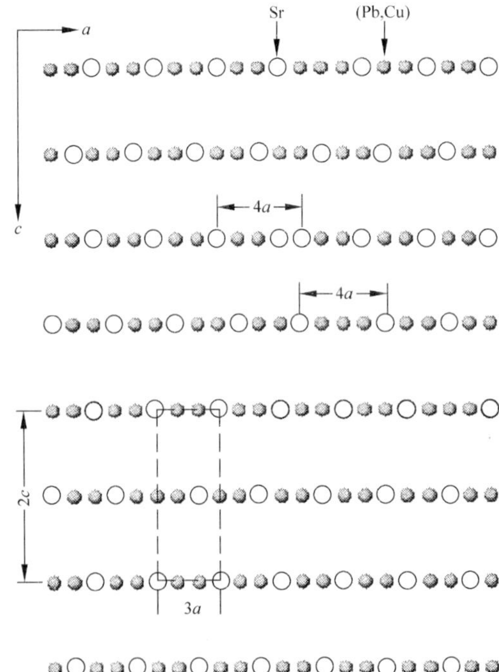

Fig. 7.21 Structural model for the present superstructure. Only cation positions in the "Pb"-O sheets are represented in this figure. Both $a_s=3.1a$ and a B-face-centered superlattice determined by the ED analysis are considered as an average result (see text)

On the basis of the HRTEM observations, the superstructural model of "Pb"-1223, namely the ordering of cations (Pb, Cu) and Sr, can be constructed as Fig.7.21. In most cases, the ratio of (Pb, Cu) to Sr in the "Pb"-O sheets is $2 : 1$, but in some local regions it becomes $3 : 1$ or $2 : 2$, which leads to the periodicity of Sr-Sr becoming $4a_{30}$ rather than $3a_{30}$. The total number of $4a_{30}$ intervals is about one ninth of that of $3a_{30}$ intervals. Therefore, the average interval becomes about $3.1a_{30}$, which corresponds to the a-axis length of superlattice determined by the ED analysis. On the other hand, such local interval variation results in the change of the relative position between adjacent rows. Hence, the Sr position in the B-plane can be either $(1/3, 0, 1/2)_{3s}$ or $(2/3, 0, 1/2)_{3s}$, as shown in Fig.7.21. Physically we can not find a B-face-centered superlattice with lattice parameters $a_{3s} = 3.1\ a_{30}$, $b_{3s} = a_{30}$ and $c_{3s} = 2\ c_{30}$ in the crystal, although we could obtain such a structure by the ED analysis. In the present case, the ED analysis reveals only an average superlattice, which somehow deviates from the real structure. Obviously, the information from the HRTEM observation reveals the real structure of "Pb"-1223. Such structure details can not be obtained by any other method.

7.5   New Developments in HRTEM

The basic theories of HRTEM were established in 1940s and 1950s, while it was used to solve the real structural problems from the 1970s. Therefore, actually HRTEM is not a new subject at all. Some researchers then suggested that as a scientific topic, there is nothing to do in this field any longer. This statement, however, is not correct.

A major advantage of HRTEM is that the atom column positions could be identified from the so-called structure image, which can reveal the structure projection directly. However, because of the effect of the objective lens on the transmission function, experimentally we know that it is not so easy to select a structure image from a series of images.

Under the WPOA, the intensity on the image plane can be expressed as:

$$I(r) = |\Psi'(r)|^2 \approx 1 - 2\sigma\varphi_{p}(r) * \mathscr{F}^{-1}[A(H)] \tag{7.45}$$

Except at the optimum conditions, the presence of the second term on the right side makes a remarkable modification for $\Psi(r)$. As a result, in most cases $I(r)$ does not have any direct agreement with the projection potential, and then the image contrast does not reflect the projection structure any longer. By doing Fourier transform for (7.45), we obtain

$$T(H) = \delta(H) - 2\sigma F(H)[A(H)] \tag{7.46}$$

350

where $T(H)=\mathscr{F}[I(r)]$, $F(H)=\mathscr{F}[\varphi_p(r)]$ and $\delta(H)$ is the δ-function which is defined as $\delta(0) = 1$ and $\delta(H) = 0$ when $H\neq0$. When $H\neq0$, (7.46) can be written as:

$$F(H) = -T(H)/2\sigma[A(H)] \tag{7.47}$$

By doing the inverse Fourier transform for both sides of (7.47), we can calculate the projection potential function as:

$$\varphi_p(r) = -\mathscr{F}^{-1}\{T(H)/2\sigma[A(H)]\} \tag{7.48}$$

$T(H)$ can be obtained experimentally from the Fourier transform of the HRTEM image which can be taken under any Δf. Therefore, if we can find $A(H)$, then the projection potential can be calculated from (7.48). The procedure to calculate $\varphi_p(r)$ by (7.48) is called image deconvolution. Obviously, the key point of the image deconvolution is to determine $A(H)$ correctly.

In recent years, the image deconvolution technique was developed. So far, several methods have been developed for doing image deconvolution. Some of them require to do this from a series of through-focus images (Kirkland et al., 1985; van Dyck and de Beeck, 1990), and the others need only one image (Unwin and Henderson, 1975; Hovmöller et al., 1984; Han et al., 1986; Tang and Li 1988; Liu et al., 1990; Hu and Li, 1991). It has been demonstrated in these studies that the structure image could be determined by the image deconvolution. This means that it is not necessary to do image simulation to obtain the structural image any longer. Consequently, the crystal structure model could be constructed directly based on the structure image. For some specimens, due to the structural damage by electron-beam irradiation, it is not easy to take a series of through focus images. In this case, the image deconvolution from one micrograph has a major advantage. The image deconvolution technique can be used not only on the structure determination, but also on the defect structure observation (He et al., 1997).

As mentioned above, the point resolving power (d_p) is one of the most important parameters for TEM. Nowadays, for most new-type conventional TEMs used in materials science research, d_p can reach 1.6–2.0 Å. Comparing with about 4.0 Å in the early 1970s, this value has been improved remarkably. But for material research, it is not good enough yet. For example, calculation indicates that in order to distinguish an individual oxygen column from the cation columns in some ceramics, it is necessary to have d_p better than 1 Å. The present resolving power of conventional TEM is still much lower than this requirement.

There are two approaches to enhance d_p. One is to make a new HRTEM with the point resolving power of 1 Å, and the second is to enhance d_p by image processing.

From the formula (1.43), it is known that d_p can be improved by (a) decreasing the spherical aberration constant C_s, or, (b) increasing the accelerating voltage E. d_p is proportion to $C_s^{1/4}$, indicating that the space for improving d_p by changing C_s is rather small. For instance, in a 200 kV TEM, when a super-high-resolution pole piece is used, C_s is 0.5mm, and d_p is 1.9Å. If C_s is reduced to a half, i.e., 0.25mm, technically this is very difficult to do, the point resolving power will become 1.6Å, being only slightly improved. This result implies that the further improvement of d_p only by decreasing C_s is very difficult. An efficient way to enhance d_p is by increasing the accelerating voltage. Actually, a new generation high-voltage TEM with an accelerating voltage of 1.2 MV could have a point resolving power of 1.0Å (Matsui et al., 1991; Horiuchi et al., 1991). However, the cost for making and maintaining such a high-voltage HRTEM is extremely high, and only a few institutions can pay for it. Besides, the HRTEM observation at a high voltage often causes serious damage to the original structure. Therefore, we need to find another way to enhance the resolving power.

The major restriction on the point resolving power of TEM is the spherical aberration of the objective lens. On the other hand, the objective lens does not have any evident influence on the diffraction plane, and there is no difficulty at all in collecting diffraction information with space frequency better than 1.0 Å^{-1} in EDP by a conventional TEM. However, EDP contains only the intensity information but the phase information of the reflections. Fortunately, such phase information can be recovered from the HRTEM images in the low-frequency range at the reciprocal space, and then be extended to the high-frequency range using a phase-extension technique that includes the direct method or the maximum entropy method. In other words, combining the information from both HRTEM and ED, it is possible to enhance the point resolving power from about 2.0 Å to about 1.0 Å by the image-processing method. Several succeeded studies have been reported in the recent years (Fan et al., 1991; Hu et al., 1992; Fu et al., 1994; Lu et al., 1997; Liu et al., 1998a), indicating that this technique has great potential for further development in the future.

The image deconvolution and the image processing combining of HRTEM and electron diffraction are still under development. At this moment, these techniques are not easy to master yet. With the progresses of these techniques, we expect that HRTEM can be used much more readily, much more efficiently and much more powerfully for the materials research in the future.

References

Cowley, J.M. and S. Iijima, Z. Natureforsch. A **27**, 445 (1972)

Er-rakho, L., C. Michel and B. Raveau. J. Sol. State Chem. **73**, 514 (1988)

Fan, H.F., S.B. Xiang, F.H. Li, Q. Pan, N. Uyeda and Y. Fujiyoshi. Ultramicroscopy. **36**, 361 (1991)

Fu, Z.Q., D.X. Huang, F.H. Li, J.Q. Li, Z.X. Zhao, T.Z. Cheng and H.F. Fan. Ultramicroscopy. **54**, 229 (1994)

Fu, W.T., D.J.W. Ijdo and R.B. Helmholdt. Mat. Res. Bull. **27**, 287 (1992)

Han, F.S., H.F. Fan and F.H. Li. Acta Cryst. A **42**, 353 (1986)

Hasegawa, M., E. Ohshima, M. Kikuchi, K. Hiraga, Y. Matsushita and H. Takei. Physica C **258**, 341 (1996)

He, W.Z., F.H. Li, H. Chen, K. Kawasaki and Oikawa. Ultramicroscopy. **70**, 1 (1997)

Horiuchi, S., Y. Matsui, Y. Kitami, M. Yokoyama, S. Suehara, X.J. Wu, I. Matsui and T. Katsuta. Ultramicroscopy. **39**, 231 (1991)

Hovmöller, S., A. Sjögren, G. Farrants, M. Sundberg and B.-O. Marinder. Nature **311**, 238 (1984)

Hu, J.J. and F.H. Li. Ultramicroscopy. **35**, 353 (1991)

Hu, J.J., F.H. Li and H.F. Fan. Ultramicroscopy. **41** , 387 (1992)

Kirkland, E.J., B.M. Siegel, N. Uyeda and Y. Fujiyoshi. Ultramicroscopy. **17** , 87 (1985)

Krekels, T., T.S. Shi, J. Reyes-gasga, G. Van Tendeloo, J. Van Landuyt and S. Amelinckx. Physica C **167**, 677 (1990)

Liu, J., F.H. Li, Z.H. Wan, H.F. Fan, X.-J. Wu, T. Tamura and K. Tanabe. Materials Transactions. JIM **39**, 920 (1998a)

Liu, Y.W., S.B. Xiang, H.F. Fan, D. Tang, F.H. Li, Q. Pan, N. Uyeda and Y. Fujiyoshi, Acta Cryst. A **46**, 459 (1990)

Lu, B., F.H. Li, Z.H. Wan, H.F. Fan and Z.Q. Mao. Ultramicroscopy. **70**, 13 (1997)

Maeda, H., Y. Tanaka, M. Fukutomi and T. Asano. Jpn. J. Appl. Phys. **27**, L209 (1988)

Matsui, Y., S. Horiuchi, Y. Bando, Y. Kitami, M. Yokoyama, S. Suehara, I. Matsui and T. Katsuta. Ultramicroscopy. **39**, 8 (1991)

Menter, J.W., Proc. Roy. Soc. A **236**, 119 (1956)

Ono, A. and S. Horiuchi. Jpn. J. Appl. Phys. **33**, 1839 (1994)

Otto, H.H., T. Zetterer and K.F. Renk. Z. Phys. B Condensed Matter. **75**, 433 (1989)

Putilin, S.N., E.V. Antipov, O. Chmaissen and M. Marezio. Nature. **362**, 226 (1993)

Radaelli, P.G., M. Marezio, M. Perroux, S. de Brion, J.L. Tholence, Q. Huang and A. Santoro. Science. **265**, 380 (1994)

Rouillon, T., J. Provost, M. Hervieu, D. Groult, C. Michel and B. Raveau. Physica C **159**, 201 (1989)

Rouillon, T., J. Provost, M. Hervieu, D. Groult, C. Michel and B. Raveau. J. Sol. State Chem. . **84**, 375 (1990)

Scherzer, O., J. Appl. Phys. **20**, 20 (1949)

Schilling, A., M. Cantoni, J.D. Gao and H.R. Ott. Nature. **363**, 56 (1993)

Shannon, R.D., Acta Cryst. A **32**, 751 (1976)

Sheng, Z.Z. and A.M. Hermann. Nature. **332**, 55 (1988a)

Sheng, Z.Z. and A.M. Hermann. Nature. **332**, 138 (1988b)

Shimakawa, Y., J.D. Jorgensen, H. Shaked, R.L. Hitterman, T. Kondo, T. Manako and Y. Kubo. Phys. Rev. **51**, 568 (1995)

Tang, D. and F.H. Li. Ultramicroscopy. **25**, 61 (1988)

Tatsuki, T., A. Tokiwa-Yamamoto, A. Fukuoka, T. Tamura, X.-J. Wu, Y. Moriwaki, R. Usami, S. Adachi, K. Tanabe and S. Tanaka. Jpn. J. Appl. Phys. **35**, L205 (1996a)

Tatsuki, T., A. Tokiwa-Yamamoto, T. Tamura, X.-J. Wu, Y. Moriwaki, S. Adachi and K. Tanabe. Physica C **273**, 162 (1996b)

Tatsuki, T., A. Tokiwa-Yamamoto, Y. Moriwaki, T. Tamura, X.-J. Wu, S. Adachi and K. Tanabe, Physica C **278**, 160 (1997)

Tokiwa-Yamamoto, A., T. Tatsuki, X.-J. Wu, S. Adachi and K. Tanabe. Physica C **257**, 36 (1996a)

Tokiwa-Yamamoto, A., T. Tatsuki, S. Adachi and K. Tanabe. Physica C **268**, 191 (1996b)

Ueda, N., T. Kobayashi, E. Suito, Y. Harada and M. Watanabe. J. Appl. Phys. **43**, 5181 (1972)

Unwin, P.N.T. and R. Henderson. J. Mol. Biol. **94**, 425 (1975)

Van Dyck, D. and M. Op de Beeck. Proc. XIIth Int. Cong. Electron Microsc. **1**, 26 (1990)

Wu, M.K., J.R. Ashburn, C.J. Torng, P.H. Hor, R.L. Meng, L. Gao, Z.J. Huang, Y.Q. Wang and C.W. Chu. Phys. Rev. Lett. **58**, 908 (1987)

Wu, X.-J., C.-J. Liu and H. Yamauchi. J. Appl. Phys. **77**, 2595 (1995a)

Wu, X.-J., T. Tamura, S. Adachi, C.-Q. Jin, T. Tatsuki and H. Yamauchi. Physica C **247**, 96 (1995b)

Wu, X.-J., T. Tamura, S. Adachi, T. Tatsuki and K. Tanabe. Physica C **266**, 261 (1996)

Wu, X.-J., T. Tatsuki, S. Adachi and K. Tanabe. Physica C **304**, 119 (1998)

Wu, X.-J., A. Tokiwa-Yamamoto, T. Tatsuki, S. Adachi and K. Tanabe. Physica C **273**, 198 (1997)

Zandbergen, H.W., W.T. Fu and J.M. van Ruitenbeek. Physica C **166**, 502 (1990)

Zandbergen, H.W., R. Gronsky, K. Wang and G. Thomas. Nature. **331**, 596 (1988)

8 Applications of Convergent-Beam Electron Diffraction in Materials Microcharacterization

Renhui Wang and Huamin Zou

8.1 Introduction

In conventional electron diffraction, a nearly parallel electron beam is incident onto a thin specimen, so the transmitted and diffracted beams form (0 0 0) transmitted and (h k l) diffracted spots on the back focal plane of the objective lens, as shown in Fig. 8.1a. In convergent-beam electron diffraction (CBED) a convergent beam with a convergent angle which is larger than 0.01 rad ($1° = 0.0175$ rad) is incident on to the thin specimen, so that the transmitted and diffracted beams are extended and form the relevant discs at the back focal plane, as shown in Fig. 8.1b. In these transmitted and diffracted discs there are intensity distributions which contain much more information than the electron diffraction spots.

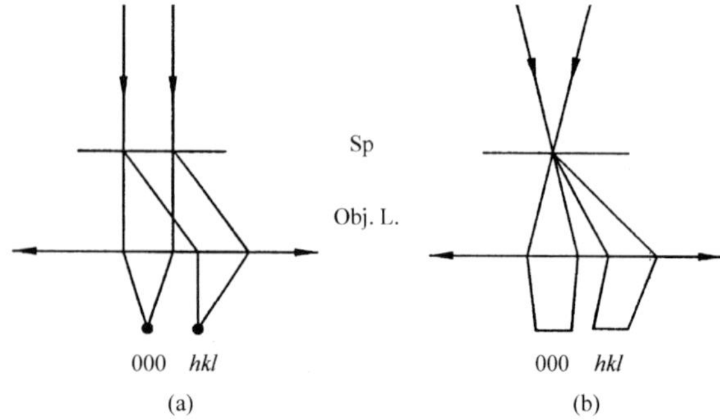

Fig. 8.1 Schematic diagrams showing (**a**) conventional electron diffraction and (**b**) CBED

In 1939 Kossel and Mölenstedt (1939) carried out, for the first time, CBED from a micro-area of 30 nm diameter of a specimen. Since 1975, with the improvement of the instruments and experimental techniques, with the deepening of the understanding of the intensity distribution in CBED patterns by the dynamic theory of electron diffraction, CBED became a rapidly growing important branch of transmission electron microscopy (TEM) and has been widely used in materials microcharacterization. In the present chapter, we will provide an introductory description about the principles, experimental technique, and some applications of CBED, such as determinations of foil thickness, lattice constants, structure factors, point groups, Burgers vectors of dislocations and interfacial residual stress fields. For details, the reader is recommended to refer to the special books (Tanaka and Terauchi, 1985; Tanaka et al., 1988; Tanaka et al., 1994; Spence and Zuo, 1992), review articles (Cowley, 1992; Spence, 1993) and original papers.

8.2 Experimental Technique

Almost all the modern transmission electron microscopes possess a function to carry out CBED experiment. Figure 8.2 shows the ray-path diagrams of (a) TEM mode, (b) energy-dispersive X-ray spectrometry (EDS) mode, (c) nanobeam electron diffraction (NBD) mode and (d) CBED mode. It is easy to switch over from one mode to another. In the TEM mode the condenser minilens is strongly excited to form a nearly parallel incident beam while in the EDS mode the condenser minilens is switched off to form the smallest incident probe. However, the switching off of this lens leads to a large convergent angle α_1. Therefore, in the NBD mode, a weakly excited condenser minilens and smaller second condenser aperture are used to obtain a smaller incident probe and smaller convergent angle α_2. In the CBED mode, an appropriate (from zero to weak) excitation of the condenser minilens and a second-condenser aperture with appropriate size are selected to obtain a convergent incident beam with appropriate, variable convergent angle.

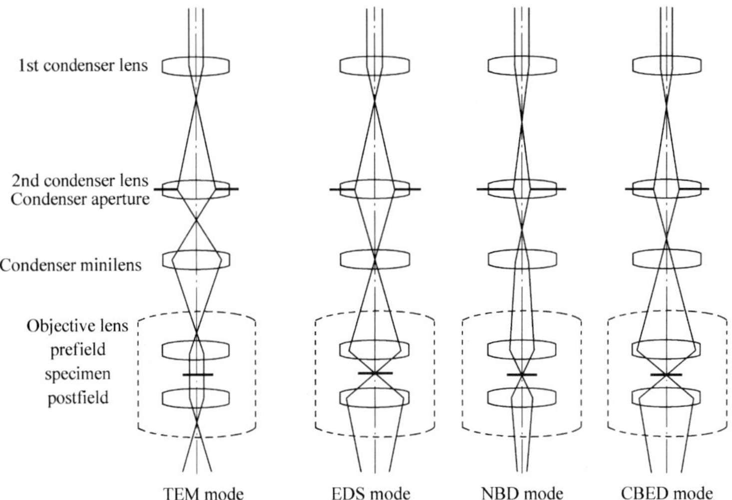

Fig. 8.2 Ray-path diagrams of (**a**) TEM mode, (**b**) EDS mode, (**c**) NBD mode and (**d**) CBED mode

In order to obtain a circular incident probe, the microscope must be accurately aligned and the astigmatism of the second condenser must be accurately corrected.

In the CBED experiment, the crossover of the incident beam may lie on the specimen (infocus CBED) as shown in Fig.8.1b or deviated from the specimen (defocus CBED) as shown in Fig.8.3. In defocus CBED, a conical incident electron beam of a half-angle of $0.4°$ $-1.0°$ converges at crossover C which is separated from the specimen Sp by a distance Δf (defocus). Thus a circular region of the specimen is illuminated by the incident beam. If the illuminated region contains a segment $D1$, $D2$ of a dislocation line u with a Burgers vector $\boldsymbol{b}$, as shown in Fig. 8.3, one observes a rather large region of the displacement field around the dislocation $D1D2$. For different illuminated points, the displacement vectors $\boldsymbol{R}$ and the incident beam directions are different, and result in different diffracted and transmitted intensities. Hence, one observes corresponding contrast in $(h\ k\ l)$ diffracted and $(0\ 0\ 0)$ transmitted disks as shown in Fig.8.3. In Fig.8.3 the points $D1$ and $D2$ in the specimen have their corresponding points $D1'$ and $D2'$ in the $(0\ 0\ 0)$ transmitted disk and points $D1''$ and $D2''$ in the $(h\ k\ l)$ diffracted disk. Therefore, a defocus CBED pattern contains both diffraction and real space information of the illuminated region.

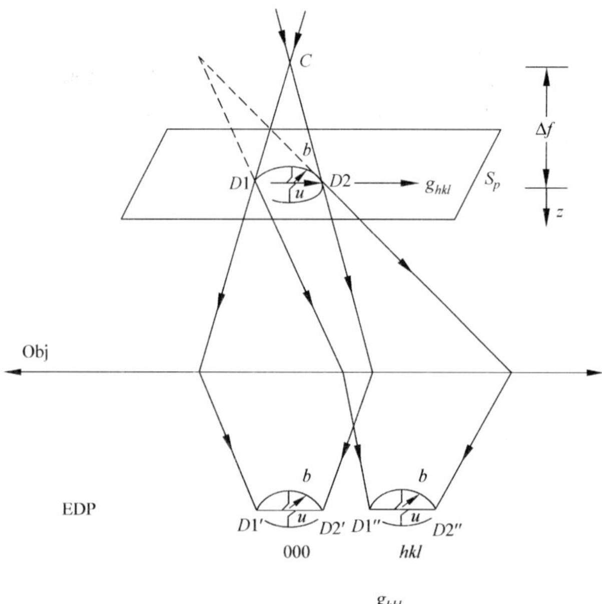

Fig. 8.3 Ray-path diagrams of defocus CBED

The angle range of a conventional CBED pattern is limited by the Bragg angle θ_{Bragg} of the nearest reflection. This angle range may be too small when a specimen with heavy elements, or small thickness, or large lattice constants is studied. A large angle CBED (LACBED) pattern consisting of nonoverlapping disks with an angle larger than $2\theta_{Bragg}$ may be obtained in a conventional TEM (Tanaka and Terauchi, 1985), as shown in Fig.8.4. In LACBED, the crossover O of the incident beam forms its first enlarged image O' at the position of the selector aperture. By using the height control of the side-entry goniometer, or adjusting the objective lens and condenser lens, the specimen is deviated from the crossover by a defocus amount of Δf and the diffracted beams will converge at the spots G, $\cdots$ which are at the same plane as the crossover O. This means the a diffraction pattern consisting of spots O, G, $\cdots$ forms at the plane which lies at the same height as the crossover O of the incident beam. This diffraction pattern is enlarged into the pattern consisting of the spots O', G', $\cdots$, which lie at the plane of the selector aperture. If the distance $O'G'=M_O 2\theta_{Bragg}\,\Delta f$, where M_O is the magnification of the objective lens, is larger than the diameter of the selector aperture, one can choose only one spot, e.g. O' or G', by selector aperture, and then switches over from magnification mode to diffraction mode. In this way a large-angle transmitted disk(0 0 0), or any large angle single diffracted disk, e.g., ($h\,k\,l$) disk marked in Fig.8.4, is observed on the fluorescent screen and recorded, without any disturbance from neighboring disks. Not that the (0 0 0) disk and its neighboring ($h\,k\,l$) disk are partly overlapped on the back focal plane of the objective lens, which is enlarged and projected onto the fluorescent screen at the diffraction mode.

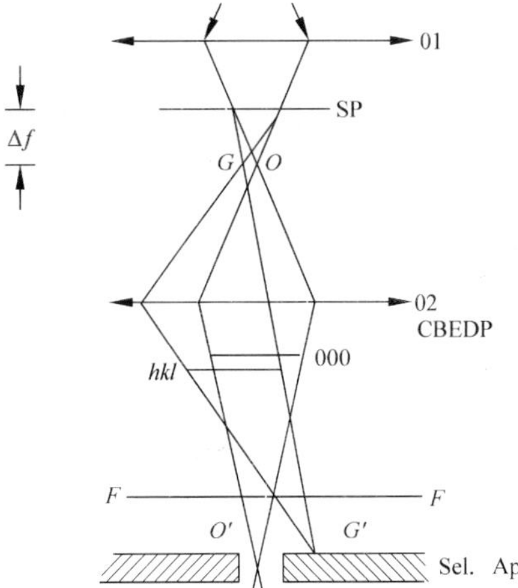

Fig. 8.4 Ray-path diagrams of large angle CBED

This LACBED technique is a special type of the defocus CBED technique. Therefore, it contains both diffraction information and real space information (shadow image) of the illuminated region. Figure 8.5 is a $[2\,1\,0]$ zone axis bright-field LACBED pattern taken from a cross-sectional specimen of a

358

Ge$_{0.5}$Si$_{0.5}$(8 nm thick) / Si (40 nm thick) strained-layer superlattice grown on Si substrate along its $[0\,0\,1]$ direction by molecular beam epitaxy (Duan, 1992). In Fig. 8.5, the diffraction pattern consists of thin higher-order reflections, of which one is marked as ($\bar{4}$ 8 0), and broad lower-order reflections. Besides, the shadow image of the GeSi / Si superlattice is also clearly seen on the upper part of the pattern, where the equidistant horizontal dark thinner bands are GeSi layers separated by broader Si layers.

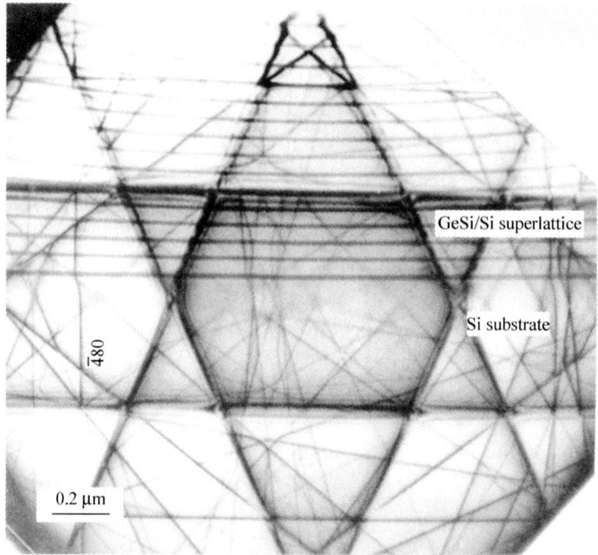

Fig. 8.5 $[2\,1\,0]$ zone axis large angle transmitted disk by using LACBED technique, taken from a cross-sectional specimen of a GeSi/Si superlattice grown on Si substrate

By using a selector aperture and at the standard diffraction mode, as described above, one obtains either a large-angle transmitted disk or a large angle single diffracted disk separately. If the excitation of the intermediate and projector lenses of the microscope can be controlled freely, one can focus the image at a plane *FF* as shown in Fig. 8.4. By this means a LACBED pattern is produced that consists of large-angle transmitted and diffracted disks without overlap. If a large flat perfect area is illuminated, one obtains a pure diffraction information with a large angle-range. For details the reader is referred to Tanaka and Terauchi (1985).

By using a newly developed slow-scan charge-coupled device (SSCCD) or imaging plate (IP), one can now record electron diffraction patterns digitally with a wide dynamic range and good linear response in intensity. Moreover, by using energy filtering techniques, either in-column type, for example, the Ω filter, equipped in the LEO 912 microscope and JEM-2010FEF microscope, or post-column type, namely Gatan imaging filter (GIF), it is possible now to obtain energy-filtered CBED patterns, for example, patterns formed mainly by elastically scattered electrons.Based on all these improvements in the instruments, one can refine the structural parameters by quantitatively comparing experimental intensity distribution in a CBED pattern with a theoretical one.

8.3 Convergent-Beam Electron Diffraction under Two-Beam Dynamic Condition—Determination of Foil Thickness and Extinction Distance

8.3.1 Two-Beam Dynamic Theory of Electron Diffraction

In the two-beam case, when only one strong reflection (hkl) is involved, the diffracted intensity I_D of the (hkl) reflection and the transmitted intensity I_T at the exit surface of a thin specimen with thickness t are expressed as (Hirsch et al., 1977)

$$I_D = 1 - I_T = \left(\frac{\pi t}{\xi_g}\right)^2 \frac{\sin^2(\pi x)}{(\pi x)^2} \tag{8.1}$$

with

$$x = t\sqrt{s^2 + 1/\xi_g^2} \tag{8.2}$$

where s is a deviation parameter describing the distance of the reciprocal point G deviated from the Ewald reflecting sphere as designated in Fig. 8.6. For the incident beam O_1O, the reciprocal point G lies exactly on the corresponding Ewald sphere S_1 and hence $s=0$. For the incident beam O_2O, which is deviated from O_1O by an angle $\Delta\theta$, the reciprocal point G lies in the interior of the corresponding Ewald sphere S_2 by a distance GG' and hence $s = GG' > 0$.

ξ_g in (8.1) and (8.2) is the extinction distance (Hirsch et al., 1977), which is related to the structure factor F_g and the Fourier component U_g of the crystal potential function for reciprocal vector g by the following expressions:

$$\xi_g = \frac{\pi V_c K \cos\theta_{\text{Bragg}}}{|F_g|} = \frac{K \cos\theta_{\text{Bragg}}}{|U_g|} \tag{8.3}$$

where V_c is the volume of the unit cell, $K \cong 1/\lambda$ is the modulus of the electron wave vector in the crystal after correction for the refractive index effect (Hirsch et al., 1977) and λ is the electron wavelength. The deviation parameter s is

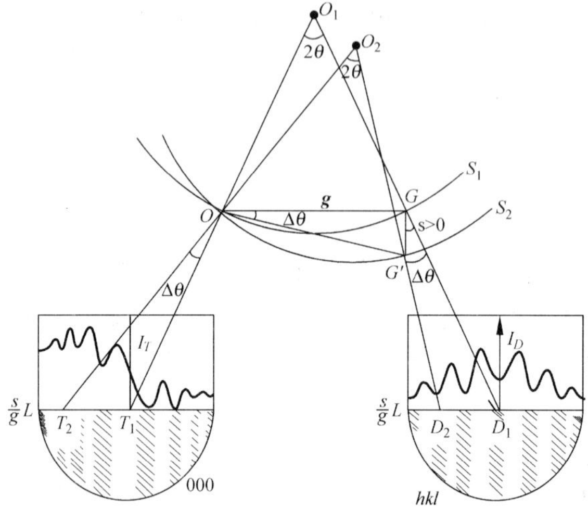

Fig. 8.6 Ewald reflecting sphere construction showing the relationship between intensity distributions in transmitted and diffracted disks and the rocking curve in two-beam case

related to the deviation angle $\Delta\theta$ from the Bragg condition by the following expression:

$$s =\mid \boldsymbol{g}\mid \Delta\theta = \lambda\mid \boldsymbol{g}\mid^{2} \Delta\theta\,/\,(2\theta_{\text{Bragg}}) \qquad (8.4)$$

The value of the function $\sin^{2}(\pi x)\,/\,(\pi x)^{2}$ in (8.1) vs. the parameter x is shown in Fig.8.7. When $x=0$, the function reaches its principal maximum value 1. When $x = n_k$ (n_k are nonzero integers), the function takes its k-th minimum value 0. When $x = x_k$ ($x_1=\pm1.431$, $x_2=\pm2.459$, $x_3=\pm3.471$, $\cdots$), the function reaches its

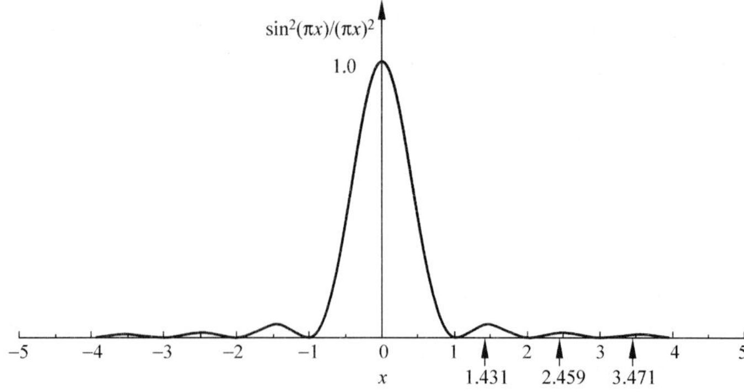

Fig. 8.7 Value of the function $\sin^{2}(\pi x)/(\pi x)^{2}$ vs. the parameter x

k-th secondary maximum value. The value of the function $\sin^2(\pi x)/(\pi x)^2$ attenuates rapidly with the increase of the modulus of x: its first secondary maximum is only 0.045 when $x_1=\pm 1.431$, its second secondary maximum is only 0.016 when $x_2=\pm 2.459$, its third secondary maximum is only 0.00834 when $x_3=\pm 3.471$, and its fourth secondary maximum is only 0.00503 when $x_4=\pm 4.477$.

The variation of the diffracted intensity I_D and the transmitted intensity I_T vs. the deviation parameter s, or equivalently vs. the deviation angle $\Delta\theta$ from the Bragg condition, is called a rocking curve. The concrete form of the rocking curve is determined by the values of the thickness t of the specimen, the extinction distance ξ_g and the modulus of the relevant reciprocal vector g. Figure 8.8 shows dark-field (diffracted beam) rocking curves for crystalline aluminum at different cases: (a) (1 1 1) reflection, $t=111.2$ nm, $\xi_g=55.6$ nm, $|g|=4.278$ nm^{-1}; (b) (2 0 0) reflection, $t=111.2$ nm, $\xi_g=67.3$ nm, $|g|=4.939$ nm^{-1}; (c) (5 3 1) reflection, $t=111.2$ nm, $\xi_g=279.8$ nm, $|g|=14.611$ nm^{-1}; (d) (5 3 1) reflection, $t=55.6$nm, $\xi_g=279.8$ nm, $|g|=14.611$ nm^{-1}. From Fig.8.8 it is clear that for higher-order reflections, when $t/\xi_g \leq 0.5$, the Bragg condition ($s=0$ and $\Delta\theta=0$) corresponds a maximal diffracted intensity and the first-, second-, $\cdots$ zero-diffracted intensity correspond to $x=1, 2, \cdots$, respectively, as shown in Figs. 8.8c and d. However, for lower-order reflections (1 1 1), (2 0 0), $\cdots$ the diffracted intensity at the Bragg condition may be maximum, as shown in Fig. 8.8b, or minimum, as shown in Fig. 8.8a, and the first-, second-, $\cdots$ zero- diffracted intensity may correspond to $x=2, 3, \cdots$ or $x=3, 4, \cdots$ depending on the value t/ξ_g. Table 8.1 lists the type (maximum or minimum) of the extreme value at $s=0$,and the sequential x values when I_D takes extreme values for different values t/ξ_g.

Table 8.1 Type of extreme values at $s=0$, and the sequential x values when I_D takes extreme values for different values of t/ξ_g

t/ξ_g	Type of extreme at $s=0$	Sequential N values when I_D takes extreme values						
$0\leq t/\xi_g<1$	max	1	1.431	2	2.459	3		$\cdots$
$1\leq t/\xi_g<1.431$	min	1.431	2	2.459	3	3.471		$\cdots$
$1.431\leq t/\xi_g<2$	max	2	2.459	3	3.471	4		$\cdots$
$2\leq t/\xi_g<2.459$	min	2.459	3	3.471	4	4.477		$\cdots$
$2.459\leq t/\xi_g<3$	max	3	3.471	4	4.477	5		$\cdots$

8.3.2 CBED Determination of Rocking Curves

Suppose O_1O in Fig.8.6 is the central beam of the incident convergent beam and satisfies the Bragg condition, namely, the reciprocal point G lies on the corresponding Ewald sphere S_1. This beam forms spot T_1 at the center of the (0 0 0) transmitted disk and its corresponding diffracted beam forms spot D_1 at the center of the ($h\,k\,l$) diffracted disk. The distance $T_1 D_1$ corresponds to the Bragg angle $2\theta_{\text{Bragg}}$. $T_1D_1 = 2\,\theta_{\text{Bragg}}L = \lambda|g|L$, where L is the effective column

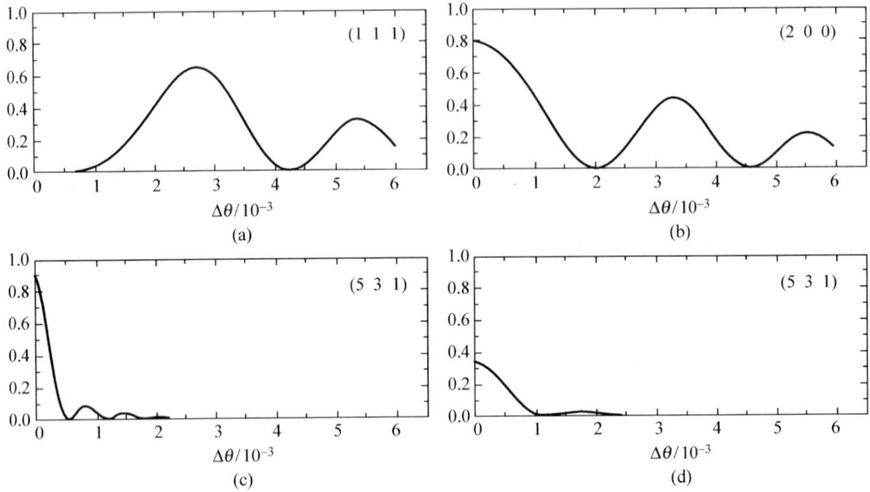

Fig. 8.8 Dark-field (diffracted beam)rocking curves for crystalline aluminum under different conditions

(**a**) (1 1 1) reflection, t=111.2 nm, ξ_g =55.6nm, $|g|$=4.278 nm^{-1};

(**b**) (2 0 0) reflection, t=111.2 nm, ξ_g =67.3 nm, $|g|$=4.939 nm^{-1};

(**c**) (5 3 1) reflection, t=111.2 nm, ξ_g =279.8 nm, $|g|$=14.611 nm^{-1};

(**d**) (5 3 1) reflection, t=55.6 nm, ξ_g =279.8 nm, $|g|$=14.611 nm^{-1}

length. Suppose O_2O is another incident beam that deviates from O_1O by an angle $\Delta\theta$ in the O_1OG plane. In this case the reciprocal point G is deviated from the Ewald sphere S_2 by $s=GG'$. The incident beam O_2O then forms the spot T_2 in the transmitted disk and spot D_2 in the diffracted disk. $T_1T_2=D_1D_2=\Delta\theta L=sL/|g|$. The above discussion indicates that the intensity variation along the straight line T_1T_2 corresponds exactly to the intensity variation of the transmitted beam versus the deviation parameter s, namely, the bright-field rocking curve. Similarly, the intensity variation along the straight line D_1D_2 corresponds exactly to the intensity variation of the diffracted beam versus the deviation parameter s, namely, the dark-field rocking curve.

Now we rotate the incident beam and its corresponding Ewald sphere around the reciprocal vector g =OG and suppose that no other strong reflection is met. Such a rotation does not change the deviation parameter s. This indicates that the transmitted intensities in the (0 0 0) disk along the direction perpendicular to T_1T_2 are equal and the diffracted intensities in the ($h\ k\ l$) disk along the direction perpendicular to D_1D_2 are also equal. Thus both the bright-field and the dark-field disks all consist of broad parallel fringes and the intensity distributions perpendicular to these fringes are exactly the bright-and dark-field rocking curves.

Figure 8.9 shows a CBED pattern (Wang et al., 1994) taken at an approximate two-beam condition in which the reflection (664400) of an icosahedral quasicrystal is strongly excited. The physical component of the (664400) reciprocal vector g possesses a modulus of $|g|=7.846$ nm^{-1} and is along a two-fold axis. As described above, this CBED pattern consists mainly of broad parallel fringes and the intensity distributions perpendicular to these fringes are exactly the bright-and-dark field rocking curves. In addition to the transmitted and (664400) diffracted disks we also see a (442200) diffracted disk lying in between. Moreover, some oblique lines belonging to other reflections can also be seen. This indicates that the so-called two-beam condition is only an approximation.

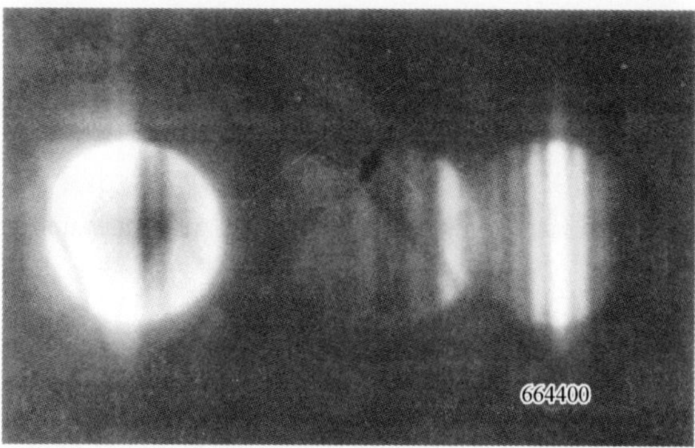

Fig. 8.9 (664400)two-beam CBED pattern from the Al$_{71}$Pd$_{19}$Mn$_{10}$ icosahedral quasicrystal

8.3.3 CBED Determination of Foil Thickness and Extinction Distance

By adjusting the related parameters, i.e., specimen thickness t, the magnitudes and phase angles of some extinction distances ξ_g and anomalous absorption distances $\xi_g{}'$, one can reach a best coincidence between the calculated and experimentally recorded quantitative rocking curves, and hence accurately determine the specimen thickness and the extinction distances. For more details see Sect. 8.7. However, for routine work, we need only to use the following much simpler method.

From (8.2) and Table 8.1, the deviation parameter s_i at which the diffracted intensity I_D reaches the i-th minimum value may be expressed as

$$\left(\frac{S_i}{n_i}\right)^2 = -\left(\frac{1}{\xi_g}\right)^2\left(\frac{1}{n_i}\right)^2 + \left(\frac{1}{t}\right)^2 \tag{8.5}$$

where the sequential values of n_1, n_2, n_3, $\cdots$ may be 1, 2, 3, $\cdots$, or 2, 3, 4, $\cdots$, or 3, 4, 5, $\cdots$, determined by the ratio of t/ξ_g, as shown in Table 8.1. The s_i values may be measured from the CBED pattern by using (8.4) where the ratio $\Delta\theta/(2\theta_{Bragg})$ is equal to the ratio of the corresponding distances measured from the CBED pattern. The sequential values of n_i may be determined by trial and error. For a set of correct values of n_i, the plot of $(s_i/n_i)^2$ against $(1/n_i)^2$ should result in a straight line with a negative slope of $-(1/\xi_g)^2$ and an intercept of $(1/t)^2$. Figure 8.10 shows a straight line plot of $(s_i/n_i)^2$ against $(1/n_i)^2$ for the CBED pattern shown in Fig. 8.9. In Fig. 8.10 the sequential values of 1, 2, 3 and 4 are taken for n_1, n_2, n_3, and n_4. From this plot the effective specimen thickness $t=105.4$ nm and the extinction distance $\xi_{(664400)}=124.7$ nm are determined. From these measured values we have the ratio $t/\xi_g<1$ which is in accordance with the selection of $n_1=1$.

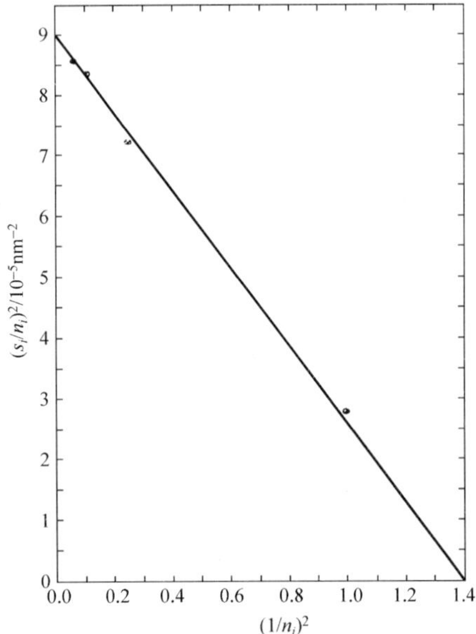

Fig. 8.10 Straight line plot of $(s_i/n_i)^2$ against $(1/n_i)^2$ for the CBED pattern shown in Fig.8.9. From this plot the effective specimen thickness $t=105.4$ nm and the extinction distance $\xi_{(664400)}=124.7$ nm are determined

8.4 Convergent-beam Electron Diffraction under Kinematic Condition—Higher-Order Laue Zone Lines

8.4.1 Formation of HOLZ Lines

There is a set of reciprocal lattice planes perpendicular to a given $[u\,v\,w]$ zone axis. The reflections produced by the reciprocal points lying on the reciprocal lattice plane passing through the origin are usually called zeroth-order Laue zone (ZOLZ) reflections. Reflections produced by the reciprocal points lying on the first, or second, ... reciprocal lattice plane are usually called first-, or second-, ... order Laue zone (FOLZ, or SOLZ, ...) reflections. They are all called higher-order Laue zone (HOLZ) reflections.

Figure 8.11 is an Ewald sphere construction showing the formation of HOLZ lines. S_1 is an Ewald sphere corresponding to the central incident beam O_1O, which is parallel to the $[u\,v\,w]$ zone axis and produces a transmitted spot T in the (0 0 0) bright-field disk. Suppose the Ewald sphere S_2 corresponding to the incident beam O_2O passes exactly through a FOLZ reciprocal point G and the beam O_2O is deviated from the central incident beam O_1O by an angle α. Obviously, the modulus $|g|$ of the reciprocal vector g corresponding to G and its extinction distance ξ_g are all large. Therefore, the ratio t/ξ_g must be less than 1, for example, $t/\xi_g \leqslant 0.5$ and hence the Bragg condition $s=0$ corresponds to the principal maximum of the rocking curve with a higher intensity than the secondary maxima, see Figs. 8.8c and d. Moreover, from (8.2) and Table 8.1 we know that the width of the principal maximum is $\Delta s = 2(1-(t/\xi_g)^2)^{1/2}/t$ in the reciprocal space which corresponds to an angle width of $\Delta\vartheta = \Delta s/|g|$. Thus the angle width of the HOLZ reflections with large $|g|$ is very narrow, and the larger the $|g|$ is, the narrower the peak of the HOLZ reflection will be. Such a situation may be seen by comparing Fig. 8.8b for small $|g|$ with Fig. 8.8c for large $|g|$. Thus the incident beam O_2O produces a bright diffracted spot D in the $(h\,k\,l)$ disk and a corresponding dark transmitted spot H in the (0 0 0) disk. Now we rotate the incident beam O_2O around the reciprocal vector $g=OG$, then the diffracted spot D is extended to a straight reflection line DD' and the transmitted spot H is extended to a deficient dark straight line HH'. Usually DD' is called a HOLZ reflection and HH' is called a HOLZ line. They are all perpendicular to the projection of the reciprocal vector on the CBED pattern.

From Fig. 8.11 we see that the distance of the $(h\,k\,l)$ HOLZ line HH' from the central point T of the (0 0 0) bright-field disk is $HT = \alpha L$ with L being the effective column length. The deviation angle α from the central incident beam O_1O may be expressed as

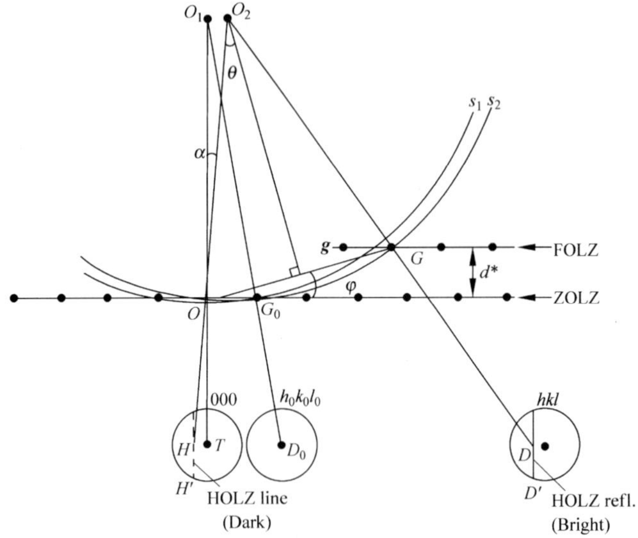

Fig. 8.11 Ewald sphere construction showing the principle of the formation of HOLZ lines

$$\alpha = \theta_{B\mathrm{ragg}} - \phi \tag{8.6}$$

where θ_{Bragg} is the Bragg angle of $(h\,k\,l)$ reflection and satisfies the following Bragg rule:

$$2\sin\theta_{\mathrm{Bragg}} \,/\, |\,\boldsymbol{g}_{hkl}\,| = \lambda \tag{8.7}$$

ϕ is the angle between the reciprocal vector $\boldsymbol{g}_{hkl}$ and the ZOLZ plane and satisfies the following equation:

$$\sin\phi = d^{*}_{uvw} \,/\, |\,\boldsymbol{g}_{hkl}\,| = \frac{\boldsymbol{r}_{uvw} \cdot \boldsymbol{g}_{hkl}}{|\,\boldsymbol{r}_{uvw}\,|\,|\,\boldsymbol{g}_{hkl}\,|} \tag{8.8}$$

where $\boldsymbol{r}_{uvw}$ is the lattice vector along the $[u\,v\,w]$ zone axis and d^{*}_{uvw} is the distance between FOLZ and ZOLZ reciprocal lattice planes measured along the direction of the $[u\,v\,w]$ zone axis.

In a CBED pattern with large convergence angle, all the FOLZ reflections nearly form a circle. From Fig. 8.12 we see that the angular radius $2\phi_{\mathrm{FOLZ}}$ of the circle is determined by the reciprocal lattice plane distance d^{*} and the radius $1/\lambda$ of the Ewald sphere: $2\phi_{\mathrm{FOLZ}} = 2d^{*}/\,|g|_{\mathrm{FOLZ}} = |g|_{\mathrm{FOLZ}}\,/\,(1/\lambda)$. Hence, we obtain the radius $|g|_{\mathrm{FOLZ}}$ of the FOLZ circle in reciprocal space,

$$|\mathbf{g}|_{\text{FOLZ}} = \sqrt{\frac{2d^*}{\lambda}} \qquad (8.9)$$

and the corresponding angular radius $2\phi_{\text{FOLZ}}$:

$$2\phi_{\text{FOLZ}} = \sqrt{2d^*\lambda} \qquad (8.10)$$

For SOLZ or other HOLZ reflections there are similar expressions like (8.9) and (8.10) by substituting relevant reciprocal plane distance for d^*.

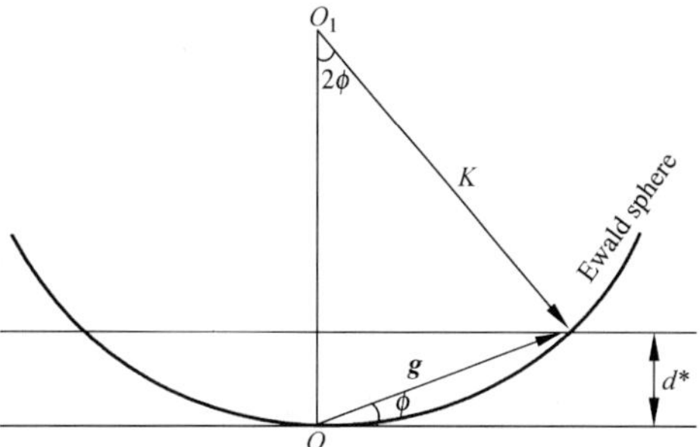

Fig. 8.12 Ewald sphere construction showing the derivation of the expressions for the radius of the FOLZ reflection circle

8.4.2 Indexing and Computer Simulation of HOLZ Line Patterns

Figure 8.13a is an experimental CBED pattern along $[1\ 1\ 3]$ zone axis of aluminum at an accelerating voltage of 120 kV. The geometrical distribution of HOLZ lines in this pattern may be calculated .We first draw a conventional electron diffraction pattern (EDP) of the $[1\ 1\ 3]$ zone axis as shown in Fig. 8.13b. In Fig. 8.13b the {2 2 0} and {4 2 2} type reflection spots belong to the ZOLZ and the {5 5 3}, {7 3 1}, {7 3 3}, {7 5 1} and {9 3 1} type reflection spots belong to FOLZ. These spots are also the projection of the reciprocal points onto the EDP plane. According to (8.9) we can estimate the approximate length of the FOLZ reciprocal vectors to be $(73/a^2)^{1/2}$ for 120 kV and $(66/a^2)^{1/2}$ for 100 kV which correspond to all the FOLZ reflections shown in Fig. 8.13b. Then we calculate the angular deviation α from the central incident beam for all the reflections shown in Fig. 13b according to (8.6), (8.7) and (8.8) and list them in Table 8.2. Now we can draw all these HOLZ lines on the principle that: (1) each HOLZ line is orthogonal to its reciprocal vector, namely, the straight line connecting the origin and the related reciprocal point in Fig. 8.13b; and (2) the HOLZ line is deviated from the center of the pattern by a distance proportional to α. The HOLZ line lies at the opposite side (or the same side) from the pattern center with respect to the reflection spot when α is positive (or negative). By this method we obtain Fig. 8.13c for 120 kV and Fig. 8.13d for 100 kV.

Table 8.2 Angular deviation α from the central incident beam for all the reflections shown in Fig. 8.13b

{hkl}	$\alpha/(2\theta_{220})$	
	100 kV	120 kV
220	0.5	0.5
422	0.866	0.866
553,731	-0.16	-0.32
733	0.02	-0.13
751	0.18	0.04
931	0.46	0.34

From (8.6), (8.7), and (8.8) it is clear that by changing accelerating voltage (and hence wavelength λ) and/or any one of the lattice constants a_0, b_0, c_0, α_0, β_0, γ_0, the deviation angle α for the given ($h\ k\ l$) reflection changes. Therefore, the geometrical distribution of the HOLZ line pattern changes by changing either the accelerating voltage or lattice constants. Figures 8.13c and d show clearly the influence of the accelerating voltage on the HOLZ line pattern.

There are some computer programs for simulating HOLZ line patterns (Tanaka and Terauchi, 1985; Wang et al., 1986; Jiao et al., 1987). First we input all the necessary data, including accelerating voltage, angular range α_{max} of the convergent incident beam, lattice constants and structural parameters of the specimen, and the zone axis $[u\ v\ w]$. Then we select indices of the reflections ($h\ k\ l$) sequentially in a given range, calculate their α angles, and remove the reflections whose α values are outside the given range ($|\alpha|>\alpha_{0max}$). If the number of the remaining reflections is too large, then calculate the structure factors and remove the weaker reflections with lower moduli of the structure factors.

Figures 8.13c and d can also be obtained by computer simulation.

The prerequisites of the above discussion are: (1) the ratio $t/\xi_g \leqslant 0.5$, (2) the modulus $|g|$ of the reciprocal vector g is large and (3) only one single reflection is excited. These are two-beam kinematical approximations. Near an intersection of two HOLZ lines with indices ($h_1k_1l_1$) and ($h_2k_2l_2$), three beams are related. If the reflection ($h_1-h_2\ k_1-k_2\ l_1-l_2$) is a strong reflection, then the coupling between reflections ($h_1k_1l_1$) and ($h_2k_2l_2$) may be strong enough to form two hyperbolas instead of two intersecting straight lines (Jones et al., 1977).

8.4.3 Factors Affecting the Quality of a HOLZ Line Pattern

It is not always possible to observe a HOLZ line pattern with good quality at any orientation for any specimen. The quality of a HOLZ line pattern is determined by the following factors.

369

(1) Intensity of the HOLZ reflection

In general, one can only observe the HOLZ lines with intensities large enough. According to (8.1) and (8.2), the peak intensity (when $s = 0$) of a HOLZ reflection with the extinction distance ξ_g is expressed as

$$I_g(s = 0) = \sin^2(\pi t / \xi_g) \tag{8.11}$$

Under the condition that the ratio $t/\xi_g < 0.5$, $I_g(s{=}0)$ increases with the thickness t of the specimen and the modulus of the structure factor $F_g = \pi V_c K \cos \theta_{\mathrm{Bragg}} / \xi_g$ of the g reflection. For a given zone axis of a given specimen, one can select the experimental conditions as follows: (a) A lower accelerating voltage to obtain a larger wavelength λ and hence smaller $|g|_{\mathrm{FOLZ}}$ value according to (8.9). Usually, the reflection with a smaller $|g|$ value possesses a larger structure factor F_g and

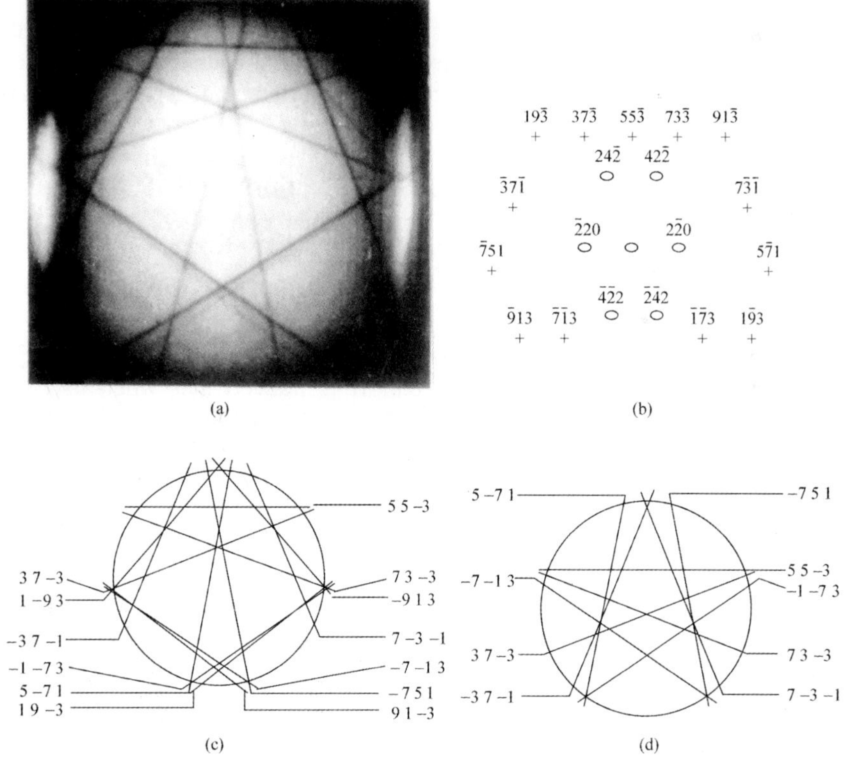

Fig. 8.13 HOLZ line pattern of Al along $[1\,1\,3]$ zone axis and its calculation.

(a) Experimental CBED pattern taken at an accelerating voltage 120 kV.

(b) Conventional EDP containing both ZOLZ and FOLZ spots. (c) Calculated HOLZ line pattern for 120 kV. (d) Calculated HOLZ line pattern for 100 kV

hence a larger intensity. (b) A lower experimental temperature to decrease the thermal vibration of the atoms in the specimen and hence to increase the reflection intensity. On the other hand, under a given accelerating voltage and a given temperature, one may select zone axes with lower symmetry or even off-zone axis orientation to obtain a CBED pattern with clearer HOLZ lines. For example, at high symmetry zone axes $\{1\ 0\ 0\}$, $\{1\ 1\ 1\}$ and $\{1\ 1\ 0\}$ of an aluminum crystal, one cannot observe clear HOLZ lines at room temperature. While at low symmetry zone axis, such as $[1\ 1\ 3]$, one observes HOLZ lines with good quality, as shown in Fig. 8.13a.

(2) Imperfection of the specimen and the beam size

According to (8.6)–(8.8), the rotation of lattice planes and the variation of lattice constants will cause the HOLZ lines to be shifted. On the other hand, a strain field in a column traveled by the electron beam will cause the HOLZ lines to be shifted and split; for details, see Sect. 8.8 .Imperfection of the illuminated area of the specimen may produce a spectrum of the local rotation of lattice planes, and of the local change of lattice constants, which may lead to a broadening and even invisibility of the HOLZ line. Obviously, a smaller size of the incident beam may decrease the broadening affect and hence improve the quality of the HOLZ line pattern.

8.4.4 Some Applications of HOLZ Reflections and HOLZ Lines

8.4.4.1 Determination of Lattice Constants of Micro-areas

Since the position of a HOLZ line is affected by the lattice constants, one can determine the lattice constants of the illuminated micro-area by comparing the experimental and simulated HOLZ line patterns. The factors affecting the HOLZ line positions are discussed in Sect. 8.8. It is noticed that when using the simulation method described in Sect. 8.4.2, one supposes the kinematical approximation. The dynamical interaction effect may be reduced by using lower symmetry zone axis patterns, e.g., $[1\ 1\ 3]$ axis for Al, or corrected by introducing an effective accelerating voltage. Take an experimental HOLZ line pattern from a standard specimen, for example, from a perfect area with known lattice constants. Then obtain a best consistency between the experimental HOLZ line pattern and the simulated one by adjusting the effective value of the wavelength λ. In order to determine unknown lattice constants in other micro-areas, experimental HOLZ line patterns must be taken at the same condition and be simulated by using the same effective wavelength value.

When the gradient of a displacement field is not too large, one can determine the strain field by taking experimental HOLZ line patterns from different micro-areas and measuring lattice constants of these micro-areas; for details see Sect. 8.8.1.

By using a known relationship between chemical composition and lattice constants one can determine the spatial variation of chemical composition on the microscale by measuring lattice constants in a series of micro-areas.

8.4.4.2 Determination of Crystallographic Symmetry

A diffraction intensity distribution contributed by ZOLZ reflections is determined by the projection of the crystal potential along the incident beam direction while the diffraction intensity distribution contributed by both ZOLZ and HOLZ reflections is determined by the three-dimensional potential of the specimen. Therefore, the symmetry of the diffraction intensity distribution contributed by ZOLZ reflections reflects the projection symmetry, while the symmetry of the diffraction intensity distribution contributed by both ZOLZ and HOLZ reflections reflects the true symmetry of the specimen. Thus one must analyze the symmetry of the HOLZ reflections and HOLZ lines when determining the specimen symmetry by using CBED pattern; for details see Sect. 8.5.

8.4.4.3 Determination of Crystal Structure Parameters

By using (8.9) and (8.10) it is possible to determine the spacing of the reciprocal lattice planes along the incident beam direction. Such information together with the information in ZOLZ pattern may help us to characterize the crystal system and crystal lattice of the specimen.

Generally, the intensities of HOLZ reflections are more sensitive to the atomic positions than those of ZOLZ reflections (Tanaka et al., 1994). Therefore, HOLZ reflections may be used to refine the atomic positions of a crystal with known structure. For details see Sect. 8.7 and the original work (Tanaka et al., 1994).

8.4.4.4 Determination of Characteristic Parameters of Defects

There is a strain field in the neighborhood of a defect, for example, a stacking fault or a dislocation line, and the strain field will induce shifting, splitting and twisting of HOLZ lines and HOLZ reflections. This influence is dependent on the indices of the relevant HOLZ reflections and the characteristic parameters of the defect, e.g., the shift vector of a stacking fault, or the Burgers vector of a dislocation line. Therefore, one can determine the characteristic parameters of defects by the behavior of the shifting, splitting and twisting of HOLZ lines and HOLZ reflections, induced by the relevant defect. In Sect. 8.6 we will describe the principle, method and some examples of determining Burgers vectors of dislocations by CBED technique. For stacking faults and twin boundaries see Tanaka et al. (1994) and Feng et al. (1995).

8.5 Convergent-Beam Electron Diffraction Determination of Crystal Symmetry

By using the reciprocity principle and the diagram illustration method, Goodman (1975) systematically studied the symmetry of a CBED pattern caused by the symmetry elements of an inversion center, a horizontal mirror, or a horizontal two-fold axis. Then, by using quantum mechanics and the diagram illustration method Buxton et al. (1976) systematically derived the symmetry of a CBED pattern caused by the possible crystallographic symmetry elements. They listed thirty one diffraction groups that may be determined by CBED at a given orientation of the specimen, and founded the method of determining crystallographic point groups by the CBED technique. Steeds and Vincent (1983), Tanaka et al. (1983), Tanaka and Terauchi (1985), and Tanaka et al. (1988) improved the method of determining crystallographic point groups by the CBED technique. Tanaka et al. (1988, 1994) further extended the CBED technique to the determination of quasicrystallographic point groups.

8.5.1 Symmetry Elements That Can Be Determined by CBED Technique

In symmetry determination, we need to observe the symmetries of the bright field (BF), dark field (DF), whole pattern (WP), and $\pm G$, for some orientations of the specimen. Thus, we need to orient the specimen so that the central incident beam is parallel to a $[u\,v\,w]$ zone axis and record the CBED pattern, which is called the zone axis pattern and consists of (0 0 0) transmitted disk and many $(h\,k\,l)$ diffracted disks. The (0 0 0) disk in a zone axis pattern is called bright field.The symmetry of the zone axis pattern is called whole-pattern (WP) symmetry, which is the same as the two-dimensional point group of the $[u\,v\,w]$ oriented specimen. A horizontal two-fold axis or a horizontal mirror will cause some new symmetry in BF symmetry, so that BF symmetry may be higher than the WP symmetry. When the center of the $(h\,k\,l)$ diffracted disk corresponds to the Bragg condition $(s=0)$, then the $(h\,k\,l)$ disk is called DF and the symmetry in the $(h\,k\,l)$ disk is called DF symmetry. The symmetry between two dark-field patterns, one corresponds to the reciprocal vector g satisfying the Bragg condition, and the other, corresponding to $-g$, is called $\pm G$ symmetry.

Table 8.3 lists BF, WP, DF and $\pm G$ symmetries in a CBED pattern caused by different symmetry elements. In Table 8.3, DF symmetry in the special case indicates the symmetry when the reciprocal vector g is parallel to the vertical mirror m or the horizontal two-fold axis 2_h. The $\pm G$ symmetry in special cases indicates the symmetry when the reciprocal vector g is perpendicular to the vertical mirror m or the horizontal two-fold axis 2_h. Now we interpret Table 8.3 by taking the horizontal two-fold axis 2_h as an example.

Table 8.3 BF, WPDF and $\pm G$ symmetries in a CBED pattern caused by different symmetry elements

Symmetry element	BF	WP	DF		$\pm G$	
			General	Special [*)]	General	Special [**)]
1	1	1	1	–	1	–
2	2	2	1	–	2	–
4	4	4	1	–	2	–
3	3	3	1	–	1	–
6	6	6	1	–	2	–
m	m_v	m_v	1	m_v	1	m_v
m_h	2	1	2	–	1	–
$\bar{1}$	1	1	1	–	2_R	–
2_h	m_2	1	1	m_2	1	m_R
$\bar{4}$	4	2	1	–	2	–

[*)] when $g // m$ or $2h$;

[**)] when $g \perp m$ or 2_h.

Figure 8.14 interprets symmetries in CBED patterns caused by a horizontal two-fold axis 2_h. Figure 8.14a illustrates the m_2 symmetry in BF, that is, in the transmitted disk caused by 2_h. T indicates an incident and transmitted beam. According to the reciprocity theorem, the intensity of such a transmitted beam T is equal to that of the transmitted beam T_R, where T_R is antiparallel to T. Then we use the 2_h symmetry, namely, rotate the T_R beam around the 2_h axis by $180°$ and obtain a transmitted beam T_{2h}, whose intensity should be the same as T_R, and hence the same as T. This is a m_R symmetry, that is, after a composite operation consisting of a reflection through the mirror m (which is parallel to the 2_h axis) and a $180°$ rotation around the center of the transmitted disk, the intensity does not change. This symmetry is called m_2 symmetry (Tanaka et al., 1983 and 1985), namely, a reflection through the mirror m_2, as shown in Fig. 8.14(a), which is perpendicular to the 2_h axis.

Figure 8.14b illustrates the m_2 symmetry in $(h\ k\ l)$ DF caused by 2_h, under a special case when the 2_h-axis is parallel to the reciprocal vector g_{hkl}. According to the reciprocity theorem, the intensity of the diffracted beam D, which corresponds to the transmitted beam T, is the same as that of the diffracted beam D_R, which is anti-parallel to the beam T. In this case, the diffracted beam D_R corresponds to the transmitted beam T_R, which is antiparallel to the beam D, as shown in Fig.8.14b. Then we use the 2_h symmetry, namely, rotate the T_R and D_R beams around the 2_h axis by $180°$ and obtain a transmitted beam T_{2h} and a diffracted beam D_{2h}. The intensity of the beam D_{2h} should be the same as D_R, and hence the same as D. The equality of the intensities of the D and D_{2h} beams is a m_R symmetry, that is, after a composite operation consisting of a reflection through the mirror m (which is parallel to the 2_h axis) and a $180°$ rotation around the center of the DF disk, the intensity does not change. This symmetry is also called m_2 symmetry (Tanaka et al., 1983; 1985), namely, a reflection through the mirror m_2 shown in Fig. 8.14b which is perpendicular to the 2_h axis.

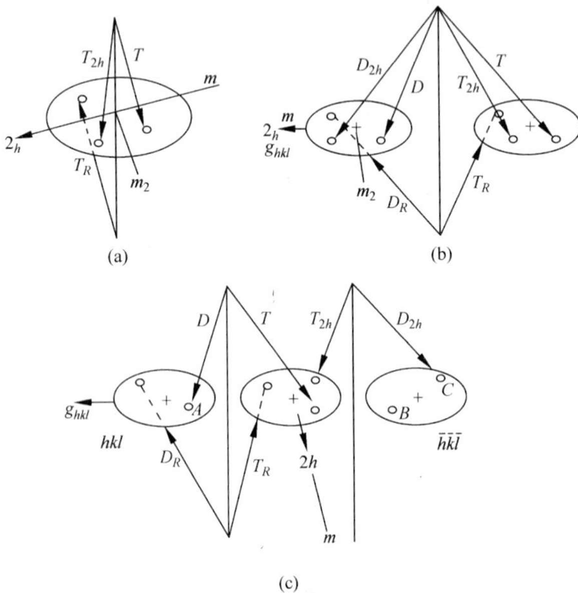

Fig. 8.14 Schematic diagrams explaining the symmetries in CBED patterns caused by 2_h axis. (**a**) general BF symmetry m_2; (**b**) special DF symmetry m_2 when $2_h // g$; (**c**) special $\pm G$ symmetry m_R when $2_h \perp g$

Figure 8.14c illustrates the m_R symmetry between $\pm G$ caused by 2_h, under a special case when 2_h axis is perpendicular to the reciprocal vector g_{hkl}. Here the meanings of T, D, T_R, D_R, T_{2h} and D_{2h} are the same as before. The equality of the intensities of the D and D_{2h} beams is a m_R symmetry, that is, after a composite operation consisting of a reflection through the mirror m (which is parallel to the 2_h axis), which transforms the spot A into spot B, and a $180°$ rotation around the center of the DF disk, which transforms the spot B into spot C, the intensity does not change.

8.5.2 Thirty One Diffraction Groups and Their CBED Determination

There are 31 groups consisting of all the symmetry elements listed in Table 8.3. These symmetry elements may be determined experimentally by the CBED technique. These 31 diffraction groups are listed in Table 8.4 and are derived as follows: From the 32 crystallographic point groups one must discard 5 cubic point groups because they possess symmetry axes oblique to the zone axis and cannot be determined by CBED technique. From the remaining 27 point groups there are four groups 2, *m*, 2*mm* and 2/*m* whose principal axis may be vertical or horizontal. Therefore, from the point of view of diffraction groups which may be experimentally determined by CBED technique, each of these groups corresponds to two groups: 2 and 2_h (i.e., m_R); *m* and m_h (i.e., 1_R); 2*mm* and $2_h mm_h$ (i.e., $m1_R$); 2/*m* (i.e., 21_R) and $2_h/m$ (i.e., $2_R mm_R$).

Table 8.4 Thirty-one diffraction groups and their BF, WP, DF, and $\pm G$ symmetries

Diffraction group	Principal symmetry elements	BF	WP	DF		$\pm G$		Projection group
				General	Special	General	Special	
1	1	1	1	1		1		
1_R	m_h	$2(1_R)$	1	$2=1_R$		1		1_R
2	2	2	2	1		2		
2_R	$\bar{1}$	1	1	1		2_R		21_R
21_R	$2/m_h$	2	2	2		21_R		
m_R	2_h	$m(m_2)$	1	1	m_2	1	m_R	
m	m	m_v	m_v	1	m_v	1	m_v	
$m1_R$	$2_h mm_h$	$2mm$ $(m_v+m_2+ (1_R))$	m_v	2	$2\,m_v\,m_2$	1	$m_v 1_R$	$m1_R$
$2m_R m_R$	$22_h 2_h$	$2mm(2+ m_2)$	2	1	m_2	2	$2\,m_R(m_2)$	
$2mm$	$2mm$	$2m_v m_v'$	$2m_v m_v'$	1	m_v	2	$2m_v'(m_v)$	
$2_R mm_R$	$2_h/m$	m_v	m_v	1	m_2 m_v	2_R	$2_R\,m_v(m_2)$ $2_R\,m_R(m_v)$	$2mm1_R$
$2mm1_R$	$2_h/mm$ m_h	$2m_v m_v'$	$2m_v m_v'$	2	$2m_v m_2$	21_R	$21_R\,m_v'(m_v)$	
4	4	4	4	1		2		
4_R	$\bar{4}$	$4(4_R)$	2	1		2		41_R
41_R	$4/m_h$	4	4	2		21_R		
$4m_R m_R$	$42_h 2_h$	$4mm$ $(4+m_2)$	4	1	m_2	2	$2m_R(m_2)$	
$4mm$	$4mm$	$4m_v m_v'$	$4m_v m_v'$	1	m_v	2	$2m_v'(m_v)$	
$4_R mm_R$	$\bar{4}\,m2_h$	$4mm$ $(2m_v m_v' +m_2)$	$2m_v m_v'$	1	m_2 m_v	2	$2m_R(m_2)$ $2m_v'(m_v)$	$4mm1_R$
$4mm1_R$	$4/m_h mm$	$4m_v m_v'$	$4m_v m_v'$	2	$2m_v m_2$	21_R	$21_R m_v'(m_v)$	
3	3	3	3	1		1		
31_R	$\bar{6}=3/m_h$	$6(3+1_R)$	3	2		1		31_R

(Continued)

Diffraction group	Principal symmetry elements	BF	WP	DF		$\pm G$		Projection group
				General	Special	General	Special	
$3m_R$	32_h	$3m(3+m_2)$	3	1	m_2	1	m_R	
$3m$	$3m$	$3m_v$	$3m_v$	1	m_v	1	m_v	$3m1_R$
	$\bar{6}\,m2_h$	$6mm$						
$3m1_R$	$=3/m_h$	$(3m_v+m_2$	$3m_v$	2	$2\,m_v\,m_2$	1	m_v1_R	
	$m2_h$	$+(1_R))$						
6	6	6	6	1		2		
6_R	$\bar{3}$	3	3	1		2_R		61_R
61_R	$6/m_h$	6	6	2		21_R		
$6m_Rm_R$	62_h2_h	$6mm$ $(6+m_2)$	6	1	m_2	2	$2m_R(m_2)$	
$6mm$	$6mm$	$6m_vm_v'$	$6m_vm_v'$	1	m_v	2	$2m_v'(m_v)$	
6_Rmm_R	$\bar{3}\,m$	$3m_v$	$3m_v$	1	m_2 m_v	2_R	$2_R\,m_v'(m_2)$ $2_R\,m_R(m_v)$	$6mm1_R$
$6mm1_R$	$6/m_hm$ m	$6m_vm_v'$	$6m_vm_v'$	2	$2m_v\,m_2$	21_R	$21_R\,m_v'(m_v)$	

In Table 8.4, the thirty-one diffraction groups, by using the symbols (Buxton et al., 1976), are listed in the first column. The second column lists their Hermann-Mauguin symbols, which are more familiar to crystallographers. The third and fourth columns are BF and WP symmetries in CBED patterns. When BF symmetry is higher than the WP symmetry, the symmetry elements that cause this additional symmetry in BF are listed in the parentheses in column 3. The fifth column lists the DF symmetries, where the symmetry m_v is a special symmetry caused by the vertical mirror m when it is parallel to the reciprocal vector g. And the symmetry m_2 is a special symmetry caused by the horizontal 2_h axis when it is parallel to g. The sixth column lists the $\pm G$ symmetries, where the special symmetry m_v is caused by the vertical mirror m when it is perpendicular to g and the special symmetry m_R is caused by the horizontal 2_h axis when it is perpendicular to g. The last column lists the projection symmetry, namely, the symmetry of the crystal potential integrated along the zone axis. Thus a projection structure possesses a horizontal mirror and, hence, a projection group is obtained by adding a horizontal mirror onto the diffraction group.

By consulting Table 8.4 one can determine the diffraction group of a specimen along a given zone axis $[u\,v\,w]$.

Table 8.5 Relationship between diffraction groups and point groups. Here, P. G. stands for the point groups, and D.G. for the diffraction groups. The entire table is separated into 3 tables, since it is too large to be put on one page

P.G.		1	$\bar{1}$	2	m	$2/m$	222	$mm2$	mmm
D.G.									
1	1	$[uvw]$		$[uvw]$	$[uvw]$		$[uvw]$	$[uvw]$	
m_h	1_R					$[010]$			
2	2			$[010]$					
$\bar{1}$	2_R		$[uvw]$			$[iuvw]$			$[uvw]$
$2/m_h$	21_R					$[010]$			
2_h	m_R			$[u0w]$			$<uv0>$	$[uv0]$	
m	m				$[u0w]$			$<u0w>$	
2_hmm_h	$m1_R$							$<100>$	
22_h2_h	$2m_Rm_R$						$<100>$		
$2mm$	$2mm$							$[001]$	
$2_h/m$	2_Rmm_R					$[u0w]$			$<uv0>$
$2/m_hmm$	$2mm1_R$								$<100>$
4	4								
$\bar{4}$	4_R								

(Continued)

P.G.	D.G.	1	$\bar{1}$	2	m	$2/m$	222	$mm2$	mmm
$4/m_h$									
	41_R								
42_h2_h	$4m_Rm_R$								
$4mm$	$4mm$								
$\bar{4}\,m2_h$	4_Rmm_R								
$4/m_hmm$	$4mm1_R$								
3	3								
$\bar{3}$	6_R								
32_h	$3m_R$								
$3m$	$3m$								
$\bar{3}\,m$	6_Rmm_R								
6	6								
$\bar{6}$	31_R								
$6/m_h$	61_R								
62_h2_h	$6m_Rm_R$								
$6mm$	$6mm$								
$\bar{6}\,m2_h$	$3m1_R$								
$6/m_hmm$	$6mm1_R$								

(Continued)

P.G. / D.G.	4	$\bar{4}$	4/m	422	4mm	$\bar{4}2m$	4/mmm	3	$\bar{3}$	32	3m	$\bar{3}m$
1	$[uvw]$	$[uvw]$		$[uvw]$	$[uvw]$	$[uvw]$		$[uvw]$		$[uvw]$	$[uvw]$	
1_R											<110>	
2										<110>		
2_R			$[uvw]$				$[uvw]$		$[uvw]$			$[uvw]$
21_R												<110>
m_R	$[uv0]$	$[uv0]$		$[u0w]$ $[uv0]$ $[uuw]$	$[uv0]$	$[u0w]$ $[uv0]$				$[u\bar{u}w]$		
m					$[u0w]$ $[uuw]$	$[uuw]$					$[u\bar{u}w]$	
$m1_R$					<100> <110>	<110>						
$2m_Rm_R$				<100> <110>		<100>						
$2mm$												
2_Rmm_R			$[uv0]$				$[u0w]$ $[uv0]$ $[uuw]$					$[u\bar{u}w]$

379

(Continued)

P.G. D.G.	4	$\bar{4}$	4/m	422	4mm	$\bar{4}2m$	4/mmm	3	$\bar{3}$	32	3m	$\bar{3}m$
$2mm1_R$							<100> <110>					
4	[001]											
4_R		[001]										
41_R			[001]									
$4m_Rm_R$				[001]								
$4mm$					[001]							
4_Rmm_R						[001]						
$4mm1_R$							[001]					
3								[001]				
6_R									[001]			
$3m_R$										[001]		
$3m$											[001]	
6_Rmm_R												[001]
6												
31_R												
61_R												
$6m_Rm_R$												
$6mm$												
$3m1_R$												
$6mm1_R$												

P.G. / D.G.	4	$\bar{6}$	6/m	622	6mm	$\bar{6}m2$	6/mmm	23	$m\bar{3}$	432	$\bar{4}3m$	$m\bar{3}m$
1	[uvw]	[uvw]		[uvw]	[uvw]	[uvw]		[uvw]		[uvw]	[uvw]	
1_R												
2												
2_R			[uvw]				[uvw]		[uvw]			[uvw]
21_R												
m_R	[uv0]			[uv0] [uuw] $[\bar{u}uw]$	[uv0]	[uuw]		<uv0>		<uv0> <uuw>	<uv0>	
m		[uv0]			[uuw] $[\bar{u}uw]$	[uv0] $[\bar{u}uw]$						<uuw>
$m1_R$					$<\bar{1}10>$ <110>	<110>						<110>
$2m_R m_R$				$<\bar{1}10>$ <110>				<100>		<100>		
$2mm$						<110>						
$2_R mm_R$			[uv0]				[uv0] [uuw] $[\bar{u}uw]$		[uv0]			[uv0] [uuw]

(Continued)

P.G. D.G.	4	$\bar{6}$	6/m	622	6mm	$\bar{6}m2$	6/mmm	23	$m\bar{3}$	432	$\bar{4}3m$	$m\bar{3}m$
$2mm1_R$							$\langle\bar{1}10\rangle$ $\langle110\rangle$		$\langle100\rangle$			$\langle110\rangle$
4												
4_R												
41_R												
$4m_Rm_R$										$\langle110\rangle$		
$4mm$												
4_Rmm_R											$\langle100\rangle$	
$4mm1_R$												$\langle100\rangle$
3								$\langle111\rangle$				
6_R									$\langle111\rangle$			
$3m_R$										$\langle111\rangle$		
$3m$											$\langle111\rangle$	
6_Rmm_R												$\langle111\rangle$
6	[001]											
31_R		[001]										
61_R			[001]									
$6m_Rm_R$				[001]								
$6mm$					[001]							
$3m1_R$						[001]						
$6mm1_R$							[001]					

8.5.3 CBED Determination of Point Groups

In some cases, a diffraction group corresponds to only one crystallographic point group. For example, if the determined diffraction group is 6, 31_R, 61_R, $6m_Rm_R$, $6mm$, $3m1_R$, or $6mm1_R$, then the point group of the specimen is 6, $\bar{6}$, $6/m$, 622, $6mm$, $\bar{6}m2$, or $6/mmm$, because there are no other crystallographic point groups that possess a six-fold axis. However, a diffraction group may correspond to several point groups. For example, if one determined a diffraction group 3, the point group of the studied crystal may be 3 or 23, because the diffraction group along a <1 1 1> axis of a crystal with a point group 23 is also 3. Table 8.5 lists the relationship between diffraction groups and point groups. From Table 8.5 one can find all the point groups that correspond to a given diffraction group, and also find all the diffraction groups that correspond to one given point group. For example, we first find the diffraction group 3 in the first column of Table 8.5, and then find $[0\,0\,0\,1]$ axis of the point group 3 and the <1 1 1> axes of the point group 23 from the same row as the diffraction group 3 belongs to. In order to distinguish these two point groups we check the columns corresponding to the point groups 3 and 23. We find that, along the <1 0 0> axis and $[u\,v\,0]$ axis of the point group 23, the diffraction groups are $2m_Rm_R$ and m_R, respectively. However, along the zone axes of the point groups 3 other than $[0\,0\,0\,1]$, the point group 3 does not correspond to any diffraction group other than diffraction group 1. Thus, by observing the possible symmetries along other zone axes one can distinguish the point group 3 from the point group 23.

Tanaka et al. (1988, 1994) extended the CBED technique to the determination of point groups of quasicrystals.

8.6 Convergent-Beam Electron Diffraction Determination of Burgers Vectors of Dislocations in Crystals and Quasicrystals

8.6.1 Twisting Direction of a Reflection Fringe Induced by a Dislocation Line

As described by Hirsch et al. (1977), kinematical theory is appropriate for weak reflections, e.g., higher-order reflections. In this case the reflected amplitude $\phi_g(x, y)$ corresponding to a reciprocal vector g at the exit point (x, y) may be expressed as

$$\phi_g(x, y) = \frac{\pi i}{\xi_g} \int_0^t \exp\left\{-2\pi i \left[s + \boldsymbol{g} \cdot \frac{d\boldsymbol{R}(x, y, z)}{dz}\right] z\right\} dz \qquad (8.12)$$

where z is the depth measured from the entrance surface and t the thickness of the specimen. ξ_g is the extinction distance of the reflection $\boldsymbol{g}$. The ξ_g value is rather large for higher-order reflections. $\boldsymbol{R}(x,y,z)$ indicates the displacement vector at the position (x,y,z) induced by a defect. The parameter s represents the deviation parameter, which is also a function of coordinates (x,y,z). Thus the diffracted intensity $I_g = |\phi_g|^2$ of the $\boldsymbol{g}$ reflection and hence the intensity $I_0 = 1 - I_g$ of the transmitted beam are all functions of the coordinates (x,y).

If a perfect region with $\boldsymbol{R}(x,y,z) \equiv 0$ is illuminated, the diffracted and transmitted intensities may be obtained by integrating (8.12):

$$I_g = 1 - I_0 = \left(\frac{\pi t}{\xi_g}\right)^2 \frac{\sin^2(\pi t s)}{(\pi t s)^2} \qquad (8.13)$$

By comparing (8.1) (8.13) it is clear that (8.1) reduces to (8.13) when s^2 is much larger than $(1/\xi_g)^2$, that is, under the kinematical approximation. Equation (8.13) indicates that I_g takes a maximum value and hence I_0 takes a minimum value at Bragg reflection condition $s=0$. However, if a defected region is illuminated, the nonvanishing term $d\boldsymbol{R}/dz$ induces local variations in both the orientations and interplanar distances of the lattice planes, and results in a local change of the deviation parameter from s into the effective deviation parameter s':

$$s' = s + \beta'_g = s + \boldsymbol{g} \cdot \frac{d\boldsymbol{R}}{dz} \qquad (8.14)$$

Thus, the region where the diffracted intensity takes maximum changes from the region with $s=0$ into the region with $s'=0$.

Figure 8.15 depicts a Burgers circuit Ci around a dislocation line $\boldsymbol{u}$ where z indicates the incident direction. The Burgers vector $\boldsymbol{b}$ of the dislocation line $\boldsymbol{u}$ is defined as the line integral of the gradient $d\boldsymbol{R}/dl$ along the Burgers circuit Ci of the dislocation: $\boldsymbol{b} = \int_{ci} (d\boldsymbol{R}/dl)dl$. Usually, $d\boldsymbol{R}/dl$ is nearly parallel to $\boldsymbol{b}$, so that we can conclude that on the $z \times \boldsymbol{u}$ (or $-z \times \boldsymbol{u}$) side of the dislocation line $\boldsymbol{u}$, $d\boldsymbol{R}/dz$ is nearly parallel to (or opposite to) the Burgers vector $\boldsymbol{b}$. In both cases the modulus of $d\boldsymbol{R}/dz$ decreases gradually to zero when the distance to the dislocation line increases.

Figure 8.16 describes the effect of the strain field induced by a dislocation line $\boldsymbol{u}$ on the defocus CBED pattern. Figures 8.16b-d correspond to the case when the incident beam crossover C lies above the specimen (see Fig. 8.16a)

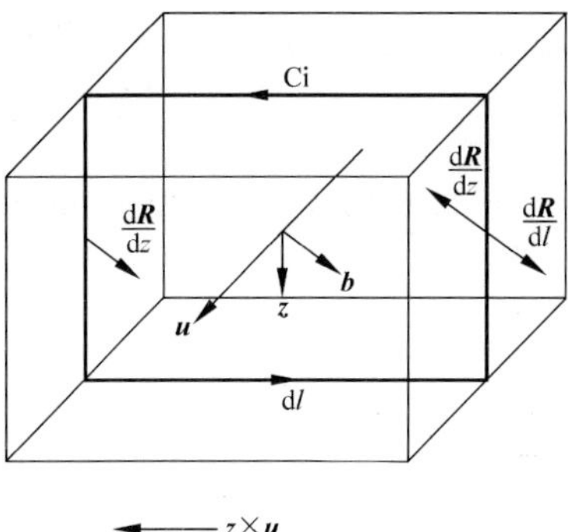

Fig. 8.15 Burgers circuit Ci around a dislocation line u showing the direction of dR/dz, which is nearly parallel to b(or$-b$) at the $z\times u$(or$-z\times u$) side of the dislocation line

and Figs. 8.16f-h correspond to the case when C is below the specimen (see Fig. 8.16(e)). Let us consider a reciprocal vector g that corresponds to a higher-order reflection. When a perfect region (R=0) of the specimen is illuminated, one observes a dark deficient fringe HL in the transmitted disk (Figs. 8.16b and f) and its corresponding bright reflection fringe in the g diffracted disk.

When the incident beam crossover C lies above the specimen (Fig. 8.16a), the side of the HL fringe pointed to by the g vector (or-g vector) corresponds to s<0 (or s>0), see Fig. 8.16b. In the specimen containing a dislocation line u with a Burgers vector b (Figs. 8.16c and d), the strain field dR/dz around the dislocation changes the Bragg condition from s=0 to $s' = s + \beta'_g = s + g \cdot dR/dz = 0$. When $g \cdot b$>0, $\cdot$ on the $z\times u$ side of the dislocation line u, $\beta'_g = g \cdot dR/dz$ >0, so that the region where $s' = 0$ is shifted toward the s<0 region, i.e., along the g direction. Similarly, on the $-z\times u$ side of the dislocation line u, $\beta'_g = g \cdot dR/dz$ <0, so that the region where $s' = 0$ is shifted toward the s>0 region, i.e., opposite to the g direction. Therefore, the fringe HL, where s=0 in the perfect region, splits into fringes HH' and $L'L$, which converge to the fringe HL when away from the dislocation line, see Fig. 8.16c. When $g \cdot b$<0, the twisting directions of the fringes HH' and $L'L$ are reversed, see Fig. 8.16d.

Now let us turn to the case when the beam crossover C lies below the specimen (Fig. 8.16e). In this case, the side of the fringe HL pointed to by the g vector corresponds to s>0 (Fig. 8.16f), hence the twisting directions of the fringes HH' and $L'L$ are all reversed with respect to the case when C is above the specimen, as can be seen by comparing Figs. 8.16c with (g) (for $g \cdot b$>0), and Figs. 8.16d with h (for $g \cdot b$<0).

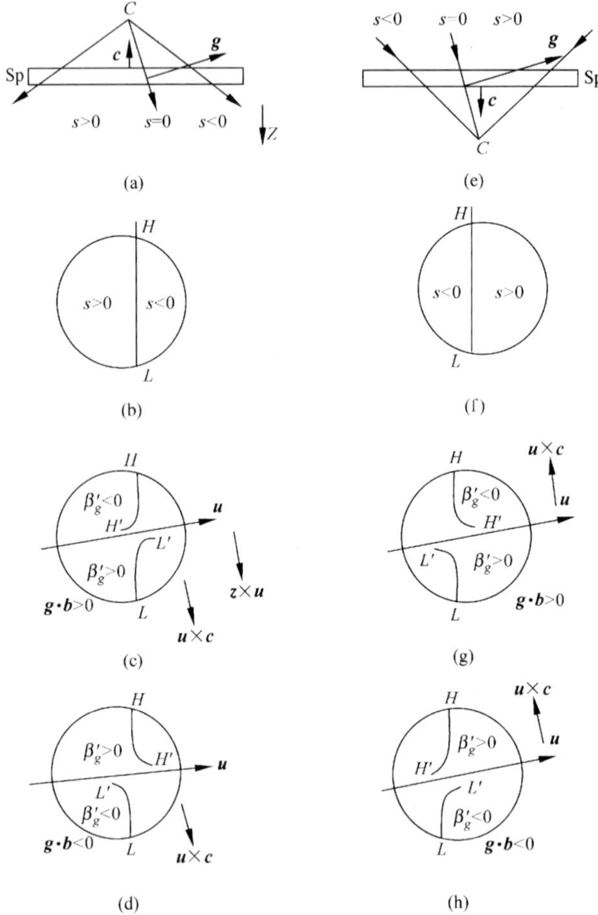

Fig. 8.16 Schematic diagram showing splitting and twisting of a weak reflection fringe *HL* in a defocus CBED pattern induced by the strain field around a dislocation line *u*

If we introduce a vector c pointing to the beam crossover C from the dislocation line (Wen et al., 1989), see Figs. 8.16a and e, then all the cases discussed above may be generally expressed as follows. A higher-order reflection g causes a bright fringe *HL* in the g diffracted disk and a corresponding dark deficient fringe *HL* in the transmitted disk (Figs. 8.16b and f) that lie in the region where Bragg condition $s=0$ is satisfied. When this fringe intersects a dislocation line u of a Burgers vector b, and $g \cdot b \neq 0$, then the fringe *HL* splits into branches *HH′* and *L′L*. The branch lying at the $u \times c$ side is twisted towards the g direction if $g \cdot b > 0$, see Figs. 8.16c and g. On the contrary, if $g \cdot b < 0$, then this branch is twisted towards the $-g$ direction, see Figs. 8.16d and h. When $g \cdot b = 0$, the fringe *HL* does not split.

Therefore, one can summarize a criterion for the relative sign of $\boldsymbol{g} \cdot \boldsymbol{b}$ determined by the defocus CBED technique as follows: the $\boldsymbol{g} \cdot \boldsymbol{b}$ value will be positive (or negative), if on the $\boldsymbol{u} \times \boldsymbol{c}$ side of the dislocation line $\boldsymbol{u}$ the reflection fringe twists towards (or opposite to) the $\boldsymbol{g}$ direction.

8.6.2 Number of Nodes in the Split Reflection Fringes Induced by a Dislocation Line

In Sect. 8.6.1 we discussed the sign criterion of the split reflection fringe induced by the strain field of an intersected dislocation line. This reflection fringe corresponds to the principal maximum of (8.13). In fact, the diffracted intensity I_g can also take secondary maxima at the positions $ts=\pm 1.431$, ± 2.459, $\cdots$, with lower intensities proportional to $1/s^2$ according to (8.13). These secondary reflection fringes will also split and twist by the strain field of the dislocation. Figure 8.17a depicts the principal reflection fringe $s_0 s_0'$ corresponding to the principal maximum of I_g at $s_0=0.0$, and four secondary reflection fringes $s_1 s_1'$, $s_{-1} s_{-1}'$, $s_2 s_2'$ and $s_{-2} s_{-2}'$ corresponding to secondary maxima of I_g at $s_1=1.431/t$, $s_{-1}=-1.431/t$, $s_2=2.459/t$ and $s_{-2}=-2.459/t$, respectively. The thickness of each fringe in Fig. 8.17 is nearly proportional to the intensity of that fringe. When these fringes intersect a dislocation line $\boldsymbol{u}$ and $\boldsymbol{g} \cdot \boldsymbol{b} \neq 0$, then these fringes split and twist. When $\boldsymbol{g} \cdot \boldsymbol{b}=1$, the branches on the $\boldsymbol{u} \times \boldsymbol{c}$ (or $-\boldsymbol{u} \times \boldsymbol{c}$) side of the dislocation line twist along the $\boldsymbol{g}$ (or $-\boldsymbol{g}$) direction, so that two branches s_0 and s_1' connect into a single fringe $s_0 s_1'$, and two branches s_{-1} and s_0' connect into a fringe $s_{-1} s_0'$, $\cdots$, see Fig. 8.17b. Among these fringes usually only two branches $s_0 h_0$ and $l_0 s_0'$ with a smaller s^2 value possess higher intensity because I_g is proportional to $1/s^2$. Thus, it seems that a reflection fringe

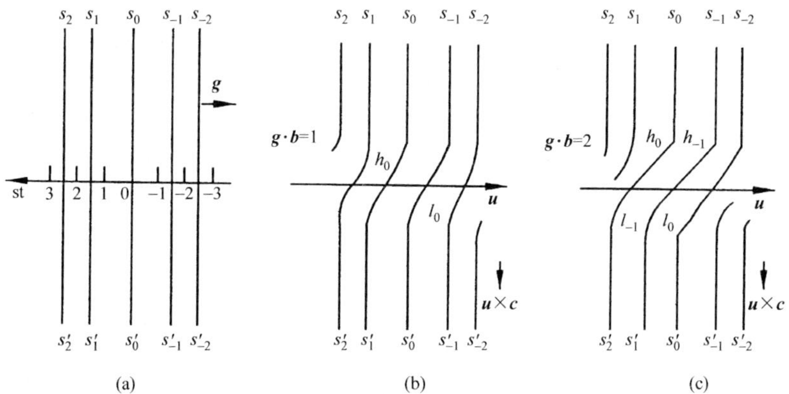

Fig. 8.17 Numbers of the nodes in the split weak reflection fringes induced by a dislocation line $\boldsymbol{u}$ under different values of $\boldsymbol{g} \cdot \boldsymbol{b}$. (a) $\boldsymbol{g} \cdot \boldsymbol{b}=0$; (b) $\boldsymbol{g} \cdot \boldsymbol{b}=1$; (c) $\boldsymbol{g} \cdot \boldsymbol{b}=2$

$s_0 s_0'$ splits into two branches $s_0 h_0$ and $l_0 s_0'$ with one node when $\boldsymbol{g} \cdot \boldsymbol{b} = 1$. When $\boldsymbol{g} \cdot \boldsymbol{b} = 2$, all the split branches twist more severely, so that connected fringes $s_0 s_2'$, $s_{-1} s_1'$, $s_{-2} s_0'$, $\cdots$ are produced, see Fig. 8.17c. Among these fringes one can observe only three strong branches $s_0 h_0$, $h_{-1} l_{-1}$ and $l_0 s_0'$ with smaller s^2 values, and it seems that a reflection fringe $s_0 s_0'$ splits into three branches with two nodes. Generally, when $\boldsymbol{g} \cdot \boldsymbol{b} = n$ with n being an integer, a higher-order reflection fringe splits into $|n|+1$ branches with $|n|$ nodes.

Therefore, in this section we proved the so-called Cherns-Preston criterion (Cherns and Preston, 1986) which may be summarized as follows. When a higher-order reflection fringe $\boldsymbol{g}$ intersects a dislocation line of a Burgers vector $\boldsymbol{b}$, this fringe will split into $|n|+1$ branches with $|n|$ nodes, if $\boldsymbol{g} \cdot \boldsymbol{b} = n$.

8.6.3 Defocus CBED Technique for Determining Burgers Vectors of Dislocations in Crystals

A Burgers vector $\boldsymbol{b} = [\,b_1, b_2, b_3\,]$ of a dislocation in a crystal is a 3-D vector with three components b_1, b_2, and b_3. Usually a defocus CBED pattern may contain several HOLZ lines that are higher-order reflection fringes. We can select three linearly independent reciprocal vectors $\boldsymbol{g}_i = [\,g_{i1}, g_{i2}, g_{i3}\,]$ $(i=1, 2, 3)$ whose fringes intersect the dislocation line. From the numbers of nodes of the split fringes, we obtain the moduli of the $\boldsymbol{g}_i \cdot \boldsymbol{b} = n_i$ $(i=1, 2, 3)$. By observing the twist direction of each split fringe, we obtain the signs of the $\boldsymbol{g}_i \cdot \boldsymbol{b} = n_i$. By indexing these reflections $\boldsymbol{g}_i$, the indices g_{ij} are obtained. Substituting g_{ij} and n_i into the following set of linear equations

$$\sum_j g_{ij} b_j = n_i \tag{8.15}$$

one can solve the indices b_1, b_2 and b_3 of the Burgers vector $\boldsymbol{b}$ of this dislocation line.

As an example of the defocus CBED determination of the Burgers vectors of dislocation lines in crystals, Fig. 8.18 shows a contrast experimental result that consists of a series of bright-field and dark-field micrographs by using various reflections. This experiment is for dislocation lines $D1$–$D5$ and $D8$–$D10$ in a $SrTiO_3$ crystal annealed at 1073 K for 74 min. The results are listed in Table 8.6 where the symbols v, i and r represent visible, invisible and residual contrast (Loretto and Smallman, 1975), respectively. If we suppose that the condition corresponds to $\boldsymbol{g} \cdot \boldsymbol{b} = 0$ when a dislocation shows weak residual contrast, then we can deduce all the directions of Burgers vectors of dislocations $D1$–$D5$ and $D8$–$D10$ as listed in the last row of Table 8.6. However, in this stage we do not know the sign and modulus of each Burgers vector.

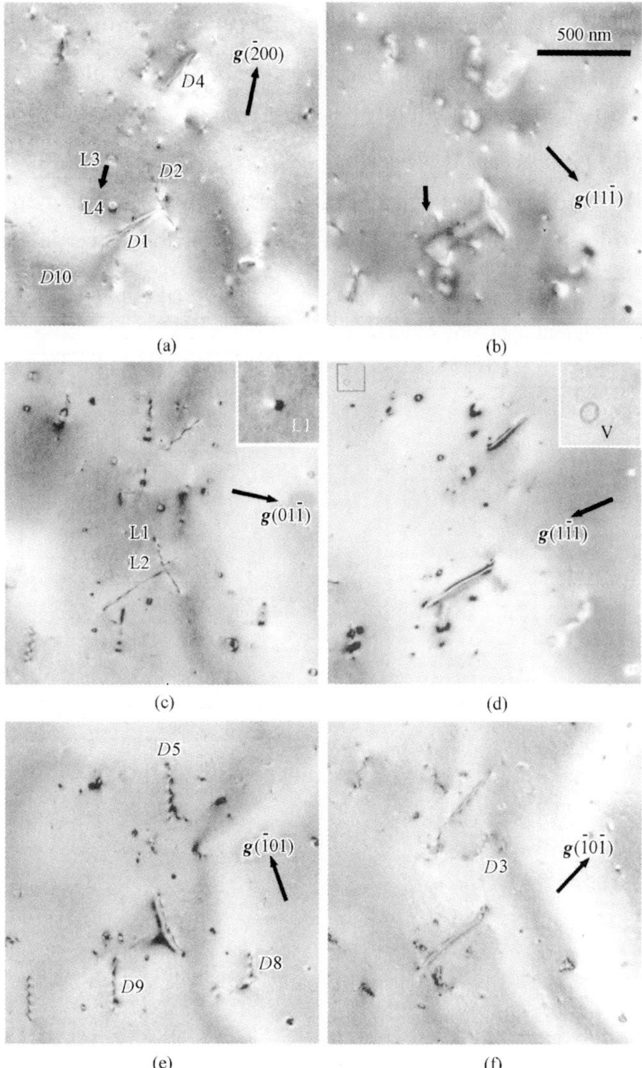

Fig. 8.18 Contrast experiment of dislocations $D1$–$D5$ and $D8$–$D10$ in a SrTiO$_3$ crystal annealed at 1073 K for 74 min by using different operating reflections. (**a**) DF, g=(–2 0 0); (**b**) DF, g=(1 1 –1); (**c**) BF, g=(0 1 –1); (**d**) BF, g=(1 –1 1); (**e**) BF, g=(–1 0 1); (**f**) DF, g=(–1 0 –1)

Figure 8.19 shows a defocus CBED pattern taken from dislocation line $D1$ shown in Fig. 8.18. In this pattern the bold arrow u indicates the line direction of dislocation $D1$ and each thin arrow $g(h\ k\ l)$ indicates the respective reciprocal vector connecting the dark deficient fringe and its corresponding bright reflection fringe. We see the fringe $g(\bar{3}\,3\,\bar{2})$ splits into 5 nodes and the branches on the $\boldsymbol{u}\times\boldsymbol{c}$ side of the dislocation line twist along the $g(\bar{3}\,3\,\bar{2})$ direction. Hence, we obtain $g(\bar{3}\,3\,\bar{2})\cdot\boldsymbol{b}=5$. Table 8.7 lists the experimental values $g_i(h_i\,k_i\,l_i)\cdot\boldsymbol{b}=n_i$ for all the $g_i(h_i\,k_i\,l_i)$ fringes appearing in Fig. 8.19 except $g(\bar{4}\,\bar{1}\,1)$, which does not intersect the dislocation line $D1$ and is marked as "ni". By substituting the values listed in Table 8.7 into (8.15) we determined the

Table 8.6 Contrast experiments of dislocation lines $D1$–$D5$ and $D8$–$D10$ shown in Fig. 8.18 by different operating reflections g

Dislocation			$D1$	$D2$	$D3$	$D4$	$D5$	$D8$	$D9$	$D10$
	g									
2	0	0	v	v	i	v	r	i	r	v
1	1	−1	r	v	i	r	r	i	r	v
0	1	−1	v	v	i	v	i	i	r	v
1	−1	1	v	i	i	v	i	i	i	r
−1	0	1	i	v	v	i	v	v	v	v
1	−1	0	v	v	v	v	v	v	v	v
1	1	0	v	v	v	v	v	v	v	v
−1	0	−1	v	i	v	v	v	v	v	i
	$b//$		[101]	[−101]	[011]	[101]	[011]	[011]	[011]	[−101]

v:visble; i:invisible; r:residual contrast.

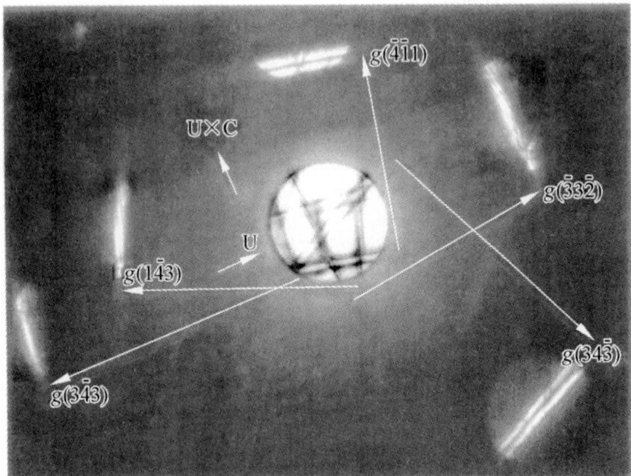

Fig. 8.19 Defocus CBED pattern taken from dislocation line $D1$ shown in Fig. 8.18

Burgers vector $\boldsymbol{b}$ of the dislocation $D1$ to be $\boldsymbol{b}=[\bar{1}\,0\,\bar{1}]$, which is consistent with the result deduced from the invisibility criterion and contains complete information of the Burgers vector including its direction, sign and modulus.

Table 8.7 Indices g_{ij} of reflection fringes g_i and the $g_i \cdot b = n_i$ values for dislocation line $D1$ shown in Fig. 8.18 measured from the defocus CBED pattern shown in Fig.8.19

g_{i1}	g_{i2}	g_{i3}	n_i
−4	−1	1	ni
−3	3	−2	5
3	4	−3	0
3	−4	3	−6
1	−4	3	−4

ni: not intersected.

8.6.4 Defocus CBED Technique for Determining Burgers Vectors of Dislocations in Quasicrystals

The Burgers vectors $B = [B_1, B_2, \cdots, B_N]$ of dislocations, the displacement vectors R and the reciprocal vectors $G = [G_1, G_2, \cdots, G_N]$ in a quasicrystal (QC) are all N-dimensional (ND) vectors with N components. For example, $N=6$ for icosahedral QCs (IQCs). In the case of QCs the discussions in Sect. 8.6.3 including (8.12) – (8.15) are all correct if we change the expression for $\beta'_g = g \cdot \mathrm{d}R/\mathrm{d}z$ into $\beta'_G = G \cdot \mathrm{d}R/\mathrm{d}z$ and change the 3-D inner product $g \cdot b$ into an inner product $G \cdot B$ of ND vectors G and B. Therefore, as pointed out by Wang and Dai (1993) and Feng and Wang (1994), the sign criterion for dislocations in QCs may be expressed as follows. The $G \cdot B$ value will be positive (or negative), if the reflection fringe on the $u \times c$ side of the dislocation line u twists along (or opposite to) the g'' direction. Here, g'' is the 3-D physical component of the ND reciprocal vector G, while its corresponding $(N-3)$-D perpendicular component is $g^{\perp}$, and $G = g'' + g^{\perp}$. Similarly, the Cherns and Preston criterion may be expressed as follows. If a higher-order reflection fringe g'' intersects a dislocation line of a Burgers vector B, this fringe will split into $|n|+1$ branches with $|n|$ nodes, if $G \cdot B = n$.

In order to solve the N components of an ND Burgers vectors $B = [B_1, B_2, \cdots, B_N]$, it is necessary to select N linearly independent reciprocal vectors $G_i = [G_{i1}, G_{i2}, \cdots, G_{iN}]$ $(i=1, 2, \cdots, N)$ whose fringes intersect the dislocation line. The indices G_{ij} of the fringes G_i may be obtained by computer simulation of the related HOLZ line patterns of QCs. Substituting G_{ij} and the values $G_i \cdot B = n_i$ into the following system of linear equations:

$$\sum_j G_{ij} B_j = n_i \tag{8.16}$$

one can solve the indices $B_1, B_2, \cdots, B_N$ of the ND Burgers vector B of the dislocation line.

Figure 8.20 shows a defocus CBED pattern taken from a perfect region of an $Al_{70.4}Pd_{21.2}Mn_{8.4}$IQC. The bright circular disk of $0.6°$ angular radius is the transmitted disk and every bright fringe labeled as 1, 2, $\cdots$, 17 is the principal reflection fringe in each diffraction disk. An arrow g_1, which connects the dark deficient fringe (HOLZ line) in the transmitted disk and its corresponding bright reflection fringe, designates the physical component $g_1^{//}$ of the 6-D reciprocal vector G_1. By simulating this pattern with a computer, we deduced all the indices of the reflections appeared in Fig. 8.20 as listed in Table 8.8.

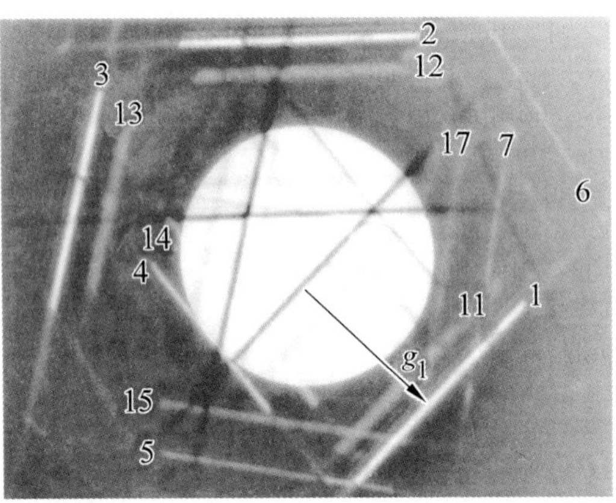

Fig. 8.20 Defocus CBED pattern taken from a perfect region of an $Al_{70.4}Pd_{21.2}Mn_{8.4}$ IQC

Figure 8.21a is a contrast image of a dislocation line in an $Al_{70.4}Pd_{21.2}Mn_{8.4}$IQC deformed plastically at high temperature. When the defocused incident beam illuminates this dislocation line, one obtains a defocus CBED pattern as shown

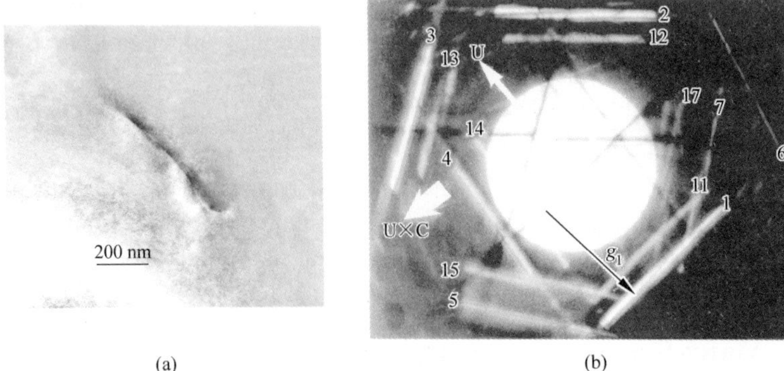

Fig. 8.21 (**a**) Contrast image of a dislocation line in an $Al_{70.4}Pd_{21.2}Mn_{8.4}$ IQC deformed plastically at high-temperature. (**b**) Defocus CBED pattern taken from the dislocation line shown in Fig. 8.21a

in Fig. 8.21b, where the arrow u designates the line direction of the dislocation. By comparing Fig. 8.21b with Fig. 8.20 we found that fringe 1 in Fig. 8.21b splits into 1 node and the branch on the $u \times c$ side of the dislocation line twists opposite to the $g_1^{//}$ direction and hence we conclude $G_1 \cdot B = n_1 = -1$. Similarly we obtain $G_2 \cdot B = 1$, $G_3 \cdot B = 0$, $\cdots$ The deficient fringes 4, 6, 12 and 14 do not intersect the dislocation line, hence they do not split and we can not obtain the corresponding experimental $G_i \cdot B = n_i$ values for i=4, 6, 12 and 14. All the information is listed in the last column of Table 8.8. There are many possibilities of selecting 6 linearly independent G_i from the 9 G_i of which the experimental $G_i \cdot B = n_i$ values exist. From any selection we solve the same Burgers vector $B = (1/2)$ $[\bar{3}\,3\,0\,2\,2\,0]$.

Table 8.8 Indices G_{ij} of reflection fringes G_i in Figs. 8.20 and 8.21b and the $G_i \cdot B = n_i$ values for a disocation line shown in Fig. 8.21a measured from the defocus CBED pattern shown in Fig. 8.21b

i	G_{i1}	G_{i2}	G_{i3}	G_{i4}	G_{i5}	G_{i6}	n_i
1	2	8	2	–6	–4	4	–1
2	2	–4	4	8	2	–6	1
3	–4	–8	–6	2	4	–2	0
4	–6	–4	–10	–6	2	4	ni
5	–4	4	–8	–12	–2	8	–2
6	8	6	12	6	–4	–4	ni
7	6	10	8	–2	–6	2	–2
11	1	5	1	–3	–3	3	0
12	1	–3	3	5	1	–3	ni
13	–3	–5	–3	1	3	–1	1
14	–4	–2	–6	–4	2	2	ni
15	–3	3	–5	–7	–1	5	1
17	3	7	5	–1	–3	1	2

ni: not intersected.

8.7 Convergent-Beam Electron Diffraction Determination of Crystal Structures

8.7.1 Many-Beam Dynamic Theory of Electron Diffraction

According to Hirsch et al. (1977), Spence and Zuo (1992), Cowley (1992), Spence (1993) and Hoier et al. (1993), the real part $V(r)$ of a periodic crystal potential can be expanded as

$$V(r) = \frac{h^2}{2me} \sum_g U_g \exp(2\pi i g \cdot r) = \sum_g V_g \exp(2\pi i g \cdot r) \qquad (8.17)$$

where U_g and V_g are Fourier coefficients of the real part of the crystal potential, which are related to the structure factor F_g and the extinction distance ξ_g by (8.3), m and e are the mass and charge of an electron. Similarly, the Fourier coefficients U_g' stem from the absorption part $iV'(r)$ of the crystal potential. U_g and U_g' are, in general, complex quantities with both amplitudes and phases:

$$U_g = |U_g| \exp(i\phi_g),$$
$$U_g' = |U_g'| \exp(i\phi_g') \qquad (8.18)$$

An incident electron beam with high energy interacts with the crystal potential and forms a wave field $\Psi(K, r)$ in the interior of the crystal:

$$\Psi(K,r) = \sum_i \varepsilon^i(K) b^i(k^i,r)$$
$$= \sum_i \varepsilon^i(K) \sum_g C_g^i \exp(2\pi i \gamma^i n \cdot r) \exp\{2\pi i (K + g) \cdot r\} \qquad (8.19)$$

where $b^i(k^i, r)$ is the i-th Bloch wave function. k^i and $\varepsilon^i(K)$ are the wave vector and excitation coefficient of the i-th Bloch wave, respectively. The Bloch wave eigenvalue γ^i is defined as $k^i = K + \gamma^i n$ with n being a unit vector along the normal of the specimen entrance surface and K is the wave vector of the incident beam corrected for the mean inner potential. The Bloch wave eigenvalues γ^i and the eigenvectors C_g^i are obtained by solving the following fundamental eigenequation:

$$A_{g,g} B_{g,g}^i + \sum_{h \neq g} A_{g,h} B_h^i = 2K_n \gamma^i B_g^i \qquad (8.20)$$

where the elements $A_{g,h}$ of the structure matrix A are expressed as

$$A_{g,h} = \frac{U_{g-h} + iU_{g-h}'}{\sqrt{1 + g_n/K_n}\sqrt{1 + h_n/K_n}} \qquad \text{for } g \neq h,$$

$$A_{g,g} = \frac{2Ks_g + iU_0'}{1 + g_n/K_n} \qquad \text{for } g = h, \qquad (8.21)$$

and the eigenvector elements B_g^i are related to C_g^i by

$$B_g^i = (1 + g_n / K_n)^{1/2} C_g^i \qquad (8.22)$$

Here $g_n = \boldsymbol{g} \cdot \boldsymbol{n}$ and $K_n = \boldsymbol{K} \cdot \boldsymbol{n}$ and the excitation error (deviation parameter) s_g is defined as

$$2Ks_g = -\boldsymbol{g}^2 - 2\boldsymbol{K} \cdot \boldsymbol{g} \qquad (8.23)$$

The total wave $\Psi(\boldsymbol{K}, \boldsymbol{r})$ is decomposed into particular reflection waves $\boldsymbol{0}, \boldsymbol{g},$ $\boldsymbol{h}, \cdots$ at the exit surface of the crystal slab with thickness t: $\varphi(\boldsymbol{K}, t) = \sum_g \Psi_g(\boldsymbol{K},t) \exp\{2\pi i (\boldsymbol{K} + \boldsymbol{g}) \cdot \boldsymbol{r}\}$, where the wave function of the $\boldsymbol{g}$ reflection is expressed as:

$$\Psi_g(\boldsymbol{K},t) = \sum_g \varepsilon^i(\boldsymbol{K}) C_g^i \exp(2\pi i \gamma^i t) \qquad (8.24)$$

and the intensity of the $\boldsymbol{g}$ reflection $I_g(\boldsymbol{K}, t) = |\Psi_g(\boldsymbol{K}, t)|^2$.

When N reflection beams (including the transmitted beam) are taken into consideration, the solution of the eigenvalue problem (8.20) gives N eigenvalues γ^i and N eigenvectors C_g^i ($i = 1, 2, \cdots, N$). Moreover, the excitation coefficients are found to be $\varepsilon^i(\boldsymbol{K}) = (C^{-1})_{i0}(\boldsymbol{K})$, where $(C^{-1})_{i0}(\boldsymbol{K})$ denotes the i-th element of the first column of the inverse eigenvector matrix C. When there is no absorption, one can prove that the structure matrix A is Hermitian, and hence the eigenvector matrix C is unitary. In this case we have $\varepsilon^i(\boldsymbol{K}) = C_0^i(\boldsymbol{K})$.

Therefore, given the Fourier components U_g, U_g', $\cdots$ of a crystal potential, the thickness t of the specimen and the incident wave vector $\boldsymbol{K}$, one can calculate the intensity $I_g(\boldsymbol{K}, t) = |\Psi_g(\boldsymbol{K}, t)|^2$ of each reflection $\boldsymbol{g}$.

8.7.2 Structure Factor Determination by Means of CBED Technique

According to Zuo and Spence (1991), Spence (1993), Hoier et al. (1993) and Tanaka et al. (1994), the structure factors U_g, U_g', $\cdots$, the thickness t of the specimen may be determined by adjusting these parameters to minimizing a weighted goodness-of-fit index χ^2 which is defined , say, as

$$\chi^2 = \sum_i \frac{f_i(cI_i^{\text{theory}} - I_i^{\text{exp}})^2}{\sigma_i^2} \tag{8.25}$$

Here, $I_i^{\text{theory}} = |\Psi_g(K_i,t)|^2$ is the calculated intensity (8.24) and I_i^{exp} is the experimental CBED intensity for the i-th pixel, and c is a normalization coefficient. f_i is a weight coefficient, and it takes larger values for the pixels that are more sensitive for the parameters to be determined. The fitting process has been automated by using a simplex method or other optimization algorithms (Zuo and Spence, 1991; Spence, 1993).

When a raw structure is to be refined, i.e., the atom positions are to be refined, it would be better to use higher-order reflections. This is because the intensities of HOLZ reflections with larger reciprocal lattice vectors g are more sensitive to positions of the atoms than those of ZOLZ reflections (Tanaka, et al., 1994). Of course, when performing dynamic calculation one must include many ZOLZ reflections. For example, Tanaka et al. (1994) obtained the rotation angle of 1.12 degrees for the oxygen octahedra in tetragonal $SrTiO_3$, a low-temperature phase, at 87 K with this method.

On the other hand, when the electron charge transfer is to be measured, one needs to determine accurately the structure factors U_g, U_g', $\cdots$ of ZOLZ reflections. For example, Spence (1993) determined the refined values of low-order reflections (2 0 0) and (4 0 0) of a BeO crystal at 80 kV to be $|U(0\ 0\ 2)|$= 3.9592(14) nm^{-2}, $\phi(0\ 0\ 2) = -0.8847\ (170)$ rad, $|U'(0\ 0\ 2)|$=0.073 (6) nm^{-2}, $\phi'(0\ 0\ 2) = -1.1(5)$ rad, $|U'\ (0\ 0\ 4)|$= 0.0002(1) nm^{-2} at t= 71.16(16) nm. From the same crystal, but at a different thickness, Spence (1993) obtained another set of parameters: $|U\ (0\ 0\ 2)|$= 3.982(13) nm^{-2}, $\phi(0\ 0\ 2) = -0.8786(170)$ rad, $|U'(0\ 0\ 2)|$= 0.090(7) nm^{-2}, $\phi'(0\ 0\ 2) = -0.4\ (5)$ rad, $|U'(0\ 0\ 4)|$= 0.004(10) nm^{-2} at t= 105.97(20) nm. The good consistency of these two sets of determined parameters shows the good confidence in the CBED technique.

CBED structure determination has some advantages compared with the X-ray and neutron diffraction methods. By the CBED method one can determine local crystal structure of the illuminated micro-area. In addition, because of the strong dynamic interaction, CBED intensities possess information on the phases of the structure factors. The main disadvantage of the CBED method is the great quantity of the calculations.

For details of the structures determined by the CBED technique, refer to Zuo and Spence (1991), Spence (1993), Hoier et al. (1993) and Tanaka et al. (1994), and the references cited by them.

8.8   CBED Study of Interfacial Strain Fields

The changes in positions of HOLZ lines in the bright-field disc of a CBED pattern can provide information on the lattice parameter changes $\Delta a/a$ to the accuracy of about 2×10^{-4} with spatial resolution of several nanometers in a modern instrument (Steeds, 1979). Such properties have been used to measure the residual strains in metallic matrix composites by detecting the shifts of HOLZ lines in conventional CBED (Rozeveld et al., 1992; Li et al., 1998). With the so-called CBED shadow-image methods, the shadow image of the illuminated area of the specimen is superimposed on the CBED pattern. The variation of lattice distortions can be revealed as shifts and splitting of the HOLZ lines. Therefore, the CBED patterns taken with such methods show directly the spatial variation of the lattice strains. According to the shifts of the HOLZ lines, the local lattice strains of cross-sectional GeSi/Si superlattice specimens have been studied with various CBED shadow-image methods, for example, Ronchigram images (Aitchison and Rodenburg, 1991), convergent beam imaging (Humphreys et al., 1988) and large-angle convergent beam electron diffraction (Duan et al., 1994). However, the surface stress relaxation, which takes place in the specimen-thinning process, can cause additional local lattice plane bending near the foil surface (Treacy et al., 1985; Perovic et al., 1991). As the electron probe is very close to the interface, HOLZ lines often become obscure or even split. In such a case, it is very difficult to measure the shifts of HOLZ lines. To avoid the influence of the surface relaxation, Cherns et al. (1988) used plan-view specimens in studying the strains in GeSi/Si superlattice with the LACBED method. The inhomogeneous strains near the coherent interfaces of spherical particles (Vincent et al., 1988) and near the incoherent interface between Al and an Al_2O_3 particle in the Al-Al_2O_3 composite (Zou et al., 1998) were studied by using the splitting of HOLZ lines in LACBED patterns.

8.8.1   Interfacial Strain Determination by Using HOLZ Line Shifting

8.8.1.1   Factors Affecting HOLZ Line Position

To determine the lattice parameters from the shifts of HOLZ lines, the HOLZ line patterns are usually simulated in kinematic approximation in order to save computing time. In such a case, dynamical effects must be taken into account (Lin et al., 1989; Okuyama et al., 1989; Zuo, 1992; Spence and Zuo, 1992; Mansfield et al., 1993).

In the kinematical approximation, the HOLZ line position is given by the Bragg law, which can be written as

398

$$K^2 - (K + g)^2 = 0 \tag{8.26}$$

where K is the mean wave vector inside the crystal. We set up an orthogonal coordinate system with the z-axis in the direction of the nearest zone axis, the x-axis parallel to one of reciprocal vectors in the ZOLZ and the y-axis being the cross-product of the z and x axes. Then the equation of the HOLZ line given by (8.26) in this coordinate system becomes

$$K_y = -\frac{g_x}{g_y}K_x + \frac{g_z}{g_y}K_z - \frac{g^2}{2g_y} \tag{8.27}$$

Here

$$K_z = \sqrt{K^2 - K_x^2 - K_y^2} \approx \sqrt{K^2 - K_{xc}^2 - K_{yc}^2} \tag{8.28}$$

in which K_{xc} and K_{yc} are the x and y components of the wave vector K at the center of the transmitted disk, respectively. Hence, K_z can be considered constant and the relationship between K_y and K_x, as shown in (8.27), is linear. In turn, the HOLZ line in a CBED disk can be well approximated by a straight line. In principle, using the lattice constants and the wave vector (accelerating voltage) as adjustable parameters, we can determine such parameters by matching the simulated HOLZ line pattern with the experimental pattern.

In other words, the HOLZ line position in the kinematic approximation is determined by the intersection of the sphere of radius K centered on the origin of the reciprocal lattice and the sphere of radius K centered on the reciprocal lattice point g in HOLZ. In the dynamical theory the HOLZ line position is approximately determined by the intersection of the sphere of radius K centered on g with the branches of ZOLZ dispersion surface. The branch of the dispersion surface that is nearest to the K-sphere centered on the origin contributes most to the intensity. Therefore, a single fringe of the HOLZ line is observed. Assuming the i-th branch is the nearest one, the condition in (8.26) changes to

$$K^2 - (K + \gamma^{(i)}\hat{z} + g)^2 = 0 \tag{8.29}$$

Here, we assume that the normal of the specimen surface is parallel to the z-axis for simplicity, $\gamma^{(i)}$ is the distance from the K sphere centered on the origin to the i-th branch of the dispersion surface along the specimen surface normal, and $\hat{z}$ is the unit vector in that z-direction. Neglecting second-order small quantities, we have

$$K_y = -\frac{g_x}{g_y}K_x + \frac{g_z}{g_y}\left(K_z - \frac{2K_z\gamma^{(i)}}{2g_z}\right) - \frac{g^2}{2g_y} \qquad (8.30)$$

This means that the HOLZ line is not at the position predicted by (8.27) and that the HOLZ line is shifted from the kinematic position by a distance shown in the second term in the right side of (8.30).In the case of strong two-beam diffraction in the ZOLZ, $2K_z\gamma^{(i)}$ can be expressed with the Bethe approximation (Spence and Zuo, 1992) as

$$2K_z\gamma^{(i)} = -\sum_{h'\neq h,0}\frac{|U_{h'}|^2}{2Ks_{h'}} \qquad (8.31)$$

where h is the strong reflection in the ZOLZ and $s_{h'}$ are the excitation errors for the weak reflections in the ZOLZ. The summation is over reflections in ZOLZ of the zone axis under consideration. For the reflections in ZOLZ, we have

$$K^2 = (\boldsymbol{K} + \boldsymbol{h'} + s_{h'})^2 \qquad (8.32)$$

and

$$2K_z s_{h'} = -2\boldsymbol{K}_t \cdot \boldsymbol{h'} - h'^2 \qquad (8.33)$$

where $\boldsymbol{K}_t$ is the component perpendicular to the zone axis. In the zone center, $\boldsymbol{K}_t=0$, $2K_z s_{h'} = -h'^2$, thus

$$2K_z\gamma^{(i)} = \sum_{h'\neq h,0}\frac{|U_{h'}|^2}{h'^2} \qquad (8.34)$$

From the discussion we can get some points as follows.

1) In practice, K_z is much larger than K_t in convergent beam diffraction of high-energy electrons. In a small area of the dispersion surface, the term $2K_z\gamma^{(i)}$ can be considered constant. It is seen from (8.30) that, to take the effect of ZOLZ diffraction into account, the HOLZ line positions can be simulated kinematically by using an adjusted accelerating voltage, which is called effective accelerating voltage.

2) As shown in (8.34), the correction term depends on the Fourier components of potential function $U_{h'}$. Hence, the dynamical correction depends on the material species.

3) Generally, the reflections in ZOLZ with higher index zone axis have

400

larger indexes and smaller $U_{h'}$. Thus $\gamma^{(i)}$ for the higher-index zone axis is smaller. Therefore, the dynamical correction becomes smaller as the index of the zone axis is getting higher, as shown experimentally by Okuyama et al. (1989).

4) One can see from (8.33) that, if the specimen is tilted away from the zone axis center, K_t becomes significant, and the correction term expressed by (8.31) is smaller than the correction term expressed by (8.34) for the case in the zone axis center.

5) The dynamical correction, as shown in (8.30), is inversely proportional to g_z, the component of HOLZ reciprocal vector g along the zone axis. Hence, the correction is least for the HOLZ lines with highest order. For example, the correction for the third-order Laue zone lines is smaller than that for the second-order Laue zone lines.

6) It has been experimentally shown by Okuyama et al. (1989) that the dynamical correction varies with the specimen thickness. Mansfield et al. (1993) theoretically analyzed the thickness dependence of HOLZ line positions. They assumed that only two branches (a and b) of the dispersion surface are significant for the ZOLZ, and the HOLZ state c intersects with the b branch. If $R = |C_0^b|^2 / |C_0^a|^2$ is much greater than 1, the HOLZ line is thickness-independent. For example, The thickness has little affect on the HOLZ line positions near the zone center of Si <1 1 4>. Also, because of the large difference between the excitations of the two branches, the center of the <1 1 4> bright-field disk has much weaker zero-order fringe structure than the outer edge of the disk. When $R \approx 1$, namely, branches a and b have comparable excitation, the HOLZ lines are highly thickness sensitive. The effect observed in Si <1 1 1> belongs to this case.

Before kinematically simulating a HOLZ line pattern to match the experimental pattern, one needs to determine the effective accelerating voltage. Usually, this is done by matching the simulated pattern, for which the accelerating voltage is the adjustable parameter, to the experimental pattern taken from a reference specimen. To increase the accuracy in lattice parameter or strain measurements, caution must be taken when choosing the reference specimen (material species and thickness), the zone axis, the order of the Laue zone, and so on. When the specimen is tilted away from higher index zone axis, the dynamical effect can be significantly reduced. A zone axis that exhibits clearly defined HOLZ lines with little or no zero-order fringe contrast is to be preferred.

8.8.1.2 Methods for Determining Lattice Strain from HOLZ Line Shifts

To extract the HOLZ line positions from a bright-field disk of a CBED pattern the pattern can be digitized with a slow-scan CCD camera or a scanner. The pixel coordinates of the intensity minima of a HOLZ line are sought across the line using a image processing software. As shown previously, the HOLZ lines can be approximated as straight lines. The line equations can be found by least squares fit to the intensity minimum coordinates. Afterwards, there are two automated methods to determine lattice parameters.

1) Method of Zuo (1992)

The distances of the intersections of the HOLZ lines are regarded as the parameters to describe the HOLZ line pattern. The intersection coordinates of HOLZ lines are calculated from the HOLZ line equations obtained above. The distances between the two intersections are then calculated. The best fit of the simulated HOLZ pattern to the experimental one is reached by using an optimization method to search for the lowest χ^2. The best fit parameter χ^2 is defined as

$$\chi^2 = \frac{1}{N-P}\sum_i \frac{1}{\sigma_i^2}(d_i^{\text{theory}} - d_i^{\text{exp}})^2 \tag{8.35}$$

Here, N is the total number of data points, P the number of adjustable parameters and d_i the i-th distance between two HOLZ line intersections. σ_i^2 is the variance of the i-th distance and equals d_i^{exp}, assuming Poisson statistics. The unknown lattice parameters are the adjustable parameters in the optimization, and are determined when χ^2 reaches its minimum. The downhill simplex method in multidimensions was used by Zuo (1992). The simulated annealing method as well as the simplex method was used by Li et al. (1998). The simplex algorithm is a rather robust algorithm but is only designed for local refinement. The simulated annealing algorithm has a high probability of finding the global minimum compared to local refinement procedures.

2) Method of Rozeveld and Howe (1993)

Triangular areas that are bounded by three HOLZ lines are defined in the CBED pattern. To suppress the effect of varying magnifications in different images, ratios between two such areas are formed. These ratios are regarded as the parameters to describe the HOLZ line pattern. The area-ratios are assumed to be linear functions of the lattice parameters. This is a valid assumption for small lattice parameter variations. By changing each lattice parameter individually by a small amount, the corresponding changes in area-ratios for that parameter can be calculated from computer simulations of the HOLZ pattern. Then, the linear functions of the area-ratios vs. lattice parameters are obtained as

$$R_i = f_{ij}a_j + c_{ij} \tag{8.36}$$

where R_i is the i-th area-ratio, a_j is one of the lattice parameters, $i, j=1,2,3,\cdots,6$. Since the lattice strains are elastic, the superposition rule may be used. Therefore, the dependence of area-ratios on lattice parameters is

$$
\begin{bmatrix} R_1 \\ R_2 \\ R_3 \\ R_4 \\ R_5 \\ R_6 \end{bmatrix} =
\begin{bmatrix}
f_{11} & f_{12} & f_{13} & f_{14} & f_{15} & f_{16} \\
f_{21} & f_{22} & f_{23} & f_{24} & f_{25} & f_{26} \\
f_{31} & f_{32} & f_{33} & f_{34} & f_{35} & f_{36} \\
f_{41} & f_{42} & f_{43} & f_{44} & f_{45} & f_{46} \\
f_{51} & f_{52} & f_{53} & f_{54} & f_{55} & f_{56} \\
f_{61} & f_{62} & f_{63} & f_{64} & f_{65} & f_{66}
\end{bmatrix}
\begin{bmatrix} a_1 \\ a_2 \\ a_3 \\ a_4 \\ a_5 \\ a_6 \end{bmatrix} +
\begin{bmatrix}
c_{11} + c_{12} + c_{13} + c_{14} + c_{15} + c_{16} \\
c_{21} + c_{22} + c_{23} + c_{24} + c_{25} + c_{26} \\
c_{31} + c_{32} + c_{33} + c_{34} + c_{35} + c_{36} \\
c_{41} + c_{42} + c_{43} + c_{44} + c_{45} + c_{46} \\
c_{51} + c_{52} + c_{53} + c_{54} + c_{55} + c_{56} \\
c_{61} + c_{62} + c_{63} + c_{64} + c_{65} + c_{66}
\end{bmatrix} .
\tag{8.37}
$$

Once the values of R_1, R_2, $\cdots$, R_6 are measured from an experimental pattern, the lattice parameters can be solved from (8.37). However, solving (8.37) would neglect all nonlinear contributions. Instead, an iterative method is used.

Starting from a first guess for the parameter set $\{a_i^0\}$ and small variations for each lattice parameters, f_{ij} and c_{ij} are obtained in the way described above. Then, new values of lattice parameters are obtained by solving (8.37). These new values are denoted by $\{a_i'\}$. The second guess for the lattice parameters are taken as $\{a_i^1 = (a_i^0 + a_i')/2\}$. Then, the operation steps are repeated. When the relative deviations of the area-ratios calculated with $\{a_i^m\}$ from the corresponding experimental values are less than, say, 0.5% and no further improvement is observed, the iteration is stopped and $\{a_i^m\}$ are the values of the lattice parameters expected.

The first problem encountered in using these methods to measure lattice parameters is that, owing to the insensitivity of certain lattice spacings to changes in lattice parameters, and to measurement errors imposed by finite HOLZ linewidths, most HOLZ line patterns can be simulated by a number of different lattice parameter combinations. It has been shown that the area-ratio precision should be greater than seven significant figures in order to solve four unknown lattice parameters simultaneously from [114] aluminum HOLZ pattern (Rozeveld and Howe, 1993). This requirement is far beyond the precision normally available in practice. To extract the reliable information on the lattice parameters, one needs either to reduce the number of lattice parameters to be determined by making physical assumptions, or to use different zone axis HOLZ patterns from the identical area of specimen and to find a unique fit of lattice parameters to all patterns.

To measure the lattice parameters of $YBa_2Cu_3O_{7-\delta}$, Zuo (1992) assumed that only lattice parameters a and b are adjustable according to the fact that the ratio of a- and b-axes in the base plane of $YBa_2Cu_3O_{7-\delta}$ is very sensitive to the oxygen deficiency. For a $K_2O.6TiO_2$/Al metallic matrix composite, there is no crystallographic orientation relationship between the whiskers and the matrix. So, one cannot assume which lattice parameters are constant and which parameters are adjustable. Li et al. (1998) proposed a method for such a case. Since the thickness of the TEM specimen is very small compared with the other two dimensions of the sample, and there are no external forces acting on the surfaces of the sample, the plane stress approximation can be used. In the plane stress approximation, only the stress components σ_x, σ_y, τ_{xy} are nonzero in the foil. In turn, only ε_{xx}, ε_{yy}, ε_{xy} are independent variables, and

$$\varepsilon_{zz} = -\frac{\nu}{1-\nu}(\varepsilon_{xx} + \varepsilon_{yy}), \quad \varepsilon_{xz} = 0, \quad \varepsilon_{yz} = 0 \tag{8.38}$$

Here, ν is Poisson's coefficient of aluminum, and x, y, z represent coordinates in the specimen coordinate system in which z is parallel to the foil normal, y is along the projection of whisker axis in the foil plane and x is parallel to the cross-product of the y- and z- axes. With this approximation, the number of independent components of the strain tensor is reduced from 6 to 3. Then, ε_{xx}, ε_{yy}, ε_{xy}, ε_{zz} are transformed from the specimen coordinate system to the crystal coordinate system according to

$$\varepsilon'_{ij} = t_{ik}t_{jl}\varepsilon_{kl} \tag{8.39}$$

where the Einstein suffix notation is used. The ε'_{ij} denotes the strain component in the crystal coordinate system in which the axes are the base vectors of the crystal lattice. The t_{ij} are direction cosines of the crystal coordinate system axes in the specimen coordinate system. Since the strains are very small, the strain components in the crystal coordinate system can be related to the changes in lattice parameters as

$$\Delta a/a_0 = \varepsilon'_{11}, \ \Delta b/b_0 = \varepsilon'_{22}, \ \Delta c/c_0 = \varepsilon'_{33}, \ \Delta\alpha = -2\varepsilon'_{23},$$
$$\Delta\beta = -2\varepsilon'_{13}, \quad \Delta\gamma = -2\varepsilon'_{12} \tag{8.40}$$

The starting values of lattice parameters are $a_0=b_0=c_0=0.404\ 958$ nm, and $\alpha=\beta=\gamma=90°$. Following the method of Zuo (1992), two optimization algorithms, the simplex method and the simulated annealing method, are used. The $[1\ 1\ 3]$

zone axis CBED patterns are taken from the regions A and C that are in the matrix and 53 nm and 158 nm away from the interface, respectively. A total of 9 HOLZ lines, 21 intersection points and 210 distances are measured and included in the refinement. The components of strain tensor in the specimen coordinate system are used as adjustable parameters. The results of the refinements with the two algorithms are shown in Tables 8.9 and 8.10. Table 8.11 gives the stress components calculated from the obtained strain components on the basis of isotropic elasticity theory (the modulus of elasticity E=62 GPa).

Table 8.9 The best fit parameter and strain components obtained with optimization

Points	Method	χ^2	ε_{xx}	ε_{yy}	ε_{xy}	ε_{zz}
A	Annealing	0.003 125	0.002 392	0.002 883	0.001 939	–0.002 778
A	Simplex	0.003 125	0.002 392	0.002 883	0.001 938	–0.002 778
C	Annealing	0.005 574	0.000 698	0.002 900	0.001 126	–0.001 895
C	Simplex	0.005 574	0.000 697	0.002 904	0.001 126	–0.001 897

Table 8.10 Lattice parameters obtained with optimization

Points	a (nm)	b (nm)	c (nm)	α (degrees)	β (degrees)	γ (degrees)
A	0.406 321	0.405 238	0.404 326	90.16	90.26	89.92
C	0.405 615	0.405 374	0.404 576	90.15	90.18	89.88

Table 8.11 Stress components calculated from the strain components given in Table 8.9

Points	σ_{xx} (MPa)	σ_{yy} (MPa)	τ_{xy} (MPa)
A	238	260	89
C	119	221	51

The second problem encountered is that, due to relaxation of internal stresses at the foil surfaces, the lattice parameters and the strains obtained in the thin specimen are not necessarily representative of bulk material. Following this, one needs to prepare the specimen and to select the zone axis carefully to minimize the effect. For example, with the LACBED pattern Duan et al. (1994) showed that, in a cross-sectional specimen of a strained-layer superlattice, relaxation and residual strains can be separated by careful selection of the zone axes used for measurement. For instance, for the (0 0 1) Si/SiGe specimen thinned along the $[1\,1\,0]$ direction, one can measure the residual strain from the shifts of diffraction lines $2\,2\,\bar{2}, 3\,\bar{3}\,\bar{3}$ in the pattern taken by tilting the specimen a few degrees away from the $[1\,1\,0]$ zone axis about $[\bar{1}\,1\,0]$ axis, and measure the relaxation in terms of the shifts of diffraction line $\bar{4}\,8\,0$ by tilting the specimen about $[0\,0\,1]$ axis. Balboni et al. (1998) discussed how to determine the bulk mismatch values from CBED results of cross sectional $Si_{1-x}Ge_x$/Si heterostructures. Orthorhombic distortion is assumed. The film is grown along the z-axis which is the $[0\,0\,1]$ direction of cubic crystal, and the y-axis is the thinning direction of the sample. Then, it is reasonable to assume that there is no lattice relaxation along the x-direction where the crystal is effectively infinite. It is demonstrated that, by using elasticity theory, the lattice mismatch along the z-direction in the bulk material can be calculated from the mismatches along y- and z-directions in thinned samples. The mismatches in thinned samples can be obtained by measuring the lattice parameters of the film along the z-and y-directions with the HOLZ line method.

8.8.2 Interfacial Strain Determination by Using HOLZ Line Splitting

The lattice plane bending can cause the HOLZ lines broadening or splitting (Vincent et al., 1988; Yan et al., 1994; Dai et al., 1994). The bending of lattice planes near the interface is due to surface relaxation in addition to the elastic mismatch between the two phases. As shown by Treacy et al.(1985) and Perovic et al.(1991), the local lattice plane rotations due to the surface relaxation are dependent on the initial strain, i.e., the unconstrained lattice misfit between the two phases. If the initial strain can be used as one of the adjustable parameters in computing the lattice displacement field caused by both the elastic mismatch and the surface relaxation, it is possible to determine the initial strain by fitting the simulated LACBED patterns to the experimental ones with respect to the shifts and splitting of HOLZ lines.

To interpret the spatial variation in HOLZ line position and intensity as being the result of strain variation at the interface by dynamical calculations, it is necessary to discuss the influence of experimental parameters on the CBED pattern. Zou et al. (1998) discussed the relerant factors. Let X be a point of the electron source with radius R_a. We assume that the crystal is thin enough for the projection approximation to be valid, and the displacement field $\boldsymbol{R}(r)$ induced by a defect is a function of the position vector $\boldsymbol{r}$ perpendicular to the surface normal. The transmission function of the sample can be written as

$$q(r) = \exp\{-i\sigma\varphi(r)\} = \sum_G F_G \exp\{-2\pi i \boldsymbol{G} \cdot \boldsymbol{R}(r)\} \exp\{2\pi i \boldsymbol{G} \cdot \boldsymbol{r}\} \quad (8.41)$$

Then, the amplitude distribution of the CBED pattern from the point X in the electron source is

$$\{\exp(-2\pi i \boldsymbol{u} \cdot \boldsymbol{X})T(\boldsymbol{u})\} * \left\{\sum_G F_G \delta(\boldsymbol{u} - \boldsymbol{G}) * \mathscr{F}[\exp(-2\pi i \boldsymbol{G} \cdot \boldsymbol{R}(r))]\right\}$$

$$(8.42)$$

where $T(\boldsymbol{u})=A(\boldsymbol{u})\exp\{i\chi(\boldsymbol{u})\}$ is the transfer function of the probe-forming lens,

406

$*$ is the convolution operation and $\mathscr{F}$ denotes the Fourier transform. The term of elastic diffuse scattering $\mathscr{F}\{\exp\left[-2\pi i \boldsymbol{G} \cdot \boldsymbol{R}(\boldsymbol{r})\right]\}$ acts as a spread function. This term is written as $W_G(\boldsymbol{u})$ for short later on. Further, we assume that the electron source is incoherent. The intensity of the diffraction beam $\boldsymbol{G}$ becomes

$$I_G(\boldsymbol{u}) = \int \left|\iint \exp(-2\pi i \boldsymbol{U} \cdot \boldsymbol{X}) T(\boldsymbol{U}) F_G \delta(\boldsymbol{V} - \boldsymbol{G}) W_G(\boldsymbol{u} - \boldsymbol{U} - \boldsymbol{V}) \mathrm{d}V \mathrm{d}U \right|^2 \mathrm{d}X$$

$$= \iint \frac{2J_1(2\pi R_a \,|\boldsymbol{U}' - \boldsymbol{U}\,|)}{2\pi R_a \,|\boldsymbol{U}' - \boldsymbol{U}\,|} T(\boldsymbol{U}) T*(\boldsymbol{U}') F_G F_G^* W_G \times$$

$$(\boldsymbol{u} - \boldsymbol{G} - \boldsymbol{U}) W_G^*(\boldsymbol{u} - \boldsymbol{G} - \boldsymbol{U}') \mathrm{d}U \mathrm{d}U'$$

$$(8.43)$$

If the electron source extends to infinity, $2J_1(X)/X$ is reduced to $\delta(\boldsymbol{U} - \boldsymbol{U}')$ which corresponds to the incoherent illumination. Then

$$I_G(\boldsymbol{u}) = \int |\, F_G A(\boldsymbol{U}) W_G(\boldsymbol{u} - \boldsymbol{G} - \boldsymbol{U})\,|^2 \,\mathrm{d}U \qquad (8.44)$$

If the electron source is a point, $2J_1(X)/X$ tends to 1, and then

$$I_G(\boldsymbol{u}) = \left|\int F_G T(\boldsymbol{U}) W_G(\boldsymbol{u} - \boldsymbol{G} - \boldsymbol{U}) \mathrm{d}U \right|^2 \qquad (8.45)$$

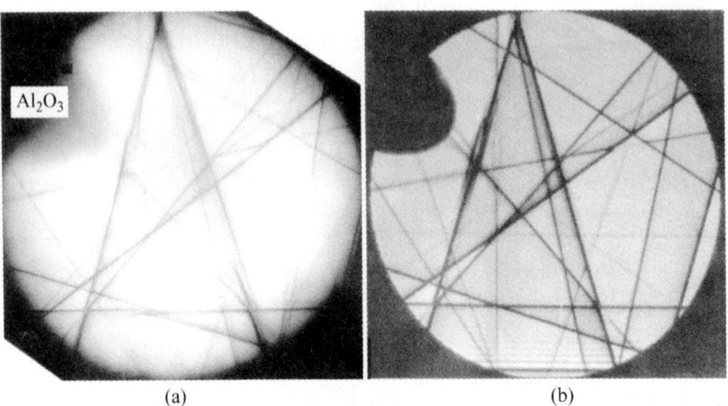

Fig. 8.22 (a) Experimental LACBED pattern taken from the area containing an alumina particle. (b) Simulated LACBED pattern. The half-lengths of the three principal axes of the ellipsoid-like particle are: a=150 nm, b=85 nm, c=60 nm. The eigenstrain used for the calculation is 3.3×10^{-3}. This pattern is calculated with 50 beams of which 40 beams are in ZOLZ

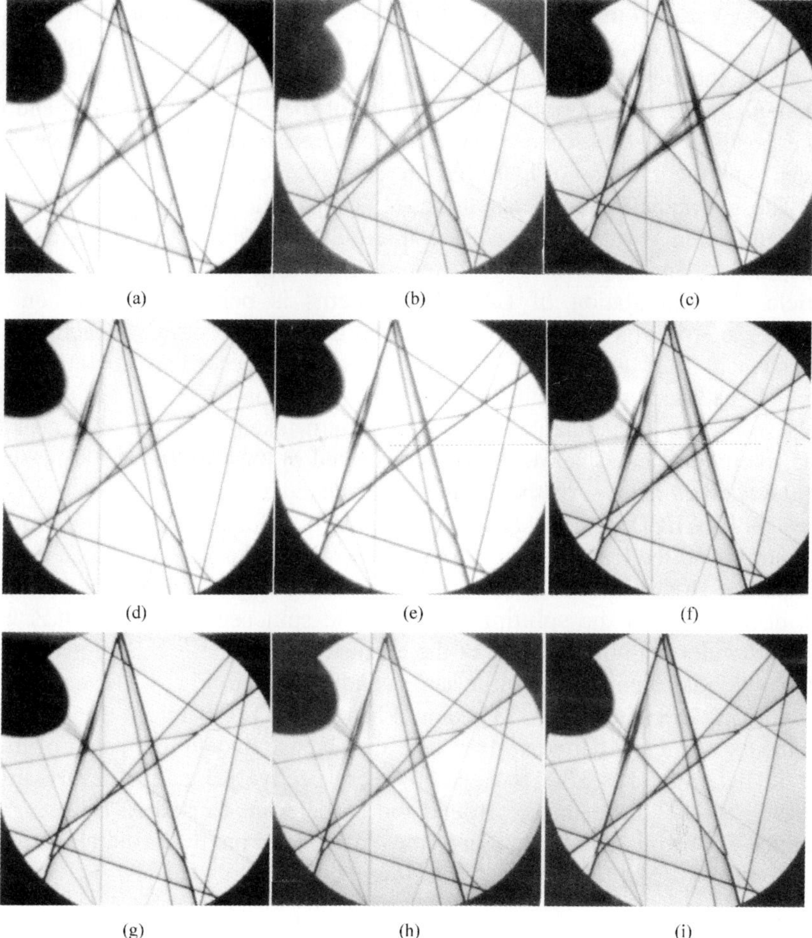

Fig. 8.23 A series of simulated LACBED patterns with the eigenstrain ε^{**} and c, the length of the shortest axis of the ellipsoid, as adjustable parameters.
(a) ε^{**}=2.7×10^{-3}, c=55 nm; (b) ε^{**}=3.3×10^{-3}, c=55 nm;
(c) ε^{**}=4.0×10^{-3}, c=55 nm; (d) ε^{**}=2.7×10^{-3}, c=60 nm;
(e) ε^{**}=3.3×10^{-3}, c=60 nm; (f) ε^{**}=4.0×10^{-3}, c=60 nm;
(g) ε^{**}=2.7×10^{-3}, c=65 nm; (h) ε^{**}=3.3×10^{-3}, c=65 nm;
(i) ε^{**}=4.0×10^{-3}, c=65 nm

For a source having a diameter of 5 μm, which is typical for a TEM instrument with a pointed, thermal electron source, one has $|U-U'|$=0.24×10^{-3} nm^{-1} for the first zero of $2J_1(X)/X$. When we simulate a LACBED pattern over semi-convergence angle of 0.0335 rad with 120 sampling points, the difference between the tangential components for two adjacent points is 0.75×10^{-1} nm^{-1}

408

(for 100-keV electrons), which is much larger than the coherence width of 0.24 $\times 10^{-3}$ nm^{-1}. Therefore, the LACBED experiments can be considered to be effectively incoherent. From (8.44), one can see that the instrument parameters such as the spherical aberration, the defocus of the probe-forming lens and the probe position (in the sense of coherent electron nanodiffraction) have little influence on the intensity of LACBED in the present case.

The interfacial strain fields around Al_2O_3 particles in Al_2O_3/Al composite were studied in this way by Zou et al.(1998). The shape of the particles is ellipsoid-like, and the two longer axes are measured from a micrograph of a particle. The simulation of LACBED patterns is performed based on the scattering matrix algorithm of the dynamical theory of electron diffraction. The displacement field around the particle is computed according to the Green's function method for an inclusion in semi-infinite space (Mura, 1987). Figure 8.22a shows the experimental LACBED pattern taken from the area near the Al_2O_3 particle whose shadow image is indicated in the pattern. The eigenstrain and the shortest axis, c, of the ellipsoid are treated as adjustable parameters. A series of LACBED patterns are calculated with different combinations of the eigenstrain and the length of the c-axis, as shown in Fig. 8.23. It is seen that the split part of HOLZ lines near the particle is getting larger as the length of the c-axis decreases, and the splitting width of the split segments of HOLZ lines becomes wider with the increase of the eigenstrain. The best fit is shown in Fig. 8.22b which is calculated with the same conditions as in Fig. 8.23e, except that Fig. 8.22b is calculated with 50 beams. When the best fit between the simulated and the experimental patterns is reached, the eigenstrain and the length of the c-axis are determined as 3.3×10^{-3} and 60 nm, respectively. Finally, the stress field and strain field in the matrix around the particle are calculated by using the obtained eigenstrain and the length of the c-axis based on the elasticity theory.

References

Aitchison, P.R., J.M. Rodenburg. In: Electron Microscopy and Analysis. 1991. Inst. Phys. Conf. Ser. No.119, Section 9, (The Institute of Physics, Bristol and London 1991) pp.409

Buxton, B.F., J.A. Eades, J.W. Steeds and G.M. Rackham. Phil. Trans. R. Soc. Lond . **A281**, 171 (1976)

Cherns, D., C.J.Kiely and A.R. Preston. Ultramicroscopy. **24**, 355 (1988)

Cherns, D. and A.R. Preston. In: Proc. XIth Int. Congr. on Electron Microscopy. (Kyoto, 1986) pp. 721

Cowley, J.M. (ed.). Electron Diffraction Techniques. Oxford Univ. Press, Oxford, (1992)

Dai, M.X., H. Zou, R. Wang, Y. Yan and J. Pan. Phys. Stat. Sol. (a) **142**, k79 (1994)

Duan, X.F. Ultramicroscopy. **41**, 249 (1992)

Duan X.F., D. Cherns and J.W. Steeds. Phil. Mag. **A70**, 1091(1994)

Feng, Jianglin and Wang, Renhui .. Phil. Mag. **A69**, 981(1994)

Feng, Jianglin, Wang, Renhui and Zou, Huamin. Phil. Mag. **A72**, 1121 (1995)

Goodman, P. Acta Cryst .. **A31**, 804 (1975)

Hirsch, P., A. Howie, R.B. Nicholson, D.W. Pashley, and M.J. Whelan. Electron Microscopy of Thin Crystals. (Robert E. Krieger, Huntington, 1977)

Hoier, R., L.N. Bakken, K.Marthinsen and R. Holmestad.Ultramicroscopy. **49**, 159 (1993)

Humphreys, C.J., D.J. Eaglesham, D.M. Maher and H.L. Fraser. Ultramicroscopy. **26**, 13(1988)

Jiao, Suilong, Zou, Huamin and Wang, Renhui.J.Chin. Electr. Microsc. Soc. **6**, No. 2, 42(1987). (in Chinese)

Jones, P.M., G.M. Rackham and J.W. Steeds.Proc. R. Soc. Lond . **A354**, 197 (1977)

Kossel, W. and G. Mölenstedt.Ann. Phys. **36**, 113 (1939)

Li, B., H. Zou and J. Pan. Scripta Materialia. **38**, 1419 (1998)

Lin, Y.P., D.M. Bird and R. Vincent. Ultramicroscopy. **27**, 233 (1989)

Loretto, M.H. and R.E. Smallman. Defect analysis in electron microscopy. (Chapman and Hall, London, 1975)

Mansfield, J., D. Bird and M. Saunders. Ultramicroscopy. **48**, 1(1993)

Mura, T. Micromechanics of Defects in Solids. (Martinus Nijhoff Publishers, Dordrecht, 1987)

Okuyama, T., S Matsumura, N. Kuwano and K. Oki. Ultramicroscopy. **31**, 309 (1989)

Perovic, D.D., G.C. Weatherly and D.C. Houghton. Phil. Mag. **A64**, 1 (1991)

Rozeveld, S.J., J.M. Howe and S. Schmauder. Acta Metall. Mater. **40**, S173 (1992)

Rozeveld, S.J., and J.M. Howe. Ultramicroscopy. **50**, 41 (1993)

Spence, J.C.H. Acta Cryst. **A49**, 231 (1993)

Spence, J.C.H. and J.M. Zuo. Electron Microdiffraction (Plenum, New York, 1992)

Steeds, J.W. In: Introduction to Analytical Electron Microscopy, (eds.) J.J.Hren, J. I . Goldstein and D.C. Joy (Plenum Press, New York, 1979)

Steeds, J.W. and R.J. Vincent. Appl. Cryst. **16**, 317 (1983)

Tanaka, M., R. Saito and H. Sekii.Acta Cryst. **A39**, 357 (1983)

Tanaka, M. and M. Terauchi. Convergent- Beam Electron Diffraction. (JEOL Ltd., Tokyo, 1985)

Tanaka, M., M. Terauchi and T. Kaneyama. Convergent-Beam Electron Diffraction II . (JEOL Ltd., Tokyo, 1988)

Tanaka, M., M. Terauchi and K. Tsuda. Convergent- Beam Electron Diffraction III. (JEOL Ltd., Tokyo, 1994)

Treacy, M.M.J., J.M. Gibson and A. Howie. Phil. Mag. **A51**, 389 (1985)

Vincent, R., A.R. Preston and M.A. King. Ultramicroscopy. **24**, 409 (1988)

Yan,Y., R.Wang, H.Zou, M.X.Dai, L.Froyen and L.Delaey. Scripta Metall.Mater. **30**, 885 (1994)

Wang, R. and M.X. Dai.Phys. Rev.. **B47**, 15326 (1993)

Wang, R., H. Zou and S. Jiao. In: Proc. XIth Int. Congr. on Electron Microscopy. Kyoto, (1986)

Wang, Z.G., R. Wang and J.L. Feng. Phil. Mag. **A70**, 577 (1994)

Wen, Jianguo, Wang, Renhui and Lu, Ganghua. Acta Cryst. **A45**, 42 (1989)

Zou, H., J. Liu, D.-H. Ding, R. Wang, L. Froyen and L. Delaey. Ultramicroscopy. **72**, 1 (1998)

Zuo, J.M. Ultramicroscopy. **41**, 211 (1992)

Zuo, J.M. and J.C.H. Spence.Ultramicroscopy. **35**, 185 (1991)

9 Numerical Simulations for Micromechanical Analyses of Fiber-Reinforced Composites, Thin Films and Coatings

Yijun Liu

9.1 Introduction

In recent years, more and more advanced materials have been developed and applied in various engineering fields with the advance of materials science and engineering. Traditional materials, such as metals, are being gradually replaced or complemented in the design of engineering structures by the advanced materials, such as composites, thin films and coatings, and the most interesting ones—smart materials. This trend will continue and will be advancing at a faster rate, well into the 21st century.

Composite materials are important modern structural materials in aerospace, civil, and most recently, automotive engineering. Composites are of light weight, high strength and long-service life. With the advance of technology in fabrication or manufacturing of composite materials, more composite materials are being developed in order to meet the demands in engineering applications (Chawla, 1998; Hyer, 1998).

Thin materials or structures, such as thin films in electronic devices and coatings on structure components, have been designed and utilized in many industries. For example, in recent years the use of coatings on machine compon ents for wear resistance, friction reduction, or as thermal barriers, has become widespread. Coatings can provide a durable and low-cost solution to many structure performance problems (Subramanian and Strafford, 1993).

Smart materials or structures (also referred to as adaptive structures or intelligent structures in the literature) are structures with attached or embedded sensors and actuators that are c ontrolled by microprocessors to both monitor the state of the structures and control their responses. In recent years, smart materials have generated a great deal of interest in various fields like aerospace, naval, automobile, civil and biomedical engineering. Smart materials can be applied, for example, in aerospace and automobile engineering for active vibration and noise controls; in civil engineering for monitoring and controlling the "health" of structures; in biomedical engineering for developing artificial biomaterials that can sense and respond; and in mechanical engineering for high precision in various robots. One of the ultimate goals of smart materials technology is to replicate biological functions in man-made structures or systems, which employ netwo rks of advanced sensors and actuators as the nerve systems and have the capabilities of sensing, responding, learning and self-healing in the working environment. Extensive reviews, regarding the concepts, development and applications of smart materials in various fields, can be found in (Rogers, 1993; Tzou, 1993; Banks, Smith et al. 1996; Culshaw, 1996; Sater, 1997; Wang and Kang, 1998).

Before any new materials can be used in an engineering structure, their overall mechanical properties (for example, the engineering material constants: equivalent Young's moduli, Poisson's ratios, thermal expansion coefficients, etc.) must be determined first so that the structure can be analyzed to meet the design requirements. Another major concern in the applications of these advanced materials in engineering is the durability issue: can the designed materials in a structure sustain the mechanical, thermal or other loads in service? Any excessive deformation or stresses in a structure can lead to malfunctions or failures of the structure. This issue should be addressed during the design stage of such materials to avoid catastrophes in later application s. Using numerical simulations in the design of the advanced materials is one of the important tools for such purposes.

9.1.1 Numerical Simulations in Materials Research

Numerical simulations are very important for the understanding, analysis and design of advanced materials, since analytical methods are usually restricted to simple geometries and boundary conditions, and experimental work is often limited by cost, time, size and other factors. Numerical modeling of the advanced materials, such as composites, thin films and coatings, and piezoelectric solids in smart materials, is versatile, cost effective, and easily attainable. With the ever-increasing power of the computer hardware and software, many of the simulation tasks can now be done on a desktop computer. The main issues, or difficulties in the simulation, are still the underlying physics of the problems, often the subjects of solid mechanics, fracture mechanics and so on. Meaningful computer simulations are very much dependent on the accuracy of the mathematical models or formulations of the problem for the material under investigation. For the same mathematics or mechanics model of the problem, there are often several numerical solution techniques available. The choice of one numerical technique over the other is situation dependent. Preferably, at least two techniques should be employed and compared against each other to validate the simulation results. The most commonly used numerical techniques in the study of the mechanical properties of materials are the finite element method (FEM) and the boundary element method (BEM),

412

which will be introduced briefly in the following two subsections.

9.1.2   The Finite Element Method (FEM)

The finite element method is a domain-based numerical method for solving the governing equations (usually partial differential equations) describing the physics of the problem (e.g., determining deformation, stress, temperature or electromagnetic fields). First, the entire domain of the problem is divided into many small subdomains (called elements) of particular shapes (triangles and quadrilaterals in 2-D, or cubes and tetrahedrons in 3-D). Distribution of the physical quantities (e.g., the displacement) is approximated over each element using the (unknown) values of the quantities at a finite number of points (called nodes) on the element. A simple relation between the physical quantities (e.g., displacement and force) at the nodes is established for the element. Then, these relations, or algebraic equations, for all the elements covering the entire domain are assembled to form the whole system of equations in terms of the nodal values of the quantities (displacements and forces). Solving this system of equations will provide the values of the physical quantities at all the nodes. Values at any point other than the nodes can be obtained from the interpolation of the nodal values over the element covering that point. The finite element procedures are simple and straightforward. The systematic nature of the method makes it very easy to implement on computers. Interested readers should consult reference books on the finite element method that are abundant in the literature (see, e.g., (Carey and Oden 1983; Cook, Malkus et al., 1989; Zienkiewicz and Taylor, 1989; Reddy, 1993; Bathe, 1996; Chandrupatla and Belegundu, 1997)).

The FEM dates back to the 1940s and has, over the years, become the most popular computational method in the engineering fields for a wide range of analyses. There are many commercial software packages available today for engineers (such as MSC/NASTRAN, ANSYS, ABAQUS, SDRC/I-DEAS, LS-DYNA3D and many others). The FEM can deal with homogeneous and isotropic materials, as well as inhomogeneous and anisotropic materials. It is especially good for large-scale modeling of mechanical systems (an automobile, spacecraft, etc.) to predict the overall performance of the structures. However, for detailed analyses, such as interface stress analysis needed for materials research, a large number of the elements (finer mesh) will be required in a very small region, which is usually in the microscale (This will be demonstrated in Sect. 9.4 using an example in stress analysis of thin films/coatings). Thus, the FEM can be inefficient for same cases in the simulations of materials.

9.1.3   The Boundary Element Method (BEM)

The boundary element method, also called boundary integral equation (BIE) method, is a boundary-based solution technique. Instead of solving the original governing equations in the problem domain, these governing equations are first transformed into boundary integral equations using the Green's identity and the fundamental solution associated with the problem. Then, elements are introduced on the boundary of the domain for the problem, for example, surface elements for 3-D problems and line elements for 2-D problems. The BIEs are subsequently reduced to algebraic equations involving the physical quantities (displacements and tractions) at a finite number of nodes (nodal values) on the boundary. By solving this system of algebraic equations, one obtains the boundary solutions (displacements and tractions), and hence the s olution at any interior point through the use of a boundary integral representat ion containing the boundary values. The BIE formulation for stress analysis (elastostatics) will be presented in the next section. Readers can consult   (Mukherjee, 1982; Cruse, 1988; Brebbia and Dominguez, 1989; Banerjee, 1994; Kane, 1994) to learn more about this interesting numerical method.

The BEM has been demonstrated to be a viable alternative or complement to the FEM, due to its features of boundary discretization and high accuracy. The high accuracy of the BEM for stress analysis, especially in fracture mechanics (Cruse, 1988; Cruse, 1996), is well recognized, mainly because of its semianalytical nature and boundary-only discretization. Many researchers have employed the BEM in the simulations of materials; see (Mackerle, 1994) for an incomple te list of the publications in composite materials, and (Chen and Lin, 1995; Lee, 1995; Shi et al., 1995; Shi et al., 1996; Luo et al., 1998) for some recent applications to other advanced materials, including piezoelectric solids for smart materials. In order to appreciate the effectiveness and efficiency of the boundary element method, we will discuss briefly the formulations of the BEM in the next section, using the elastostatic problem as an example.

9. 2 Formulation of the Boundary Element Method

9.2.1 The Governing Equations

Consider the linear elastostatic problem in a 2-D or 3-D domain V (Fig.9.1). The governing equations are given as follows (Index notation is used here. Refer to textbooks on elasticity for more details):

Equilibrium equations $\qquad \sigma_{ij,j} + f_i = 0, \qquad \forall P \in V$ $\qquad\qquad$ (9.1)

Constitutive equations $\qquad \sigma_{ij} = E_{ijkl}\varepsilon_{kl}, \qquad \forall P \in V$ $\qquad\qquad$ (9.2)

Strain-displacement relations $\quad \varepsilon_{ij} = \dfrac{1}{2}(u_{i,j} + u_{j,i}), \qquad \forall P \in V$ $\qquad$ (9.3)

where σ_{ij}, ε_{ij} and u_i are the stresses, strains and displacements, respectively, at a point P inside the body V; f_i the body force and E_{ijkl} the elastic modulus tensor,

414

which satisfies

$$E_{ijkl} = E_{jikl} = E_{ijlk} = E_{klij} \tag{9.4}$$

For isotropic materials, we have

$$E_{ijkl} = \frac{2\mu\nu}{1-2\nu}\delta_{ij}\delta_{kl} + \mu(\delta_{ik}\delta_{jl} + \delta_{il}\delta_{jk}) \tag{9.5}$$

where μ is the shear modulus, ν the Poisson's ratio and δ_{ij} the Kronecker delta.

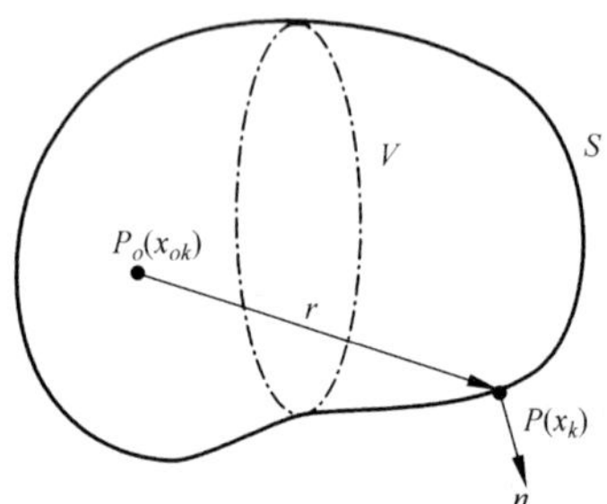

Fig.9.1　3-D closed domain V in R^3 with boundary S

The set of equations in (9. 1)-(9. 3) (15 equations in total for 3-D problems) must be solved for the given domain and material properties to fully determine the displacement, strain and stress fields, under the following boundary conditions (BCs):

Displacement BC　　　　$u_i = \bar{u}_i,$　　　　　$\forall P \in S_1$ 　　　(9.6)

Traction BC　　　　　$t_i = \sigma_{ij}n_j = \bar{t}_i,,$　　　$\forall P \in S_2$ 　　　(9.7)

where the quantities with the over bar indicate the given values, n_i the directional cosines of the normal n and $S=S_1 \cup S_2$. Only in very simple casescan the governing equations be solved exactly using the analytical approach (see, e. g., (Timoshenko and Goodier, 1987)). In general, numerical methods, such as the BEM, are the only way to solve the elastostatic problems.

9.2.2　The Boundary Integral Equation Formulation

To establish the boundary integral equations for a given problem, one needs two ingredients: the generalized Green's identity and the fundamental solution for the type of problem. For the elastostatic problem, the generalized Green's identity, or reciprocal theorem, is

$$\int_V (\sigma_{jk,k} u_j^* - \sigma_{jk,k}^* u_j)\,\mathrm{d}V = \int_S (t_j u_j^* - t_j^* u_j)\,\mathrm{d}S, \tag{9.8}$$

where $(\sigma_{jk},\, \varepsilon_{jk},\, u_j)$ and $(\sigma_{jk}^*,\, \varepsilon_{jk}^*,\, u_j^*)$ are two sets of solutions to (9.1)-(9.3). This identity can be verified easily by using the Gauss theorem. The fundamental solution for the elastostatic problem is the response (displacement, strain and stress fields) at a field point P due to a concentrated unit force at a source point P_0 in an infinite elastic medium. The stress tensor $\sum_{ijk}(P,P_0)$ of the fundamental solution satisfies

$$\sum_{ijk,k}(P,P_0) + \delta_{ij}\delta(P,P_0) = 0, \qquad \forall P, P_0 \in R^3 \tag{9.9}$$

where the first subscript i indicates the direction of the unit force and $\delta(P,P_0)$ is the Dirac-δ function, which is the mathematical representation of the unit force.

Choosing σ_{jk}, t_j, u_j as the solution of the given problem governed by (9.1) – (9.3) and BCs (9.6)–(9.7), and $\sigma_{jk}^* = \sum_{ijk}$, $u_j^* = U_{ij}$, $t_j^* = T_{ij}$ (the fundamental solution), in the reciprocal theorem (9.8), one obtains the following, after some simple manipulations:

$$u_i(P_0) = \int_S \left[U_{ij}(P,P_0) t_j(P) - T_{ij}(P,P_0) u_j(P) \right] \mathrm{d}S(P)$$
$$+ \int_V U_{ij}(P,P_0) f_j(P)\,\mathrm{d}V(P), \qquad \forall P_0 \in V \tag{9.10}$$

This is an integral representation of the displacement at any point inside the domain V, once the boundary values of the traction and displacement are found on S. The volume integral in (9.10) involves no unknown functions. To obtain the unknown boundary values of the traction and displacement (only half of them are given by the boundary conditions), let the source point P_0 go to the boundary S in the representation (9.10) and one obtains

$$C_{ij}(P_0) u_j(P_0) = \int_S \left[U_{ij}(P,P_0) t_j(P) - T_{ij}(P,P_0) u_j(P) \right] \mathrm{d}S(P)$$
$$+ \int_V U_{ij}(P,P_0) f_j(P)\,\mathrm{d}V(P), \qquad \forall P_0 \in S \tag{9.11}$$

where $C_{ij}(P_0) = \dfrac{1}{2}\delta_{ij}$ for a smooth boundary. Equation (9.11) is the desired boundary integral equation (BIE) for the elastostatic problem, which relates the unknown boundary data (traction or displacement) to the given boundary data. By solving (9.11), numerically in most cases, one can find the unknown boundary values of the traction or displacement.Then, by using (9.10), one can determine the displacement or stress at any point inside the domain V. Detailed derivations of the BIE (9.11) and the expressions for the fundamental solution (also called kernel functions) $U_{ij}(P, P_0)$ and $T_{ij}(P, P_0)$ can be found in any BEM reference (see, e.g., (Mukherjee, 1982; Cruse, 1988; Brebbia and Dominguez, 1989; Banerjee, 1994; Kane, 1994)).This BIE for 2-D elastostatic problems in the form of (9.11) was first established by Rizzo in 1967 (Rizzo, 1967), based on his Ph.D. dissertation research, which was completed earlier (in 1964).This BIE formulation has formed the bases of the BEM applications for not only elastostatics, but also elastodynamics, fracture mechanics and many other problems (see, (Cruse and Rizzo, 1968; Cruse, 1969; Cruse and Rizzo, 1975; Rizzo and Shippy, 1977; Rizzo, Shippy et al., 1985; Cruse, 1988; Liu, 1992; Liu and Rizzo, 1993; Liu and Rizzo, 1997) and many others).

It should be emphasized that, up to (9.11), one has not introduced any approximations beyond the assumptions in elasticity.Errors are introduced only when one tries to discretize BIE (9.11) using the boundary elements.These errors are local, meaning that they are mostly confined to near the boundary of the body.Also, these errors are, in general, not propagating in the numerical (BEM) procedure (errors for values inside the domain are, in general, smaller than those on the boundary) due to the desirable features of the integration.All these are in strong contrast to other differentiation and domain-based numerical methods, such as the FEM or FDM (finite difference method).These integral and boundary based features are the main sources of accuracy and efficiency of the boundary element method in stress analysis, especially when stress concentration occurs, such as near cracks, voids, or interfaces in materials.

For the problems in materials research, there are often several different materials, each of which may be modeled as a homogeneous and isotropic material, as in the case of composite materials.In these cases, the BIE given in (9.11) will be applied to each material domain and the resulting BEM equations from each domain will be coupled through the enforcement of the interface conditions (perfect bonding, imperfect bonding or debonding).The next section will demonstrate the effectiveness and efficiency of the BEM for the micromechanical analysis of fiber-reinforced composites with the presence of interphases.

9.3 Micromechanical Analysis of Fiber-Reinforced Composites with Interphases

9.3.1 Review of the Research

Interphases, or interfacial zones, in fiber-reinforced composite materials are the thin layers between the fiber and matrix, Fig. 9.2. These interphases are formed by, for example, chemical reactions between the fiber and matrix materials, or the use of protective coatings on the fiber during manufacturing. The fiber, which is employed to reinforce the matrix material in the fiber direction, is usually much stiffer and stronger than the matrix material.Different levels of stresses and deformations can develop in the fiber and matrix materials, because of this mismatch in the material properties.It is the interphases that bond the fiber and matrix together to ensure the desired functionality of the composite material under external loads. Although small in thickness, interphases have a large area-to-volume ratio in the composites, and significantly affect the overall mechanical properties of the fiber-reinforced composites, as observed in many studies (Achenbach and Zhu, 1989; Achenbach and Zhu, 1990; Hashin, 1991; Zhu and Achenbach, 1991; Yeh, 1992; Jasiuk and Kouider, 1993; Lagache et al., 1994; Chawla, 1998; Wacker, Bledzki et al., 1998).It is the weakest link in the load path, and consequently most failures in fiber-reinforced composites, such as debonding, fiber pullout and matrix cracking, occur in or near this region. Thus, it is crucial to fully understand the mechanism and effects of the interphases in the fiber-reinforced composites.

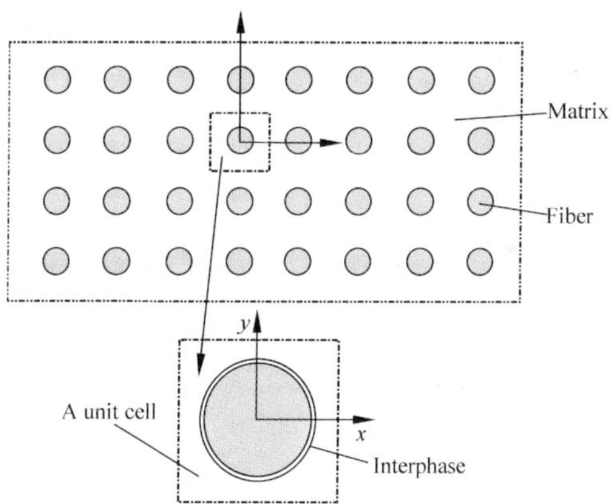

Fig. 9.2　Interphase in a fiber-reinforced composite

Modeling of the fiber-reinforced composite materials presents great challenges to both the FEM and BEM, especially for the analysis at the microstructural level. The main issue in the micromechanical analysis of the fiber-reinforced composites is to predict the interface stresses for durability assessment, and to determine the engineering properties, such as the effective Young's modulus, Poisson's ratios and thermal expansion coefficients, which are needed for structural analysis.Idealized models using the unit cell (or RVE-representative volume element) concept are usually employed in the micromechanics analysis, in which the fibers are assumed to be infinitely long and packed in a square or hexagonal pattern (see, for example, (Aboudi, 1989; Sun and Vaidya, 1996; Hyer, 1998)), Fig. 9.2.Although only one fiber and the surrounding matrix materials are modeled in the unit cell approach, the presence of the interphase between the fiber and matrix still makes the FEM and BEM modeling difficult, simply because of the thicknesses of the interphases, which are in the micrometer scale or below.

Many finite element models based on 2-D elasticity theory have been developed to study the micromechanical properties of the fiber-reinforced composites under transverse loading and with the presence of an interphase, for example, in (Adams, 1987; Yeh, 1992; Nassehi et al., 1993; Lagache et al., 1994) and most recently in (Wacker et al., 1998).In all these FEM models, layers of very fine finite elements were used between the fiber and matrix to model the interphase.Because of the thinness of the interphase, a large number of very fine finite elements are needed in these models, in order to avoid elements with large aspect ratios which can deteriorate the FEM solutions.This, in turn, causes a large number of elements in the fiber and matrix regions because of the connectivity requirement.For instance, in Wacker et al. (1998), more than 3500 finite elements were used to model one quarter of the chosen unit cell.With further smaller thickness of the interphase as compared with the diameter of the fiber, even more elements will be needed in the FEM model.Thus, using finite elements based on elasticity theory for the modeling of interphases can be costly and inefficient, especially in 3-D analyses that are still very scarce in the literature.

For the analysis of micromechanical behaviors of fiber-reinforced composites using the BEM, there are very few publications in the literature, and all of the BEM models developed so far are two-dimensional and account only for transverse loading. Perfect bonding interface conditions, that is, continuous displacements and tractions across the interface of the fiber and matrix, are often assumed in these studies. No BEM models have been attempted earlier to model the interphases.Nevertheless, the BEM has been demonstrated to be an accurate and efficient, alternative numerical tool for the micromechanical analysis of fiber-reinforced composites.

Achenbach and Zhu (1989) developed a 2-D model of a square unit cell using the BEM.To study the effect of the interphase, the continuity of the tractions across the interface of fiber and matrix is maintained, while a linear relation between the displacement differences and the tractions across the interface is introduced.This simple relation represents a spring-like model of the interphase. The proportionality constants used in this model characterize the stiffness and strength of the interphase.Although this spring-like model of the interphase in fiber-reinforced materials is very simple, it can reveal many basic characteristics of an interphase.For example, the study in (Achenbach and Zhu, 1989) shows that the variations of the interphase parameters can cause pronounced changes in the stress distributions in the fiber and matrix.The initiation, propagation and arrest of the interface cracks were also analyzed in (Achenbach and Zhu, 1989).The same approach to the interphase modeling was extended in (Achenbach and Zhu, 1990) to study hexagonal-array fiber composites, and in (Zhu and Achenbach, 1991) to study the micromechanical behaviors of a cluster of fibers.Oshima and Watari (1992) calculated the transverse effective Young's modulus using the 2-D BEM for a square unit cell model.No interphase was modeled and perfect bonding between the fiber and the matrix was assumed.The BEM results using constant elements were shown to be in very good agreement with the experimental data.Gulrajani and Mukherjee (1993) studied the sensitivities and optimal design of composites with a hexagonal array of fibers.A 2-D BEM model with the same, spring-like interphase model as in (Achenbach and Zhu, 1989) was used.The sensitivities of stresses at the interphase were calculated and employed to optimize the value of the stiffness of an interphase in order to minimize the possibility of failure of a composite.Most recently, Pan et al. (1998) developed a similar 2-D BEM model using the same interphase relation as in (Achenbach and Zhu, 1989) to study the perfectly bonded as well as imperfectly bonded fiber-reinforced composites.A main component in this research was the development of a library of Green's functions (or matrices of BEM equations) for fiber-reinforced composite materials, for use by engineers in the design of such composites. Although successful to some extent, the BEM models of the unit cells for fiber-reinforced composites with the spring-like interphase relations are incapable of providing other important information about the properties of composites, such as changes of the thickness and nonuniform distribution of the interphases. An improved BEM model of the interphases based on the elasticity theory is needed.

Interphases as in the fiber-reinforced composites are thin, shell-like structures.For this class of structures, there have been two major concerns in applying the BEM.The first concern is whether or not the conventional boundary integral equation (CBIE) for elasticity problems can be applied successfully to thin structures.It is well known in the BIE/BEM literature that the CBIE will degenerate when it is applied to cracks or thin voids in structures because of the closeness of the two crack surfaces (see, e.g., (Cruse, 1988; Krishnasamy et al., 1994)). One of the remedies to such degeneracy in the CBIE for crack-like problems (exterior-like problems), is to employ the hypersingular boundary integral equation (HBIE), or the traction BIE (see, e.g., (Gray et al., 1990; Krishnasamy et al., 1991; Cruse, 1996; Liu and Rizzo, 1997)).Does this degeneracy occur when the CBIE is applied to thin structures (interior-like problems), such as thin shells?It was not clear in the BEM literature and the BEM based on elasticity had been avoided in analyzing thin shell-like structures for a long time due to this concern.Recently, it was shown in (Liu, 1998; Luo et al., 1998), both analytically and numerically, that the CBIE will not degenerate when it is applied to thin shell-like structures, contrary to the case of crack-like problems.Thus the degeneracy issue should no longer be a concern when the CBIE is applied to thin structures, once the second concern, i.e., the numerical difficulty is addressed.

420

The numerical difficulty in the BIE is how to compute the nearly singular integrals that arise in thin-structure problems.The integrals in BIE, which determine the influence matrices, contain singular kernels of the order $O(1/r)$ and $O(\ln r)$ in the 2-D elasticity case, where r is the distance between the source point and the integration point on the boundary element.When the source point is very close to, but not on the element (with small r), although the kernels are regular in the mathematical sense, the values of the kernels change rapidly in the neighborhood of the source point.The standard Gauss quadrature is no longer practical in this case since a large number of integration points are needed in order to achieve the required accuracy.In the last two decades, numerous research studies have been published on this subject in the BEM literature. Detailed studies on the behaviors of the nearly singular integrals and comprehensive reviews of the earlier work in this regard can be found in (Cruse and Aithal, 1993) and (Huang and Cruse, 1993).One of the most efficient and accurate approaches to deal with the nearly singular integrals in the BEM for 3-D problems is to transform these (surface) integrals to line integrals analytically and then carry out the numerical integration for these line integrals (Liu et al., 1993; Krishnasamy et al., 1994; Liu, 1998).A similar approach can be established for 2-D elasticity problems (Luo et al., 1998). It has been demonstrated in (Luo et al. 1998) that very accurate numerical solutions can be obtained for thin structures with the thickness to length ratio in the micro- and even nano- scales, using the newly developed BEM approach, without seeking refinement of the meshes as the thickness decreases.

With the degeneracy issue for the CBIE in thin-structure problems having been clarified and the nearly singular integrals dealt with accurately and efficiently, it is believed that the BEM can now be applied to a wide range of problems in engineering, including simulations of thin shell-like structures (Liu, 1998), thin-film and coatings on the micro-or nano-scales (Luo et al. 1998), and in particular, the interphases in fiber-reinforced composite materials.

A 2-D detailed model for the interphases in fiber-reinforced composite materials has been developed to study their micromechanical behaviors.All the regions–the fiber, matrix and interfacial zone contained in a unit cell, are modeled using an advanced boundary element method with thin-body capabilities and based on the 2-D elasticity theory.Effective elastic moduli in the transverse directions and stresses in the interphases are computed using this approach. This 2-D model of the interphases can provide more accurate interface stresses and therefore a more accurate account of the micromechanical behaviors of fiber-reinforced composites than the current spring-like models.Extension of this work to 3-D models is underway in order to investigate the fiber-pullout failure modes.

9.3.2 Two Unit Cell Models with the Interphase

Two unit cell models are used in this study, namely, the concentric cylinder model and the square model (see, e.g., (Hyer, 1998)), both of which include the interphase (Figs.9.3–9.5). For the cylinder model, analytical solutions are given for the displacement and stress fields, which can be employed to validate the BEM results.For the square model, many numerical (FEM and BEM) solutions are available in the literature for the effective elastic moduli that will be compared with the data from the present BEM approach.

a) Concentric cylinder model

For the concentric cylinder model, Fig.9.3, the response of the composite in the x-y plane is axisymmetric if the applied load or displacement on the boundary S_3 is also axisymmetric. Here it is assumed that a radial displacement δ is given on S_3 (at $r = c$, Fig. 9.3). Applying the theory of elasticity for the plane strain case in the polar coordinate system (r, θ), one can derive the following expressions for the radial displacement and stress fields in the fiber, interphase and matrix, respectively:

$$u^{(f)}(r) = A^{(f)}r, \qquad (0 \leqslant r \leqslant a)$$

$$u^{(i)}(r) = A^{(i)}r + \frac{B^{(i)}}{r}, \qquad (0 \leqslant r \leqslant b) \qquad (9.12)$$

$$u^{(m)}(r) = A^{(m)}r + \frac{B^{(m)}}{r}, \qquad (0 \leqslant r \leqslant c)$$

and,

$$\sigma_r^{(f)} = \sigma_\theta^{(f)} = k^{(f)}A^{(f)}, \qquad (0 \leqslant r \leqslant a)$$

$$\sigma_r^{(i)} = k^{(i)}\left[A^{(i)} - (1 - 2v^{(i)})\frac{B^{(i)}}{r^2} \right],$$

$$\sigma_\theta^{(i)} = k^{(i)}\left[A^{(i)} + (1 - 2v^{(i)})\frac{B^{(i)}}{r^2} \right], \qquad (a \leqslant r \leqslant b) \qquad (9.13)$$

$$\sigma_r^{(m)} = k^{(m)}\left[A^{(m)} - (1 - 2v^{(m)})\frac{B^{(m)}}{r^2} \right],$$

$$\sigma_\theta^{(m)} = k^{(m)}\left[A^{(m)} + (1 - 2v^{(m)})\frac{B^{(m)}}{r^2} \right], \qquad (b \leqslant r \leqslant c)$$

where the constants $A^{(\beta)}$, $B^{(\beta)}$ are $k^{(\beta)}$ (β=f, i and m)are determined by using

422

the boundary and interface conditions. From the above expressions, one can compute the radial displacement and stress components at any point in the three domains within the cylinder model for any small values of the interphase thickness.

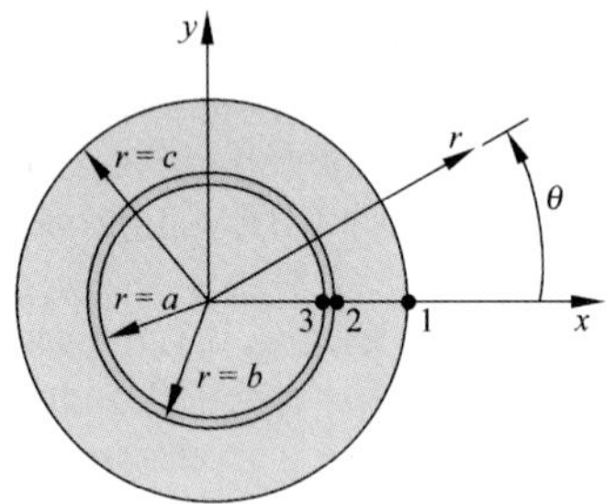

Fig. 9.3 Concentric cylindrical model

b) Square model

As shown in Fig. 9.4, the boundary conditions for the square model under tension are

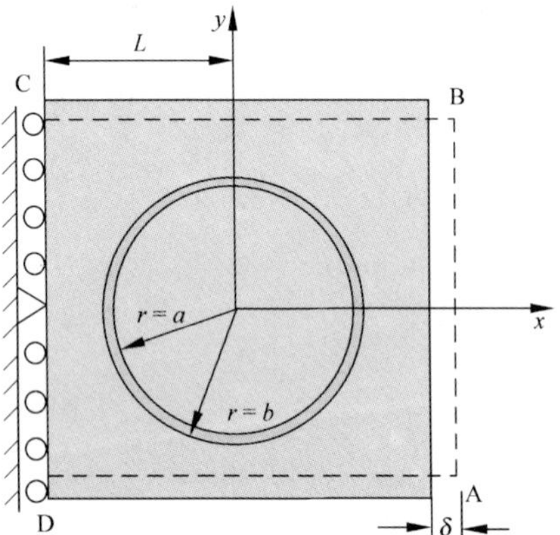

Fig. 9.4 Square model under tension

along AB:	$u_x = \delta$,	$t_y = 0$	
along BC:	$u_y = -C_0$,	$t_x = 0$	
along CD:	$u_x = 0$,	$t_y = 0$,	except at $y=0$ where $u_x = u_y = 0$
along DA:	$u_y = C_0$	$t_x = 0$	

$$(9.14)$$

where u_x, u_y, t_x and t_y are the displacement and traction components, respectively; δ the given displacement (Fig. 9.4) and C_0 an unknown constant. This unknown constant is meant to keep the edges BC and DA straight after the deformation. This represents the constrain of the neighboring cells to the one under study.In the literature, there are several ways of dealing with these subtle boundary conditions along the top and bottom edges (Achenbach and Zhu, 1989; Oshima and Watari, 1992; Pan et al., 1998; Wacker et al., 1998).

Once stresses on the boundary are determined, the average tensile stress along the edge AB is evaluated by:

$$\overline{\sigma}_x = \frac{1}{2L}\int_{-L}^{L}\sigma_x(L,y)dy \tag{9.15}$$

The effective Young's modulus in the transverse direction and under the plane-strain condition is thus determined by:

$$E'_x = \frac{\overline{\sigma}_x}{\overline{\varepsilon}_x} = \frac{\int_{-L}^{L}\sigma_x(L,y)dy}{\delta} \tag{9.16}$$

where $\overline{\varepsilon}_x = \delta/2L$ is the average tensile strain.The effective Poisson's ratio under the plane-strain condition can be determined by:

$$V'_{xy} = -\frac{\overline{\varepsilon}_y}{\overline{\varepsilon}_x} \tag{9.17}$$

in which $\overline{\varepsilon}_y$ is the average strain in the y-direction.

For the square model under shear deformation, Fig. 9.5, the boundary conditions are

$$
\begin{array}{lll}
\text{along AB:} & u_x = 0, u_y = \eta & \\
\text{along BC:} & u_x = 0, t_y = 0 & \\
\text{along CD:} & u_x = 0, u_y = 0 & \\
\text{along DA:} & u_x = 0, t_y = 0 &
\end{array} \tag{9.18}
$$

The average shear stress along edge AB can be evaluated by:

$$\overline{\tau}_{xy} = \frac{1}{2L}\int_{-L}^{L}\tau_{xy}(L,y)dy \tag{9.19}$$

and the effective shear modulus in the transverse plane and under the plane strain condition is:

$$G'_{xy} = \frac{\overline{\tau_{xy}}}{\overline{\gamma_{xy}}} = \frac{\int_{-L}^{L} \tau_{xy}(L, y)\,dy}{\eta} \tag{9.20}$$

where $\overline{\gamma_{xy}} = \eta / 2L$ is the average shear strain.

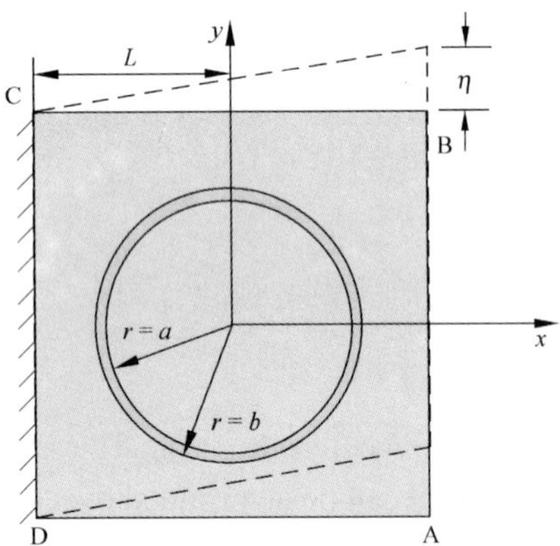

Fig. 9.5 Square model under shear deformation

Finally, one recognizes that the material constants E'_x, V'_{xy} and G'_{xy} given in (9.16), (9.17) and (9.20), respectively, are determined under the plane-strain condition that accounts for the constraint in the z-direction ($\varepsilon_z = 0$). These constants are related to the intrinsic material properties by the following relations (cf., e.g., (Timoshenko and Goodier, 1987; Pan et al., 1998)):

$$E_x = \frac{1 + 2v'_{xy}}{(1 + v'_{xy})^2} E'_x, \qquad v'_{xy} = \frac{v'_{xy}}{1 + v'_{xy}}, \qquad G_{xy} = G'_{xy} \tag{9.21}$$

which are the effective Young's modulus, Poisson's ratio and shear modulus, respectively, in the transverse direction for the composite.

9.3.3 Numerical Examples

a) Cylinder model

The cylinder model (Fig. 9.3) is studied first to validate the developed BEM formulation and the solution strategy. The specified radial displacement on the boundary (at $r = c$) is δ. The following material constants for a glass/epoxy composite are used:

for fiber: $\quad E^{(f)} = 72.4$ GPa$(10.5 \times 10^6$ psi$), v^{(f)} = 0.22;$

for interphase: $E^{(i)} = 36.2$ GPa$(5.25 \times 10^6$ psi$), v^{(i)} = 0.30;$

for matrix: $\quad E^{(m)} = 3.45$ GPa$(0.5 \times 10^6$ psi$), v^{(m)} = 0.35;$

where the Young's modulus of the interphase has been taken as half of that of the fiber; and the dimensions used are:

$$a = c/2, \quad b = a+h,$$

with h being the thickness of the interphase, which is varying.

Quadratic line elements are employed in the discretization and two meshes are tested, one with 24 elements (eight on each circle) and another one with 48 elements (16 on each circle).Differences in the results from the two meshes are less than 5% and the results from the refined mesh (48 elements) are reported.The radial displacements and stresses at selected points (Fig. 9.3) are given in Table 9.1 and Table 9.2, respectively.It is observed that the maximum errors of the displacement and stress using the developed BEM are less than 0.05% in all the cases with different thickness of the interphase. These results demonstrate that the developed BEM approach is extremely accurate and effective in modeling the interphases with any small thickness, as has been confirmed in the context of single material problems (Luo et al. 1998).

Table 9.1 Results of the radial displacement u $(\times 10^{-2}\delta)$ for the cylinder model

	$h=0.1a$		$h=0.01a$		$h=0.001a$	
	Point 2	Point 3	Point 2	Point 3	Point 2	Point 3
BEM	8.2961	7.0833	6.7379	6.6225	6.5909	6.5794
Analytical	8.2958	7.0830	6.7378	6.6224	6.5925	6.5810
Error(%)	0.0036	0.0042	0.0015	0.0015	0.0243	0.0243

Table 9.2 Results of the radial stress σ_r $(\times E^{(m)}\delta/c)$ for the cylinder model

	$h=0.1a$			$h=0.01a$			$h=0.001a$		
	Point 1	Point 2	Point 3	Point 1	Point 2	Point 3	Point 1	Point 2	Point 3
BEM	3.6515	4.2806	4.3550	3.4215	4.0636	4.0714	3.4009	4.0442	4.0449
Analytical	3.6513	4.2803	4.3544	3.4214	4.0633	4.0711	3.4006	4.0461	4.0458
Error(%)	0.0055	0.0070	0.0138	0.0029	0.0074	0.0074	0.0088	0.0470	0.0222

b) Square model

(i) Calculation of effective Young's modulus with varying interphase property

First, the square model under a stretch in the x-direction is considered, Fig.9.4.The properties of the constituent materials considered are for fiber: $\quad\quad\quad E^{(f)} = 84.0 \text{ GPa}, v^{(f)} = 0.22;$

for interphase: $\quad E^{(i)} = 4.0 \sim 12.0 \text{ GPa}, v^{(i)} = 0.34;$

for matrix: $\quad\quad E^{(m)} = 4.0 \text{ GPa}, v^{(m)} = 0.34;$

and a=8.5 μm, b=a+h, L=21.31 μm (fiber-volume fraction V_f =50%).Young's modulus for the interphase is changing in the range between 4.0 and 12.0 GPa. The effect of the variations in the interphase material on the effective Young's modulus of the composite is of the primary interest here.A total of 64 quadratic boundary elements are used, with 16 elements on each of the two circular interfaces and 32 elements on the outer boundary. Table 9.3 shows the effective Young's moduli obtained from the BEM stress data using (9.15) and then (9.20), and compared with those from the FEM quarter model with 3518 linear triangular elements in (Wacker et al., 1998) for the thickness h= 1.0 μm. The BEM results are slightly lower than those from the FEM data.This may be caused by the use of linear triangular elements in the FEM, which tend to overestimate the stiffness of the structure. It is noticed that the different boundary conditions along the top and bottom edges of the square model (free-traction or straight-line conditions) have very little influence on the final effective Young's modulus. It should also be pointed out that for the FEM results in (Wacker et al., 1998), the only thickness considered is h=1.0 μm, which is relatively large compared with the fiber radius (a = 8.5 μm). If a smaller thickness were used in the FEM model, a much larger number of elements would have been needed in order to avoid large aspect ratios in the FEM mesh, as demonstrated in a similar study (see, (Luo et al., 1998)). However, for the BEM employed here, the same number of elements can be used no matter how small the thickness of the interphase.

Table 9.3 Effective transverse elastic modulus (GPa) using the square unit cell model

Interphase property $E^{(i)}$ (GPa)	Current BEM with traction-free conditions on BC and DA	Current BEM with straight-line conditions on BC and DA	FEM (Wacker et al., 1998)
4.0	11.61	11.61	12.25
6.0	13.18	13.02	13.71
8.0	13.97	13.89	14.68
12.0	15.04	14.93	15.91

(ii) Effect of the interphase thickness

Figure 9.6 shows the Effect of different interphase thicknesses on the effective Young's modulus.It is found that the effect of the thickness is not significant on the effective Young's moduli when the fiber-volume fraction V_f is small (50% and less), while a significant effect is observed when V_f is large (70%).This may be due to the fact that the effective elastic moduli are obtained by evaluating the average stress on the outer boundary of the matrix (edge AB, Fig. 9.4).When the fiber-volume fraction is small, the interphase is away from the matrix outer boundary and thus changing the interphase thickness does not affect considerably the stresses on the edge AB.This will change if the fiber-volume fraction is large (e.g., 70%) when the interphase becomes closer to the outer boundary of the matrix. It should also be pointed out that when the fiber-volume fraction is large, it will present an additional difficulty in the modeling using the FEM and earlier BEM formulation, because of the thinness of the matrix region.However, for the current BEM formulation, this additional thickness of the matrix domain does not present any problem.

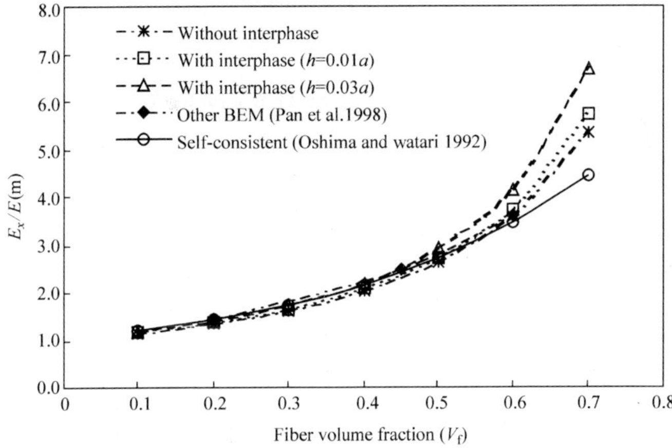

Fig. 9.6 Influence of the thickness on the effective Young's modulus

(iii) Effect of nonuniform thickness

Next, the effect of nonuniform thickness of the interphase on the interface stresses and effective elastic moduli is investigated.The starting model (Fig. 9.4) is the same as the one used for Table 9.3 with the material constants listed at the beginning of this subsection b) (with $E^{(i)}$=42.0 GPa).To form the nonuniform distribution of the interphase, the outer boundary of the interphase is shifted to the left slightly (see Fig. 9.7).When the offset Δ is close to h (the initial, uniform thickness), the change of the interphase thickness in the x-direction is the largest ($\Delta = 0$ corresponds to the uniform interphase).The interface normal stresses at points 1 and 2 (Fig. 9.7), normalized by those in the uniform case, are plotted in Fig. 9.8. Due to the misalignment of the fiber and interphase centers, the interface stress at point 1 (which is the maximum interface normal stress) increases by about 30%, while the stress at point 2 (the second largest interface stress) increases by about 50%.However, the effect of the nonuniform thickness of the interphase on the effective Young's modulus is found to be less than 3%.This, again, is largely due to the averaging process on the edge AB, which is

428

away from the interphase.

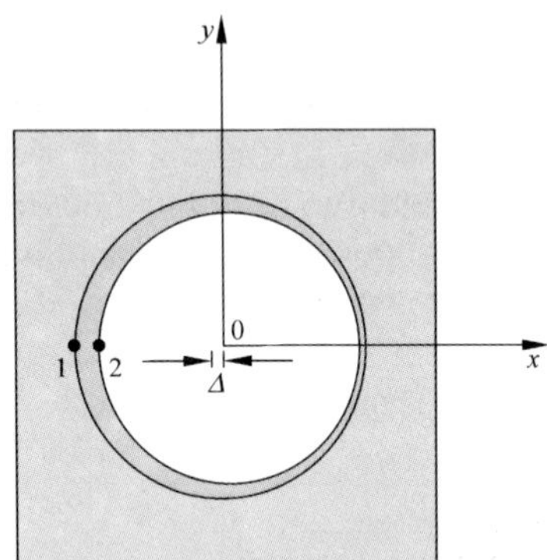

Fig. 9.7   Interphase with nonuniform thickness

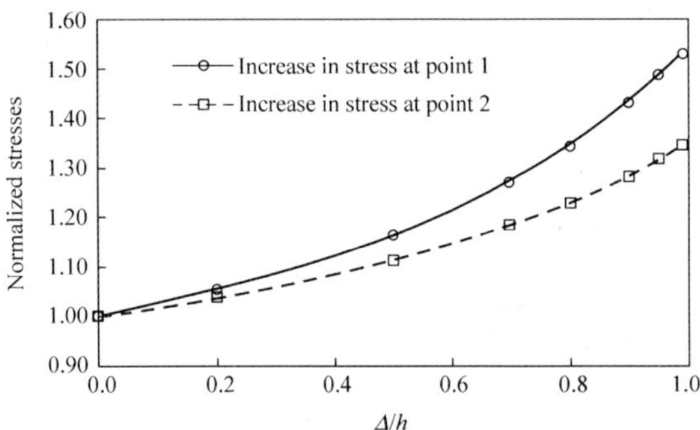

Fig. 9.8   Effect of nonuniform thickness on the interface stress

(iv) Calculation of shear modulus with varying interphase thickness
Finally, the effective shear modulus in the transverse direction is calculated using the square unit cell model shown in Fig. 9.5. The boundary conditions applied are listed in (9.18) and (9.20)–(9.21) are used to compute the shear modulus. In order to compare the results with those in the literature, the following materials properties for a Kevlar/epoxy composite are used in the current BEM calculation:

for fiber:	$E^{(f)} = 7.0 \text{ GPa}, v^{(f)} = 0.30;$
for interphase:	$E^{(i)} = 5.0 \text{ GPa}, v^{(i)} = 0.35;$
for matrix:	$E^{(m)} = 3.0 \text{ GPa}, v^{(m)} = 0.35.$

Table 9.4 shows the results of the effective shear modulus by the current BEM with and without the presence of the interphase. The data without the

interphase ($h = 0$) agrees very well with the results from the FEM (Yeh, 1992); and the other BEM (Pan et al., 1998), both of which used the perfect bonding condition and did not model the interphase.With the increase of the thickness of the interphase, the shear modulus deviates from the perfect bonding case slightly, with the largest change (about 8%) occurring at the fiber volume fraction V_f=0.6, for the interphase property considered.When V_f =0.7 and h= 0.1a, the interphase will be outside the boundary of the unit cell.This is not permissible and thus no BEM data is generated.

Table 9.4   Effective transverse shear modulus (GPa)

	Fiber volume fraction (V_f)			
	0.2	0.4	0.6	0.7
FEM (Yeh,1992)	1.297	1.529	1.826	–
Other BEM (Pan et al.,1998)	1.294	1.513	1.798	–
Current BEM (h=0.0)	1.2939	1.5133	1.7981	1.9866
Current BEM (h=0.001a)	1.2941	1.5139	1.7994	1.9885
Current BEM (h=0.01a)	1.2964	1.5196	1.8111	2.0059
Current BEM (h=0.1a)	1.3205	1.5807	1.9446	–

9.3.4   Discussion

The advanced BEM formulation with thin-body capabilities for elastostatic problems (Liu, 1998; Luo et al., 1998) has been extended to multidomain problems and applied to model the interphases in fiber-reinforced composites under transverse loading.Compared with the current spring-like models for the interphases in the BEM literature, this new interphase model is based on the elasticity theory and thus provides a more accurate account of the interphases in fiber-reinforced composites within the linear theory.The developed BEM code using the object-oriented programming language (C++) can be utilized in analyzing the micromechanical properties of fiber-reinforced composites with the presence of the interphases of any arbitrarily small thickness (uniform or nonuniform).The approach is very accurate, as is validated using the concentric cylinder model for which the analytical solution has been derived.It is also very efficient as only a small number (less than one hundred) of boundary elements are needed to model the whole unit cells for the BEM analysis, compared with the large number (more than a few thousands) of finite elements often needed for a quarter model in the FEM analysis.The approach provides greater flexibility in parametric study of the interphases as well, since the geometry, size or material property of the interphases can be changed very easily to investigate their effect on the micromechanical behaviors of the fiber-reinforced composites.

Numerical studies in this section show that the thickness, nonuniform distribution and material property of the interphase can have significant influences on the micromechanical behaviors of the composites, such as effective elastic moduli and interface stresses, especially when the fiber volume fractions are large.These observations are consistent with the findings in both the FEM and BEM literatures on this subject.

Considerations of interface cracks in the present BEM model and extension of the BEM code to three dimensions to study the fiber-pullout failure modes will be interesting and challenging next steps, both of which will further demonstrate the robustness of the developed BEM approach as compared with the FEM or previous BEM approaches to the micromechanical analysis of fiber-reinforced composites.

9.4 Interface Stress Analysis of Thin Films and Coatings

To further demonstrate the accuracy and efficiency of the BEM in the simulations of various materials, thin films and coatings were modeled using the developed BEM formulation in which the nearly singular integrals have been dealt with accurately.More details and examples can be found in (Luo et al., 1998; Luo et al., 2000).

9.4.1 A Thin Film Model

First, a thin film model with pressure load p shown in Fig. 9.9 is studied. The plane-strain condition is assumed. The length L of the film is constant in this study, while the thickness h changes from L to $10^{-9}L$. Note that the thickness is changing from macroscale to microscale, and eventually to nanoscale, which may be already outside of the limits of the continuum mechanics assumptions for many materials.However, it is of more interest here to verify the validity and effectiveness of the developed BEM approach for such 2-D thin materials.

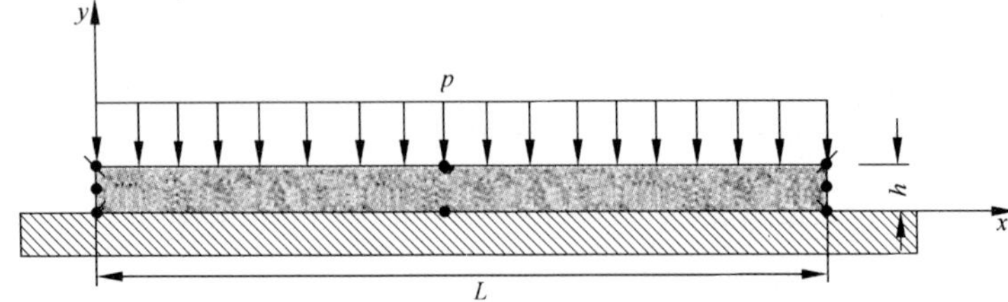

Fig. 9.9 Thin flim model under constant pressure load p (2-D plane strain model, shear modulus $\mu = 8.0 \times 10^{10}$ Pa, Poisson's ratio $\nu = 0.2$)

In the BEM model, the boundary of the film is discretized with only four quadratic boundary elements, two elements with length $L = 4$ m, and two other elements with size (thickness h) changing from L to $10^{-9}L$. Figure 9.10 shows the displacement of the upper-right node in the x-direction by the advanced BEM.We see that even for thickness-to-length ratio h/L in the nanoscale, results are still very good.Results for the stress components are even more accurate for this example, almost reproducing the exact values for all values of the thickness.This proves that the developed analytical work in the BEM procedure is effective.The finite element analysis of this simple problem was also attempted, but it was soon found that the number of the 2-D finite elements was so large that the task quickly exceeds the capacity of the computer used.A simple calculation reveals that the total number of the 2-D finite elements required for this example is determined by $M = 4L / 5h$, if only two layers of elements are used across the thickness and an aspect ratio of five is maintained.This gives $M = 3200000$ finite elements if $h = 10^{-6}$ m ($L = 4$ m) and an even larger number of nodes, depending on the type of finite elements employed.

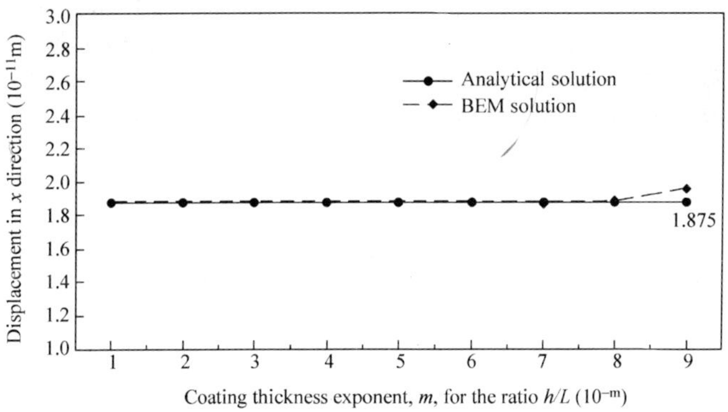

Fig. 9.10 Displacement at the upper-right node in x-direction as a function of the thickness of the film

The example problem shown here is a very simple one in the study of thin films. The purpose of using this example is to verify the correctness of the singularity subtraction and nonlinear coordinate transformation in dealing with nearly singular integrals developed in (Luo et al., 1998).Because of the simplicity of the boundary condition (constant pressure), a model with much smaller length can actually be applied. However, more cases exist where the length of the model can not be decreased.For example, in this simple problem if the pressure p is not constant but varies along the length of the film, then the model with length = 4m should be used.In such cases, with the 2-D BEM, accurate results can be obtained by using just a few boundary elements, as we have seen in this example.

9.4.2 Thin Coatings on a Shaft

The BEM developed is applied to model a shaft with a thin coating, which can be uniform or nonuniform, Fig. 9.11.Plane strain is assumed, and the shaft is considered to be rigid when compared to the coating.The shaft and coating have radii r_s and r_c, respectively. For the case of nonuniform thickness, both the shaft and coating profiles remain circular, but their centers are misaligned, producing some normalized eccentricity $\delta = x_c /(r_c - r_s)$, where x_c is the center offset, as shown in Fig. 9.11b. The coated shaft is loaded by pressure p, which is uniformly distributed around the circumference.The BEM discretization uses only 16 quadratic line elements, regardless of the thickness of the coating.The FEM is also used for comparison.The FEM discretization varies according to the coating thickness and element aspect ratio requirements. The FEM mesh in the nonuniform thickness case must consider the entire structure, which significantly increases the number of elements.In addition, as the eccentricity δ increases, more finite elements are required in the thinnest section in order to maintain a reasonable element aspect ratio. In this case, the shaft radius is held constant at $r_s = 0.1$ m and the coating outer radius is also held constant at $r_c = 0.11$ m. However, the eccentricity has been systematically varied over the entire range $0 \leqslant \delta < 1$. For each case of eccentricity, the stress solution has been obtained by using the BEM, while the FEM solution breaks down as $\delta \rightarrow 1$.

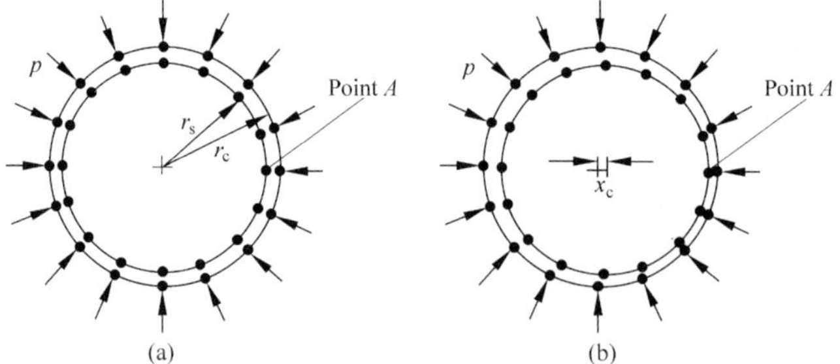

Fig. 9.11 Cross section of a shaft with coatings of (**a**) uniform and (**b**) nonuniform thickness

Figure 9.12 shows the normalized radial stress σ_{rr} at point A, and also the number of nodes needed to achieve solution for both the BEM and FEM.Note first the asymptotic behavior of the BEM solution, which approaches the analytical value of the sample problem as $\delta \rightarrow 0$ (case (a)). Also note that the same stress value at point A approaches the applied pressure p as $\delta \rightarrow 1$, which is consistent with the physical interpretation.Again, the BEM mesh does not change across the entire range of normalized eccentricity.The solution time and memory requirements are therefore quite modest for the BEM procedure. The FEM solution, however, demonstrates a very different behavior.While the low eccentricity cases are easily solved with the FEM, with a fairly small number of elements, the solution requires significantly more effort for $\delta \rightarrow 1$. The element number for the FEM increases dramatically at approximately $\delta = 0.9$, corresponding to an absolute coating thickness of $h = 0.001$ m at the thinnest point on the structure. Indeed, for $\delta > 0.99$, the FEM solution becomes infeasible due to memory limitations on a PC. On the other hand, the BEM can continue to provide results for $\delta = 0.999$, 0.9999 and 0.99999 without any difficulty, because of the effective approach developed in (Luo et al., 1998) to deal with the nearly singular integrals. Clearly, the BEM solution, which does not require a refined mesh, regardless of the coating thickness, is preferable.

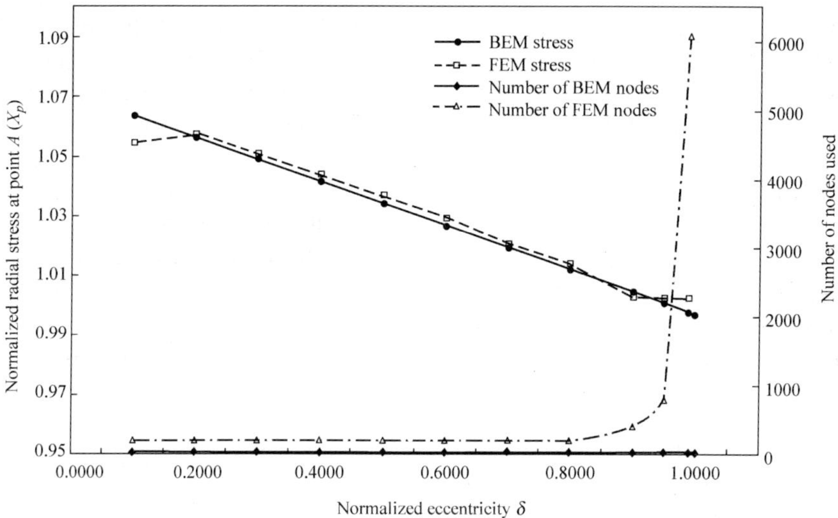

Fig. 9.12   Radial stress and numbers of nodes used for the nonuniform coating

9.4.3   Discussion

The applicability of the conventional BIE for 2-D elasticity problems to the analysis of 2-D thin materials or structures in the micro- and nano-scales is investigated in this subsection. It is shown that the BEM can be very accurate and efficient in modeling thin films and coatings. The numerical verification on two sample problems supports this assertion.

For all the test problems of thin materials studied in this subsection, the BEM provides a computationally efficient solution for both displacement and stress fields.While the FEM performs well over certain parameter ranges, in general it is not suitable for thin materials such as thin films and coatings due to the huge number of finite elements required to accurately model such materials.The BEM approach to thin materials analysis described here can solve the stress field problem accurately for (in theory) arbitrarily thin materials, because of the following attractive features: The BEM uses only boundary discretization (e.g., quadratic line elements) instead of domain discretization (area elements).Thus the BEM does not suffer from thickness or aspect-ratio difficulties related to FEM shell or 2-D continuum elements.The element number using the BEM can be held constant for thin materials problems as described in the examples, without any loss in accuracy as the thickness decreases.The BEM procedure therefore requires no remeshing as the thickness decreases.The BEM procedure seldom confronts memory or storage issues in the numerical solution, and can be implemented on desktop computers (PCs).

The developed method of using the BEM based on elasticity theory for analyzing 2-D thin materials or structures can be extended readily to model layered coatings, thin films or other layered structures to study contact stresses, interface cracks, thermal effects and nonlinear deformations. Some work along this line is already underway (Luo et al., 1999).

9.5 Closing Remarks

Two commonly applied numerical methods, the finite element method and the boundary element method, in the simulations for materials research are introduced briefly in this chapter.The formulation of the BEM is presented to illustrate the accuracy and efficiency of the method for certain problems in materials research, such as studies of the interphases in fiber-reinforced composites, interface stresses in thin films and coatings.Numerical examples in these areas clearly demonstrate the superior features of the BEM approach over the FEM approach.However, it should be emphasized that the FEM is still the dominant numerical tool in the computational materials science for general purposes.The BEM should be used as a special tool or complementary method to the FEM in cases where the FEM is inaccurate or inefficient, as illustrated by example problems presented in this chapter.Both the FEM and BEM will evolve, and more and more materials science problems will be simulated on computers by using these numerical methods.

The developed BEM is being extended to model the piezoelectric solids used in smart materials, such as for sensors and actuators.The BEM formulations for piezoelectric solids are much more involved than those for the composite, thin film or coating materials, because of the coupling of the electric field and the elastic field.The effectiveness and accuracy of the BEM in the analysis of piezoelectric materials have been demonstrated in (Chen 1993; Chen and Lin, 1995) for bulky piezoelectric solids.However, piezoelectric sensors and actuators are in general made in very thin shapes (patches) for which the general BEM formulations will fail due to the nearly singular integrals.Efforts are underway to extend the capabilities for analyzing thin materials in (Liu, 1998; Luo et al., 1998; Luo et al., 2000) to thin piezoelectric materials.Results from this research can be utilized to predict the performance of piezoelectric sensors and actuators in the design stage, including the durability issues regarding interface cracks or voids and high cyclic electric and mechanical loads.

9.6   Acknowledgment

This research is supported by the U.S. National Science Foundation under the grant CMS 9734949.The author would like to thank Professor Edward Berger, and graduate students Mr Jianfeng Luo and Ms Nan Xu for their contributions in various stages of the research reported in this chapter.

References

Aboudi, J. Micromechanical analysis of composites by the method of cells. Appl. Mech. Rev.. **42**(7), 193–221(1989)

Achenbach, J. D. and H.Zhu. Effect of interfacial zone on mechanical behavior and failure of fiber-reinforced composites. J. Mech. Phys. Solids. **37**(3), 381–393 (1989)

Achenbach, J. D. and H.Zhu. Effect of interphases on micro and macromechanical behavior of hexagonal-array fiber composites. J. Appl. Mech. **57**, 956–963 (1990)

Adams, D. F. A micromechanics analysis of the influence of the interface on the performance of polymer-matrix composites. J. Reinforced Plastics and Composites. **6**, 66–88 (1987)

Banerjee, P. K. The Boundary Element Methods in Engineering. (New York, McGraw-Hill, 1994)

Banks, H. T.R.C. Smith. and Y.Wang. Smart Material Structures—Modeling, Estimation and Control. (John Wiley & Sons, Chichester, 1996)

Bathe, K. J. Finite Element Procedures. (Prentice Hall, Englewood Cliffs, NJ, 1996)

Brebbia, C. A. and J.Dominguez. Boundary Elements—An Introductory Course. (McGraw-Hill, New York, 1989)

Carey, G. F. and J.T. Oden. Finite Elements: A Second Course. Englewood Cliffs, (Prentice-Hall, New Jersey, 1983)

Chandrupatla, T. R. and. A.D.Belegundu. Introduction To Finite Elements in Engineering. (Prentice-Hall , Upper Saddle River, NJ, 1997)

Chawla, K. K. Composite Materials : Science and Engineering. (Springer-Verlag , New York, 1998)

Chen, T. Green's functions and the non-uniform transformation problem in a piezoelectric medium . Mech. Res. Comm. **20**, 271–278 (1993)

Chen, T. and F.Z.Lin. Boundary integral formulations for three-dimensional anisotropic piezoelectric solids . Comput. Mech. **15**, 485–496 (1995)

Cook, R. D., D.S.Malkus. and M.E. Plesha. Concepts and Applications of Finite Element Analysis. (John Wiley & Sons, Inc. New York, 1989)

Cruse, T. A. Numerical solutions in three dimensional elastostatics. Int. J. Solids & Structures **5**, 1259–1274 (1969)

Cruse, T. A. Boundary Element Analysis in Computational Fracture Mechanics. (Kluwer Academic Publishers , Dordrecht, Boston, 1988)

Cruse, T. A. BIE fracture mechanics analysis: 25 years of developments. Comput. Mech. **18**, 1–11 (1996)

436

Cruse, T. A. and R. Aithal. Non-singular boundary integral equation implementation. Int. J. Numer. Methods Engrg. **36**, 237–254 (1993)

Cruse, T. A. and F.J.Rizzo. A direct formulation and numerical solution of the general transient elastodynamic problem, I.J. Math. Anal. Appl. **22**, 244–259 (1968)

Cruse, T. A. and F.J. Rizzo. Eds. Boundary-Integral Equation Method: Computational Applications in Applied Mechanics, AMD-ASME (1975)

Culshaw, B. Smart Structures and Materials. (Artech House, Boston, 1996)

Gray, L. J., L.F.Martha. and A.R.Ingraffea. Hypersingular integrals in boundary element fracture analysis.Int. J. Num. Meth. Eng. **29**, 1135–1158 (1990)

Gulrajani, S. N. and S. Mukherjee. Sensitivities and optimal design of hexagonal array fiber composites with respect to interphase properties. Int. J. Solids Structures. **30**(15), 2009–2026 (1993)

Hashin, Z. Composite materials with interphase: thermoelastic and inelastic effects. Inelastic Deformation of Composite Materials. G. J. Dvorak. (Springer, New York, 1991) 3–34

Huang, Q. and T.A.Cruse.Some notes on singular integral techniques in boundary element analysis.Int. J. Numer. Methods Engrg . **36**, 2643–2659 (1993)

Hyer, M. W. Stress Analysis of Fiber-Reinforced Composite Materials. (McGraw-Hill, Boston, 1998)

Jasiuk, I. and M.W.Kouider. The effect of an inhomogeneous interphase on the elastic constants of transversely isotropic composites. Mechanics of Materials **15**, 53–63 (1993)

Kane, J. H. Boundary Element Analysis in Engineering Continuum Mechanics. (Prentice Hall, Englewood Cliffs, NJ, 1994)

Krishnasamy, G., F.J.Rizzo. and Y.J.Liu. Boundary integral equations for thin bodies. Int. J. Numer. Methods Eng. **37**, 107–121 (1994)

Krishnasamy, G., F.J.Rizzo. and T.J.Rudolphi. Hypersingular boundary integral equations: their occurrence, interpretation, regularization and computation. Developments in Boundary Element Methods. P. K. Banerjee et al. (Elsevier Applied Science Publishers , London, 1991). **VII**: Chap.7

Lagache, M., A.Agbossou, J.Pastor. et al. Role of interphase on the elastic behavior of composite materials: theoretical and experimental analysis.J. Composite Materials **28**(12), 1140 –1157 (1994)

Lee, J. S. Boundary element method for electroelastic interaction in piezoceramics. Eng. Anal. Boundary Elements. **15**(4), 321–328 (1995)

Liu, Y. J. Development and applications of hypersingular boundary integral equations for 3-D acoustics and elastodynamics. Ph.D. Dissertation, University of Illinois at Urbana-Champaign. (1992)

Liu, Y. J. Analysis of shell-like structures by the boundary element method based on 3-D elasticity: formulation and verification. Int. J. Numer. Methods Eng. **41**, 541–558 (1998)

Liu, Y. J. and F.J.Rizzo. Hypersingular boundary integral equations for radiation and scattering of elastic waves in three dimensions. Comput. Methods Appl. Mech. Eng. **107**, 131–144 (1993)

Liu, Y. J. and F.J.Rizzo. Scattering of elastic waves from thin shapes in three dimensions using the composite boundary integral equation formulation. J. Acoust. Soc. Am. **102**(2)

(Pt.1, August), 926–932 (1997)

Liu, Y. J., D.Zhang. and F.J.Rizzo. Nearly singular and hypersingular integrals in the boundary element method. in: Boundary Elements XV, (Computational Mechanics Publications, Worcester, MA, 1993) 453– 468

Luo, J., Y.J. Liu. and E. Berger. Analysis of two-dimensional thin structures (from micro- to nano-scales) using the boundary element method. Comput. Mech. **22**, 404 – 412 (1998)

Luo, J., Y.J.Liu. and E.Berger. Interfacial stress analysis for multi-coating systems using an advanced boundary element method. Comput. Mech. **24**, 448– 455 (2000)

Mackerle, J. Finite element and boundary element library for composites—a bibliography (1991–1993).Finite Elements in Analysis and Design **17**, 155–165 (1994)

Mukherjee, S. Boundary Element Methods in Creep and Fracture. (Applied Science Publishers, New York, 1982)

Nassehi, V., J.Dhillon. and L. Mascia. Finite element simulation of the micromechanics of interlayered polymer/fibre composites: a study of the interactions between the reinforcing phases.Composites Science and Technology. **47**, 349–358 (1993)

Oshima, N. and N.Watari. Calculation of effective elastic moduli of composite materials with dispersed parallel fibers. Theoretical and Applied Mechanics. M. Hori and G. Yagawa, Japan National Committee for Theoretical and Applied Mechanics, Science Council of Japan. **41**, 181–188 (1992)

Pan, L., D.O.Adams. and F.J. Rizzo. Boundary element analysis for composite materials and a library of Green's functions. Comput. & Struct . **66**(5), 685–693 (1998)

Reddy, J. N. An Introduction to the Finite Element Method. (McGraw-Hill, New York, 1993)

Rizzo, F. J. An integral equation approach to boundary value problems of classical elastostatics.Quart. Appl. Math. **25**, 83–95 (1967)

Rizzo, F. J. and D.J. Shippy. An advanced boundary integral equation method for three-dimensional thermoelasticity.Int. J. Numer. Methods Eng. **11**, 1753–1768 (1977)

Rizzo, F. J.,D.J.Shippy. and M. Rezayat. A boundary integral equation method for radiation and scattering of elastic waves in three dimensions. Int. J. Numer. Methods Eng. **21**, 115–129 (1985)

Rogers, C. A. Intelligent material systems—the dawn of a new materials age. J. Intell. Mater. Syst. Struct. **4**(1), 4–12 (1993)

Sater, J. M., Ed. Smart Structures and Materials 1997—Industrial and Commercial Applications of Smart Structures Technologies. (SPIE, San Diego, 1997)

Shi, F., P. Ramesh. and S. Mukherjee. Simulation methods for micro-electro-mechanical structures (MEMS) with application to a microtweezer. Comput. & Struct. **56**(5), 769–783 (1995)

Shi, F., P.Ramesh. and S. Mukherjee. Dynamic analysis of micro-electro-mechanical systems (MEMS). Int. J. Numer. Methods Eng. **39**(24), 4119– 4139 (1996)

Subramanian, C. and K.N.Strafford. Review of multicomponent and multilayer coatings for tribological applications. Wear, 85–95 (1993)

Sun, C. T. and R.S.Vaidya. Prediction of composite properties from a representative volume element.Composites Science and Technology **56**, 171–179 (1996)

Timoshenko, S. P. and J.N. Goodier. Theory of Elasticity. (McGraw-Hill, New York, 1987)

Tzou, H. S. Piezoelectric Shells: Distributed Sensing and Control of Continua. (Kluwer Academic Publishers , Dordrecht, 1993)

Wacker, G., A.K.Bledzki. and A.Chate. Effect of interphase on the transverse Young's modulus of glass/epoxy composites. Composites Part A **29A,** 619–626 (1998)

Wang, Z. L. and Z.C. Kang. Functional and Smart Materials—Structural Evolution and Structure Analysis. (Plenum, New York, 1998)

Yeh, J. R. The effect of interface on the transverse properties of composites. Int. J. of Solids & Structures **29**, 2493–2505 (1992)

Zhu, H. and J.D.Achenbach. Effect of fiber-matrix interphase defects on microlevel stress states at neighboring fibers. J. Composite Materials. **25**, 224–238 (1991)

Zienkiewicz, O. C. and R.L. Taylor.The Finite Element Method. (McGraw-Hill, New York, 1989)

Index

Springer Series in
MATERIALS SCIENCE

Editors: R. Hull   R. M. Osgood, Jr.   J. Parisi

Springer Series in
MATERIALS SCIENCE

Editors: R. Hull   R. M. Osgood, Jr.   J. Parisi

The Springer Series in Materials Science covers the complete spectrum of materials physics, including fundamental principles, physical properties, materials theory and design. Recognizing the increasing importance of materials science in future device technologies, the book titles in this series reflect the state-of-the-art in understanding and controlling the structure and properties of all important classes of materials.

Series homepage – http://www.springer.de/phys/books/ssms/

Volumes 1–50 are listed at the end of the book.